# Plant Membrane and Vacuolar Transporters

### EDITORS

**PAWAN K. JAIWAL**
*Professor*
Advanced Centre for Biotechnology
Maharishi Dayanand University
Rohtak - 124 001
India

**RANA P. SINGH**
*Professor*
Department of Environmental Science
B.B. Ambedkar (Central) University
Rae Bareily Road
Lucknow - 226 025
India

**OM PARKASH DHANKHER**
*Assistant Professor*
Department of Plant, Soil and Insect Sciences
University of Massachusetts
Amherst, MA 01002
USA

www.cabi.org

**CABI is a trading name of CAB International**

CABI Head Office
Nosworthy Way
Wallingford
Oxfordshire OX10 8DE
UK

CABI North American Office
875 Massachusetts Avenue
7th Floor
Cambridge, MA 02139
USA

Tel: +44 (0)1491 832111
Fax: +44 (0)1491 833508
E-mail: cabi@cabi.org
Website: www.cabi.org

Tel: +1 617 395 4056
Tel: +1 617 354 6875
E-mail: cabi-nao@cabi.org

© CAB International 2008. All rights reserved. No part of this publication may be reproduced in any form or by any means, electronically, mechanically, by photocopying, recording or otherwise, without the prior permission of the copyright owners.

A catalogue record for this book is available from the British Library, London, UK.

**Library of Congress Cataloging-in-Publication Data**

Plant membrane and vacuolar transporters/editors, Pawan K. Jaiwal, Rana P. Singh, Om Parkash Dhankher.

p. cm.

Includes bibliographical references and index.

ISBN 978-1-84593-402-6 (alk. paper)

1. Plant cell membranes. 2. Plant translocation. I. Jaiwal, Pawan K. II. Singh, Rana P. III. Dhankher, Om Parkash.

QK725.P4756 2008

571.6'42—dc22

2007051246

ISBN-13: 978 1 84593 402 6

Printed and bound in the UK from copy supplied by the editors by Biddles Ltd., King's Lynn.

# CONTENTS

**PREFACE**

**Chapters**

1. **Mechanisms of potassium uptake and transport in higher plants**    1-50
   Tracey A. Cuin[1], Igor I. Pottosin[2] and Sergey N. Shabala[1]
   [1]*School of Agricultural Science, University of Tasmania, Private Bag 54, Hobart, Tasmania 7001, Australia*
   [2]*Centro de Investigaciones Biomedicas, Universidad de Colima, Mexico*

2. **Calcium transporters: from fields to the table**    51-82
   Jay Morris[1] and Kendal Hirschi[1,2]
   [1]*The Vegetable and Fruit Improvement Center, Texas A&M University, College Station, Texas 77845, USA*
   [2]*Children's Nutrition Research Center, Department of Pediatrics, Baylor College of Medicine, Houston, Texas 77030, USA*

3. **Nitrate and ammonium transporters in plants**    83-103
   Rana P. Singh[1], Manish Sainger[1], D.P. Singh[1] and Pawan K. Jaiwal[2]
   [1]*Department of Environmental Science, Baba Saheb Bhimrao Ambedkar (Central) University, Lucknow - 226 025, India*
   [2]*Advanced Centre for Biotechnology, M.D. University, Rohtak - 124 001, India*

4. **Plant sulfate transporters**    105-130
   Peter Buchner
   *Plant Science Department, Rothamsted Research, Harpenden AL5 2JQ, UK*

5. **Phosphate uptake and transport to plant cells**    131-147
   Toshio Sano[1] and Toshiyuki Nagata[2]
   [1]*Graduate School of Frontier Sciences, The University of Tokyo, Tokyo, Japan*
   [2]*Graduate School of Science, The University of Tokyo, Tokyo, Japan*

6. **Iron uptake and transport in plants**    149-172
   Tzvetina Brumbarova and Petra Bauer
   *Department of Biological Sciences – Botany, Saarland University, PO Box 151150, D-66041 Saarbrücken, Germany*

7. **Mechanisms of manganese accumulation and transport**    173-204
Jon K. Pittman
*Faculty of Life Sciences, University of Manchester, 3.614 Stopford Building, Oxford Road, Manchester, M13 9PT, UK*

8. **Silicon uptake and transport in higher plants**    205-212
Yongchao Liang[1,2]
[1]*Institute of Soil and Fertilizer, and Ministry of Agriculture Key Laboratory of Plant Nutrition and Nutrient Cycling, Chinese Academy of Agricultural Sciences, Beijing - 100 081, P.R. China*
[2]*Key Laboratory of Eco-agriculture Shihezi University, Shihezi - 832 003, P.R. China*

9. **Heavy metal transporters in plants**    213-238
Bibin Paulose[1], Pawan K. Jaiwal[2] and Om Parkash Dhankher[1]
[1]*Department of Plant, Soil and Insect Sciences, University of Massachusetts, Amherst, MA 01002, USA*
[2]*Advanced Centre for Biotechnology, Maharshi Dayanand University, Rohtak - 124 001, India*

10. **Sugar and polyol transporters in plants**    239-266
Katsuhiro Shiratake
*Graduate School of Bioagricultural Sciences, Nagoya University, Chikusa, Nagoya 464-8601, Japan*

11. **Amino acid transporters in plants**    267-282
Uwe Ludewig and Wolfgang Koch
*Center for Plant Molecular Biology, Plant Physiology, University of Tübingen, Auf der Morgenstelle 1, D-72076 Tübingen, Germany*

12. **Membrane transport of secondary metabolites in plants**    283-300
Nobukazu Shitan and Kazufumi Yazaki
*Research Institute for Sustainable Humanosphere, Kyoto University, Gokasho, Uji 611-0011, Japan*

13. **Proteomic analysis of the vacuolar membrane**    301-343
Tetsuro Mimura[1], Miwa Ohnishi[1], Taise Shimaoka[2] and Ken-ichi Tomizawa[2]
[1]*Department of Biology, Graduate School of Science, Kobe University, Nada, Kobe 657-8501, Japan*
[2]*Plant Research Group, Research Institute of Innovative Technology for the Earth, Kizugawadai, Kizu-cho, Soraku-gun, Kyoto 619-0292, Japan*

14. **Elemental biofortification of crop plants**  345-371
Savita Dahiya[1], Darshna Chaudhary[1], Ranjana Jaiwal[1],
Om Parkash Dhankher[2], Rana P. Singh[3] and Pawan K. Jaiwal[1]
[1]*Advanced Centre for Biotechnology, M.D. University, Rohtak - 124 001, India*
[2]*Department of Plant, Soil and Insect Sciences, University of Massachusetts, Amherst, MA 01002, USA*
[3]*Department of Environmental Science, BBA University, Lucknow - 226 025, India*

**SUBJECT INDEX**

# PREFACE

Plant membrane and tonoplast bound transporter proteins are involved in uptake, transport, partitioning and redistribution of different ions and metabolites and play a wider role in various cellular activities. These transporter proteins and their corresponding genes have been identified, isolated, characterized, cloned and over-expressed to ascertain their precise functions and specific roles in mineral nutrition, carbon and nitrogen metabolism, cell signaling, osmoregulation, cell homeostasis, storage and stress responses. Recent advances in functional genomics have provided new insights into molecular transport biology based on mutant and down-stream gene regulation. Plant as a continuous system needs a very well-coordinated and finely tuned genetic programme for uptake, transport, assimilation, storage and redistribution of ions and metabolites. Disruption in any of these processes results in impairment of vital functions. Therefore, a large number of studies have been conducted in the recent past on the localization, characterization and regulation of ion and metabolite transport systems. This title is a sincere effort to bring an updated critical account of the various transporter systems to one place.

Potassium, the second most abundant mineral nutrient, is essential in plant growth. Numerous families of $K^+$ transporters have been discussed by Tracey A. Cuin and co-authors (pp. 1-50). An understanding of the underlying mechanisms for the regulation of $K^+$ uptake and utilization by crop species will help to develop crop species with higher $K^+$ use efficiency. The chapter by Jay Morris and Kendal Hirschi (pp. 51-82) focuses on the overview of different types of $Ca^{2+}$-transporters in plants. This review presents a summary of recent advances in cloning and characterization of $Ca^{2+}$-transporters and their putative roles in $Ca^{2+}$-signaling. They have extensively discussed how these transporters are being engineered to improve $Ca^{2+}$ content of agriculturally important crops.

Nitrate and ammonium are major available forms of nitrogen which are taken up from the rhizosphere and transported to the various plant parts before their assimilation into organic N-molecules. Low and high affinity membrane transporters for both $NO_3^-$ and $NH_4^+$ have been characterized in plants. Tonoplast transport of the these nutrient ions has been reported to be involved in storage of unassimilated N in the vacuoles. Rana P. Singh and co-authors (pp. 83-103) have discussed the role and regulation of various molecular species of N transporters which have recently been augumented by mutant studies and genetically engineered plants. Sulfur transport across the plasma membrane and tonoplast is mediated by members of a single gene family which is divided into 5 groups. Different sulfate transporters are involved in its uptake, vascular transport and vacuolar efflux which has been discussed by Peter Buchner in his chapter (pp. 105-130). Phosphate (Pi), the available form of P for plants and other living organisms, exists in very low concentration in soil. The plant responses to Pi starvation, the physiological and molecular characterization and the regulation of plant Pi-transporters have been analyzed by T. Sano and T. Nagata (pp. 131-147).

Manganese is an essential mineral nutrient which is taken up by the roots, and transported to various other parts through the membrane bound transporters. The characterization of putative $Mn^{2+}$ transporters and their potential roles in various physiological processes as well as the genetic manipulation of $Mn^{2+}$ transport and homoeostasis proteins have been discussed by Jon K. Pittman in chapter 7 (pp. 173-204). There is a potential for genetic manipulation as a strategy to overcome $Mn^{2+}$ toxicity or $Mn^{2+}$ deficiency stress in plants. Iron transport is based either on reduction of Fe, its uptake or chelation of ions by phytosidephores. The physiological and molecular aspects of Fe uptake and transport in plants have been described by T. Brumbarova and P. Bauer (pp. 149-172).

Silicon is the second most abundant element after oxygen in the earth's crust and soil. Although the essentiality of Si in higher plants has not been recognized, Si has been proved to be beneficial for healthy growth of many plant species. The better understanding of Si-uptake mechanisms and its role in eleviation of stress tolerance and plant biology have been evaluated by Y. Liang (pp. 205-212).

Heavy metal toxicity is a serious environmental hazard affecting human health. Uptake, transport and partitioning of these toxic metals are mediated by various groups of membrane bound transporters (located in both plasma membrane and tonoplast). Bibin Paulose and co-authors (pp. 213-238) have critically analyzed the heavy metal transport in plants so as to draw a cost effective, eco-friendly approach to detoxification of the metal contaminated sites.

The transport and compartmentalization of sugars and polyols are important in controlling plant growth and development. The characters, regulations and functions of their transporters in plants have been reviewed by K. Shiratake (pp. 239-266). A greater understanding of these transporters will produce crops with greater yield and quality.

Inorganic nitrogen taken up by the plants is assimilated into amino acids which are translocated to the sink tissue via phloem and xylem. The loading and unloading steps require amino acid transport proteins. Several amino acid transporters have been identified and characterized. The potential role of these transporters in the plant adaptation to meet the nitrogen requirements of different plant tissues and under varying environmental conditions has been discussed by Uwe Ludewig and W. Koch (pp. 267-282).

In addition to the primary metabolites, plants produce a vast number of secondary metabolites such as alkaloids, terpenoids, phenolic compounds etc., which get accumulated in particular sink organs and some are translocated from source cells via long distance transport. Recent advances in functional genomic research and analysis of many mutants have identified membrane transporters of plant secondary metabolites. Characterization of such transporters has been presented by N. Shitan and K. Yazaki (pp. 283-300).

Molecular analysis of various proteins in the vacuolar membrane including pumps, carriers, ion channels and receptors is an essential step to understand how vacuoles function. T. Mimura and co-authors (pp. 301-343) have made a comprehensive proteomic analysis of purified vacuolar membrane and identified well characterized proteins such as V-type, $H^+$-ATPases and V-type PPases, along with a number of novel proteins. Plants serve as a major source for all essential minerals required for good health of humans and animals. Unfortunately the micronutrient concentrations are often low in many staple food crops. Therefore, the diets of populations dependent on cereals are deficient in regions where soil mineral imbalances occur. Micronutrient deficiency (Fe, Zn and Se) affects a large portion of the global population. Savita Dahiya and co-authors (pp. 345-371) have presented the various strategies to overcome the deficiencies of micronutrients in staple food crops.

The contents of the book have been carefully organized in a planned sequence to give a detailed insight of the subject to the beginners as well as to the experts. The chapters critically evaluate the current knowledge, state of art and future prospects of plant membrane and vacuolar transporters in relation to plant productivity and quality improvement. The book is valuable for the scientists, academicians, researchers, students, planners and industrialists working in the area of biotechnology, plant agriculture, agronomy, horticulture, plant physiology, molecular biology, nutritional biology, plant sciences and environmental sciences.

The readers will immediately recognize that each chapter in this volume has been written by the world's leading experts, group in plant membrane and vacuolar transporters. A special effort has been made to concentrate on new findings in these areas. It is through this vision that a new lead will develop to a more complete understanding of the complexities of plant metabolism and their regulation for the fruitful improvement in human health and the environment in sustainable way.

We are grateful to the contributors for their efforts in preparing insightful and authoritative accounts of various aspects of the knowledge in this area. We express our sincere thanks and gratitude to all these colleagues and warm appreciation and thanks to CABI, UK for their keen interest in bringing out this title with quality work. We are also thankful to our family members and Ph.D. students for their understanding and patience during planning and preparation of this title. Efforts made by Mr. Ashok Datta and Ms. Reema of *LaPrints*, New Delhi, for preparing a camera-ready version deserve high appreciation.

Rohtak, India  
October, 2007

Pawan K. Jaiwal  
Rana P. Singh  
Om Parkash Dhankher

# Chapter 1

## MECHANISMS OF POTASSIUM UPTAKE AND TRANSPORT IN HIGHER PLANTS

TRACEY A. CUIN[1], IGOR I. POTTOSIN[2] AND SERGEY N. SHABALA[1]
[1]*School of Agricultural Science, University of Tasmania, Private Bag 54, Hobart, Tasmania 7001, Australia*
[2]*Centro de Investigaciones Biomedicas, Universidad de Colima, Mexico*
E-mail: Tracey.Cuin@utas.edu.au

### Abstract

Potassium, the second most abundant mineral nutrient, is essential to plant growth. Over millions of years, plants have evolved a sophisticated network of potassium transport systems. These include Shaker-type and "two-pore" potassium channels; various types of potassium-permeable non-selective cation channels; and KUP/HAK/KT, HKT and $K^+/H^+$ transporters. This chapter gives an overview of these transport systems and discusses their regulation and functional expression in plant membranes under potassium deficiency conditions. Molecular and electrophysiological data are discussed in the context of physiology and agronomy of the regulation of potassium nutrition at the whole plant level grown under various soil types and environmental conditions. An overview of modern methods for studying the mechanisms of potassium transport in plants is also discussed.

**Keywords:** potassium; membrane; ion channels; transporters; homeostasis

## 1. INTRODUCTION

Potassium is the second most abundant mineral nutrient in plants. It is essential to plant growth and, consequently, to crop production. Humans also require potassium to maintain the osmotic equilibrium within the body, as well as for neuronal signalling, muscle activity, heartbeat and activation of a large number of enzymes involved in various metabolic processes. This vital potassium is obtained either directly from plants or indirectly through the animal products in the diet.

Although concentrations of potassium in the soil solution are not very high (between 0.1 to 6 mM, depending on the soil type; Adams, 1971), plants are able to accumulate the large quantities required; usually between 2% and 10% of the plant dry weight (Leigh and Wyn Jones, 1984; Tisdale *et al.*, 1993). Within the plant, potassium plays crucial roles in cell elongation, leaf movement, tropisms,

---

© CAB International 2008. *Plant Membrane and Vacuolar Transporters* (eds P.K. Jaiwal, R.P. Singh and O.P. Dhankher)

metabolic homeostasis, germination, osmoregulation, stomatal function and in numerous biochemical processes. The importance of potassium to crop production has long been recognised, with potassium fertilisers, traditionally derived from livestock manure and wood ashes (hence the name "potash" for the oxide of potassium), being supplied for over 250 years. Today, farmers apply over 20 million tonnes of potassium fertiliser annually (FAOSTAT, 2003). Because of its crucial importance in crop production, and that it is often a limiting factor in crop production, potassium nutrition and transport has been studied intensively for over fifty years. As a result, we know more about the transport of potassium by plants than about any other mineral or organic solute.

## 1.1. Potassium content of soils

Potassium is the fourth most abundant mineral in the earth crust, constituting about 2.5% of the lithosphere (Sparks and Huang, 1985). This relatively large amount does not however reflect the availability of potassium to plants. Potassium can be present in the soil in four different pools (Fig. 1): (i) mineral potassium; (ii) fixed potassium; (iii) exchangeable potassium, and (iv) soil solution potassium (Mclaren and Cameron, 1996; Syers, 1998). The largest fraction is mineral potassium that is almost completely unavailable to plants (Pal et al., 1999). Physiologically relevant pools are the soil solution and the exchangeable potassium; their size dependent upon the nutrient dynamics within the soil rather than the actual total potassium content of the soil. Importantly for crop production, the release of exchangeable potassium is often slower than the rate of potassium acquisition by plants (Sparks and Huang, 1985). Consequently, potassium content in soils is often very low (Pretty and Stangel, 1985; Johnston, 2005). Furthermore, the exchange of potassium between different pools is strongly dependent upon the concentration of other macronutrients in the soil solution (Yanai et al., 1996). The plant potassium status may also further deteriorate in the presence of high levels of other monovalent cations such as sodium and ammonium that interfere with potassium uptake (Spalding et al., 1999; Rus et al., 2004; Qi and Spalding, 2004).

Potassium is delivered to the root surface predominantly by diffusion (Seiffert et al., 1995; Oliveira et al., 2004), with only a small fraction (ca 4%) delivered via mass flow (Oliveira et al., 2004). Potassium in the soil is rapidly removed by the plant at the root surface (Rosolem et al., 2003). Thus, potassium depletion around the root is the most frequently observed phenomenon associated with plant-evoked soil potassium perturbations. Variations in soil density may also affect potassium availability. Soil compaction is associated with higher volumetric water content and therefore tends to facilitate potassium transport to the root surface (Kuchenbuch et al., 1986). However, dense soil may also cause a reduction in the root length, so the higher bulk density does not necessarily result in increased potassium availability to, and accumulation by the plant (Seiffert et al., 1995). The spatial heterogeneities in potassium distribution encountered by a root are often superimposed on the temporal variations in potassium availability resulting from continuously changing soil moisture content. For example, in dry soils, bulk

Potassium uptake and transport

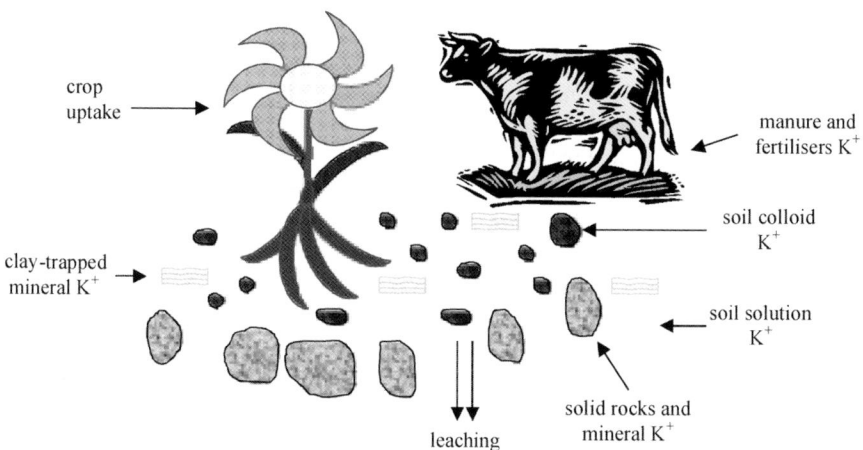

**Figure 1:** The potassium cycle in the soil-plant system. Most of the potassium is in a fixed form (structural potassium) and is not directly available to plants. Potassium content in the soil slowly increases due to the breakdown of potassium-containing rocks and this is released into the soil solution. Part of this released potassium is bound to clay particles and can be replaced relatively easily by other positively charged ions, making it available to the plant (exchangeable pool). Potassium within the soil solution is immediately available for uptake by the roots, and moves through the soil to the plant root surface predominantly by diffusion. Potassium is added to the soil through fertilisation and manure addition and is removed by harvesting of crops and through the leaching of potassium through the soil.

potassium content is normally higher, but mass flow and diffusion are restricted (Kuchenbuch *et al.*, 1986; Seiffert *et al.*, 1995; Vetterlein and Jahn, 2004). Thus, the negative effects of drought on potassium transport in soil are likely to be more significant than the increases in soil potassium content and therefore these environmental conditions lead to reduced availability of the nutrient (Kuchenbuch *et al.*, 1986; Seiffert *et al.*, 1995; Liebersbach *et al.*, 2004).

### 1.2. Physiological and agronomical aspects of potassium nutrition

Due to the high demand, potassium as a nutrient is frequently in short supply in many agricultural systems (Öborn *et al.*, 2005). This results in potassium deficiency which causes a severe reduction in net photosynthesis (Bednarz and Oosterhuis, 1999; Peoples and Koch, 1979; Cakmak and Engels, 1999) and, ultimately, in crop yield. Such an impairment of photosynthesis leads to increased reactive oxygen species production, resulting in photooxidative damage (Cakmak, 2005). Importantly, potassium appears to play an important role in contributing to the survival of crop plants under environmental stress conditions and there is increasing evidence that plants suffering from environmental stresses have a larger potassium requirement (Cakmak and Engels, 1999). For example, plants exposed to high light intensities or grown under long-term sunlight conditions, may have larger potassium requirements than plants grown under lower light intensities. In addition, when plants are grown under a low supply of potassium, drought stress induced reactive oxygen species production is additionally enhanced, possibly due to such disturbances in stomatal opening that affect water relations and photosynthesis (Marschner, 1995; Mengel and Kirkby, 2001). High levels of potassium also appear to provide protection against oxidative damage caused by chilling or frost (Grewal and Singh, 1980; Hakerlerler *et al.*, 1997). Accumulation of sodium and impairment of potassium nutrition is a major characteristic of salt stressed plants, and potassium nutrition is known to be crucial in the tolerance of crop plants to saline soils (Maathuis and Amtmann, 1999). All these facts call upon a need for the better understanding of potassium uptake and transport in plants. Such knowledge will eventually lead to the development, either by traditional breeding, or by genetic engineering, of crop species with higher potassium use efficiency.

## 2. POTASSIUM HOMEOSTASIS AND COMPARTMENTATION

### 2.1. Potassium compartmentation at the tissue and whole-plant levels

Under normal growth conditions, potassium is the most abundant cation in plant cells. The maximum concentration however differs significantly between root and leaf cells. Potassium concentrations in the leaf cell vacuoles are approximately 230 mM (Cuin *et al.*, 2003) compared to 120 mM in the root cell vacuoles (Walker *et al.*, 1996) (Fig. 2f, g). At the same time, the cytosolic potassium concentrations (ca. 100 mM; see below) in root and leaf cells are comparable (Walker *et al.*, 1996; Cuin *et al.*, 2003).

There is also a certain degree of heterogeneity between the vacuolar (but not cytosolic) potassium content of different cell types in leaves under potassium-replete conditions (Cuin *et al.*, 2003). This heterogeneity however, is not as extensive as has been found for other solutes (Fricke *et al.*, 1994). As such, in barley, potassium concentrations are only slightly lower in the mesophyll cells than the epidermal cells (Fricke *et al.*, 1994; Cuin *et al.*, 2003). Slight differences in vacuolar potassium content between abaxial and adaxial epidermal cells have been reported in *Lupinus* (Treeby and Van Steveninck, 1988) and *Sorghum* (Boursier and Läuchli, 1989). However, the substantial heterogeneity in potassium concentration between different cell types only becomes pronounced under potassium-limiting conditions, where concentrations are maintained in the mesophyll cells, while decreasing in the epidermal cells (Fricke *et al.*, 1994).

### 2.2. Intracellular potassium compartmentation

Potassium is present in all compartments of plant cells and occurs in two major pools, one in the vacuole and one in the cytosol (Fig. 2f, g) as revealed by a variety of experimental techniques such as compartmental analysis, NMR, energy-dispersive X-ray microanalysis and multi-barrelled microelectrodes (Pitman *et al.*,

Potassium uptake and transport

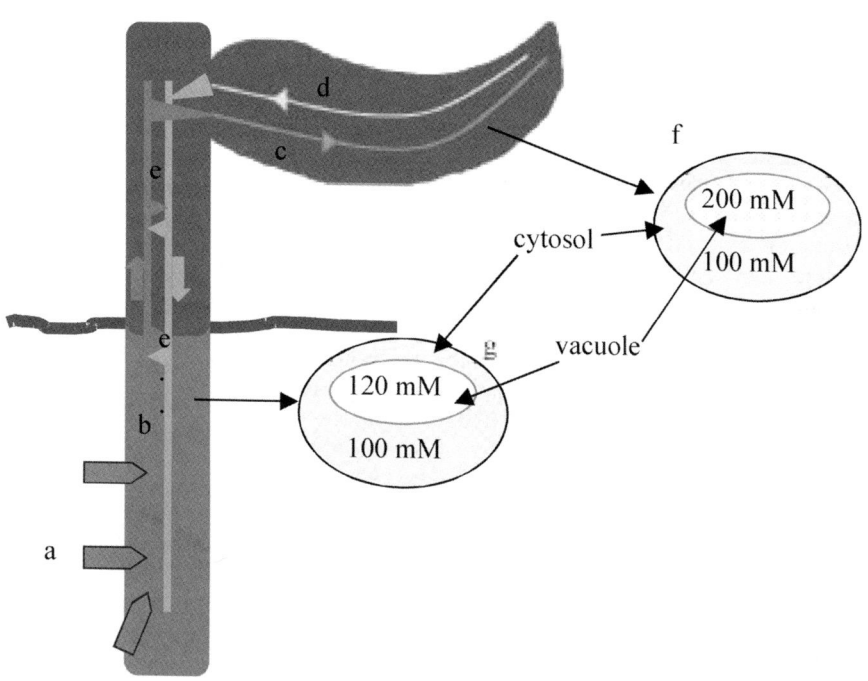

**Figure 2:** Diagrammatic representation of the long-distance transport of potassium in higher plants. (a) Potassium is absorbed across the root cell plasma membrane and moved radially towards the root vascular system. (b) Potassium is released into the xylem vessels from where it is transported through the plant to the shoot. (c) In the shoot, potassium moves from the xylem vessels to the apoplast and is absorbed into the leaf cells. (d) Potassium is loaded into the phloem of the leaf cells whereupon it can be exported from the leaf and unloaded for subsequent use, or be recycled back to the root (e). Typical concentrations within the cytosol and vacuole of leaf (f) and root (g) cells from a potassium-replete plant are also shown.

1981; Rona *et al.*, 1982; Fricke *et al.*, 1994; Walker *et al.*, 1996). Other pools of potassium are also present within the cell, although in much smaller quantities.

The function of potassium in enzyme activation and protein biosynthesis relies on high and stable concentrations within the cytoplasm. Therefore, cytoplasmic potassium homeostasis is essential for metabolic processes to proceed and is strictly controlled at approximately 100 mM (Maathuis and Sanders, 1994; Walker *et al.*, 1996; Leigh *et al.*, 1999; Cuin *et al.*, 2003). Two major mechanisms are involved in maintaining this homeostasis: (i) control of potassium influx across the plasma membrane and (ii) mobilisation of potassium from vacuolar reserves (Fernando and Glass, 1992; Walker *et al.*, 1996).

The vacuolar potassium content is regulated less strictly and large fluctuations in vacuolar content may occur depending on the potassium supply (Leigh and

Wyn Jones, 1984; Malone et al., 1991; Walker et al., 1996; Leigh et al., 1999). Under potassium-limiting conditions, other cations (such as sodium, magnesium and calcium) or organic solutes (e.g. sugars) may substitute for the osmotic functions of potassium in vacuoles. Under potassium-replete conditions, the concentration of vacuolar potassium is typically around 200-250 mM but can reach 500 mM in open stomatal guard cells (MacRobbie, 1998). On the other hand, extreme potassium deficiencies can lead to very low levels of potassium within this compartment (Leigh and Wyn Jones, 1984; Huang and van Steveninck, 1989; Walker et al., 1996) and there appears to be no lower limit to vacuolar potassium concentrations (Walker et al., 1996).

Apoplastic potassium concentrations rarely exceed 20 mM (Karley et al., 2000; Roelfsema and Hedrich, 2002), with the exception of specialised cells or tissues (such as stomata and pulvini), where the peak potassium level may reach 100 mM, depending on environmental conditions (Roelfsema and Hedrich, 2002).

Potassium plays a key role in charge balancing in thylakoid membranes (Pottosin and Schönknecht, 1996; Hinnah and Wagner, 1998), as well as in enzymatic control of leaf photochemistry in stroma. This explains the relatively high level (50 to 100 mM) of potassium found in the chloroplast stroma (Demmig and Gimmler, 1983; Pier and Berkowitz, 1987). The extent to which chloroplasts are able to maintain potassium homeostasis in a changing chloroplast environment is unclear. Also unclear is the nature of the transport systems mediating potassium transport across chloroplast envelopes and thylakoid membranes. In addition to several types of cation-permeable channels (reviewed by Shabala, 2003), there are also suggestions that various secondary active transport systems are also present at the chloroplast envelope where their physiological role may be in controlling stromal pH via an $K^+/H^+$ antiport (Demmig and Gimmler, 1983; Wu and Berkowitz, 1992; Song et al., 2004).

Underlying potassium homeostasis within the plant at all levels is potassium transport. Both the uptake of potassium from the soil and internal potassium homeostasis rely on the presence of potassium transport systems in different subcellular membranes and in different cell types. These systems are likely to have different affinities and modes of function, depending on the prevailing electrochemical potentials. Thus, in order to understand how the requirements for potassium are met by the plant and how potassium homeostasis within the plant is maintained, it is necessary to fully understand both the uptake of potassium into the plant and its internal transport.

Studies into potassium transport mechanisms in plants have a long history (e.g. Epstein et al., 1963; Welch and Epstein, 1968). This early research has evolved into a vast amount of detailed studies on the various mechanisms of potassium transport at all levels within the plant. At present, more than fifty genes, potentially encoding potassium transporter have been identified in the model plant: *Arabidopsis thaliana* (Mäser et al., 2001). However, although regulation and location of some potassium transporters have been studied in detail, much still remains to be elucidated before we can obtain a complete understanding of potassium uptake, transport and regulation of homeostasis within a plant. Our present understanding of these transporters involved in the uptake of potassium and transport within the plant is the subject of this review.

## 3. METHODS TO STUDY POTASSIUM UPTAKE AND TRANSPORT MECHANISMS

Both the uptake of potassium from the soil and internal potassium homeostasis rely on the presence of a large number of potassium transport systems in different subcellular membranes. These systems are likely to have different affinities and modes of function, depending on the prevailing electrochemical potentials. Summarised below are basic methods for the studying and analysing of potassium uptake and transport systems in plants.

### 3.1. Tracer measurements

The radiotracer technique is a non-invasive method, used for over fifty years to study absorption or retention of radiolabelled ions (e.g. potassium or its analogue; rubidium) by plant organs, tissues and isolated cells such as stomatal guard cells. This method can also be used to study ion (in this case, potassium) long-distance transport and re-distribution *in planta*. A suitable isotope needs to be stable and, for safety reasons, mainly beta-emitting. Unfortunately, $^{42}$K has a relatively short decay halftime (12.36 hrs) compared to $^{86}$Rb (18.66 days). Therefore, $^{86}$Rb is generally used as potassium-tracer. As some potassium transporters and channels are strongly selective for potassium over rubidium, the latter might be considered an inappropriate analogue for certain situations (Kronzucker *et al.*, 2003). The radiotracer technique studies *unidirectional*, and not necessarily electrogenic, fluxes of ions and is suitable for measuring steady-state ion fluxes. By its nature, tracer flux measurement is not linked to the membrane potential difference, which is, however, an important factor in determining the flux magnitude. Also, while working with intact tissues, the absorption of a cation radioisotope by the cell wall occurring during the initial phase, has to be taken into the account (Kochian, 2000). Despite these methodological issues, the application of radiotracers has resulted in several major breakthroughs, such as the identification of low- and high-affinity potassium uptake mechanisms by roots (Epstein *et al.*, 1963), of the pivotal role of potassium in the development of action potentials in plants (Gaffey and Mullins, 1958) and in stomatal opening (Fischer and Hsiao, 1968). The tracer technique is perhaps the only option for investigating vacuolar potassium uptake and release *in vivo* (MacRobbie, 1995; 1998), because, once isolated from the cell, the vacuole looses its natural environment (cytoplasm and adjacent macromolecules/ organelles), which is impossible to mimic in a satisfactory manner.

A drawback of this technique, as with any other non-invasive technique, is the problem of interpreting the results, as the overall unidirectional potassium influx is studied via a sum of the large number of different transporters/channels. Dissection of these transporters is no trivial task. Already in early studies (Epstein and Rains, 1965) a spectrum of transport states for potassium and sodium was observed, challenging the notion of a simple dual mechanism of absorption by roots. The application of inhibitors, such as *N*-ethylmaleimide for high-affinity transporter-mediated flux and tetraethylammonium (TEA$^+$) as a general potassium channel blocker, provides one option for separating different transport mechanisms. However, recent studies have shown that some transporters (e.g. AtKUP1) are also sensitive to both TEA$^+$ and other non-specific potassium channel blockers. Moreover, this particular transporter displayed a dual, high- and low-affinity for potassium, with $K_m$ values similar to those observed for potassium uptake by whole roots (Fu and Luan, 1998). Furthermore, the AKT1 channel has been shown to mediate not only low, but also high-affinity potassium uptake, especially when high-affinity transporters are impaired (Hirsch et al., 1998). Therefore, the classical but oversimplified Michaelis-Menten analysis of potassium transport mechanisms on the complex system may be counterproductive. Studies of individual transporters, complemented to a heterological system (e.g. yeast cell line, defective in potassium transport) could be considered as a useful alternative. However, results appear to be rather sensitive to the nature of the heterological system *per se*. For instance, AtKUP1 expressed in Arabidopsis cell culture displayed predominantly high-affinity uptake, whereas its expression in yeast results in domination by the low-affinity component (Kim et al., 1998; Fu and Luan, 1998). Switching between the two activity modes of a transporter could be produced by post-translational mechanisms such as site-specific phosphorylation (Liu and Tsay, 2003). Also, knockout of a certain high-affinity transporter often results in an adjustment in the apparent affinities of other transporters in a way that minimises the alteration (Gierth et al, 2005). This could occur, at least partly, via lowering of the membrane potential and/or enhancing the driving force for potassium ($^{86}$Rb$^+$). To verify this however, requires additional techniques such as microelectrodes.

### 3.2. Fluorescent dyes: visualisation of intracellular potassium and potassium-transporting proteins

Another non-invasive technique to measure dynamic changes of potassium concentrations relies on a use of potassium-sensing fluorescent dyes. PBFI (Potassium Binding Bensofurane Isophtalate) was synthesised in Tsien's laboratory together with SBFI for sodium, by increasing the size of central macrocycle (diazacrown in PBFI), conferring potassium over sodium selectivity. Unfortunately, this selectivity is far from perfect ($K_d$ = 8 and 21 mM for potassium and sodium, respectively), implying poorer performance for potassium sensing on a background of high sodium (Minta and Tsien, 1989). A more sophisticated phase-modulation fluorometry needs to be used to measure potassium in the presence of high sodium concentrations instead of simple excitation measurements (Szmacinski and Lakowicz, 1999). Also, the sensitivity to potassium decreases greatly at low pH (<6.0, Mühling and Sattelmacher, 1997). Another problem is dye loading. Ideally, a simple infiltration should be used to load the indicator into the apoplast (Mühling and Sattelmacher, 1997; Mühling and Läuchli, 2000). Successful loading of both PBFI and SFBI has been reported for Arabidopsis root hairs (Halperin and Lynch, 2003), but this method does not work in many other tissues. A more efficient way of introducing the dye into the cell is the use of the membrane-permeable ester form; PBFI-AM. Operation of endogenic esterases converts it into less permeable PBFI, achieving an efficient intracellular loading (Lindberg and Strid, 1997). To date, no reports of PBFI loading into plant cell organelles have been published. However, this is possible in principle. For instance, to detect selectively the fluorescence from mitochondria, PBFI-loaded cells were simply permeabilised by digitonion, washing out the dye from the cytosol (Zoeteweij et al., 1994).

Ion transporters and especially ion channels are not the most abundant membrane components. However, modern sensors can detect the fluorescence emission from a single fluorophore molecule. The most popular fluorescent reporter (tag) is jellyfish (*Aequorea victoria*) green fluorescent protein (GFP) and its derivatives. This protein has a beta-barrel structure, with a chromophore inside, shielded from environment perturbations. Removal of the cryptic intron in animal GFP resulted in the achievement of a stable expression of GFP in Arabidopsis (Haseloff et al., 1997). Tagging by GFP could be used not only for protein localisation, but also for studying protein and membrane dynamics and interactions, and for marking and identification of cellular compartments or tissues (Hanson and Köhler, 2001; Dixit et al., 2006). For instance, for Arabidopsis, marking by GFP was used to separate stele and cortex protoplasts in roots, and to identify protoplasts originating from the phloem; thus enabling investigations of the specific pattern of potassium channel expression within these tissues (Maathuis et al., 1998; Ivashikina et al., 2003). Another successful application in relation to potassium transport is concerned with subcellular (vacuolar) localisation of the novel potassium channel; KCO1, (Czempinski et al, 2002) the OsHAK10 potassium transporter (Bañuelos et al., 2002) and the plasma membrane localisation of the cyclic-nucleotide gated channel family member; CNGC3 (Gobert et al., 2006). GFP-tagging can also be used to study channel protein and vesicle trafficing, for instance; KAT1 insertion into and retrieval from the guard cell plasma membrane (Hurst et al., 2004; Meckel et al., 2004). Multiple GFP-derivatives are now available, covering the excitation spectrum from blue to red light (Zhang et al., 2002; Dixit et al., 2006). One important application of different FPs is the creation of donor-acceptor pairs for Förster non-radiative energy transfer, to study intra- and intermolecular interactions (e.g. between two proteins and functional groups within a protein). As an example of such a pair, fulfilling the mandatory requirement that the donor emission spectrum overlaps acceptor excitation spectrum, is a combination of CFP/YFP, cyan and yellow fluorescent proteins. FRET (fluorescence resonance energy transfer) is a quantum mechanical phenomenon, occurring at typical distances of 2-6 nm. There are several available intensity-based methods, such as, for instance, measuring the ratio of donor-acceptor intensity or lifetime-based methods (e.g. FLIM, fluorescent-life time imaging microscopy, monitoring the decrease of donor lifetime). Details, advantages and pitfalls of these technique, as well as comments on alternative fluorescent probes can be found in recent reviews (Hanson and Köhler, 2001; Zhang et al., 2002; Hink et al., 2002; Dixit et al., 2006).

Finally, fluorescence techniques could be used to measure the change of concentration of ions other than potassium whose transport is coupled to the transport of potassium. A relevant example is the vacuolar $K^+/H^+$ antiport, where fluorescence quenching of acridine orange has been used to monitor potassium coupled movement of protons (Zhang and Blumwald, 2001; Apse et al., 2003). It is worth mentioning, that this cation/$H^+$ exchanger operates at 1:1 stoichiometry, so that, in this case, no net current movement can be detected by electrophysiological techniques.

### 3.3. Ion-selective microelectrodes

Use of ion-selective electrodes allows the investigation of potassium concentrations and fluxes *in situ* or even *in vivo*. A combination of a potassium-selective electrodes with one measuring the membrane potential allows the separation of the chemical and electrical components of the potassium gradient across the membrane. Moreover, this knowledge of the whole electrochemical gradient gives information regarding the thermodynamics of potassium transport, i.e. whether it is passive or active. As an example, under the condition of high-affinity potassium uptake by Arabidopsis roots (10 µM external potassium), measurements of membrane potential together with external and internal potassium activities revealed that the driving force for potassium movement is in an outward direction, hence any influx potassium has to be energised (Maathuis and Sanders, 1993). Very large changes of intracellular potassium were measured by potassium-selective electrodes upon stomata closure and opening (Penny and Bowling, 1974). Also, studies with triple-barrelled microelectrodes, incorporating pH and potassium and membrane potential electrodes, enabled the separation on the basis of differential intracellular pH, vacuolar and cytosolic microelectrode impalements (Walker et al., 1995). Systematic studies on barley roots and leaf cells allowed the changes in potassium activity and pH in vacuolar and cytosolic compartments under conditions of potassium deficiency and salt stress to be followed (Walker et al., 1996; Cuin et al., 2003). Introduction of a potassium-selective barrel into a multifunctional probe (including a pressure probe and a membrane potential electrode) enabled simultaneous real-time measurements of xylem solute transport, xylem pressure and trans-root potential (Wegner and Zimmermann, 2002).

As the external potassium concentration is usually a couple of orders of magnitude lower than the intracellular one, extracellularly positioned potassium-selective microelectrodes can detect relatively larger changes in potassium activity due to potassium fluxes across the plasma membrane. Combination of an extracellular potassium-selective electrode, conventional intracellular potential electrode, patch-clamp measurements and RT-PCR on Arabidopsis root hairs reported elicitor-induced potassium efflux mediated by the potassium outward-rectifying channel; GORK (Ivashikina et al., 2001). More convenient information on potassium fluxes may be obtained by a combination of a potassium-selective microelectrode with a vibration probe. A stepwise movement of an ion-selective microelectrode between two distinct points in close proximity to the plant tissue allows the determination, with a high (few seconds) temporal resolution, of the diffusion gradient for a given ion, which can then be transformed into the flux for that ion.

This approach is employed in the so-called MIFE technique (for ion-selective Microelectrode Ion Flux Estimation, see Newman, 2001 and Shabala 2006 for reviews). A few examples of MIFE application include demonstrating the role of potassium fluxes; in charge-balancing during light-induced transient potential change in photosynthetic tissue (Shabala and Newman, 1999); in osmotic adjustment upon hypertonic challenge (Shabala et al., 2000; Shabala et al. 2003); and in salt sensitivity (Chen et al., 2005; Shabala et al. 2005). The power of this technique is its non-invasive nature and versatile application for studying transport processes *in planta*. Although the deconvolution of potassium gradient into flux rate is not an absolutely accurate procedure due to complications in diffusion calculations, a parallel measurement of ion fluxes by MIFE and patch-clamp techniques supports the validity of MIFE in quantitative terms (Tyerman et al., 2001).

## 3.4. Voltage- and patch-clamp techniques: sorting membrane currents

The voltage-clamp technique was designed to measure transmembrane currents under fully controlled voltage conditions (Cole, 1949; Hodgkin et al., 1949). Currents, mediated solely by electrogenic channels and transporters can be classified by their authentic patterns and hitherto dissected, using specific recording protocols, solutions and pharmacology. Classic voltage-clamp amplifiers use two intracellular electrodes, one, a high impedance, zero current flowing, to measure the actual voltage difference across the membrane, while another, a current electrode, is used to pass the current in such a way as to maintain the desired potential. The latter current is the outcome of the measurement. In higher plant tissues, most cells are interconnected via plasmodesmata, so voltage-clamp, due to non-uniform current spread and resulting space clamp problems *in situ*, can be applied only to specialised (separate) cells such as guard cells (Blatt et al., 1990) or root hairs (Lew, 1991). Recent developments of this technique allow investigations of potassium currents in intact stomata in their unaltered environment, in combination with other single-cell techniques (Roelfsema et al., 2001; Goh et al., 2002) Classic voltage-clamp is characterised by a relatively low sensitivity. However, this technique is indispensable when large or very large currents are measured, which is common for ion channels and transporters, heterologously expressed in *Xenopus* oocytes. Patch-clamp is a modification of the voltage-clamp technique, when the functions of current-measuring and voltage-clamping electrodes are advocated to a single patch electrode via resistance feedback of the amplifier. Patch-clamp amplifiers are suited to readily detect currents as low as ~100 fA (1 fA= 6,000 elementary charges per second) up to the range of 100 nA. As currents mediated by a single ionic channel are normally of an order of few pA (Fig. 3), patch-clamp is an excellent technique to study the kinetics of individual ion channel proteins. Individual porters and pumps transport rates are normally 10-100 ions/second, thus only currents mediated by large (e.g. whole cell) populations of these proteins can be measured. Similar to voltage-clamp, patch-clamp is dealing with *net* electric currents, therefore, only the activity of *electrogenic* transporters can be registered by this method.

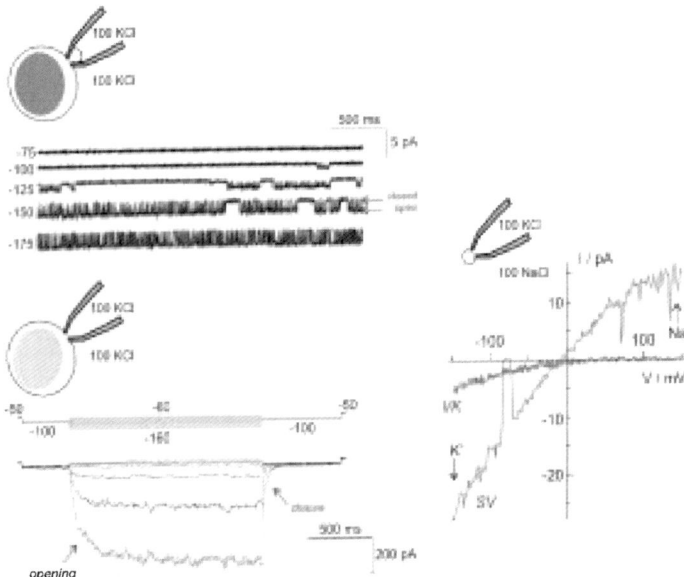

**Figure 3:** Different patch-clamp configurations. Top left panel. A small Ω-shaped patch of a wheat (*Triticum sativum* L.) root protoplast membrane is pulled into the pipette by suction. After obtaining a giga-Ohm seal, the *cell-attached* configuration is reached. The activity of individual inward-rectifying potassium channels is detected in this patch. As can be seen from the record, the channel *open probability* is greatly enhanced when the membrane potential is stepped to more negative values (<-100 mV). Breaking the patch by strong suction/voltage pulse leads to so-called *whole-cell* configuration (bottom left panel). In this configuration the tight seal between the cell membrane and patch-pipette is preserved and the cell interior is perfused by pipette solution. A low resistance access to the cell via the pipette tip allows current recording from the whole membrane surface. In this example, the whole protoplast current was determined by many (~250) potassium channels, identical to that measured in the attached patch. Pulling the pipette away from the cell results in an excised patch in either the *inside-out* or the *outside-out* configuration, depending on the initial configuration, *cell-attached* or *whole-cell*. Such approaches can be used also to study the characteristics of ion channels in intracellular membranes such as the vacuolar membrane. *Right panel*. Properties of single channel currents in an excised cytosolic side-out tonoplast patch from *Beta vulgaris* L. taproot vacuole. This patch contained a copy of the potassium selective channel (VK) and a large non-selective cation channel (SV). In this example, the clamped voltage was changed in a ramp-wave manner, yielding *current-voltage (I/V) relations* of single open VK and SV channels. Exchange of the bath to pure NaCl solution reveals that sodium can not pass through the VK channel (no outward current could be seen) but freely permeates through the SV channel pore. A measure of the *relative permeability* ($P_{K+}/P_{Na+}$) is the *reversal potential* of the single channel current. As the VK-mediated current asymptotically reaches the voltage axis at high positive potentials, one may conclude that $P_{K+} >> P_{Na+}$ for this channel. The SV channel-mediated current changes its sign at ~-10 mV. This implies that the SV channel is slightly selective for sodium over potassium.

Potassium uptake and transport

To take advantage of the high resolution of the patch-clamp, an electrically tight seal is required between patch-electrode and membrane surface. Enzymatic cleaning of the membrane surface and experimentation with glass microelectrode tips (geometry, fire polishing) plus some specific tricks (suction, etc) leads to seal values in the range of GΩ (Hamill et al., 1981). Although up to now, the mechanism of GΩ seal formation remains a mystery, many consequences of this phenomenon are utilized. The contact between glass and membrane is mechanically stable, which permits isolation of cell-free patches of either orientation, inside or outside-out (Fig. 3). These configurations are suitable for assaying single channel activity under controlled conditions (voltage, solution composition, mechanical pressure). Alternatively, currents mediated by whole cell population of channels and transporters can be studied in the whole-cell mode. In this configuration one can also evaluate other electric parameters such as membrane capacitance (hence, membrane surface - to study secretion, protein traffic and/or transporters and channels expression/density) and membrane voltage (in current-clamp mode). To make plant cell membrane accessible to patch-pipette, the most common technique is the enzymatic digestion of the cell wall, leading to a release of protoplasts. There are several improvements on this, making the overall procedure shorter and the enzyme treatment minimal (Elzenga et al., 1991); indeed, the use of plasmolysis/ deplasmolysis and mechanical treatment enables, in some instances, the complete avoidance of enzymatic treatment (Trebacz and Schönknecht, 2000). Alternatively, laser microsurgery can be employed. This approach is especially important in studies of polarised cells and/or cells within tissue *in situ*, as a laser beam cuts only a small window in a cell wall of selected cell (Kurkdjian et al., 1993; Henriksen et al., 1996). Several preparative techniques have also been developed to study channels in plant endomembranes (Keller and Hedrich, 1992). By this means, potassium permeable channels were discovered by patch-clamp in plant cell plasma membranes (Moran et al., 1984, Schroeder et al., 1984), in vacuoles (Hedrich et al., 1986; 1988), chloroplast thylakoid (Pottosin and Schönknecht, 1996) and envelope (Pottosin, 1992) membranes, as well as in nuclei (Grygorczyk and Grygorczyk, 1998). Recently, using giant liposomes formed on the basis of native membrane fraction (Keller and Hedrich, 1992) single channels in plant endoplasmic reticulum were characterised. Some developments and problems of patch-clamp technique in its application to plant membranes are described elsewhere (White et al., 1998).

Some plant cell potassium currents are very distinct in their characteristics, e.g. inward-and outward-rectifier potassium channels. However, in a single plant or even in the same cell, distinct channels with apparently similar characteristics can be present, as for instance two inward rectifiers KAT1 and KAT2 in guard cells (Table 1). So far, there is no specific pharmacology targeted against plant potassium channels. Therefore, in order to characterise individual potassium channels and transporters, heterological expression is used.

There are several heterological expression systems proved to be suitable for patch- and/or voltage-clamp analysis. The following potassium channels and transporters have been analysed: AKT2 in yeast *Saccharomyces cerevisiae* plasma membrane (Bertl et al., 1995), KCO1 in yeast vacuolar membrane (Bihler et al.,

Table 1. The major channels involved in mediating potassium transport in Arabidopsis.

| Name | Type | Location | Function | References |
|---|---|---|---|---|
| **Shaker-type channels** | | | | |
| AKT1 | IR | Root cortex, epidermis, endodermis and hair, leaf mesophyll | $K^+$ uptake | Sentenac et al., 1992 |
| AKT2/3 | WIR | Phloem, xylem, leaf mesophyll, guard cell | Phloem loading/ unloading; mesophyll uptake | Marten et al., 1999; Deeken et al., 2000; Lacombe et al., 2000 |
| AKT6 | – | Flower | – | Lacombe et al., 2000 |
| KAT1 | IR | Guard cell | stomatal movement | Anderson et al., 1992; Szyroki et al., 2001 |
| KAT2 | IR | Guard cell, leaf phloem, flowers | Stomatal movement? | Pilot et al., 2001; Szyroki et al., 2001 |
| SPIK | IR | Pollen | Tube development | Mouline et al., 2002 |
| SKOR | OR | Root pericycle; stelar parenchyma cells | Xylem loading | Gaymard et al., 1998 |
| GORK | OR | Guard cell, stem, root hair and epidermis | Stomatal movement, $K^+$ efflux, $K^+$ sensing | Ache et al., 2001; Ivashikina et al., 2001; Hosy et al., 2003 |
| KC1 | – | Meristem, epidermis, cortex, endodermis | $K^+$ uptake | Ivashikina et al., 2001: Reintanz et al., 2002: Pilot et al., 2003 |
| **"Two-pore" channels** | | | | |
| KCO2 | OR? | - | - | Czempinski et al., 2002 |
| KCO3 | OR? | – | – | Czempinski et al., 2002 |

**Table 1.** Continued

| Name | Type | Location | Function | References |
|---|---|---|---|---|
| KCO4 | OR? | Pollen | Pollen tube $K^+$ homeostasis | Czempinski et al., 2002; Becker et al., 2004 |
| KCO5 | OR? | – | – | Czempinski et al., 2002 |
| KCO6 | OR? | Roots, mesophyll, guard cell | – | Czempinski et al., 2002 |

IR = inward rectifier; OR = outward rectifier; WIR = weakly inward rectifying channel

2005), KAT1, SKOR, KZM1, AtCNGC1 and AtCNGC2, as well as HKT1 in *Xenopus* oocytes (Hoshi, 1995; Gaymard et al., 1998; Philippar et al., 2003; Hua et al., 2003; Golldack et al., 2002), SKT1, KST1, KCO1 in baculovirus-infected insect cells (Zimmermann et al., 1998; Czempinski et al., 1999) and in transfected mammalian cell lines, such as hamster ovary cells (KDC1, Downey et al., 2000) or HEK293 cells (OsAKT1, Fuchs et al., 2005). The choice of the adequate system depends on several considerations. For instance, yeast cells are a very attractive system for genetic manipulations, but could become a nightmare for a patch-clamper. *Xenopus* oocytes are much easier to handle, but only transient expression is observed in this system. For example, all attempts to express AKT1 in *Xenopus* oocytes have failed, but this channel can be successfully expressed in baculovirus-infected cell lines (Gaymard et al., 1996). Once the wild type channel or transporter is characterised in detail, structure-function studies can be initiated. As an example, the use of site-directed mutagenesis to identify the groups responsible for pH dependence in KAT1 and KST1 channels (Hoth and Hedrich, 1999) can be mentioned. This approach showed that deletions in C-terminal domain of SKT1 prevents its hetero-oligomerisation with KST1 (Zimmermann et al., 2001) and gave a clue to the origin of the opposite rectification in the two related *Shaker* channels, KAT1 and SKOR (Porée et al., 2005).

### 3.5. Molecular methods

Some molecular biology techniques (such as the expression of GFP-constructs, transfection into a heterologous system, site-directed mutagenesis) have been already mentioned in combination with other methods. However, molecular biology methods by themselves provide a powerful tool to identify genes encoding potassium-transporting proteins, their expression (level and profile), function of a particular gene product etc. The first two plant potassium channel sequences, KAT1 (Anderson et al., 1992) and AKT1 (Sentenac et al., 1992), were discovered by screening of constructed Arabidopsis cDNA libraries for animal Shaker potassium channel homologues. These two channels were cloned in yeast, defective in potassium transport, so that the expression of foreign channel genes restored the growth of colonies in a low potassium medium. Alternative approaches have emerged thanks to expanding genetic databases. The polymerase chain reaction coupled with reverse transcription (RT-PCR), was used to identify a multi-gene family that in barley and other cereals encodes proteins homologous to high affinity potassium transporters SoHAKl from fungi *Schwanniomyces occidentalis* and KUP from *E. coli* (Santa-María *et al.*, 1997). In Arabidopsis, the homologues of these potassium transporters were found simply by computer-aided searches in gene banks (Kim *et al.*, 1998). As the sequencing of the Arabidopsis and rice genomes is now completed, bioinformatics has become the most important tool for identification of new transporters and channels in plants (Mäser *et al.*, 2001). Functional characterisation is commonly done by complementation of growth of potassium transport deficient yeast (Anderson *et al.*, 1992; Sentenac *et al.*, 1992; Santa-María *et al.*, 1997; Gobert *et al.*, 2006) or *E. coli* (Kim *et al.*, 1998). However, only heterological expression with a subsequent electrophysiological characterisation may provide the firm evidence that a particular gene product is responsible for a certain current *in planta*. With a heterological system, however, the problem of mistargeting of the protein arises. This is particularly important for organelle (e.g. vacuolar) channels. Analysis of knock-out mutants and plants with the over-expression of a given gene product may be a viable alternative, especially, when the respective family is represented by a single gene as, for instance, AtNHX1 and TPC1 in Arabidopsis, encoding a vacuolar cation/$H^+$ antiporter and non-selective channel, respectively (Zhang and Blumwald, 2001; Peiter *et al.*, 2005). It should be noted here, that, although phenotypic analysis of knock-out mutants and over-expressors is a widely used approach to investigate a possible protein function *in planta*, these results should be treated *cum grano salis*, bearing in mind the redundancy of protein function in plants as well as the enormous adaptive plasticity of plant responses.

To further specify authentic functions of potassium transporters *in planta*, the profile of expression is important. Use of reporter gene constructs is the most common way to unravel the potential preferential expression of transporter genes within different tissues in a plant. Besides the already described GFP-fusions, GUS constructs are widely used; most of the data on the location of channels/transporters in Tables 1 and 2 were obtained in this way. This method is based on the histochemical activity of the β-glucuronidase enzyme from *E. coli*. This enzyme converts the substrate (5-bromo-4-chloro-3-indolyl-β-D-glucuronide) into the reactive indoxyl molecule, which, under oxidative conditions (ferricyanide added), forms indigo blue, by dimerisation. This is visualised as blue precipitate granules. Measurements in semi thin sections of Arabidopsis explants embedded in plastic allows spatial resolution up to the single cell level (De Block and Lijsebettens, 1998). *In situ* hybridisation and autoradiography is another option to determine the spatial distribution of RNA or DNA species in the whole plant (Kong and Simon, 1998) The rapidly developing technique of single-cell RT-PCR (Brandt, 2005) which enables information to be obtained about the temporal and spatial expression of specific transporter genes (Brandt *et al.*, 1999; 2002; Kehr, 2001) will lead to a greater specification of putative potassium transporters within the plant. Although little information using this technique to relate potassium gene expression to potassium transporter activity is currently available, detailed procedures for this technique are available (Karrer *et al.*, 1995; Brandt *et al.*, 1999; Gallagher *et al.*, 2001) and have been recently reviewed (Brandt, 2005).

As an alternative to FRET analysis, a more robust method to study protein interactions utilises a yeast two-hybrid system. The idea of this method is to use the transcription of a certain yeast reporter gene, which is activated by a two-domain factor (Gal4), containing the DNA-binding domain and the activator domain. Expression in yeast of these domains, each fused with the protein of interest (or its fragment), allows us to screen them for direct interaction, because only in this case will the transcription of reporter gene be activated. Using this approach, it was found that the guard cell inward-rectifier potassium channels form functional oligoheteromers both in Arabidopsis; KAT1 and KAT2 (Pilot *et al.*, 2001), and in potato; KST1, SKT2, SKT3 (Erhardt *et al.*, 1997). Moreover, potato guard cell channels were shown to interact via their C-termini, which was also confirmed by "Green Western" blot on the basis of *in vitro* cross-interaction of the channel proteins C-termini fused with GFP and native ones.

## 4. BASIC FEATURES OF POTASSIUM UPTAKE AND TRANSPORT IN PLANTS

Potassium is moved from the soil solution to the root surface predominantly by diffusion and enters the root symplast via the cell plasma membrane (Fig. 2a). From there, it can travel through the symplast to the vascular tissues (Fig. 2b), where it is unloaded from the xylem vessels for long-distance transport to the leaves. Potassium is reabsorbed from the xylem into leaf cells (Fig. 2c). Being a highly mobile element (Marschner, 1995), it can be easily loaded into the phloem for translocation to actively growing sink tissues (e.g. shoot and root apices) and can be unloaded by way of symplasmic or apoplastic pathways (Fig. 2d). Potassium can also cross the tonoplast membrane for storage in vacuoles of both root and shoot cells. The integration and regulation of potassium transport systems at different sites along the long-distance pathway allows the plant to direct the partitioning and circulation of potassium. Such an integration of transport systems plays a central role in plant growth and development and in the allocation of potassium in response to changes in potassium availability and plant potassium requirements.

### 4.1. General features of membrane transporters in plants

Solute and ion transport across cellular membranes is dependent on metabolically coupled proton pumps that generate an electrochemical potential gradient across cellular membranes (Fig. 4a). Two major electrogenic proton pumps, $H^+$-ATPase (present at both the plasma membrane and tonoplast) and $PP_i$-ase (tonoplast only), hydrolyse ATP and inorganic pyrophosphate ($PP_i$), respectively, to power the transport of protons out of the cytosol. This establishes a significant electrical gradient across the membrane, as well as a pH gradient. Together, these two gradients constitute the electrochemical potential or proton motive force across the membrane. This proton motive force is used in secondary transport processes to drive the transport of many other solutes and ions (including potassium) against their electrochemical gradient.

Transport systems that couple the downhill flow of one ion (such as $H^+$) to the uphill flow of another ion (such as $K^+$) or solutes are called carriers. This is termed secondary active transport. An antiport (e.g. $H^+/K^+$ antiporter) refers to coupled transport in which the downhill movement of protons drives the uphill transport of potassium in the opposite direction (Fig. 4b). Co-transport or proton symport (Fig. 4c) involves the influx of a cation alongside the inward movement of $H^+$ such as is found for high-affinity potassium transport (e.g. $K^+/H^+$ co-transporter) which shows a 1:1 transport stoichiometry (Maathuis and Sanders, 1993). Such carriers are generally high-affinity transport systems with a high specificity for a particular solute or ion and a low capacity of movement, in the order of $10^3$ molecules of solute transport per second.

Ion channels (Fig. 4d) are integral membrane proteins facilitating ion movement along the electrochemical gradient. Transport through channels is always passive and they are thought to be low-affinity ion transporters with a high capacity of about $10^8$ solute molecules transported per second. These channels do not tend to stay open for long periods, but have "gates" that open and close the channel pore in response to external signals, such as voltage, pH, mechanical stretch, G-proteins, cytosolic calcium etc. Many such channels show rather steep rectification, e.g. are capable of conducting ions only in a certain range of membrane potentials and/or in a particular direction. Inward-rectifying channels mediate current flow into the cell, and outward-rectifiers - out of the cell. Ion currents through such channels may be either time-dependent (e.g. having current that increases over a period of several hundred milliseconds) upon a change in the membrane voltage, or time-independent (instantaneous). For potassium channels, outward-rectifiers are opened upon membrane depolarisations, resulting in a loss of potassium from the cell, while inward-rectifiers are activated by hyperpolarisation of the membrane potential resulting in the movement of potassium into the cell.

Within a plant, the uptake of potassium from the growth medium and transport of potassium within the plant utilises both potassium carriers and potassium mediating channels. These will be discussed below.

### 4.2. Potassium uptake in plant roots

Plant roots absorb potassium over a wide range of soil concentrations. The pioneering tracer flux studies by Epstein and co-workers (Epstein *et al.*, 1963) indicated that at least two potassium transport mechanisms exist. One of them (so-called System *I*) mediates high-affinity potassium uptake at external potassium concentrations below 1 mM. The other (so-called System *II*) operates at higher external potassium concentrations as a low-affinity uptake mechanism. While the

Potassium uptake and transport

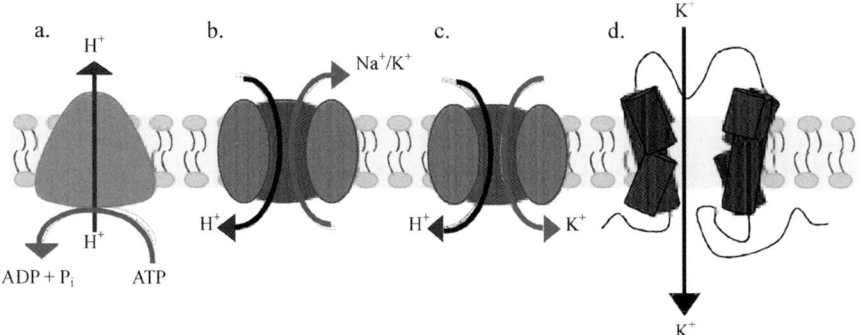

**Figure 4:** A summary model of the various transporters responsible for mediating potassium transport at an idealised membrane. The proton pumping $H^+$-ATPase (a) transports protons through the membrane from the internal to the external surface. The generated electrochemical gradient is used for the uphill transport of potassium or other cations via secondary active co-transport mechanisms (b, c). Two major types of symporters are known: (b) the antiport in which the downhill movement of a proton is coupled to the uphill movement of a cation and (c) a symport in which both the proton and cation are moved in the same direction. (d) An inward-rectifying potassium channel in which the movement of potassium is passive, but dependent on the electrochemical driving forces on that ion. Outward-rectifying channels, weakly-voltage rectifying and stretch-activated channels (not shown) are also involved in mediating potassium transport across the membrane. At the tonoplast, but not the plasma membrane, the $PP_i$-ase also contributes, along with the $H^+$-ATPase, to the electrochemical gradient.

high-affinity potassium uptake; System *I* is strongly selective for potassium over other alkali cations and shows increased gene expression or transport activity under potassium starvation conditions, the low-affinity potassium transport; System *II* is less selective for potassium over sodium and less influenced by changes in the potassium status of the plant (Marschner, 1995). At the molecular level, these components are conventionally attributed to the activities of transporters and channels respectively (Maathuis and Sanders, 1994; 1997). Patch-clamp studies suggest that System *I* is indeed an active transport mechanism (Maathuis and Sanders, 1993), most likely via a $K^+/H^+$ symporter (Maathuis and Sanders, 1994). The inward-rectifying potassium-selective (KIR) channels have been found in root cells of various species (White and Tester, 1992; Gassmann and Schroeder, 1994; Maathuis and Sanders, 1995; Roberts and Tester, 1995) and are believed to mediate potassium uptake within the concentration range of System *II* (above 1 mM). Importantly, KIR channels are also found in root hairs (Gassmann and Schroeder, 1994), suggesting their important role in potassium acquisition beyond the root depletion zone. However, recent findings indicate that, at least in terms of affinity to potassium, the boundaries between transporters and channels are not clearly defined (Fu and Luan, 1998; Hirsch *et al.*, 1998; Spalding *et al.*, 1999). It is now apparent that there is functional overlap between the high- and low-affinity uptake mechanisms for potassium acquisition in plants (Hirsh *et al.*, 1998; Santa-María *et al.*, 2000).

### 4.3. Xylem loading and potassium transport into the shoot

To be delivered to the shoot, potassium absorbed by plant roots has to be transported to the vascular tissues where it is unloaded from xylem parenchyma into xylem vessels. Significant evidence has accumulated demonstrating the presence of both anion and cation-permeable channels at the xylem/parenchyma boundary. These channels are likely to be responsible for loading of solutes into the xylem for transport to the shoot (Wegner *et al.*, 1994; Wegner and de Boer, 1997; Maathuis *et al.*, 1998; Köhler and Raschke, 2000). Patch-clamp experiments have revealed a significant difference in KIR and KOR channel densities between epidermal and stellar root tissues (Roberts and Tester 1995), supporting the important role of potassium outward-rectifying channels in xylem loading. More supporting evidence comes from experiments with Arabidopsis mutants. Several authors have shown that SKOR (outward-rectifying Shaker channel in Arabidopsis xylem parenchyma tissue) contributes to about 50% of the potassium translocated towards the shoot (Gaymard *et al.*, 1998; Lacombe *et al.*, 2000). However, some authors have argued against the role of outward potassium channels in xylem sap potassium loading (de Boer, 1999), suggesting instead that potassium secretion into the xylem occurs against the potassium electrochemical gradient in a process mediated by active transport systems (Kochian and Lucas, 1988; Moshelion *et al.*, 2002). The major source of disagreement results from the absence of reliable estimates of membrane potential and apoplastic ion activities in inner root tissues (Grignon and Sentenac, 1991).

### 4.4. Remobilisation and recycling

After delivery to the leaf tissue, potassium can be loaded into phloem cells for translocation to actively growing sink tissues (e.g. shoot and root apices), where it can be unloaded by way of symplasmic or apoplastic pathways. Such a redistribution of potassium also involves the remobilisation from the older, senescing leaves to the younger parts of the plant (Marschner 1995). It is estimated that in wheat and rice, up to 35% of xylem potassium originates from the phloem (Grignon and Sentenac 1991).

## 5. IONIC MECHANISMS OF POTASSIUM ACQUISITION AND TRANSPORT IN PLANTS

### 5.1. General features of potassium transporters in plants

Since the classical work of Epstein and co-workers (Epstein *et al.*, 1963), many advanced electrophysiological and molecular techniques, as described above, have become available which have allowed considerable progress in the analysis of potassium transport in plants at both the molecular and physiological level. However, although channel mediated transport has been studied in great detail because of the availability of advanced electrophysiological techniques and the relative ease of expression in heterologous systems (i.e. expressed in *Xenopus* oocytes, insect and mammalian cell lines, or yeast), less information is available on transport systems characterised by a lower rate of potassium transport. To date, most of our current information regarding potassium transport in plant cells is from the model plant species, *Arabidopsis thaliana*. While information concerning potassium transport systems from other plant species, including crop species is slowly becoming available, it still lags severely behind that of Arabidopsis.

The completion of the Arabidopsis genome sequencing project has enabled the identification of putative plant transporter proteins (Ward, 2001). Seven major families of Arabidopsis cation transporters (75 genes in total), proposed to be involved in mediating potassium transport across plant membranes have been identified (Mäser *et al.*, 2001; Véry and Sentenac, 2002; 2003). The major potassium channels known to be involved in mediated potassium efflux and influx across cellular membranes are listed in Table 1, and important potassium transporters in Table 2.

## 5.2. Potassium permeable channels

### 5.2.1. Shaker-type potassium channels

***Classification***: Plant Shaker-like potassium channels are related to the animal potassium channels initially cloned from *Drosophila* (hence the name "Shakers"). The Shaker-type potassium channels KAT1 (Anderson *et al.*, 1992) and AKT1 (Sentenac *et al.*, 1992) were the first potassium transporting proteins cloned from plants. Since then, nine members of the Shaker family of potassium channels have been identified in Arabidopsis (Mäser *et al.*, 2001). These include, in addition to the previously found AKT1 (Sentenac *et al.*, 1992) and KAT1 (Anderson *et al.*, 1992), SKOR (Gaymard *et al.*, 1998), KAT2 (Pilot *et al.*, 2001), SPIK (Mouline *et al.*, 2002), AKT2/3 (Marten *et al.*, 1999; Deeken *et al.*, 2000; Lacombe *et al.*, 2000), GORK (Ache *et al.*, 2001; Ivashikina *et al.*, 2001; Hosy *et al.*, 2003); AtKC1 (Ivashikina *et al.*, 2001; Reintanz *et al.*, 2002; Pilot *et al.*, 2003) and AKT6 (Lacombe *et al.*, 2000). Shaker-like potassium channels are present in numerous cell types within plants. They operate at millimolar potassium concentrations and are involved in a low-affinity transport.

Comparison of the functional properties of potassium Shaker-type channels in heterologous expression systems with channel activity recorded *in planta*, indicates that these channels mediate most of the potassium-selective voltage-gated currents that dominate the membrane potassium conductance at both hyperpolarised and depolarised membrane potentials (Véry and Sentenac, 2002). Based on this voltage dependency, these channels are grouped into three functional subfamilies (Véry and Sentenac, 2002; 2003):

(i) Inward-rectifying channels (AKT1, KAT1, KAT2 and SPIK). These channels are mainly involved in potassium uptake and are activated by membrane hyperpolarisation, from a threshold generally more negative than the potassium equilibrium potential ($E_k$);

**Table 2.** Potassium transporters cloned in Arabidopsis.

| Name | Location | Function | References |
|---|---|---|---|
| **KUP/HAK/KT family** | | | |
| KUP1 | Flower, leaf, stem, root | Root $K^+$ uptake | Quintero & Blatt., 1997; Kim et al., 1998; Fu & Luan, 1998 |
| KUP2 | Flower, leaf, stem, root | $K^+$ transport, regulation of cell elongation | Kim et al., 1998; Elumalai et al., 2002 |
| KUP3 | Flower, leaf, stem, root | – | Kim et al., 1998 |
| KUP4 | Flower, leaf, stem, root | Root hair elongation | Kim et al., 1998; Rigas et al., 2001 |
| KUP5-12 | Generally, ubiquitously expressed | $K^+$ transport | Ahn et al., 2004 |
| HAK5 | Roots, shoots | High affinity $K^+$ uptake | Ahn et al., 2004; Shin & Schachtman, 2004; Gierth et al., 2005 |
| **HKT family** | | | |
| HKT1 | Root | Root $Na^+$ and/or $K^+$ uptake | Uozumi et al., 2000; Rus et al., 2001 |
| **$K^+/H^+$ antiporters** | | | |
| NHX1 | Leaf mesophyll | $Na^+$ and $K^+$? transport at tonoplast | Zhang and Blumwald, 2001; Venema et al., 2002; Apse et al., 2003 |
| CHX17 | Cortex, epidermis | $K^+$ uptake | Cellier et al., 2004 |

(ii) Weakly-inward-rectifying channels (AKT2/3) are able to mediate both potassium uptake and release depending on the local potassium electrochemical gradients; and

(iii) Outward-rectifying channels (SKOR and GORK). These channels activate at membrane potentials more positive than $E_k$ and thus mediate potassium release (Véry and Sentenac, 2003).

***Functional expression***: The physiological roles and functional expression of the Shaker family channels are diverse. AKT1 is expressed in roots (Bertl et al., 1994;

## Potassium uptake and transport

Lagarde *et al.*, 1996) where it plays a major role in potassium acquisition by the plant (Lagarde *et al.*, 1996; Hirsch *et al.*, 1998). Interestingly, this channel facilitates transport within a very broad range of external potassium concentrations, including micromolar concentrations (Hirsch *et al.*, 1998; Spalding *et al.*, 1999). AtKC1 is also expressed in root periphery cells (Ivashikina *et al.*, 2001; Pilot *et al.*, 2003) but does not form functional ion channels when expressed in heterologous systems. However, disruption of this protein does affect the biophysical characteristics of the AKT1-mediated inward current in root hairs (Dreyer *et al.*, 1997; Reintanz *et al.*, 2002). This has lead to the proposal that the role of AtKC1 is as part of a heteromeric functional channel protein, along with AKT1 (Reintanz *et al.*, 2002). This hypothesis is supported by two-hybrid testing, confirming a physical interaction between AtKC1 and AKT1 (Pilot *et al.*, 2003).

Several orthologues of *AKT1* have been identified in a number of other plant species, such as *SKT1* in potato (*Solanum tuberosum*; Zimmerman *et al.*, 1998). *LKT1* in tomato (*Lycopersicon esculentum*; Hartje *et al.*, 2000), *MKT1* in common ice plant (*Mesembryanthemum crystallinum*; Su *et al.*, 2001), *TaAKT1* in wheat (*Triticum aestivum*; Buschmann *et al.*, 2000) and *OsAKT1* in rice (*Oryza sativa*; Golldack *et al.*, 2003). In total, ten Shaker-like potassium channels have now been reported in rice (Bridges *et al.*, 2005).

AKT2 and AKT3 are differentially initiated transcripts from a single gene. As based on the promoter reporter experiments for *AKT2* and *in situ* hybridisation for *AKT3*, this gene is predominantly expressed in phloem and xylem parenchyma (Marten *et al.*, 1999; Lacombe *et al.*, 2000). The products of *AKT2/AKT3* expression in *Xenopus* oocytes and COS cells are characterised as weakly-rectifying potassium channels. Such a weakly-rectifying channel would enable bi-directional potassium transport. Thus AKT2/AKT3 may be involved in phloem loading and/or unloading (Marten *et al.*, 1999; Lacombe *et al.*, 2000; Deeken *et al.*, 2002). AKT2 has also been shown to physically interact with AKT1 and AtKC1 (Pilot *et al.*, 2003). Furthermore, low-level expression of *AKT2* has also been detected in the leaf lamina (Lacombe *et al.*, 2000). Hence, it may also be an important contributor, along with AKT1, to potassium uptake into leaf mesophyll cells (Dennison *et al.*, 2001).

KAT1 and KAT2 are guard cell specific channels (Szyroki *et al.*, 2001) mediating potassium influxes for turgor dependent regulation of stomatal aperture (Nakamura *et al.*, 1995; Pilot *et al.*, 2001). During stomatal closure, the potassium depolarisation-activated, outward-rectifier GORK mediates potassium release from the guard cells (Ache *et al.*, 2001; Hosy *et al.*, 2003). As well as being expressed in guard cells, GORK is also expressed in root hairs and root epidermal cells where it mediates potassium efflux from root epidermal cells and root hairs into the root medium in response to external stimuli (Ivashikina *et al.*, 2001). Because its channel gating is potassium sensitive, GORK has been proposed to function as a potassium sensor within the root cells (Ache *et al.*, 2000; Ivashikina *et al.*, 2001). Like GORK, SKOR is also a depolarisation-activated outward-rectifying potassium channel, but it is expressed in the root pericycle and stelar parenchyma cells. Here, it is proposed to be involved in xylem loading, mediating the delivery of up to 50% of the potassium to the xylem sap (Gaymard *et al.*, 1998).

The inward-rectifier SPIK is specifically expressed in pollen where it mediates potassium uptake. Pollen tube growth and, therefore, pollen competitive ability are dependent on the activity of this channel (Mouline *et al.*, 2002). For the remaining known Shaker-like potassium channel; AKT6, much less is known. Only localisation data in flowers have so far been obtained (Lacombe *et al.*, 2000).

*5.2.2. "Two-pore" potassium channels*

The first plant two-pore potassium channel was cloned from Arabidopsis and named *AtKCO1* for $K^+$ channel, $Ca^{2+}$-activated, outward rectifying 1 (Czempinski *et al.*, 1997). KCO channels display a hydrophobic core composed of either four transmembrane segments and two P domains (KCO-2P family), or two transmembrane segments and one P domain (KCO-1P family). The channels exhibit a predicted 14-3-3 binding motif at their N-terminus (Becker *et al.*, 2004) and are characterised by the presence of one or two EF hands localised to the C-terminal domains of the respective channel proteins. These domains are thought to be involved in calcium regulation of the channel (Czempinski *et al.*, 1999; Moshelion *et al.*, 2002; Véry and Sentenac, 2002). Although this EF hand distinguishes them from their animal relatives of the KCNK family (Goldstein *et al.*, 2001), the channels do share some structural homologies with 4TMS-2P (leak-like) and 2TMS-1P (inward-rectifying) animal potassium channels, respectively (Jan and Jan, 1997; Doupnik *et al.*, 1995).

In Arabidopsis, the KCO-2P family has five members; KCO1 - 5, and the KCO-1P family has a single member (Czempinski *et al.*, 2002). The tonoplast located *AtKCO1* (Czempinski *et al.*, 2002; Schönknecht *et al.*, 2002), has been characterised by heterologous expression in insect cells, where it has been shown to encode a potassium selective outward-rectifying channel activated by cytosolic calcium (Czempinski *et al.*, 1997), while showing voltage independence (O'Connell *et al.*, 2002; Bihler *et al.*, 2005).

It has recently been demonstrated, however, that some members of the KCO family, such as AtKCO4, cannot be uniformly classified as outward-rectifiers, leading to the suggestion that this group of potassium channels should be renamed the *TPK* family for Tandem-Pore $K^+$ (Becker *et al.*, 2004). Moreover, some TPK/KCO channels also appear to differ in their regulation. As well as different rectification properties, AtKCO4/AtTPK4 show a lack of the EF hands and the sequence for AtKCO5/AtTPK5 exhibits only weak similarity to the EF-consensus motif, suggesting that direct regulation of these particular channels by cytoplasmic calcium is unlikely (Becker *et al.*, 2004). Furthermore, AtKCO4/AtTPK4 lacks the 14-3-3 binding site (Becker *et al.*, 2004). This indicates that these particular channels share greater functional properties with members of the animal potassium channels of the KCNK family (Goldstein *et al.*, 2001) than other members of this family.

The location of many members of the AtKCO family has not yet been identified within Arabidopsis. Among members that have been localised, *KCO1* is found to be expressed throughout the plant, with strongest expression in the seedlings, leaves and flowers (Czempinski *et al.*, 2002; Schönknecht *et al.*, 2002) as well as roots (Schönknecht *et al.*, 2002). Similarly *AtKCO6* demonstrates strongest

Potassium uptake and transport

expression in roots and leaves (Schönknecht *et al.*, 2002). *AtKCO1-GUS* (β-glucuronidase) expression is detected in mitotically active tissue (Czempinski *et al.*, 2002) and *AtTPK4* (*AtKCO4*); predominantly in pollen (Becker *et al.*, 2004). At the subcellular level, *AtKCO1* has been localised to the tonoplast (Czempinski *et al.*, 2002; Schönknecht *et al.*, 2002), while AtTPK4 functions in the plasma membrane of pollen, where it may play a role in potassium homeostasis and membrane potential regulation of the growing pollen tube (Becker *et al.*, 2004).

Within other plant species KCO1-like channels has been cloned from the rain tree *Samanea samem* (named *SPOCK1*; Moshelion *et al.*, 2002) and potato (*StKCO1α* and *StKCO1β*; Czempinski *et al.*, 2002).

*5.2.3. Non-selective cation channels (NSCCs)*

Non-selective cation channels (NSCCs) are widely present in plants, with forty putative NSCCs revealed in the Arabidopsis genome sequence (Mäser *et al.*, 2001; Demidchik *et al.*, 2002; Véry and Sentenac, 2002; 2003). NSCCs appear to form a large, heterogeneous group of channels which typically show a high selectivity for cations over anions. Some NSCCs do not permit permeation of divalent cations, while other NSCCs are relatively selective for divalent over monovalent cations. Monovalent cations are usually similarly permeable through NSCCs, and these channels can conduct cations which are impermeable or poorly permeable through potassium-selective channels. In general the $K^+$:$Na^+$ selectivity ratios of NSCCs are between 0.3 and 3 (Demidchik *et al.*, 2002). Non-selective cation movements appear to be ubiquitous in the plasma membrane of plant cells as well as in other endomembranes and they likely dominate tonoplast ion transport.

Research into NSCCs is still in its infancy, and the contribution of these channels to potassium acquisition and transport in the plant is not yet fully known. Furthermore, the proteins responsible for non-selective cation currents have not been identified. The characterisation of the non-selective cation currents is also still in its initial stages and the physiological roles of these currents are not well understood, although they have been proposed to function in low-affinity uptake (see Demidchik *et al.*, 2002 for a review). Many plant NSCCs are thought to be voltage-independent, but again, little is known about the mechanisms that control gating in this type of channel. Numerous methods of activation of these channels have been reported including depolarisation activated, hyperpolarisation activated, voltage insensitive, calcium activated, mechanosensitive NSCCs, cyclic nucleotide-gated and glutamate-gated (Demidchik *et al.*, 2002).

***Cyclic nucleotide-gated (CNG) channels***: Plant cyclic nucleotide-gated channels (CNG channels) were first identified in a screen for calmodulin binding partners in barley (Schuurink *et al.*, 1998). In the same year, two cDNAs encoding proteins similar to the barley CNG channel; HvCBT1, were cloned from Arabidopsis; *AtCNGC1* and *AtNGCC2* (Köhler *et al.*, 1999). CNG channels are defined functionally as ligand-gated channels that are activated by either 3'5'-cyclic AMP (cAMP) or 3'5' cyclic GMP (cGMP) which bind to the channel protein (Leng *et al.*, 2002). Such cyclic nucleotides have been shown to activate channel opening in all CNG channel proteins in plants examined so far, resulting in the influx of cations into the cell (Leng et al., 1999; 2002; Balague et al., 2003; Hua et al., 2003). There are now known to be twenty members of the CNG family in Arabidopsis (Köhler et al., 1999; Mäser et al., 2001). Sixteen CNG channel proteins have also been identified in rice (Yuan et al., 2003) and several more in barley (e.g. HvCBT1; Schuurink et al., 1998) and tobacco (NtCBP4 and NtCBP7; Arazi et al., 1999; Leng et al., 1999; 2002). CNG channels are also represented in expressed sequence tags from other plant species so it is now obvious that they are widely present across the plant kingdom (Talke et al., 2003).

The CNG channels in Arabidopsis share structural homologies with the cyclic nucleotide-gated channels (CNGCs), characterised and cloned from animal cells (Zagotta and Siegelbaum, 1996). The gating in Arabidopsis by the intracellular nucleotides; cGMP and cAMP is similar to the gating in their animal counterparts (Broillet and Firestein, 1999; Maathuis and Sanders, 2001), although plant CNGs are also gated by calcium and calmodulin (Kleene, 1999; Mäser et al., 2001; Véry and Sentenac, 2002). Furthermore, in contrast to animal CNGCs, the domains binding cyclic nucleotide and calmodulin overlap in plants (Köhler et al., 1999; Arazi et al., 2000; Köhler and Neuhaus, 2000) enabling cross-talk between cyclic nucleotides and calmodulin signalling (Arazi et al., 2000). Moreover, the primary amino acid sequence homology between this family of plant transporter proteins and their presumed animal homologues is not very large at only 22% (Leng et al., 1999).

The structure of the Arabidopsis CNG channels is also somewhat similar to that of the Shaker-type channels, suggesting that they are related to this family. For instance, similar to the Shaker-type potassium channels structure, CNG channels have six membrane-spanning regions (S1-S6) and a P-loop region with the P-loop region comprised of the S5 and S6 transmembrane segments and a pore region, or P-loop. However, in contrast to the Shaker family, the CNG family does not contain the high potassium selectivity hallmark motif in their P domains. This likely results in the non-selective characteristics of this channel; they readily conduct both monovalent and divalent cations (Véry and Sentenac, 2002) similar to their animal counterparts (Gamel and Torre, 2000). Indeed, electrophysiological studies have shown that plant CNG channels are permeable to sodium and calcium as well as potassium (Leng et al., 1999; 2002; White et al., 2002; Balague et al., 2003). More specifically for potassium transport, the Arabidopsis CNG channels, AtCNGC1 and AtCNGC4, show equal permeability for potassium and sodium (Hua et al., 2003; Balague et al., 2003; Bridges et al., 2005). In addition, one member of the family, AtCNGC2, is characterised by a unique Ala-Asn-Asp selectivity filter and this filter may result in this protein's much higher selectivity for potassium over sodium (Leng et al., 2002; Hua et al., 2003).

Cyclic nucleotide gated channels appear to be expressed ubiquitously in all tissues in Arabidopsis (Talke et al., 2003), although to date, no detailed studies have been performed on the cell specificity of CNG channel expression. Nonetheless, the subcellular localisation of CNG channels has been analysed in only two cases - for barley, *HvCBT1* and tobacco, *NtCBP4* (Schuurink et al., 1998; Arazi et al.,

1999), where they are localised to the plasma membrane. Like all non-selective cation channels, questions also remain about the physiological role of plant CNG channels, although the large size of the plant CNG channel family indicates a wide diversity of physiological functions. Initially the proposed role of plant CNG channels was in cell signalling (Demidchik et al., 2002; Véry and Sentenac, 2002), similar to their role in sensory neurons in animals (Lee et al., 2001). However, as Arabidopsis CNG channels are shown to be expressed in roots, they may also provide a physiologically significant pathway for cation uptake into the root system, (Maathuis and Sanders, 2001; White et al., 2002). In this role, it has been suggested that some individual CNG channels, because of their poor cation selectivity, may also impact upon a plant's response to soil salinity and heavy metals (Leng et al., 2002). On the other hand, AtCNGC2, because of their high potassium permeability (Leng et al., 2002; Hua et al., 2003) and high level of expression in roots (Talke et al., 2003), CNG channels may be directly involved in potassium uptake. There is an evolving picture indicating that plant CNG channels may be involved in a broad array of mechanisms impacting upon growth, development and response to environmental stresses. However, their contribution towards potassium acquisition and transport within plants still remains to be quantified.

***Glutamate receptors (iGluRs)***: Ionotropic glutamate receptors (iGluRs) represent a class of cation-selective, ligand gated ion channels mediating excitatory synaptic transmission in nerve cells in animals (Madden, 2002). In such animal systems, binding of the agonist; glutamate, leads to channel opening and, depending on the electrochemical gradients and channel selectivity, to membrane depolarisation and cation entry. These animal iGluRs are non-selective cation channels that function predominantly as glutamate-gated sodium and calcium influx pathways at neuronal (vertebrate) or neuromuscular (invertebrate) junctions (Dingledine et al., 1999).

In plants, a family of polypeptides that appear to be related to these animal ionotropic glutamate receptors has been found. Two complementary DNA clones; *GLR1* and *GLR2*. and two genomic sequences; *GLR3* and *GLR4*, both of which have sequence similarities and similarity in terms of secondary structures to animal iGluRs, have been identified (Lam et al., 1998). In Arabidopsis, twenty genes are reported to encode putative glutamate receptor subunits (Lacombe et al., 2001). However, only six of the twenty genes have been cloned as full-length cDNAs, although a further eleven have been identified as transcribed. These putative GLR genes have been assigned to three subfamilies (so-called "Clades") on the basis of parsimony analysis (Lacombe et al., 2001). Clades I and III have been shown to be expressed ubiquitously throughout the plants, and genes from Clade II are largely root specific (Chiu et al., 2002; Meyerhoff et al., 2005).

However, the precise physiological role of the plant glutamate receptor family is unknown (Lacombe et al., 2001; Davenport, 2002; White et al., 2002; Bouche et al., 2003). It is not yet clear whether Arabidopsis glutamate receptors, like their animal counterparts, are actually capable of forming functional ion channels, although evidence is emerging for a non-selective cation channel function of AtGLRs

(Dennison and Spalding, 2000) mediating both calcium and monovalent cation transport (Lacombe *et al.*, 2001; Kim *et al.*, 2001). The strong expression levels of all *AtGLR* genes in Arabidopsis roots imply that they are important in regulating ion uptake from the soil (Chiu *et al.*, 2002). In addition, as GLR genes are ubiquitously expressed in Arabidopsis organs (Meyerhoff *et al.*, 2005), they could be involved in many different signalling and physiological processes. For example the *AtGLR* genes have been implicated in processes such as light signal transduction (Lam *et al.*, 1998) and calcium homeostasis (Kim *et al.*, 2001).

### 5.3. Potassium transporters

*5.3.1. KUP/HAK/KT transporters*

Along with the potassium Shaker-type channels, KUP/HAK/KT transporters play central roles in potassium homeostasis in plant cells including a role in both high- and low-affinity potassium uptake (Santa-María *et al.*, 1997; Rigas *et al.*, 2001; Elumalai *et al.*, 2002; Vallejo *et al.*, 2005). They are homologous to the $H^+$-$K^+$ symporters first identified in *E. coli* (Schleyer and Bakker, 1993) and *Schwanniomyces occidentalis* (Bañuelos *et al.*, 1995) and form a large family in plants, with thirteen members in Arabidopsis (Mäser *et al.*, 2001), at least seventeen members in rice (Bañuelos *et al.*, 2002) and five in barley (Santa-María *et al.*, 1997; Rubio *et al.*, 2000). Little is known about the structure of these transporters. Hydrophobicity profiles suggest that they posses twelve transmembrane-spanning (TMS) domains with a long cytosolic loop between the second and third TMS (Kim *et al.*, 1998; Rubio *et al.* 2000; Bañuelos *et al.*, 2002). Current evidence indicates that KUP/HAK/KT transporters are ubiquitous throughout the plant (Kim *et al.*, 1998; Rubio *et al.*, 2000; Ahn *et al.*, 2004) and function both in the plasma membrane and at the tonoplast (Rodríguez-Navarro, 2000; Serrano and Rodríguez-Navarro, 2001; Senn *et al.*, 2001; Bañuelos *et al.*, 2002).

Plant KUP/HAK/KT transporters form four phylogenetic clusters (Rubio *et al.*, 2000). Representatives of Cluster I are the barley *HvHAK1* and the Arabidopsis *AtHAK5* which have been characterised in yeast as high-affinity potassium transporters. Representatives of Cluster II are the Arabidopsis *AtKT1/KUP1*, *AtKT2/KUP2* and *AtKT3/TRH1* and the barley *HvHAK2*. The characterisation of all these transporters from Cluster II presents serious problems in yeast mutants defective for the endogenous potassium transporters, and the results from different laboratories do not always concur (Fu and Luan, 1998; Kim *et al.*, 1998). From the data available, this cluster appears to include the low-affinity transporter KT5/KUP2 (Quintero and Blatt, 1997), the high-affinity transporter KT3/TRH1 (Rigas *et al.*, 2001) and the biphasic (high- and low-affinity) transporter KT1/KUP1 (Fu and Luan, 1998; Kim *et al.*, 1998). Less detail is known about the members of the remaining two clusters. Cluster III contains three genomic sequences from Arabidopsis and Cluster IV holds the translated sequence of an incomplete cDNA clone from rice; OsHAK4 (Rubio *et al.*, 2000).

Potassium uptake and transport

Expression studies of all thirteen members of the KUP/HAK/KT family in Arabidopsis has revealed that most of the genes are expressed in all of the plant tissues tested under all experimental conditions (Ahn et al., 2004). Some of these however were found to have a more limited range of expression (*AtKT/KUP9* to *12*). Specifically, *AtKT/KUP10* and *12* were mainly expressed in leaves, whereas *AtKT/KUP11* expression was barely detectable (Ahn et al., 2004).

The *AtHAK5* gene is constitutively expressed in roots and it is induced in both root and shoots of potassium-starved plants and down-regulated upon potassium re-supply (Ahn et al., 2004; Armengaud et al., 2004; Shin and Schachtman, 2004; Gierth et al., 2005). This is in contrast to the other KUP/HAK/KT genes in Arabidopsis which are unresponsive to changes in external potassium (Maathuis et al., 2003; Ahn et al., 2004). Similar to *AtHAK5*, the exclusively root expressed barley *HvHAK1* gene is also induced under potassium-starved conditions (Santa-María et al., 1997). *AtKUP3* has also been shown to be expressed in Arabidopsis roots and, along with *AtHAK5*, is proposed to contribute to potassium uptake (Fu and Luan, 1998; Kim et al., 1998; Ahn et al., 2004). However, experimental evidence for the proposed potassium transport functions of other KUP/HAK/KT transporters is still under investigation. Only for one KUP/HAK/KT transporter; *AtKUP2*, has an unequivocal function in potassium-dependent cell expansion of growing tissues has been shown (Elumalai et al., 2002).

### 5.3.2. HKT transporters

Plant HKT transporters are related to the fungal Trk transporters and prokaryote KtrB and TrkH potassium transporter subunits (Durell and Guy, 1999; Rodríguez-Navarro, 2000). They display a core structure with eight TMS and four P-forming domains, four repeats of 1TMS-1P-1TMS, with the four P loops lining a central P, and C- terminal cytosolic regions (Durell and Guy, 1999; Durell et al., 1999; Kato et al., 2001).

Wheat *TaHKT1*, which facilitates $K^+/Na^+$ symport, was the first potassium transporter to be cloned from plants (Schachtman and Schroeder, 1994; Rubio et al., 1995). Further studies suggest that its role appears not to be restricted to potassium transport, but also includes low affinity sodium transport into the roots (Laurie et al., 2002), especially under low potassium to sodium ratios. Available information suggests that these HKT homologues operate in two transport modes with differing selectivity for potassium and sodium: (i) high-affinity sodium-potassium symporter (in the presence of low potassium and sodium concentrations), and (ii) low-affinity sodium-sodium (co)-transporter (when the sodium-potassium concentration ratio in the external solution is high; Rubio et al., 1995; Gassmann et al., 1996). Potassium starvation induces *TaHKT1* expression (Wang et al., 1998) as well as stimulating inward sodium currents within the root cortical cells (Buschmann et al., 2000). However, as HKT1 appears to play a role in net sodium accumulation (Uozumi et al., 2000; Laurie et al., 2002; Mäser et al., 2001), its decreased expression under salt stress often correlates with plant salt tolerance

(Golldack et al., 1997). Not surprisingly, HKT1 has sometimes been proposed to be a determinant of salt sensitivity in plants (Rubio et al., 1995).

Although HKT homologues have been isolated or detected in many species, they do not constitute multigene families. Wheat only has one member (*TaHKT1*; Schachtman and Schroeder, 1994) and there is also only one member of this group in Arabidopsis (*AtHKT1*; Uozumi et al., 2000). Eucalyptus and rice each have two HKT paralogues (Fairbairn et al., 2000; Horie et al. 2001). The only exception is japonica rice in which the genome shows the presence of up to nine *OsHKT* genes (Garciadeblás et al., 2003). All HKT transporters identified so far are expressed predominantly in roots.

### 5.3.3. $K^+/H^+$ antiporters

Among the genes encoding putative proton-coupled transporters in Arabidopsis, thirty-eight genes encode proteins homologous to the sodium-proton exchangers described in mammals or micro-organisms (Mäser et al., 2001). Phylogenic analysis indicates that these antiporters fall into three families, the monovalent Cation:Proton Antiporter-1 (CPA1) family, consisting of eight members, the monovalent Cation:Proton Antiporter-2 (CPA2) family, also referred to as the CHX (Cation/H$^+$ eXchanger) family which contains twenty-eight members, and the NhaD family with two members (Mäser et al., 2001). Our knowledge of the role of these transporters in potassium homeostasis is severely limited.

Several members of the CPA1 family have been functionally characterised in yeast where they display Na$^+$/H$^+$ antiporter activity (Gaxiola et al., 1999; Darley et al., 2000; Shi et al., 2002; Yokoi et al., 2002; Aharon et al., 2003), and, depending on their subcellular location, they exclude sodium at the plasma membrane (Qiu et al., 2002; 2003), or compartmentalise sodium into the cell vacuole (Apse et al., 1999; 2003). Importantly, for cellular potassium homeostasis, the tonoplast located Na$^+$/H$^+$ antiporter, *AtNHX1* (Apse et al., 1999), also mediates potassium transport (Zhang and Blumwald, 2001; Venema et al., 2002; Apse et al., 2003). It has been proposed that other members of the CPA1 family may also be involved in the regulation of potassium homeostasis within plant cells (Cellier et al., 2004), although to date no other members have been characterised for such a role.

A role in potassium acquisition and homeostasis, as opposed to sodium transport has been proposed for one member of the CPA2 family, *AtCHX17* (Cellier et al., 2004). *AtCHX17* is induced by potassium starvation and is expressed in epidermal and cortical cells of the mature root zone. Experimental evidence indicates that it contributes towards potassium acquisition and homeostasis under conditions of environmental stress such as high salinity and potassium deficiency (Cellier et al., 2004).

A final group of ion transporters in plants which may potentially contribute to potassium acquisition and homeostasis are the KEA transporters. KEA transporters, (called KEA for K$^+$ Efflux Antiporter) show substantial sequence similarities (up to 35% identity) with bacterial Ker (K$^+$ Efflux) antiporters regulated by glutathione (Munro et al., 1991; Yao et al., 1997). This gene family consists of six members, but their role in potassium nutrition is largely unknown. Furthermore, their tissue and subcellular localisations are also unknown and they are the least studied class of the plant transporters.

*CCC family*: A few putative members of the cation chloride co-transporter family (CCC) have been found in plants (Véry and Sentenac, 2003). In animal cells the CCC family comprises potassium-chloride, sodium-chloride and sodium-potassium-chloride transporters (Gamba *et al.*, 1993; Gillen *et al.*, 1996; Isenring and Forbush, 1997). Members of this family have important roles in cellular ionic and osmotic homeostasis in animal cells, but their role in potassium transport in plants is unknown.

*LCT1*: LCT1 is low-affinity transporter mediating uptake of a wide range of monovalent cations, including potassium and sodium (Schachtman *et al.*, 1997; Clemens *et al.*, 1998; Amtmann *et al.*, 2001) and is expressed in both roots and leaves (Schachtman *et al.*, 1997). So far, *LCT1* has been found only in wheat, has no counterpart in Arabidopsis and shares no sequence homology with any other gene to date.

## 6. RESPONSES OF PLANT POTASSIUM TRANSPORTERS TO LOW POTASSIUM STATUS

Numerous physiological experiments have revealed that potassium uptake by the root is tuned in response to shoot demand. Potassium deficiency is known to result in both quantitative and qualitative changes in the mechanisms of potassium uptake (Siddiqi and Glass, 1987; Benlloch *et al.*, 1989; Fernando *et al.*, 1990; Fernando and Glass, 1992; Maathius and Sanders, 1995; Kochian and Lucas, 1988; Shin and Schachtman, 2004). This activation has conventionally been associated with the induction of expression of high-affinity transporters, traditionally considered to be major mechanisms of adaptation to potassium starvation (Drew *et al.*, 1984; Fernando *et al.*, 1990). Surprisingly, very little is known about how these transporter systems are regulated and coordinated with the potassium status of the plant and the available potassium supply. Interestingly, it appears that there is only a very limited transcriptional response to potassium deprivation in plants (Maathuis *et al.*, 2003; Gierth *et al.*, 2005). Indeed, very few Arabidopsis genes encoding potassium channels and transporters have been reported to be regulated directly by external potassium concentration, although many of these genes have been shown to be induced or repressed by other stresses and hormones (Pilot *et al.*, 2003).

### 6.1. Responses of potassium transporters to potassium deprivation

As mentioned above, very few potassium transporter appear to be activated by potassium-starvation in Arabidopsis (Maathuis *et al.*, 2003; Gierth *et al.*, 2005). One however (*AtHAK5*) has found to be consistently and strongly up-regulated in response to potassium deprivation (Ahn *et al.*, 2004; Armengaud *et al.*, 2004; Shin and Schachtman, 2004; Gierth et al., 2005). Elevated transcription of *AtHAK5* is also down-regulated by potassium re-supply (Ahn et al., 2004; Armengaud et al., 2004; Gierth et al., 2005). This activation and down-regulation of *AtHAK5* indicates that this transporter may be of physiological significance in maintaining potassium homeostasis under varying levels of potassium supply. Importantly, *AtHAK5* expression is localised to the epidermis of main and lateral roots of Arabidopsis (Ahn et al., 2004; Gierth et al., 2005) e.g. in the tissues functionally specialised in potassium uptake. This, along with its activation under potassium-deprivation, indicates that the *AtHAK5* gene encodes an important component that functions in potassium starvation-induced high-affinity potassium uptake in roots (Gierth et al., 2005). Interestingly, *AtHAK5* has also been localised to the mature leaves of Arabidopsis where it is also upregulated by potassium starvation (Ahn et al., 2004). The significance of this is not yet known.

To date, it has been demonstrated that *AtHAK5* orthologues are also induced by low external potassium in barley *HvHAK1* (Santa-María et al., 1997), tomato *LeHAK5* (Wang et al., 2002), and rice *OsHAK1* (Bañuelos et al., 2002). Thus, activation of this high-affinity transporter *HAK5* in response to potassium deprivation is probably a common feature of this gene across various plant species.

It is possible that other KT/KUP/HAK transporters may complement the activation of *AtHAK5* under potassium-restricted conditions, but evidence for this is more limited. For instance, activation or de-repression of *AtKUP3* expression has been shown in roots of plants grown for two to three weeks on solidified potassium-depleted media (Kim et al., 1998), although this response has been found to be absent from mature Arabidopsis (Maathuis et al., 2003). *AtKUP2* has also been proposed to play a role in acclimation to mineral deficiencies as it is down-regulated in shoots after potassium re-supply (Armengaud et al., 2004).

Similar to *AtKUP3* and *AtKUP2*, evidence for regulation by potassium supply is also unclear for *HKT1*. In roots of both barley and wheat, rapid up-regulation at the transcriptional level has been reported for HKT1 following the withdrawal of potassium (Wang et al., 1998); potassium re-supply reverses this trend. This up-regulation of *HKT1* transcript levels in barley roots corresponds to the increase of high-affinity potassium uptake in roots following potassium removal (e.g. Glass, 1975; 1976; 1978; Kochian and Lucas, 1982; Fernando et al., 1990), which has led to the implication of HKT1 in potassium-supply regulated potassium transport (Wang et al., 1998). However, no evidence for HKT1 upregulation by potassium-deprivation was found in Arabidopsis (Maathuis et al., 2003).

Amongst members of the KEA potassium transporter family, *KEA5* has been shown to be transiently induced after potassium deprivation (Shin and Schachtman, 2004). However, as the location of this transporter and its role in potassium transport is not known, it is difficult to assess the contribution of KEA5 in plant acclimation to potassium deficiencies.

Transcription of *AtCHX17*, the member of the *CHX* family involved in potassium acquisition and homeostasis, has also been shown to be induced by potassium starvation in Arabidopsis (Cellier et al., 2004). The localisation of *AtCHX17* to the epidermal, root hairs and cortical cells of the mature root zone, and lack of expression in the shoots (Cellier *et al.*, 2004), indicate its role in potassium acquisition by the plant. Interestingly, in addition to a strong induction under potassium limiting conditions, *AtCHX17* was also found to be strongly induced by salinity (Kreps *et al.*, 2002; Cellier *et al.*, 2004). This could be due, in addition to the direct effects of sodium toxicity, to the reduced potassium availability resulting from the competition at the root level between potassium and sodium. In contrast, Maathuis *et al.* (2003) did not observe *AtCHX17* up-regulation in response to either salt stress or potassium deprivation. A possible explanation for this inconsistency may be the difference in the developmental stage of the plant, implying that at different developmental stages of the plant, different potassium transporters come into play.

## 6.2. Potassium deprivation and functional expression of potassium-conducting channels

Unlike the high-affinity transporters, low affinity transport systems appear to be much less sensitive to the potassium status of the plant. However, there is evidence that the activities of some potassium-mediating channels may also be modified by the potassium status of the plant in some species. While in barley, ryegrass (Glass and Dunlop, 1978) and maize (Kochian and Lucas, 1982), the low-affinity potassium transporter systems appears to be insensitive to the plant potassium status, in Arabidopsis (Maathuis and Sanders, 1995), wheat (Buschmann *et al.*, 2000) and sunflower (Benlloch *et al.*, 1989), low affinity potassium uptake and channel activity have been shown to be slightly increased by potassium starvation.

An increase in the activity of the root periphery cell expressed, dominant inward potassium; AKT1 (Hirsch *et al.*, 1998) has been reported in Arabidopsis root cells in response to changes in potassium availability (Maathuis and Sanders, 1995). Such activation is most likely to occur post-transcriptionally because neither RNA blot nor microarray experiments reveal an alteration in *AKT1* transcription in potassium starved plants (Pilot *et al.*, 2003; Maathuis *et al.*, 2003). On the other hand, enhanced inward-rectifying currents have been reported under potassium starved conditions in wheat (Buschmann *et al.*, 2000). This increased current was associated with increased transcription of *TaAKT1* suggesting that in this species, in contrast to Arabidopsis, up-regulation of AKT1 under potassium starved conditions is controlled at the transcriptional level.

As well as potentially up-regulating potassium transporters, potassium starvation has also been shown to down-regulate the transcription of a small number of potassium conducting channels. Potassium starvation results in decreased levels of *SKOR* and *AKT2* transcripts (Maathuis *et al.*, 2003; Pilot *et al.*, 2003). SKOR is thought to be involved in delivery of potassium to the shoot (Gaymard *et al.*, 1998) and AKT2 in potassium transport in the phloem tissue (Marten *et al.*, 1999; Lacombe *et al.*, 2000). Regulation in the level of expression of *SKOR* and *AKT2* would enable the plant to control potassium secretion into the xylem sap, thus retaining it in the root and would also regulate potassium movement within the plant via the phloem sap. Integration of the xylem and phloem fluxes has been proposed to play a central role in the adjustment of root potassium transport activity to both the potassium availability and the shoot demand for potassium (Kochian and Lucas, 1988; White, 1997).

## 7. CONCLUSIONS AND FUTURE PROSPECTS

Considerable progress has been made in our understanding of the mechanisms of potassium transport in plant species. Numerous families of potassium transporters have now been identified, some of which have now been functionally characterised. In addition, the completion of the Arabidopsis genome sequencing project has enabled the identification of numerous putative plant transporter proteins involved in potassium transport (Ward, 2001). However, most of our knowledge concerning potassium transport within plant species is based on the model plant; Arabidopsis and is not applicable directly to crops. Furthermore, we have just scratched the surface in our understanding of how these potassium transport mechanisms are regulated and coordinated. The rice genome, which is likely to be followed by other crops, will give us a more comprehensive understanding of the mechanisms underlying regulation of potassium uptake and utilisation by crop species. This could potentially lead to better targeting of potassium within agricultural systems and also increase the possibility to develop, either by traditional breeding, or by genetic engineering, crop species with higher potassium use efficiency.

## 8. ACKNOWLEDGEMENTS

This work was supported by the ARC Discovery and DEST grants to S. Shabala.

## 9. LITERATURE CITED

Ache, P., Becker, D., Deeken, R., Dreyer, I., Weber, H., Fromm, J. and Hedrich, R. (2001) VFK1, a *Vicia faba* $K^+$ channel involved in phloem unloading. *Plant J.*, **27**: 571-580

Ache, P., Becker, D., Ivashikina, N., Dietrich, P., Roelfsema, M.R.G. and Hedrich, R. (2000) GORK, a delayed outward rectifier expressed in guard cells of *Arabidopsis thaliana*, is a $K^+$-selective, $K^+$-sensing ion channel. *FEBS Lett.*, **486**: 93-98

Adams, F. (1971) Soil solution. In: *The Plant Root and its Environment* (Ed. EW Carson). University Press of Virginia, Charlottesville, VA, pp 441-481

Aharon, G.S., Apse, M.P., Duan, S.L., Hua, X.J. and Blumwald, E. (2003) Characterization of a family of vacuolar $Na^+/H^+$ antiporters in *Arabidopsis thaliana*. *Plant Soil*, **253**: 245-256

Ahn, S.J., Shin, R. and Schachtman, D.P. (2004) Expression of KT/KUP genes in Arabidopsis and the role of root hairs in $K^+$ uptake. *Plant Physiol.*, **134**: 1135-1145

Amtmann, A., Fischer, M., Marsh, E.L., Stefanovic, A., Sanders, D. and Schachtman, D.P. (2001) The wheat cDNA LCT1 generates hypersensitivity to sodium in a salt-sensitive yeast strain. *Plant Physiol.*, **126**: 1061-1071

Anderson, J.A., Huprikar, S.S., Kochian, L.V., Lucas, W.J. and Gaber, R.F. (1992) Functional expression of a probable *Arabidopsis thaliana* potassium channel in *Saccharomyces cerevisiae*. *Proc. Natl. Acad. Sci. USA*, **89**: 3736-3740

Apse, M.P., Aharon, G.S., Snedden, W.A. and Blumwald, E. (1999) Salt tolerance conferred by overexpression of a vacuolar $Na^+/H^+$ antiport in Arabidopsis. *Science*, **285**: 1256-1258

Apse, M.P., Sottosanto, J.B. and Blumwald, E. (2003) Vacuolar cation/$H^+$ exchange, ion homeostasis, and leaf development are altered in a T-DNA insertional mutant of AtNHX1, the Arabidopsis vacuolar $Na^+/H^+$ antiporter. *Plant J.*, **36**: 229-239

Arazi, T., Kaplan, B. and Fromm, H. (2000) A high-affinity calmodulin-binding site in a tobacco plasma-membrane channel protein coincides with a characteristic element of cyclic nucleotide-binding domains. *Plant Mol. Biol.*, **42**: 591-601

Arazi, T., Sunkar, R., Kaplan, B. and Fromm, H. (1999) A tobacco plasma membrane calmodulin-binding transporter confers $Ni^{2+}$ tolerance and $Pb^{2+}$ hypersensitivity in transgenic plants. *Plant J.*, **20**: 171-182

Armengaud, P., Breitling, R. and Amtmann, A. (2004) The potassium-dependent transcriptome of Arabidopsis reveals a prominent role of jasmonic acid in nutrient signaling. *Plant Physiol.*, **136**: 2556-2576

Balague, C., Lin, B.Q., Alcon, C., Flottes, G., Malmstrom, S., Köhler, C., Neuhaus, G., Pelletier, G., Gaymard, F. and Roby, D. (2003) HLM1, an essential signaling component in the hypersensitive response, is a member of the cyclic nucleotide-gated channel ion channel family. *Plant Cell*, **15**: 365-379

Bañuelos, M.A., Garciadeblas, B., Cubero, B. and Rodríguez-Navarro, A. (2002) Inventory and functional characterization of the HAK potassium transporters of rice. *Plant Physiol.*, **130**: 784-795

Bañuelos, M.A., Klein, R.D., Alexander-Bowman, S.J. and Rodríguez-Navarro, A. (1995) A potassium transporter of the yeast *Schwanniomyces occidentalis* homologous to the Kup system of *Escherichia coli* has a high concentrative capacity. *EMBO J.*, **14**: 3021-3027

Becker, D., Geiger, D., Dunkel, M., Roller, A., Bertl, A., Latz, A., Carpaneto, A., Dietrich, P., Roelfsema, M.R.G., Voelker, C., Schmidt, D., Mueller-Roeber, B., Czempinski, K. and Hedrich, R. (2004) AtTPK4, an Arabidopsis tandem-pore $K^+$ channel, poised to control the pollen membrane voltage in a pH- and $Ca^{2+}$-dependent manner. *Proc. Natl. Acad. Sci. USA*, **101**: 15621-15626

Bednarz, C.W. and Oosterhuis, D.M. (1999) Physiological changes associated with potassium deficiency in cotton. *J. Plant Nutr.*, **22**: 303-313

Benlloch, M., Moreno, I. and Rodríguez-Navarro, A. (1989) Two modes of Rubidium uptake in sunflower plants. *Plant Physiol.*, **90**: 939-942

Bertl, A., Anderson, J.A., Slayman, C.L. and Gaber, R.F. (1995) Use of *Saccharomyces cerevisiae* for patch-clamp analysis of heterologous membrane proteins: Characterization of KAT1, an inward-rectifying $K^+$ channel from *Arabidopsis thaliana*, and comparison with endogenous yeast channels and carriers. *Proc. Natl. Acad. Sci. USA*, **62**: 2701-2705

Bertl, A., Anderson, J.A., Slayman, C.L., Sentenac, H. and Gaber, R.F. (1994) Inward and outward rectifying potassium currents in *Saccharomyces cerevisiae* mediated by endogenous and heterologously expressed ion channels. *Folia Microbiol.*, **39**: 507-509

Bihler, H., Eing, C., Hebeisen, S., Roller, A., Czempinski, K. and Bertl, A. (2005) TPK1 is a vacuolar ion channel different from the slow vacuolar cation channel. *Plant Physiol.*, **197**: 417-424

Blatt, M.R., Thiel, G. and Trentham, D.R. (1990) Reversible inactivation of $K^+$ channels of *Vicia* stomatal guard cells following the photolysis of caged inositol 1,4,5-triphosphate. *Nature*, **346**: 766-769.

Bouche, N., Lacombe, B. and Fromm, H. (2003) GABA signaling: a conserved and ubiquitous mechanism. *Trend Cell Biol.*, **13**: 607-610

Boursier, P. and Läuchli, A. (1989) Mechanism of chloride partitioning in leaves of salt-stressed *Sorghum bicolor* L. *Physiol. Plant.*, **77**: 537-544

Brandt, S., Kehr, J., Walz, C., Imlau, A., Willmitzer, L. and Fisahn, J. (1999) A rapid method for detection of plant gene transcripts from single epidermal, mesophyll and companion cells of intact leaves. *Plant J.,* **20**: 245-250.

Brandt, S., Kloska, S., Altmann, T. and Kehr, J. (2002) Using array hybridization to monitor gene expression at the single cell level. *J. Exp. Bot.,* **53**: 2315-2323

Brandt, S.P. (2005) Microgenomics: gene expression analysis at the tissue-specific and single-cell levels. *J. Exp. Bot.,* **56**: 495-505

Bridges, D., Fraser, M.E. and Moorhead, G.B.G. (2005) Cyclic nucleotide binding proteins in the *Arabidopsis thaliana* and *Oryza sativa* genomes. *BMC Bioinformatics,* **6**

Broillet, M.C. and Firestein, S. (1999) Cyclic nucleotide-gated channels - Molecular mechanisms of activation. In: *Molecular and Functional Diversity of Ion Channels and Receptors,* Vol. 868, pp 730-740

Buschmann, P.H., Vaidyanathan, R., Gassmann, W. and Schroeder, J.I. (2000) Enhancement of $Na^+$ uptake currents, time-dependent inward-rectifying $K^+$ channel currents, and $K^+$ channel transcripts by $K^+$ starvation in wheat root cells. *Plant Physiol.,* **122**: 1387-1397

Cakmak, I. (2005) The role of potassium in alleviating detrimental effects of abiotic stresses in plants. *J. Plant Nutr. Soil Sci.,* **168**: 521-530

Cakmak, I. and Engels, C. (1999) Role of mineral nutrition in photosynthesis and yield formation. In: *Mineral Nutrition of Crops: Mechanisms and Implications* (Ed. Rengel, Z.). The Haworth Press, New York, pp 141-168

Cellier, F., Conejero, G., Ricaud, L., Luu, D.T., Lepetit, M., Gosti, F. and Casse, F. (2004) Characterization of AtCHX17, a member of the cation/$H^+$ exchangers, CHX family, from *Arabidopsis thaliana* suggests a role in $K^+$ homeostasis. *Plant J.,* **39**: 834-846

Chen, Z., Newman, I., Zhou, M., Mendham, N., Zhang, G. and Shabala, S. (2005) Screening plants for salt tolerance by measuring $K^+$ flux: a case study for barley. *Plant Cell Environ.,* **28**: 1230-1246

Chiu, J.C., Brenner, E.D., DeSalle, R., Nitabach, M.N., Holmes, T.C. and Coruzzi, G.M. (2002) Phylogenetic and expression analysis of the glutamate-receptor-like gene family in *Arabidopsis thaliana. Mol. Biol. Evolut.,* **19**: 1066-1082

Clemens, S., Antosiewicz, D.M., Ward, J.M., Schachtman, D.P. and Schroeder, J.I. (1998) The plant cDNA LCT1 mediates the uptake of calcium and cadmium in yeast. *Proc. Natl. Acad. Sci. USA,* **95**: 12043-12048

Cole, K.S. (1949) Dynamic electrical characteristics of the squid axon membrane. *Arch. Sci. Physiol.,* **3**: 253-258

Cuin, T.A., Miller, A.J., Laurie, S.A. and Leigh, R.A. (2003) Potassium activities in cell compartments of salt-grown barley leaves. *J. Exp. Bot.,* **54**: 657-661

Czempinski, K., Frachisse, J.M., Maurel, C., Barbier-Brygoo, H. and Mueller-Roeber, B. (2002) Vacuolar membrane localization of the Arabidopsis 'two-pore' $K^+$ channel KCO1. *Plant J.,* **29**: 809-820

Czempinski, K., Gaedeke, N., Zimmermann, S. and Mueller-Roeber, B. (1999) Molecular mechanisms and regulation of plant ion channels. *J. Exp. Bot.,* **50**: 955-966

Czempinski, K., Zimmermann, S., Ehrhardt, T. and Mueller-Roeber, B. (1997) New structure and function in plant $K^+$ channels: KCO1, an outward rectifier with a steep $Ca^{2+}$ dependency. *EMBO J.,* **16**: 2565-2575

Darley, C.P., van Wuytswinkel, O.C.M., van der Woude, K., Mager, W.H. and de Boer, A.H. (2000) *Arabidopsis thaliana* and *Saccharomyces cerevisiae* NHX1 genes encode amiloride sensitive electroneutral $Na^+/H^+$ exchangers. *Biochem. J.,* **351**: 241-249

Davenport, R. (2002) Glutamate receptors in plants. *Ann. Bot.,* **90**: 549-557

Potassium uptake and transport

De Block, M. and Van Lijsebettens, M. (1998) Glucouronidase enzyme histochemistry on semithin sections of plastic-embedded Arabidopsis explants. In: *Methods in Molecular Biology, vol. 82: Arabidopsis Protocols* (Eds. Martinez-Zapater, J., Salinas, J.). Humana Press Inc, Totowa, NJ
de Boer, A.H. (1999) Potassium translocation into the root xylem. *Plant Biol.*, 1: 36-45
Deeken, R., Geiger, D., Fromm, J., Koroleva, O., Ache, P., Langenfeld-Heyser, R., Sauer, N., May, S.T. and Hedrich, R. (2002) Loss of the AKT2/3 potassium channel affects sugar loading into the phloem of Arabidopsis. *Planta*, 216: 334-344
Deeken, R., Sanders, C., Ache, P. and Hedrich, R. (2000) Developmental and light-dependent regulation of a phloem-localised $K^+$ channel of *Arabidopsis thaliana*. *Plant J.*, 23: 285-290
Demidchik, V., Davenport, R.J. and Tester, M. (2002) Nonselective cation channels in plants. *Annu. Rev. Plant Biol.*, 53: 67-107
Demmig, B. and Gimmler, H. (1983) Properties of the isolated intact chloroplast at cytoplasmic $K^+$ concentrations. I. Light-induced cation uptake into intact chloroplasts is driven by an electric potential difference. *Plant Physiol.*, 73: 169-174
Dennison, K.L., Robertson, W.R., Lewis, B.D., Hirsch, R.E., Sussman, M.R. and Spalding, E.P. (2001) Functions of AKT1 and AKT2 potassium channels determined by studies of single and double mutants of Arabidopsis. *Plant Physiol.*, 127: 1012-1019
Dennison, K.L. and Spalding, E.P. (2000) Glutamate-gated calcium fluxes in Arabidopsis. *Plant Physiol.*, 124: 1511-1514
Dingledine, R., Borges, K., Bowie, D. and Traynelis, S.F. (1999) The glutamate receptor ion channels. *Pharmacol. Rev.*, 51: 7-61
Dixit, R., Cyr, R. and Gilroy, S. (2006) Using intrinsically fluorescent proteins for plant cell imaging. *Plant J.*, 45: 599-615
Doupnik, C.A., Davidson, N. and Lester, H.A. (1995) The inward rectifier potassium channel family. *Curr. Opin. Neurobiol.*, 5: 268-277
Downey, P., Szabo, I., Ivashikina, N., Negro, A., Guzzo, F., Ache, P., Hedrich, R., Terzi, M. and Lo Schiavo, F. (2000) *KDC1*, a novel carrot root hair $K^+$ channel. Cloning, characterization, and expression in mammalian cells. *J. Biol. Chem.*, 275: 39420-39426
Drew, M.C., Saker, L.R., Barber, S.A. and Jenkins, W. (1984) Changes in the kinetics of phosphate and potassium absorption in nutrient-deficient barley roots measured by a solution-depletion technique. *Planta*, 160: 490-499
Dreyer, I., Antunes, S., Hoshi, T., Mulle-rRober, B., Palme, K., Pongs, O., Reintanz, B. and Hedrich, R. (1997) Plant $K^+$ channel alpha-subunits assemble indiscriminately. *Biophys. J.*, 72: 2143-2150
Durell, S.R. and Guy, H.R. (1999) Evaluation of structural models of $K^+$ channels. *Biophys. J.*, 76: A333
Durell, S.R., Hao, Y.L., Nakamura, T., Bakker, E.P. and Guy, H.R. (1999) Evolutionary relationship between $K^+$ channels and symporters. *Biophys. J.*, 77: 775-788
Elumalai, R.P., Nagpal, P. and Reed, J.W. (2002) A mutation in the Arabidopsis KT2/KUP2 potassium transporter gene affects shoot cell expansion. *Plant Cell*, 14: 119-131
Elzenga, J.T.M., Keller, C. and Van Volkenburgh, E. (1991) Patch-clamping protoplasts from higher plants: A method for the quick isolation of protoplasts having a high success rate of gigaseal formation. *Plant Physiol.*, 97: 1573-1575
Epstein, E., Rains, D.W. and Elzam, O.E. (1963) Resolution of dual mechanisms of potassium absorption by barley roots. *Proc. Natl. Acad. Sci. USA*, 49: 684-692

Epstein, E. and Rains, D.W. (1965) Carrier-mediated cation transport in barley roots: kinetic evidence for a spectrum of active sites. *Proc. Natl. Acad. Sci. USA*, **53**: 1320-1324

Erhardt, T., Zimmermann, S. and Müller-Röber, B. (1997) Association of plant $K^+$(in) channels is mediated by conserved C-termini and does not affect subunit assembly. *FEBS Lett.*, **409**: 166-170

Fairbairn, D.J., Liu, W.H., Schachtman, D.P., Gomez-Gallego, S., Day, S.R. and Teasdale, R.D. (2000) Characterisation of two distinct HKT1-like potassium transporters from *Eucalyptus camaldulensis*. *Plant Mol. Biol.*, **43**: 515-525

FAOSTAT (2003) Fertilizer usage. http://apps.fao.org/lim500/nph-wrap.pl? Fertilizers&Domain=LUI&servlet=1

Fernando, M. and Glass, A.J.M. (1992) Homeostatic processes for the maintenance of the $K^+$ content of plant cells: a model. *Israel J. Bot.*, **41**: 145-166

Fernando, M., Kulpa, J., Siddiqi, M.Y. and Glass, A.D.M. (1990) Potassium-dependent changes in the expression of membrane-associated proteins in barley roots .1. Correlations with $K^+(^{86}Rb^+)$ influx and root $K^+$ concentration. *Plant Physiol.*, **92**: 1128-1132

Fischer, R. and Hsiao, T. (1968) Stomatal opening in isolated epidermal strips of *Vicia faba*. II: Response to KCl concentrations and role of $K^+$ absorption. *Plant Physiol.*, **43**: 1953-1958

Fricke, W., Leigh, R.A. and Tomos, A.D. (1994) Concentrations of inorganic and organic solutes in extracts from individual epidermal, mesophyll and bundle-sheath cells of barley leaves. *Planta*, **192**: 310-316

Fu, H.H. and Luan, S. (1998) AtKUP1: A dual-affinity $K^+$ transporter from Arabidopsis. *Plant Cell*, **10**: 63-73

Fuchs, I., Stölze, S., Ivashikina, N. and Hedrich, R. (2005) Rice $K^+$ uptake channel OsAKT1 is sensitive to salt stress. *Planta*, **221**: 212-221

Gaffey and Mullins, L.J. (1958) Ion fluxes during the action potential in *Chara*. *J. Physiol. London*, **144**: 505-524

Gallagher, J.A., Koroleva, O.A., Tomos, D.A., Farrar, J.F. and Pollock, C.J. (2001) Single cell analysis technique for comparison of specific mRNA abundance in plant cells. *J. Plant Physiol.*, **158**: 1089-1092

Gamba, G., Saltzberg, S.N., Lombardi, M., Miyanoshita, A., Lytton, J., Hediger, M.A., Brenner, B.M. and Hebert, S.C. (1993) Primary structure and functional expression of a cDNA-encoding the thiazide-sensitive, electroneutral sodium-chloride cotransporter. *Proc. Natl. Acad. Sci. USA*, **90**: 2749-2753

Gamel, K. and Torre, V. (2000) The interaction of $Na^+$ and $K^+$ in the pore of cyclic nucleotide-gated channels. *Biophys. J.*, **79**: 2475-2493

Garciadeblás, B., Senn, M.E., Bañuelos, M.A. and Rodríguez-Navarro, A. (2003) Sodium transport and HKT transporters: the rice model. *Plant J.*, **34**: 788-801

Gassmann, W., Rubio, F. and Schroeder, J.I. (1996) Alkali cation selectivity of the wheat root high-affinity potassium transporter HKT1. *Plant J.*, **10**: 869-882

Gassmann, W. and Schroeder, J.I. (1994) Inward-rectifying $K^+$ channels in root hairs of wheat - a mechanism for aluminum-sensitive low-affinity $K^+$ Uptake and membrane-potential control. *Plant Physiol.*, **105**: 1399-1408

Gaxiola, R.A., Rao, R., Sherman, A., Grisafi, P., Alper, S.L. and Fink, G.R. (1999) The *Arabidopsis thaliana* proton transporters, AtNHX1 and AVP1, can function in cation detoxification in yeast. *Proc. Natl. Acad. Sci. USA*, **96**: 1480-1485

Potassium uptake and transport

Gaymard, F., Cerutti, M., Horeau, C., Lemaillet, G., Urbach, S., Ravallec, M., Devauchelle, G., Sentenac, H. and Thibaud, J.B. (1996) The baculovirus/insect cell system as an alternative to *Xenopus* oocytes - First characterization of the AKT1 K⁺ channel from *Arabidopsis thaliana*. *J. Biol. Chem.*, **271**: 22863-22870
Gaymard, F., Pilot, G., Lacombe, B., Bouchez, D., Bruneau, D., Boucherez, J., Michaux-Ferriere, N., Thibaud, J.B. and Sentenac, H. (1998) Identification and disruption of a plant Shaker-like outward channel involved in K⁺ release into the xylem sap. *Cell*, **94**: 647-655
Gierth, M., Mäser, P. and Schroeder, J.I. (2005) The potassium transporter AtHAK5 functions in K⁺ deprivation-induced high-affinity K⁺ uptake and AKT1 K⁺ channel contribution to K⁺ uptake kinetics in Arabidopsis roots. *Plant Physiol.*, **137**: 1105-1114
Gillen, C.M., Stirewalt, V.L., Bryant, D.A. and Forbush, B. (1996) Cloning, sequencing, and initial characterization of a putative ion cotransport protein from cyanobacterium *Synechococcus* sp PCC 7002. *Biophys. J.*, **70**: MP293
Glass, A.D.M. (1978) Regulation of potassium influx into intact roots of barley by internal potassium levels. *Can. J. Bot.*, **56**: 1759-1764
Glass, A.D.M. (1976) Regulation of potassium absorption in barley roots - allosteric model. *Plant Physiol.*, **58**: 33-37
Glass, A.D.M. (1975) Inhibition of phosphate uptake in barley roots by hydroxy-benzoic acids. *Phytochem.*, **14**: 2127-2130
Glass, A.D.M. and Dunlop, J. (1978) Influence of potassium content on kinetics of potassium influx into excised ryegrass and barley roots. *Planta*, **141**: 117-119
Gobert, A., Park, G., Amtmann, A., Sanders, D. and Maathuis, F.J.P. (2006) *Arabidopsis thaliana* Cyclic Nucleotide Gated Channel 3 forms a non-selective ion transporter involved in germination and cation transport. *J. Exp. Bot.*, **57**: 791-800
Goh, C.H., Dietrich, P., Steinmeyer, R., Schreiber, U., Nam, H.-G. and Hedrich, R. (2002) Parallel recordings of photosynthetic electron transport and K⁺- channel activity in single guard cells. *Plant J.*, **32**: 623–630.
Goldstein, S.A.N., Bockenhauer, D., O'Kelly, I. and Zilberberg, N. (2001) Potassium leak channels and the KCNK family of two-P-domain subunits. *Nature Rev. Neurosci.*, **2**: 175-184
Golldack, D., Kamasani, U.R., Quigley, F., Bennett, J. and Bohnert, H.J. (1997) Salt stress-dependent expression of a HKT1-type high affinity potassium transporter in rice. *Plant Physiol.*, **114**: 529-529
Golldack, D., Quigley, F., Michalowski, C.B., Kamasani, U.R. and Bohnert, H.J. (2003) Salinity stress-tolerant and -sensitive rice (*Oryza sativa* L.) regulate AKT1-type potassium channel transcripts differently. *Plant Mol. Biol.*, **51**: 71-81
Golldack, D., Su, H., Quigley, F., Kamasani, U.R., Muñoz-Garay, C., Balderas, E., Popova, O.V., Bennett, J., Hans, J., Bohnert, H.J. and Pantoja, O. (2002) Characterization of a HKT-type transporter in rice as a general alkali cation transporter. *Plant J.*, **31**: 529-542
Grewal, J.S. and Singh, S.N. (1980) Effect of potassium nutrition on frost damage and yield of potato plants on alluvial soils of the Punjab (India). *Plant Soil.*, **57**: 105-110
Grignon, C. and Sentenac, H. (1991) pH and ionic concentrations in the apoplast. *Annu. Rev. Plant Physiol. Plant Mol. Biol.*, **42**: 103-128
Grygorczyk, C. and Grygorczyk, R. (1998) A Ca²⁺ and voltage-dependent cation channel in the nuclear envelope of red beet. *Biochim. Biophys. Acta.*, **1375**: 117-130

Hakerlerler, H., Oktay, M., Eryuce, N. and Yagmur, B. (1997) Effect of potassium sources on the chilling tolerance of some vegetable seedlings grown in hotbeds. In: *Food Security in the WANA Region, the Essential Need for Balanced Fertilization* (Ed. Johnston, A.E.). International Potash Institute, Basel, pp. 317-327

Halperin, S.J. and Lynch, J.P. (2003) Effects on salinity on cytosolic $Na^+$ and $K^+$ in root hairs of *Arabidopsis thaliana* : *in vivo* measurements using fluorescent dyes SBFI and PBFI. *J. Exp. Bot.*, **54**: 2035-2043

Hamill, O.P., Marty, A., Neher, E., Sakmann, B. and Sigworth, F.J. (1981) Improved patch-clamp techniques for high-resolution current recording from cells and cell-free membrane patches. *Pflügers Arch.*, **391**: 85-100

Hanson, M.R. and Köhler, R.H. (2001) GFP imaging: methodology and application to investigate cellular compartments in plants. *J. Exp. Bot.*, **52**: 529-539

Hartje, S., Zimmermann, S., Klonus, D. and Mueller-Roeber, B. (2000) Functional characterisation of LKT1, a $K^+$ uptake channel from tomato root hairs, and comparison with the closely related potato inwardly rectifying $K^+$ channel SKT1 after expression in *Xenopus* oocytes. *Planta*, **210**: 723-731

Haseloff, J., Siemering, K.R., Prasher, D.C. and Hodge, S. (1997) Removal of a cryptic intron and subcellular localization of green fluorescent protein are required to mark transgenic Arabidopsis plants brightly. *Proc. Natl. Acad. Sci. USA*, **94**: 2122-2127

Hedrich, R., Flügge, U.I. and Fernandez, J.M. (1986) Patch-clamp studies of ion transport in isolated plant vacuoles. *FEBS Lett.*, **204**: 228-232

Hedrich, R., Barbier-Brygoo, H., Felle, H., Fluegge, U.I., Luettge, U., Maathuis, F.J.M., Marx, S., Prins, H.B.A., Raschke, K., Schnabl, H., Schroeder, J.I., Struve, I., Taiz, L. and Zeigler, P. (1988) General mechanisms for solute transport across the tonoplast of plant vacuoles: A patch-clamp survey of ion channels and proton pumps. *Bot. Acta*, **101**: 7–13

Henriksen, G.H., Taylor, A.R., Brownlee, C. and Assmann, S.A. (1996) Laser microsurgery of plant cell walls permits patch-clamp access. *Plant Physiol.*, **110**: 1063-1068

Hink, M.A., Bisseling, T. and Visser, A.J.W.G. (2002) Imaging protein-protein interactions in living cells. *Plant Mol. Biol.*, **50**: 871-883

Hinnah, S.C. and Wagner, R. (1998) Thylakoid membranes contain a high-conductance channel. *Eur. J. Biochem.*, **253**: 606-613

Hirsch, R.E., Lewis, B.D., Spalding, E.P. and Sussman, M.R. (1998) A role for the AKT1 potassium channel in plant nutrition. *Science*, **280**: 918-921

Hodgkin, A.L., Huxley, A.F. and Katz, B. (1949) Ionic currents underlying activity of the giant axon of the squid. *Arch. Sci. Physiol.*, **3**: 129-150

Horie, T., Yoshida, K., Nakayama, H., Yamada, K., Oiki, S. and Shinmyo, A. (2001) Two types of HKT transporters with different properties of $Na^+$ and $K^+$ transport in *Oryza sativa*. *Plant J.*, **27**: 129-138

Hoshi, T. (1995) Regulation of voltage dependence of the KAT1 channel by intracellular factors. *J. Gen. Physiol.*, **105**: 309-328

Hosy, E., Vavasseur, A., Mouline, K., Dreyer, I., Gaymard, F., Poree, F., Boucherez, J., Lebaudy, A., Bouchez, D., Very, A.A., Simonneau, T., Thibaud, J.B. and Sentenac, H. (2003) The Arabidopsis outward $K^+$ channel GORK is involved in regulation of stomatal movements and plant transpiration. *Proc. Natl. Acad. Sci. USA*, **100**: 5549-5554

Hoth, S. and Hedrich, R. (1999) Distinct molecular bases for pH sensitivity of the guard cell $K^+$ channels KST1 and KAT1. *J. Biol. Chem.*, **274**: 11599-11603

Hua, B.G., Mercier, R.W., Leng, Q. and Berkowitz, G.A. (2003) Plants do it differently. A new basis for potassium/sodium selectivity in the pore of an ion channel. *Plant Physiol.*, **132**: 1353-1361

Huang, C.X. and Van Steveninck, R.F.M. (1989) Longitudinal and transverse profiles of K+ and Cl- concentration in low-salt and high-salt barley roots. *New Phytol.*, **112**: 475-480

Hurst, A.C., Meckel, T., Tayefeh, S., Thiel, G. and Homann, U. (2004) Trafficking of the plant potassium inward rectifier KAT1 in guard cell protoplasts of *Vicia faba*. *Plant J.*, **37**: 391-397

Isenring, P. and Forbush, B. (1997) Ion and bumetanide binding by the Na-K-Cl cotransporter - importance of transmembrane domains. *J. Biol. Chem.*, **272**: 24556-24562

Ivashikina, N., Becker, D., Ache, P., Meyerhoff, O., Felle, H.H. and Hedrich, R. (2001) K+ channel profile and electrical properties of Arabidopsis root hairs. *FEBS Lett.*, **508**: 463-469

Ivashikina, N., Deeken, R., Ache, P., Kranz, E., Pommerrenig, B., Sauer, N. and Hedrich, R. (2003) Isolation of *AtSUC2* promoter-GFP-marked companion cells for patch-clamp studies and expression profiling. *Plant J.*, **36**: 931-945

Jan, L.Y. and Jan, Y.N. (1997) Cloned potassium channels from eukaryotes and prokaryotes. *Annu. Rev. Neurosci.*, **20**: 91-123

Johnston, A.E. (2005) Understanding potassium and its use in agriculture. Brussels: EFMA.

Karley, A.J., Leigh, R.A. and Sanders, D. (2000) Differential ion accumulation and ion fluxes in the mesophyll and epidermis of barley. *Plant Physiol.*, **122**: 835-844

Karrer, E.E., Lincoln, J.E., Hogenhout, S., Bennet, A.B., Bostock, R.M., Martineau, B., Lucas, W.J., Gilchrist, D.G. and Alexander, D. (1995) *In situ* isolation of mRNA from individual plant cells: creation of cell-specific cDNA libraries. *Proc. Natl. Acad. Sci. USA*, **92**: 3814-3818

Kato, Y., Sakaguchi, M., Mori, Y., Saito, K., Nakamura, T., Bakker, E.P., Sato, Y., Goshima, S. and Uozumi, N. (2001) Evidence in support of a four transmembrane-pore-transmembrane topology model for the *Arabidopsis thaliana* Na+/K+ translocating AtHKT1 protein, a member of the superfamily of K+ transporters. *Proc. Natl. Acad. Sci. USA*, **98**: 6488-6493

Kehr, J. (2001) High resolution spatial analysis of plant systems. *Curr. Opin. Plant Biol.*, **4**: 197-201

Keller, B.U. and Hedrich, R. (1992) Patch-clamp techniques to study ion channels from organelles. In: *Methods in Enzymology*, vol. 207 Ion Channels, Academic Press, NY pp 673-681

Kim, E.J., Kwak, J.M., Uozumi, N. and Schroeder, J.I. (1998) AtKUP1: An Arabidopsis gene encoding high-affinity potassium transport activity. *Plant Cell*, **10**: 639-639

Kim, S.A., Kwak, J.M., Jae, S.-K., Wang, M.-H. and Nam, H.G. (2001) Overexpression of the AtGluR2 gene encoding an Arabidopsis homolog of mammalian glutamate receptors impairs calcium utilisation and sensitivity to ionic stress in transgenic plants. *Plant Cell Physiol.*, **42**: 74-84

Kleene, S.J. (1999) Both external and internal calcium reduce the sensitivity of the olfactory cyclic-nucleotide-gated channel to cAMP. *J. Neurophysiol.*, **81**: 2675-2682

Kochian, L.V. and Lucas, W.J. (1988) Potassium transport in roots. *Adv. Bot. Res.*, **15**: 93-178

Kochian, L.V. and Lucas, W.J. (1982) Potassium transport in corn roots. I. Resolution of kinetics into a saturable and linear component. *Plant Physiol.*, **70**: 1723-1731

Kochian, L.V. (2000) Molecular physiology of mineral nutrient acquisition, transport, and utilization. In: *Biochemistry and Molecular Biology of Plants* (Eds. Buchanan. B., Gruissem, W., Jones, R.) American Society of Plant Biologists, pp 1204-1249

Köhler, B. and Raschke, K. (2000) The delivery of salts to the xylem. Three types of anion conductance in the plasmalemma of the xylem parenchyma of roots of barley. *Plant Physiol.*, **122**: 243-254

Köhler, C., Merkle, T. and Neuhaus, G. (1999) Characterisation of a novel gene family of putative cyclic nucleotide- and calmodulin-regulated ion channels in *Arabidopsis thaliana*. *Plant J.*, **18**: 97-104

Köhler, C. and Neuhaus, G. (2000) Characterisation of calmodulin binding to cyclic nucleotide-gated ion channels from *Arabidopsis thaliana*. *FEBS Lett.*, **471**: 133-136

Kong, Q. and Simon, A.E. (1998) *In situ* hybridization to RNA in whole Arabidopsis plants In: *Methods in Molecular Biology, vol. 82: Arabidopsis Protocols* (Eds. Martinez-Zapater, J., Salinas, J.) Humana Press Inc, Totowa, NJ pp 409-415

Kreps, J.A., Wu, Y.J., Chang, H.S., Zhu, T., Wang, X. and Harper, J.F. (2002) Transcriptome changes for Arabidopsis in response to salt, osmotic, and cold stress. *Plant Physiol.*, **130**: 2129-2141

Kronzucker, H.J., Szczerba, M.W. and Britto, D.T. (2003) Cytosolic potassium homeostasis revisited: $^{42}$K-tracer analysis in *Hordeum vulgare* L. reveals set-point variations in [K$^+$]. *Planta*, **217**: 540-546

Kuchenbuch, R., Classen, N. and Jungk, A. (1986) Potassium availability in relation to soil moisture. I. Effect of soil moisture on potassium distribution, root growth and potassium uptake of anion plants. *Plant Soil.*, **95**: 221-231

Kurkdjian, A., Leitz, G., Manigault, P., Harim, A. and Greulich, K.O. (1993) Non-enzymatic access to the plasma membrane of *Medicago* root hairs by laser microsurgery. *J. Cell Sci.*, **105**: 263-268

Lacombe, B., Becker, D., Hedrich, R., DeSalle, R., Hollmann, M., Kwak, J.M., Schroeder, J.I., Le Novere, N., Nam, H.G., Spalding, E.P., Tester, M., Turano, F.J., Chiu, J. and Coruzzi, G.M. (2001) The identity of plant glutamate receptors. *Science*, **292**: 1486-1487

Lacombe, B., Pilot, G., Michard, E., Gaymard, F., Sentenac, H. and Thibaud, J.B. (2000) A Shaker-like K$^+$ channel with weak rectification is expressed in both source and sink phloem tissues of Arabidopsis. *Plant Cell*, **12**: 837-851

Lagarde, D., Basset, M., Lepetit, M., Conejero, G., Gaymard, F., Astruc, S. and Grignon, C. (1996) Tissue-specific expression of Arabidopsis AKT1 gene is consistent with a role in K$^+$ nutrition. *Plant J.*, **9**: 195-203

Lam, M., Bhat, M.B., Nunez, G., Ma, J.J. and Distelhorst, C.W. (1998) Regulation of Bcl-xl channel activity by calcium. *J. Biol. Chem.*, **273**: 17307-17310

Laurie, S., Feeney, K.A., Maathuis, F.J.M., Heard, P.J., Brown, S.J. and Leigh, R.A. (2002) A role for HKT1 in sodium uptake by wheat roots. *Plant J.*, **32**: 139-149

Lee, H.J., Xiong, L.M., Gong, Z.Z., Ishitani, M., Stevenson, B. and Zhu, J.K. (2001) The Arabidopsis HOS1 gene negatively regulates cold signal transduction and encodes a RING finger protein that displays cold-regulated nucleo-cytoplasmic partitioning. *Genes Develop.*, **15**: 912-924

Leigh, R. and Wyn Jones, R.G. (1984) A hypothesis relating critical potassium concentrations for growth to the distribution and functions of this ion in the plant cell. *New Phytol.*, **97**: 1-13

Leigh, R.A., Walker, D.J., Fricke, W., Tomos, A.D. and Miller, A.J. (1999) Patterns of potassium compartmentation in plant cells as revealed by microelectrodes and

microsampling. In: *Frontiers in Potassium Nutrition: New Perspectives on the Effects of Potasium on Physiology of Plants* (Eds. Oosterhuis, D.M., Berkowitz, G.A.). The Potash and Phosphate Institute, Norcross, Georgia, pp 63-70
Leng, Q., Mercier, R.W., Hua, B.G., Fromm, H. and Berkowitz, G.A. (2002) Electrophysiological analysis of cloned cyclic nucleotide-gated ion channels. *Plant Physiol.*, **128**: 400-410
Leng, Q., Mercier, R.W., Yao, W.Z. and Berkowitz, G.A. (1999) Cloning and first functional characterization of a plant cyclic nucleotide-gated cation channel. *Plant Physiol.*, **121**: 753-761
Lew, R.R. (1991) Electrogenic transport properties of growing Arabidopsis root hairs: the plasma membrane proton pump and potassium channels. *Plant Physiol.*, **97**: 1527-1534
Liebersbach, H., Steingrobe, B. and Claassen, N. (2004) Roots regulate ion transport in the rhizosphere to counteract reduced mobility in dry soil. *Plant Soil*, **260**: 79-88
Lindberg, S. and Strid, H. (1997) Aluminium induces rapid changes in cytosolic pH and free calcium and potassium concentrations in root protoplasts of wheat (*Triticum aestivum*). *Physiol. Plant.*, **99**: 405-414
Liu, K.-H. and Tsay, I.-F. (2003) Switching between the two action modes of the dual-affinity nitrate transporter CHL1 by phosphorylation. *EMBO J.*, **22**: 1005-1013
Maathuis, F.J.M. and Amtmann, A. (1999) $K^+$ nutrition and $Na^+$ toxicity: The basis of cellular $K^+/Na^+$ ratios. *Ann. Bot.*, **84**: 123-133
Maathuis, F.J.M., Filatov, V., Herzyk, P., Krijger, G.C., Axelsen, K.B., Chen, S.X., Green, B.J., Li, Y., Madagan, K.L., Sanchez-Fernandez, R., Forde, B.G., Palmgren, M.G., Rea, P.A., Williams, L.E., Sanders, D. and Amtmann, A. (2003) Transcriptome analysis of root transporters reveals participation of multiple gene families in the response to cation stress. *Plant J.*, **35**: 675-692
Maathuis, F.J.M., May, S.T., Graham, N.S., Bowen, H.C., Jelitto, T.C., Trimmer, P., Bennett, M.J., Sanders, D. and White, P.J. (1998) Cell marking in *Arabidopsis thaliana* and its application to patch-clamp studies. *Plant J.*, **15**: 843-851
Maathuis, F.J.M. and Sanders, D. (1993) Energization of potassium uptake in *Arabidopsis thaliana*. *Planta*, **191**: 302-307
Maathuis, F.J.M. and Sanders, D. (1994) Mechanism of high-affinity potassium uptake in roots of *Arabidopsis thaliana*. *Proc. Natl. Acad. Sci. USA*, **91**: 9272-9276
Maathuis, F.J.M. and Sanders, D. (1995) Contrasting roles in ion-transport of 2 $K^+$-channel types in root-cells of *Arabidopsis thaliana*. *Planta*, **197**: 456-464
Maathuis, F.J.M. and Sanders, D. (1997) Regulation of $K^+$ absorption in plant root cells by external $K^+$: Interplay of different plasma membrane $K^+$ transporters. *J. Exp. Bot.*, **48**: 451-458
Maathuis, F.J.M. and Sanders, D. (2001) Sodium uptake in Arabidopsis roots is regulated by cyclic nucleotides. *Plant Physiol.*, **127**: 1617-1625
MacRobbie, E.A.C. (1995) Effects of ABA on $^{86}Rb^+$ fluxes at plasmalemma and tonoplast of stomatal guard cells. *Plant J.*, **7**: 835-843
MacRobbie, E.A.C. (1998) Signal transduction and ion channels in guard cells. *Phil. Trans. Royal Soc. London Series B*, **353**: 1475-1488
Madden, D.R. (2002) The structure and function of glutamate receptor ion channels. *Nature Rev. Neurosci.* **3**: 91-101
Malone, M., Leigh, R.A. and Tomos, A.D. (1991) Concentrations of vacuolar inorganic-ions in individual cells of intact wheat leaf epidermis. *J. Exp. Bot.*, **42**: 305-309
Marschner, H. (1995) *The mineral nutrition of higher plants*, 2$^{nd}$ edition. Academic Press, San Diego, USA

Marten, I., Hoth, S., Deeken, R., Ache, P., Ketchum, K.A., Hoshi, T. and Hedrich, R. (1999) AKT3, a phloem-localized K$^+$ channel, is blocked by protons. *Proc. Natl. Acad. Sci. USA*, **96**: 7581-7586

Mäser, P., Thomine, S., Schroeder, J.I., Ward, J.M., Hirschi, K., Sze, H., Talke, I.N., Amtmann, A., Maathuis, F.J.M., Sanders, D., Harper, J.F., Tchieu, J., Gribskov, M., Persans, M.W., Salt, D.E., Kim, S.A. and Guerinot, M.L. (2001) Phylogenetic relationships within cation transporter families of Arabidopsis. *Plant Physiol.*, **126**: 1646-1667

Mclaren, R.G. and Cameron, K.C. (1996) *Soil Science: Sustainable Production and Environmental Protection.* Oxford University Press. Auckland 304 pp

Meckel, T., Hurst, A.C., Thiel, G. and Homann, U. (2004) Endocytosis against high turgor: intact guard cells of *Vicia faba* constitutively endocytose fluorescently labelled plasma membrane amd GFP-tagged K$^+$ channel KAT1. *Plant Cell*, **39**: 182-193

Mengel, K. and Kirkby, E.A. (2001) *Principles of Plant Nutrition*, 5th edition. Kluwer Academic Press, Dordrecht, The Netherlands

Meyerhoff, O., Muller, K., Roelfsema, M.R., Latz, A., Lacombe, B., Hedrich, R., Dietrich, P. and Becker, D. (2005) AtGLR3.4, a glutamate receptor channel-like gene is sensitive to touch and cold. *Planta*, **222**: 418-427

Minta, A. and Tsien, R.Y. (1989) Fluorescent indicators for cytosolic sodium. *J. Biol. Chem.*, **264**: 19449-19457

Moran, N., Ehrenstein, G., Iwasa, K., Bare, C. and Mischke, C. (1984) Ion channels in plasmalemma of wheat protoplasts. *Science*, **226**: 835-838

Moshelion, M., Becker, D., Czempinski, K., Mueller-Roeber, B., Attali, B., Hedrich, R. and Moran, N. (2002) Diurnal and circadian regulation of putative potassium channels in a leaf moving organ. *Plant Physiol.*, **128**: 634-642

Mouline, K., Véry, A.A., Gaymard, F., Boucherez, J., Pilot, G., Devic, M., Bouchez, D., Thibaud, J.B. and Sentenac, H. (2002) Pollen tube development and competitive ability are impaired by disruption of a Shaker K$^+$ channel in Arabidopsis. *Genes Develop.*, **16**: 339-350

Mühling, K.H. and Sattelmacher, B. (1997) Determination of apoplastic K$^+$ in intact leaves by ratio imaging of PBFI fluorescence. *J. Exp. Bot.*, **48**: 1609-1614

Mühling, K.H. and Läuchli, A. (2000) Light-induced pH and K$^+$ changes in the apoplast of intact leaves. *Planta*, **212**: 9-15

Munro, A.W., Ritchie, G.Y., Lamb, A.J., Douglas, R.M. and Booth, I.R. (1991) The cloning and DNA-sequence of the gene for the glutathione-regulated potassium-efflux system KefC of *Escherichia coli*. *Mol. Microbiol.*, **5**: 607-616

Nakamura, R.L., McKendree, W.L., Hirsch, R.E., Sedbrook, J.C., Gaber, R.F. and Sussman, M.R. (1995) Expression of an Arabidopsis potassium channel gene in guard-cells. *Plant Physiol.*, **109**: 371-374

Newman, I.A. (2001) Ion transport in roots: measurement of fluxes using ion-selective microelectrodes to characterize transporter function. *Plant Cell Environ.*, **24**: 1-14

Öborn, I., Andrist-Rangel, Y., Askegaard, M., Grant, C.A., Watson, C.A. and Edwards, A.C. (2005) Critical aspects of potassium management in agricultural systems. *Soil Use Manag.*, **21**: 102-112

O'Connell, A.D., Morton, M.J. and Hunter, M. (2002) Two-pore domain K$^+$ channels-molecular sensors. *Biochim. Biophys. Acta*, **1566**: 152-161

Oliveira, R.H., Rosolem, C.A. and Trigueiro, R.M. (2004) Importance of mass flow and diffusion on the potassium supply to cotton plants as affected by soil water and potassium. *Revista Brasileira De Ciencia Do Solo*, **28**: 439-445

Pal, Y., Wong, M.T.F. and Gilkes, R.J. (1999) The forms of potassium and potassium adsorption in some virgin soils from south-western Australia. *Austral J. Soil Res.*, **37**: 695-709

Peiter, E., Maathuis, F.J.M., Mills, L.N., Knight, H., Pelloux, J., Hetherington, A.M. and Sanders, D. (2005) The vacuolar $Ca^{2+}$-activated channel regulates seed germination and stomatal movements. *Nature*, **434**: 404-408

Penny, M.G. and Bowling, D.J.F. (1974) A study of potassium gradients in the epidermis of intact leaves of *Commelina communis* in relation to stomatal opening. *Planta*, **119**: 17-25

Peoples, T.R. and Koch, D.W. (1979) Role of potassium, in carbon dioxide assimilation in *Medicago sativa* L. *Plant Physiol.*, **63**: 878-881

Philippar, K., Büchsenschütz, K., Abshagen, M., Fuchs, I., Geiger, D., Lacombe, B. and Hedrich, R. (2003) The $K^+$ channel KZM1 mediates potassium uptake into the phloem and guard cells of the C4 grass *Zea mays*. *J. Biol. Chem.*, **278**: 16973-16981

Pier, P.A. and Berkowitz, G.A. (1987) Modulation of water-stress effects on photosynthesis by altered leaf $K^+$. *Plant Physiol.*, **85**: 655-661

Pilot, G., Gaymard, F., Mouline, K., Cherel, I. and Sentenac, H. (2003) Regulated expression of Arabidopsis Shaker $K^+$ channel genes involved in $K^+$ uptake and distribution in the plant. *Plant Mol. Biol.*, **51**: 773-787

Pilot, G., Lacombe, B., Gaymard, F., Cherel, I., Boucherez, J., Thibaud, J.B. and Sentenac, H. (2001) Guard cell inward $K^+$ channel activity in Arabidopsis involves expression of the twin channel subunits KAT1 and KAT2. *J. Biol. Chem.*, **276**: 3215-3221

Pitman, M.G., Wellfare, D. and Carter, C. (1981) Reduction of hydraulic conductance during inhibition of exudation from excised maize and barley roots. *Plant Physiol.*, **61**: 802-808

Porée, F., Wulfetang, K., Naso, A., Carpaneto, A., Soller, A., Natura, G., Bertl, A., Sentenac, H., Thibaud, J.-B. and Dreyer, I. (2005) Plant $K_{in}$ and $K_{out}$ channels: Approaching the trait of opposite rectification by analyzing more than 250 KAT1–SKOR chimeras. *Biochem. Biophys. Res. Comm.*, **332**: 465-473

Pottosin, I.I. (1992) Single channel recording in the chloroplast envelope. *FEBS Lett.*, **308**: 87-90

Pottosin, I.I. and Schönknecht, G. (1996) Ion channel permeable for divalent and monovalent cations in native spinach thylakoid membranes. *J. Membr. Biol.*, **152**: 223-233

Pretty, K.M. and Stangel, P.J. (1985) Current and future use of world potassium. In: *Potassium in Agriculture* (Ed. Munson, R.D.) American Society of Agronomy, Madison, Wisconsin, USA, pp 99-128

Qi, Z. and Spalding, E.P. (2004) Protection of plasma membrane $K^+$ transport by the salt overly sensitive $Na^+$-$H^+$ antiporter during salinity stress. *Plant Physiol.*, **136**: 2548-2555

Qiu, Q.S., Barkla, B.J., Vera-Estrella, R., Zhu, J.K. and Schumaker, K.S. (2003) $Na^+/H^+$ exchange activity in the plasma membrane of Arabidopsis. *Plant Physiol.*, **132**: 1041-1052

Qiu, Q.S., Guo, Y., Dietrich, M.A., Schumaker, K.S. and Zhu, J.K. (2002) Regulation of SOS1, a plasma membrane $Na^+/H^+$ exchanger in *Arabidopsis thaliana*, by SOS2 and SOS3. *Proc. Natl. Acad. Sci. USA*, **99**: 8436-8441

Quintero, F.J. and Blatt, M.R. (1997) A new family of KC transporters from Arabidopsis that are conserved across phyla. *FEBS Lett.*, **415**: 206-211

Reintanz, B., Szyroki, A., Ivashikina, N., Ache, P., Godde, M., Becker, D., Palme, K. and Hedrich, R. (2002) AtKC1, a silent Arabidopsis potassium channel alpha-subunit modulates root hair $K^+$ influx. *Proc. Natl. Acad. Sci. USA*, **99**: 4079-4084

Rigas, S., Debrosses, G., Haralampidis, K., Vicente-Agullo, F., Feldmann, K.A., Grabov, A., Dolan, L. and Hatzopoulos, P. (2001) TRH1 encodes a potassium transporter required for tip growth in Arabidopsis root hairs. *Plant Cell*, **13**: 139-151

Roberts, S.K. and Tester, M. (1995) Inward and outward $K^+$-selective currents in the plasma membrane of protoplasts from maize root cortex and stele. *Plant J.*, **8**: 811-825

Rodríguez-Navarro, A. (2000) Potassium transport in fungi and plants. *Biochim. Biophys. Acta Biomembr.*, **1469**: 1-30

Roelfsema, M.R.G., Steinmeyer, R., Staal, M. and Hedrich, R. (2001) Single guard cell recordings in intact plants: light-induced hyperpolarization of the plasma membrane. *Plant J.*, **26**: 1-13.

Roelfsema, M.R.G. and Hedrich, R. (2002) Studying guard cells in the intact plant: modulation of stomatal movement by apoplastic factors. *New Phytol.*, **153**: 425-431

Rona, J.-P., Cornel, D., Grignon, C. and Heller, R. (1982) The electrical potential difference across the tonoplast of *Acer pseuoplatanus*. *Physiol. Veg.*, **20**: 459-463

Rosolem, C.A., Mateus, G.P., Godoy, L.J.G., Feltran, J.C. and Brancaliao, S.R. (2003) Root morphology and potassium supply to pearl millet roots as affected by soil water and potassium contents. *Revista Brasileira De Ciencia Do Solo*, **27**: 875-884

Rubio, F., Gassmann, W. and Schroeder, J.I. (1995) Sodium-driven potassium uptake by the plant potassium transporter HKT1 and mutations conferring salt tolerance. *Science*, **270**: 1660-1663

Rubio, F., Santa-María, G.E. and Rodríguez-Navarro, A. (2000) Cloning of Arabidopsis and barley cDNAs encoding HAK potassium transporters in root and shoot cells. *Physiol. Plant.*, **109**: 34-43

Rus, A., Lee, B.H., Munoz-Mayor, A., Sharkhuu, A., Miura, K., Zhu, J.K., Bressan, R.A. and Hasegawa, P.M. (2004) AtHKT1 facilitates $Na^+$ homeostasis and $K^+$ nutrition *in planta*. *Plant Physiol.*, **136**: 2500-2511

Rus, A., Yokoi, S., Sharkhuu, A., Reddy, M., Lee, B.H., Matsumoto, T.K., Koiwa, H., Zhu, J.-K., Bressan, R.A. and Hasegawa, P.M. (2001) AtHKT1 is a salt tolerance determinant that controls $Na^+$ entry into plant roots. *Proc. Natl. Acad. Sci. USA*, **98**: 14150-14155

Santa-María, G.E., Danna, C.H. and Czibener, C. (2000) High-affinity potassium transport in barley roots. Ammonium-sensitive and -insensitive pathways. *Plant Physiol.*, **123**: 297-306

Santa-María, G.E., Rubio, F., Dubcovsky, J. and Rodríguez-Navarro, A. (1997) The HAK1 gene of barley is a member of a large gene family and encodes a high-affinity potassium transporter. *Plant Cell*, **9**: 2281-2289

Schachtman, D.P., Kumar, R., Schroeder, J.I. and Marsh, E.L. (1997) Molecular and functional characterization of a novel low-affinity cation transporter (LCT1) in higher plants. *Proc. Natl. Acad. Sci. USA*, **94**: 11079-11084

Schachtman, D.P. and Schroeder, J.I. (1994) Structure and transport mechanism of a high-affinity potassium uptake transporter from higher-plants. *Nature*, **370**: 655-658

Schleyer, M. and Bakker, E.P. (1993) Nucleotide-sequence and 3'-end deletion studies indicate that the $K^+$-uptake protein KUP from *Escherichia coli* is composed of a hydrophobic core linked to a large and partially essential hydrophilic-C terminus. *J. Bacteriol.*, **175**: 6925-6931

Schönknecht, G., Spoormaker, P., Steinmeyer, R., Bruggeman, L., Ache, P., Dutta, R., Reintanz, B., Godde, M., Hedrich, R., Palme, K. (2002) KCO1 is a component of the slow-vacuolar (SV) ion channel. *FEBS Lett.*, **511**: 28-32

Potassium uptake and transport

Schroeder, J.I., Hedrich, R. and Fernandez, J.M. (1984) Potassium-selective single channels in guard cell protoplasts of *Vicia faba*. *Nature,* **312**: 361-362

Schuurink, R.C., Shartzer, S.F., Fath, A. and Jones, R.L. (1998) Characterization of a calmodulin-binding transporter from the plasma membrane of barley aleurone. *Proc. Natl. Acad. Sci. USA,* **95**: 1944-1949

Seiffert, S., Kaselowsky, J., Jungk, A. and Claassen, N. (1995) Observed and calculated potassium uptake by maize as affected by soil water content and bulk density. *Agron. J.,* **87**: 1070-1077

Senn, M.E., Rubio, F., Banuelos, M.A. and Rodriguez-Navarro, A. (2001) Comparative functional features of plant potassium HvHAK1 and HvHAK2 transporters. *J. Biol. Chem.,* **276**: 44563-44569

Sentenac, H., Bonneaud, N., Minet, M., Lacroute, F., Salmon, J.M., Gaymard, F. and Grignon, C. (1992) Cloning and expression in yeast of a plant potassium-ion transport-system. *Science,* **256**: 663-665

Serrano, R. and Rodríguez-Navarro, A. (2001) Ion homeostasis during salt stress in plants. *Curr. Opin. Cell Biol.,* **13**: 399-404

Shabala, L., Cuin, T.A., Newman, I.A. and Shabala, S. (2005) Salinity-induced ion flux patterns from the excised roots of Arabidopsis SOS mutants. *Planta,* **222**: 1041-1050

Shabala, S. and Newman, I. (1999) Light-induced changes in hydrogen, calcium, potassium, and chloride ion fluxes and concentrations from the mesophyll and epidermal tissues of bean leaves. Understanding the ionic basis of light-induced bioelectrogenesis. *Plant. Physiol.,* **119**: 1119-1124

Shabala, S., Babourina, O. and Newman, I. (2000) Ion-specific mechanisms of osmoregulation in bean mesophyll cells. *J. Exp. Bot.,* **51**: 1243-1253

Shabala, S., Shabala, L. and Van Volkenburgh, E. (2003) Effect of calcium on root development and root ion fluxes in salinised barley seedlings. *Functional Plant Biol.,* **30**: 507-514

Shabala, S. (2003) Regulation of potassium transport in leaves: from molecular to tissue level. *Ann. Bot.,* **92**: 627-634

Shabala, S. (2006) Non-invasive microelectrode ion flux measurements in plant stress physiology. In: *Plant Electrophysiology - Theory and Methods* (Ed. Volkov, A.). Springer, Heidelberg, pp. 35-71

Shi, H.Z., Quintero, F.J., Pardo, J.M. and Zhu, J.K. (2002) The putative plasma membrane $Na^+/H^+$ antiporter SOS1 controls long-distance $Na^+$ transport in plants. *Plant Cell,* **14**: 465-477

Shin, R. and Schachtman, D.P. (2004) Hydrogen peroxide mediates plant root cell response to nutrient deprivation. *Proc. Natl. Acad. Sci. USA,* **101**: 8827-8832

Siddiqi, M.Y. and Glass, A.D.M. (1987) Regulation of $K^+$ influx in barley - evidence for a direct control of influx by $K^+$ concentration of root-cells. *J. Exp. Bot.,* **38**: 935-947

Song, C.P., Guo, Y., Qiu, Q.S., Lambert, G., Galbraith, D.W., Jagendorf, A. and Zhu, J.K. (2004) A probable $Na^+(K^+)/H^+$ exchanger on the chloroplast envelope functions in pH homeostasis and chloroplast development in *Arabidopsis thaliana*. *Proc. Natl. Acad. Sci. USA,* **101**: 10211-10216

Spalding, E.P., Hirsch, R.E., Lewis, D.R., Qi, Z., Sussman, M.R. and Lewis, B.D. (1999) Potassium uptake supporting plant growth in the absence of AKT1 channel activity - Inhibition by ammonium and stimulation by sodium. *J. Gen. Physiol.,* **113**: 909-918

Sparks, D.L. and Huang, P.M. (1985) Physical chemistry of soil potassium. In: *Potassium in Agriculture* (Ed. Munson, R.D.). American Society of Agronomy, Madison, Wisconsin, USA, pp 201-276

Su, H., Golldack, D., Katsuhara, M., Zhao, C.S. and Bohnert, H.J. (2001) Expression and stress-dependent induction of potassium channel transcripts in the common ice plant. *Plant Physiol.*, **125**: 604-614

Syers, J.K. (1998) *Soil and Plant Potassium in Agriculture*. The Fertiliser Society, York, UK

Szmacinski, H. and Lakowitz, J.R. (1999) Potassium and sodium measurements at clinical concentrations using phase-modulation fluorometry. *Sensors Actuators*, B **60**: 8-18

Szyroki, A., Ivashikina, N., Dietrich, P., Roelfsema, M.R.G., Ache, P., Reintanz, B., Deeken, R., Godde, M., Felle, H., Steinmeyer, R., Palme, K. and Hedrich, R. (2001) KAT1 is not essential for stomatal opening. *Proc. Natl. Acad. Sci. USA*, **98**: 2917-2921

Talke, I.N., Blaudez, D., Maathuis, F.J.M. and Sanders, D. (2003) CNGCs: prime targets of plant cyclic nucleotide signalling? *Trend Plant Sci.*, **8**: 286-293

Tisdale, S.L., Nelson, W.L., Beaton, J.D. and Havlin, J.L. (1993) *Soil Fertility and Fertilizer*. Macmillan, New York.

Trebacz, K. and Schönknecht, G. (2000) Simple method to isolate vacuoles and protoplasts for patch-clamp experiments. *Protoplasma*, **213**: 39-45

Treeby, M.T. and Van Steveninck, R.F.M. (1988) The influence of salinity on phosphate-uptake and distribution in lupin roots. *Physiol. Plant.*, **72**: 617-622

Tyerman, S.D., Beilby, M., Whittington, J., Juswonon, U., Newman, I. and Shabala, S. (2001) Oscillations in proton transport revealed from simultaneous measurements of net current and net proton fluxes from isolated root protoplasts: MIFE meets patch-clamp. *Aust. J. Plant Physiol.*, **28**: 591-604

Uozumi, N., Kim, E.J., Rubio, F., Yamaguchi, T., Muto, S., Tsuboi, A., Bakker, E.P., Nakamura, T. and Schroeder, J.I. (2000) The Arabidopsis HKT1 gene homolog mediates inward $Na^+$ currents in *Xenopus laevis* oocytes and $Na^+$ uptake in *Saccharomyces cerevisiae*. *Plant Physiol.*, **122**: 1249-1259

Vallejo, A.J., Peralta, M.L. and Santa-María, G.E. (2005) Expression of potassium-transporter coding genes, and kinetics of rubidium uptake, along a longitudinal root axis. *Plant Cell Environ.*, **28**: 850-862

Venema, K., Quintero, F.J., Pardo, J.M. and Donaire, J.P. (2002) The Arabidopsis $Na^+/H^+$ exchanger AtNHX1 catalyzes low affinity $Na^+$ and $K^+$ transport in reconstituted liposomes. *J. Biol. Chem.*, **277**: 2413-2418

Véry, A.A. and Sentenac, H. (2002) Cation channels in the Arabidopsis plasma membrane. *Trend Plant Sci.*, **7**: 168-175

Véry, A.A. and Sentenac, H. (2003) Molecular mechanisms and regulation of $K^+$ transport in higher plants. *Annu. Rev. Plant Biol.*, **54**: 575-603

Vetterlein, D. and Jahn, R. (2004) Gradients in soil solution composition between bulk soil and rhizosphere - In situ measurement with changing soil water content. *Plant Soil*, **258**: 307-317

Walker, D.J., Leigh, R.A. and Miller, A.J. (1995) Simultaneous measurements of intracellular pH and $K^+$ or $NO_3^-$ in barley roots using triple barreled, ion-selective microelectrodes. *Plant Physiol.*, **108**: 743-751

Walker, D.J., Leigh, R.A. and Miller, A.J. (1996) Potassium homeostasis in vacuolate plant cells. *Proc. Natl. Acad. Sci. USA*, **93**: 10510-10514

Wang, T.B., Gassmann, W., Rubio, F., Schroeder, J.I. and Glass, A.D.M. (1998) Rapid up-regulation of HKT1, a high-affinity potassium transporter gene, in roots of barley and wheat following withdrawal of potassium. *Plant Physiol.*, **118**: 651-659

## Potassium uptake and transport

Wang, Y.H., Garvin, D.F. and Kochian, L.V. (2002) Rapid induction of regulatory and transporter genes in response to phosphorus, potassium, and iron deficiencies in tomato roots. Evidence for cross talk and root/rhizosphere-mediated signals. *Plant Physiol.*, **130**: 1361-1370

Ward, J.M. (2001) Identification of novel families of membrane proteins from the model plant *Arabidopsis thaliana. Bioinformatics*, **17**: 560-563

Wegner, L.H. and de Boer, A.H. (1997) Two inward $K^+$ channels in the xylem parenchyma cells of barley roots are regulated by G-protein modulators through a membrane-delimited pathway. *Planta*, **203**: 506-516

Wegner, L.H., Deboer, A.H. and Raschke, K. (1994) Properties of the $K^+$ inward rectifier in the plasma-membrane of xylem parenchyma cells from barley roots - Effects of TEA(+), $Ca^{2+}$, $Ba^{2+}$ and $La^{3+}$. *J. Membr. Biol.*, **142**: 363-379

Wegner, L.H. and Zimmermann, U. (2002) On-line measurements of $K^+$ activity in the tensile water of the xylem conduit of higher plants. *Plant J.*, **32**: 409-417

Welch, R.M. and Epstein, E. (1968) Dual mechanisms of alkali cation absorption by plant cells - their parallel operation across plasmalemma. *Proc. Natl. Acad. Sci. USA*, **61**: 447-453

White, P.J. (1997) Cation channels in the plasma membrane of rye roots. *J. Exp. Bot.*, **48**: 499-514

White, P.J., Biskup, B., Elzenga, J.T.M., Homann, U., Thiel, G., Wissing, F. and Maathuis, F.J.M. (1998) Advanced patch-clamp techniques and single channel analysis. *J. Exp. Bot.*, **50**: 1037-1054

White, P.J., Bowen, H.C., Demidchik, V., Nichols, C. and Davies, J.A. (2002) Genes for calcium-permeable channels in the plasma membrane of plant root cells. *Biochim. Biophys. Acta Biomembr.*, **1564**: 299-309

White, P.J. and Tester, M.A. (1992) Potassium channels from the plasma-membrane of rye roots characterized following incorporation into planar lipid bilayers. *Planta*, **186**: 188-202

Wu, W.H. and Berkowitz, G.A. (1992) Stromal pH and photosynthesis are affected by electroneutral $K^+$ and $H^+$ exchange through chloroplast envelope ion channels. *Plant Physiol.*, **98**: 666-672

Yanai, J., Linehan, D.J., Robinson, D., Young, I.M., Hackett, C.A., Kyuma, K. and Kosaki, T. (1996) Effects of inorganic nitrogen application on the dynamics of the soil solution composition in the root zone of maize. *Plant Soil*, **180**: 1-9

Yao, Z.H., Mizumura, T., Mei, D.A. and Gross, G.J. (1997) K-ATP channels and memory of ischemic preconditioning in dogs: Synergism between adenosine and K-ATP channels. *Amer. J. Physiol.*, **41**: H334-H342

Yokoi, S., Quintero, F.J., Cubero, B., Ruiz, M.T., Bressan, R.A., Hasegawa, P.M. and Pardo, J.M. (2002) Differential expression and function of *Arabidopsis thaliana* NHX $Na^+/H^+$ antiporters in the salt stress response. *Plant J.*, **30**: 529-539

Yuan, Q.P., Ouyang, S., Liu, J., Suh, B., Cheung, F., Sultana, R., Lee, D., Quackenbush, J. and Buell, C.R. (2003) The TIGR rice genome annotation resource: annotating the rice genome and creating resources for plant biologists. *Nucleic Acids Res.*, **31**: 229-233

Zagotta, W.N. and Siegelbaum, S.A. (1996) Structure and function of cyclic nucleotide-gated channels. *Annu. Rev. Neurosci.*, **19**: 235-263

Zhang, H.X. and Blumwald, E. (2001) Transgenic salt-tolerant tomato plants accumulate salt in foliage but not in fruit. *Nature Biotechnol.*, **19**: 765-768

Zhang, J., Campbell, R.E., Ting, A.Y. and Tsien, R.Y. (2002) Creating new fluorescent probes for cell biology. *Nat. Rev. Cell Mol. Biol.*, **3**: 906-918

Zimmermann, S., Talke, I., Ehrhardt, T., Nast, G. and Muller-Rober, B. (1998) Characterization of SKT1, an inwardly rectifying potassium channel from potato, by heterologous expression in insect cells. *Plant Physiol.*, **116**: 879-890

Zimmermann, S., Hartje, S., Ehrhardt, T., Plesch, G. and Bernd Mueller-Roeber, B. (2001) The $K^+$ channel SKT1 is co-expressed with KST1 in potato guard cells - both channels can co-assemble via their conserved $K_T$ domains. *Plant J.*, **28**: 517-527

Zoeteweij, P.J., Van de Water, B., De Bont, H.J.G.M. and Nagelkerke, J.F. (1994) Mitochondrial $K^+$ as modulator of $Ca^{2+}$-dependent cytotoxicity in hepatocytes. Novel application of the $K^+$-sensitive dye PBFI ($K^+$-binding benzofuran isophthalate) to assess free mitochondrial $K^+$ concentrations. *Biochem. J.*, **299**: 539-543

# Chapter 2

## CALCIUM TRANSPORTERS: FROM FIELDS TO THE TABLE

### JAY MORRIS[1] AND KENDAL HIRSCHI[1,2]
[1]*The Vegetable and Fruit Improvement Center, Texas A&M University, College Station, Texas 77845, USA*
[2]*Children's Nutrition Research Center, Department of Pediatrics, Baylor College of Medicine, Houston, Texas 77030, USA*
E-mail: jaym@bcm.tmc.edu

**Abstract**

Calcium transporters regulate calcium fluxes within cells. Plants, like all organisms, contain channels, pumps and exchangers to carefully modulate intracellular calcium levels. This review presents a summary of the recent advances in cloning and characterizing of these transporters and highlights their putative roles in calcium signaling. Two themes pervade this review: Yeast molecular genetic analysis is a robust tool to study the function of plant transporters and there is a lack of evidence directly linking plant calcium transport mutants with alterations in calcium signaling. We conclude this review by discussing how plant transporters are being engineered to improve the calcium content of agriculturally important plants.

**Keywords:** calcium, transport, channels, pumps, exchangers, nutrition

## 1. INTRODUCTION

Calcium ($Ca^{2+}$) has various important roles in both plants and animals. Minute changes in cytosolic $Ca^{2+}$ regulate biological responses. While humans need $Ca^{2+}$ to build strong bones, plants use $Ca^{2+}$ to provide stress protection and to strengthen cell walls. Thus $Ca^{2+}$ serves a dual purpose as both a nutrient and a signal. To help modulate $Ca^{2+}$ levels, plants have $Ca^{2+}$ transporters with specific kinetic properties located on membranes of various organelles. These $Ca^{2+}$ transporters can be characterized into three major types: pumps, which require energy to move $Ca^{2+}$ across the membrane, exchangers, which are driven by proton gradients, and channels, which allow large concentrations of $Ca^{2+}$ to traverse membranes (Pittman and Hirschi, 2003). Understanding the unique roles these transporters have in $Ca^{2+}$ homeostasis is an important area of plant biology.

Factors that cause cytosolic $Ca^{2+}$ spikes range from temperature stress, hormones, to mechanical or tactile stimuli (Bothwell and Ng, 2005; Braam, 2005). These spikes

---

© CAB International 2008. *Plant Membrane and Vacuolar Transporters* (eds P.K. Jaiwal, R.P. Singh and O.P. Dhankher)

are complex and are judiciously regulated. Coupled with the amplitude of the $Ca^{2+}$ spike, the duration and location are hypothesized to be crucial to the specificity of the subsequent signal (Sanders et al., 2002; Scrase-Field and Knight, 2003; Plieth 2005). Calcium can increase up to 3µM in concentration in the cytosol during a spike (signaling event), which is up to a 1000 to 10000 fold increase from resting cytosolic $Ca^{2+}$ concentrations (Reddy and Reddy, 2004; Medvedev, 2005). To facilitate the localization and duration of these spikes, various types of $Ca^{2+}$ transporters are located on membranes throughout the cell. Arabidopsis has $Ca^{2+}$ channels on the plasma (PM), vacuole (VM), and endoplasmic reticulum (ER) membranes. $Ca^{2+}$ pumps can be found on the PM, VM, ER, Golgi, small vacuole (sVM) and chloroplast membranes (CM). Finally, $Ca^{2+}$ exchangers can be found predominantly on the VM in Arabidopsis plants (Figure 1; Reddy and Reddy, 2004). There are other calcium transporters present at mitochondrial, plastid and nuclear membranes for $Ca^{2+}$ partitioning in and out of these organelles (Chigri et al., 2006; Oldroyd and Downie, 2006; Xiong et al., 2006). Activation and regulation of these transporters are thought to shape the dynamics of a $Ca^{2+}$ spike and insure the specificity of the stimulus dependent responses (Scrase-Field and Knight, 2003; Harper et al., 2004; Bothwell and Ng, 2005; Medvedev, 2005). At one level, these $Ca^{2+}$ signaling events appear simple: cells at rest have a low level of cytosolic $Ca^{2+}$ that rises during a signal transduction event (Hirschi, 2004). Signal transductions related to $Ca^{2+}$ oscillations can be thought of as a four-part process: (1) $Ca^{2+}$ is mobilized by response triggers to become active, followed by the movement of $Ca^{2+}$ into the cytosol; (2) this increase activates $Ca^{2+}$-regulated proteins, which induce the response; (3) $Ca^{2+}$ then functions as a messenger to activate $Ca^{2+}$ sensitive processes, which are mediated by proteins such as CDPKs (calcium-dependent protein kinases) and calmodulin; (4) binding proteins and transporters remove the $Ca^{2+}$ from the cytoplasm, essentially turning "off" the $Ca^{2+}$ mediated signaling event (Berridge et al., 2000). These signaling events, which transiently alter cytosolic $Ca^{2+}$ concentrations, must be regulated with a high degree of fidelity.

Several model systems have been used to characterize the biology of $Ca^{2+}$ movement and transporters which help modulate these oscillations. Heterologous expression of plant transporters in the budding yeast *Saccharomyces cerevisiae* is a well developed and widely used technique (Ton and Rao, 2004). Yeast offers advantages such as simple and inexpensive growth conditions, tractable genetics and a conservation of basic cellular machinery and signal transduction pathways with higher eukaryotes (Ton and Rao, 2004). In plants, $Ca^{2+}$ signals can be studied in highly coordinated events such as stomatal closure and pollen tube formation (Ng et al., 2001; Golovkin and Reddy 2003; Lemtiri-Chlieh and Berkowitz, 2004; Levchenko et al., 2005; Shang et al., 2005). These processes involve highly specific $Ca^{2+}$ spikes to induce directional growth in pollen tube formation (Wang et al., 2004) and the opening/closing of guard cells (Schiøtt et al., 2004). The directional growth of the pollen tube is important to ensure delivery of male gametophytic DNA to the female ovules. Pollen granules and guard cells are both single cell plant tissues and thus make ideal model systems to study $Ca^{2+}$ transporters. These systems allow us to better understand the biology behind the

## Calcium transporters

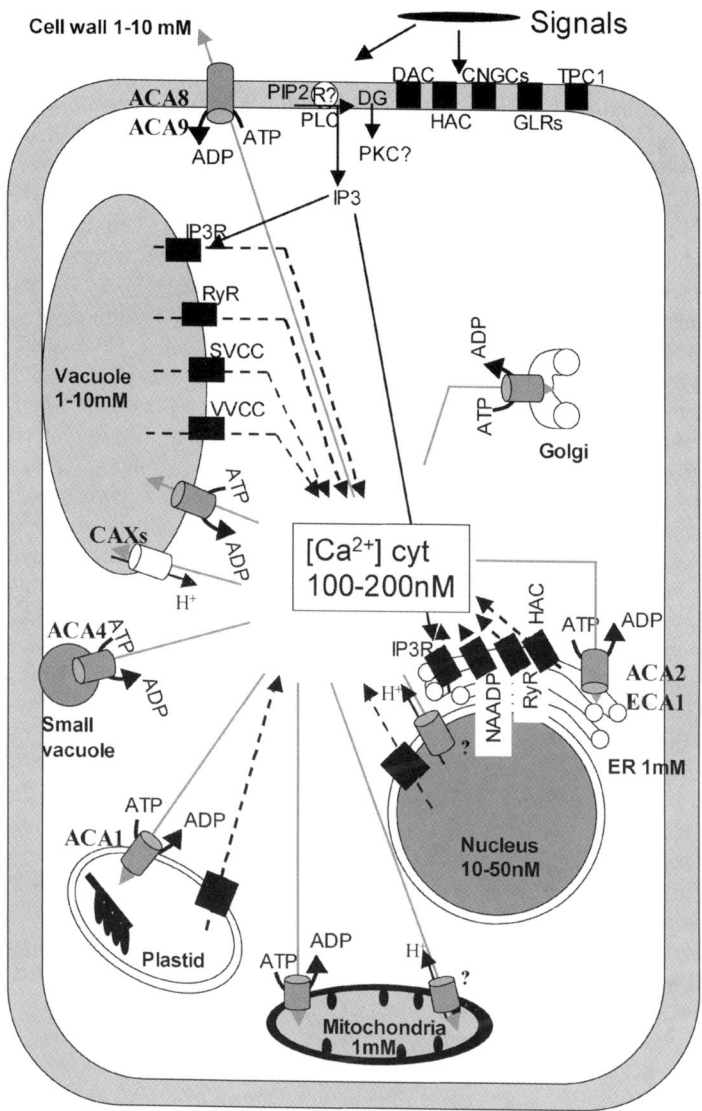

**Figure 1:** Diagram of $Ca^{2+}$ in a plant cell. Representation of $Ca^{2+}$-permable channels, pumps and transporters on various membranes of a plant cell thought to be involved in the initiation and termination of $Ca^{2+}$ specific signals. $Ca^{2+}$ channels (black boxes) allow $Ca^{2+}$ entry from high concentration sources into low concentration cytosol. Maintenance is done by $Ca^{2+}$-ATPase (gray cylinders) and $Ca^{2+}$ exchangers (white cylinders). Black (dashed) and gray arrows indicate influx and efflux sequestration of the $Ca^{2+}$ stream. PLC, phospholipase C; ER, endoplasmic reticulum (Adapted from Reddy and Reddy, 2004).

highly organized system of transporters that ensures elaborate and effective regulation of the free $Ca^{2+}$ level in cellular compartments (Yang and Poovaiah, 2003; Medvedev, 2005).

To understand the function of any specific transporter among the collection of plant proteins is difficult. For instance, the Arabidopsis genome contains 855 open reading frames which encode for transporters (Sze et al., 2004; Shigaki and Hirschi, 2006). To study any transporter one must take a fundamental approach, which includes localization, expression and gene disruption studies. The determination of tissue and cellular localization is often done by the use of reporter constructs, mainly green fluorescent protein (GFP) and $\beta$-glucuronidase (GUS). Often specific transport studies are initially conducted in a heterologous host such as yeast and then function further clarified in planta (Kamiya et al., 2005; Ali et al., 2006). Another greatly utilized resource is gene knockout plants and subsequent forward/reverse genetics. Screening for phenotypes in mutant plants or identifying a phenotype in a known gene disruption can be beneficial in characterizing the function of a gene product. Along with biochemical studies in yeast, the changes in plant biochemistry related to specific gene knockouts help elucidate the function of individual gene products. However, many of the transports are members of multigene families (Mäser et al., 2001) and functional redundancy can mask the changes seen by disruption of a specific gene. The use of chameleon (ratiometric fluorescent proteins) $Ca^{2+}$ indicators is also useful to measure local changes in $Ca^{2+}$ concentrations in specific tissues due to their ability to compare the different $Ca^{2+}$ oscillation patterns between the wild-type and the mutated gene product (Scrase-Field and Knight, 2003; Thuleau et al., 2003). These changes in $Ca^{2+}$ concentrations, subtle or substantial, can be useful in trying to determine the biological function of a specific transporter. Experimental approaches using $Ca^{2+}$ reporters to measure these types of changes have not yet been reported.

To date, numerous articles and reviews have been published on $Ca^{2+}$ transporters and their specific functions in plants (Scrase-Field and Knight, 2003; Harper et al., 2004; Reddy and Reddy, 2004; Bothwell and Ng, 2005; Braam, 2005; Hepler, 2005; Medvedev, 2005). Here we will review the most current research, which shows the role $Ca^{2+}$ transporters have in controlling and maintaining $Ca^{2+}$ movement within the cell as well as in various tissues. First, we will briefly discuss the use of yeast as a model system used to study plant $Ca^{2+}$ transporters. Then, we will discuss the biology of $Ca^{2+}$ channels and their related roles in $Ca^{2+}$ signaling. The next section will detail the role $Ca^{2+}$-ATPases have in $Ca^{2+}$ transport, followed by a section discussing $Ca^{2+}$ exchangers. We conclude with a section dealing with how manipulating $Ca^{2+}$ transporters can alter $Ca^{2+}$ content and be used to improve plant production and benefit human nutrition.

## 2. YEAST AS A MODEL

$Ca^{2+}$ has been shown to be involved in numerous physiological processes in yeast, including adaptation to environmental stress, cell cycle control, mating response and processing of proteins in the secretory pathway (Kellermayer et al., 2003). A

## Calcium transporters

recent review by Ton and Rao (2004) described the advances in heterologous expression of proteins in yeast and their applications in the study of $Ca^{2+}$ homeostasis. The greatest benefit to using yeast is the conservation of the essential components of cellular $Ca^{2+}$ machinery, termed the "calciome" (Ton and Rao, 2004). The calciome includes $Ca^{2+}$ channels and transporters, $Ca^{2+}$ sensors and signal transducers (Figure 2). In yeast, $Ca^{2+}$ enters the cytosol via the PM channel complex Cch1p/Mid1p or the vacuolar transient receptor-like channel Yvc1p (Ton and Rao, 2004). This increase in $Ca^{2+}$ is in response to diverse environmental cues such as endoplasmic reticulum (ER) stress, osmotic shock or mating pheromone. The P-type ATPases Pmr1p, located in the Golgi, and Pmc1p, located in the vacuole, are primarily responsible for removing $Ca^{2+}$ from the cytosol (Kellermayer *et al.*, 2003; Aiello *et al.*, 2004; Vanoevelen *et al.*, 2005). Along with these pumps, the vacuolar $Ca^{2+}/H^+$ exchanger Vcx1p transports $Ca^{2+}$ into the vacuole by the use of a proton gradient (Miseta *et al*, 1999). It has been suggested that the ER P-type ATPase Cod1p (or Spf1p) is also involved in $Ca^{2+}$ transport (Cronin *et al.*, 2002). The yeast $Ca^{2+}$ signaling and transport pathway is similar to that of plants and

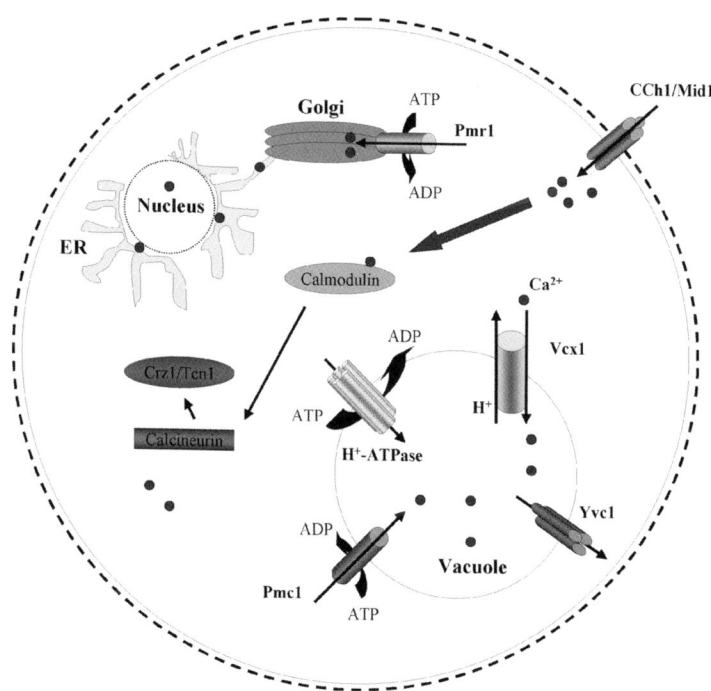

**Figure 2:** $Ca^{2+}$ signaling and transport pathways in *Saccharomyces cerevisiae*. There are calcium transporters on all membranes; small dark circles indicate $Ca^{2+}$ ions (Adapted from (Ton and Rao, 2004).

makes yeast a simple experimental model system to study the function of plant $Ca^{2+}$ transporters.

Yeast lacks the redundancy of multiple isoforms and different splice variants found in mammalian cells. This allows for the deletion of individual or multiple genes with predictable effects (Ton and Rao, 2004). Conversely, plants have calcium transporter families consisting of numerous genes (Mäser et al., 2001; Cai and Lytton, 2004; Shigaki et al., 2006). Designer yeast strains allow for the study of heterologous expression of plant genes without interference from endogenous host genes (Ton and Rao, 2004). For instance, a mutant strain lacking Vcx1p, the vacuolar $Ca^{2+}$ pump Pmc1p and the regulatory subunit of the protein calcineurin (cnb1) is sensitive to high calcium and has been used to study the putative $Ca^{2+}$ transport properties of plant proteins (Hirschi et al., 1996; Shigaki et al., 2001; Cheng et al., 2002). Rice $Ca^{2+}/H^+$ exchangers were also screened for the ability to suppress both the $Ca^{2+}$ and $Mn^{2+}$ sensitivity of a yeast strain devoid of Pmc1p and Vcx1p (Kamiya et al., 2005). The deletion of both vacuolar $Ca^{2+}$ transporters disrupts the yeast cell's ability to remove $Ca^{2+}$ from the cytosol.

Numerous yeast mutants have been used to study the function of plant transporters. A mutant lacking both $Ca^{2+}$-ATPases (Pmr1p & Pmc1p) along with cnb1 can be used to study the biochemistry of yeast plasmids encoding $Ca^{2+}$ pumps (Ton and Rao, 2004) and plant $Ca^{2+}$ pumps (Wu et al., 2002; Bonza et al., 2004; Schiøtt et al., 2004; Schiøtt and Palmgren, 2005; Baekgaard et al., 2006). Additionally, yeast strains defective in Pmr1p and Pmc1p have perturbed cellular secretion (Kellermayer et al., 2003) and can be used to study plant genes thought to be involved in cellular secretion. Similarly, the yeast strains lacking a functional Cch1p PM channel have been used to study the biochemistry of plant $Ca^{2+}$ channels (Kurusu et al., 2004; Peiter et al., 2005). Calcium channels from Arabidopsis, rice and wheat have all been studied in the yeast *Cch1* mutant background (Kurusu et al., 2004; Peiter et al., 2005; Wang et al., 2005). In sum, these yeast mutants are a powerful and simplistic tool used in studying plant $Ca^{2+}$ transporters.

Yeast also has other advantages in the study of plant transporters. The first advantage is to use yeast mutants to screen multiple variations in a specific plant protein which might have an effect on the function (Shigaki et al., 2001; Ton and Rao, 2004; Baekgaard et al., 2006). This allows rapid analysis of multiple variants of the plant protein to determine structure/function relationships by assessing their ability to complement the mutant phenotypes. These types of screens are markedly more difficult to do *in planta* due to the inherent growth difficulties and endogenous background activities (Ton and Rao, 2004). An example of the effectiveness of this approach was the screening of multiple chimeric constructs of $Ca^{2+}/H^+$ exchangers to identify functional domains (Shigaki et al, 2001). Along with the ability to study structure/function relations, one can also precisely regulate the expression level of the plant proteins in yeast cells. The yeast plasmid "tool kit" contains various promoters which allow for the control of gene expression levels. Some strong promoters have been shown to have high-level expression reaching up to 10% of the PM proteins (Ton and Rao, 2004). These promoters have been used to study the function of many heterologously expressed plant genes

(Nakamura et al., 2001; Shigaki et al., 2001; Cheng et al., 2002). This ability to control expression is useful because it may be necessary to express abundant amounts of the protein to delineate function (Ton and Rao, 2004). These methodologies, combined with the ability to quickly and efficiently transform yeast, make it a useful tool in studying the structure/function relationships of plant transporters.

One versatile tool to use with yeast to study $Ca^{2+}$ transport function is the aequorin reporter system. The aequorin reporter protein system is a calcium receptor coupled to the aequorin luminescent protein. This protein has the ability to bioluminesce in the presence of $Ca^{2+}$. As more $Ca^{2+}$ binds, the light intensity given off by the aequorin reporter protein increases (Allen et al., 1977), much like the chameleon reporter system used in plants (Allen et al., 1999) and animals (Thuleau et al., 2003). This calcium reporter system was used to show that Vcx1p plays an important role in the maintenance of resting cytosolic $Ca^{2+}$ levels in a yeast strain deficient in Pmr1p and Pmc1p function (Kellermayer et al., 2003). The ability of this reporter system to detect subtle changes in $Ca^{2+}$ is a major benefit which when coupled with genetic analysis creates a valuable tool to dissect the role of transporters in the modulation of cellular $Ca^{2+}$ levels.

Despite all the previously mentioned benefits, sometimes using yeast to study the function of plant transporters can be problematic. Plant genes expressed in yeast encounter several problems such as localization to the proper membrane, mis-folding of the protein and a heterologous host which may contain foreign proteins that alter function (Ton and Rao, 2004). For example, a plant $Na^+$ transporter expressed in yeast resides on the vacuole whereas the native yeast transport, that may have similar function, is localized on the prevacuole. Only after the plant transporter was targeted to the prevacuole could functionality be obtained (Darley et al., 2000). Another illuminating example is the expression of barley $Na^+$ (or $K^+$) uniporter or $Na^+$, $K^+$ symporter, HvHKT1 (Haro et al., 2005). Only the uniport function was observed in the barley roots and it is suspected that symport function resulted from a different translation of HKT1 when expressed in yeast. Another speculation is that $K^+$ inhibition of $Na^+$ uptake processes in roots cannot be reproduced in a heterologous yeast system (Haro et al., 2005). Expression in yeast can also be a problem for plant plasma membrane proteins where there may be difficulty targeting to the yeast PM. Exogenous proteins can also be retained in the ER due to delays or problems in the correct folding of the protein (Kauffman et al., 2002). Despite these issues, which we will discuss throughout the review, yeast is still a simple and valuable tool used to study the biology of plant transporters.

## 3. $Ca^{2+}$ CHANNELS

One component in the generation and duration of $Ca^{2+}$ signals in the plant cell is a functional system of $Ca^{2+}$ channels. Channels are activated by either a change in membrane polarization or binding of a ligand (hormone or second messenger).

These channels account for low affinity transport of $Ca^{2+}$ across the membrane and are an important portion of the generation and propagation of the $Ca^{2+}$ signal (Medvedev, 2005). These channels are also thought to maintain the spatial and temporal increase in cytosolic $Ca^{2+}$ as well as the duration and frequency of the $Ca^{2+}$ spike (Nayyar, 2003; Ng and McAnish, 2003; Miedema et al., 2003; Reddy and Reddy, 2004). In some plant tissues, these $Ca^{2+}$ channels create $Ca^{2+}$ gradients in the tissues to create polar or directional growth of the tissue (Demidchik et al., 2003; Mori and Schroeder, 2004; Wang et al., 2004; Prokić et al., 2005). For example, $Ca^{2+}$ gradients may be involved in the polarized growth of developing pollen tubes (Samaj et al., 2006; Qu et al., 2007). Hyper-polarization of $Ca^{2+}$ permeable channels have been reported in growing pollen tubes from Arabidopsis (Schiøtt et al., 2004, Wang et al., 2004) and were first seen in the growing apex from wheat roots (White et al., 2000). Calcium permeable channels play an important role in stomatal closure in response to abscisic acid (ABA), a plant hormone (Prokić et al., 2005). At pH 7 or higher, concentrations of $Ca^{2+}$ are needed to induce stomatal closure compared to an acidic pH (Prokić et al., 2005), suggesting that cytosolic pH concentrations are important in regulation of $Ca^{2+}$ channel activity. Arabidopsis contains 51 potential channels, which imply a complex role of channels in the regulation of cytosolic $Ca^{2+}$ levels. This high number of channels is part of the complex coordinated system of events maintaining the spatial temporal relationship of $Ca^{2+}$ signaling. To accomplish this, these channels are located in the plasma and tonoplast membranes along with the ER, chloroplasts and the nuclear envelope (Medvedev, 2005). This allows for $Ca^{2+}$ entry from these endomembrane stores and from extracellular spaces through numerous channels to maintain the specificity of the signal (Harper et al., 2004; Hetherington and Brownlee, 2004; Bothwell and Ng, 2005). $Ca^{2+}$ permeable channels, cyclic nucleotide gated channels (CNGCs) and glutamine receptors (GLRs) are located on the PM. Various other channels from different endomembranes also play a role in the spatial temporal increase in cytosolic $Ca^{2+}$ (Reddy and Reddy, 2004). As we will discuss below, these channels may play an important role in the initiation and termination steps of a calcium spike.

### 3.1. TPC1 is a 2-pore channel

A recently characterized two-pore channel from Arabidopsis AtTPC1 was shown to play an important role in $Ca^{2+}$ signaling (Furuichi et al., 2001; Peiter et al., 2005). AtTPC1 is the only annotated cation channel in Arabidopsis and was first characterized by heterologous expression in yeast (Furuichi et al., 2001). This protein consists of 12 predicted transmembrane (TM) domains, with two EF hand $Ca^{2+}$ motifs, and two pores, which allows for the transport of $Ca^{2+}$ (Figure 3A). AtTPC1 contains two Shaker domains with a basic region in the fourth TM of each Shaker domain. Shaker domains are named after $K^+$ channels characterized in *Drosophila melanogaster* (Tanouye et al., 1981). These Shaker domains are structures which form the pore of the transporter and contain one region with numerous basic residues. This protein conformation is similar to yeast and mammalian two pore $Ca^{2+}$ channels.

Calcium transporters

**A.** Arabidopsis Two-pore Calcium Channel (TPC1)

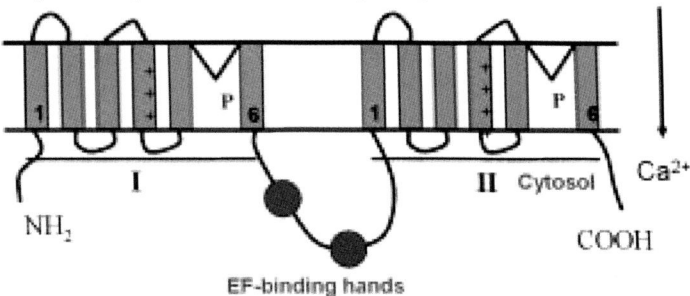

EF-binding hands

**B.** Arabidopsis Cyclic Nucleotide Gated Channel (CNCG)

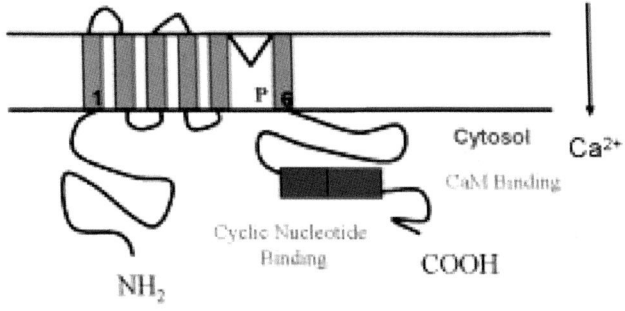

**C.** Arabidopsis Glutamate receptor protein (AtGLR)

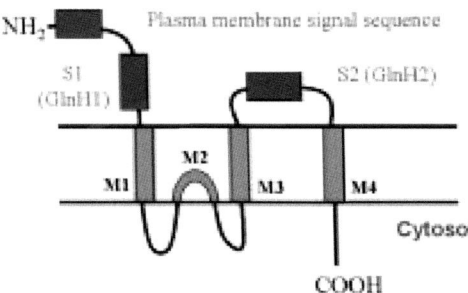

**Figure 3:** Topology Models of Putative PM proteins involved in $Ca^{2+}$ influx into the Arabidopsis cytosol. (A) TPC1 is a 2-pore channel with 2 EF calcium binding hands, the pore loop (P) located between the 5th and 6th TM domains. The 4th TM domain is enriched in basic residues. (B) CNGC structure contains a P loop, and overlapping calmodulin and cyclic nucleotide binding domains (C) GLR have 2 glutamate binding domains (GlnH) located on the outside of the membrane with 4 TM domains (Adapted from (Sanders et al., 2002).

Expression of *AtTPC1* in a mutant yeast strain, deficient in the $Ca^{2+}$ channel *CCH1p* enhances the growth rate by suppressing the mating-induced cell death and low basal $Ca^{2+}$ uptake phenotypes (Hetherington and Brownlee, 2004; Peiter et al., 2005) suggesting a similar function to the yeast $Ca^{2+}$ permeable channel. A GFP fusion of *At*TPC1 shows the protein localizes to the vacuolar membrane of Arabidopsis mesophyll protoplasts. Vacuolar proteomics and western blots reveal that *At*TPC1 is the only $Ca^{2+}$ permeable channel located on the plant vacuole (Peiter et al., 2005). Whole plant analysis shows expression of AtTPC1 in leaves, stems, root tissue, and green and developing seed pods (Medvedev, 2005). This suggests *At*TPC1 plays an important role in $Ca^{2+}$ transport from the plant vacuole.

When mesophyll vacuolar membranes from *At*TPC1 mutants were analyzed by patch clamp electrophysiology, complete absence of the slow voltage currents occurred. This suggests that TPC1 encodes a slow-vacuolar (SV) $Ca^{2+}$ channel (Peiter et al., 2005). This lack of current in the presence of $Ca^{2+}$ was restored in the mutants by transformation with the wild-type gene (Peiter et al., 2005). Suppression of the phenotype from the mutant mesophyll vacuolar membranes by the wild-type *At*TPC1 confirms it is a $Ca^{2+}$ permeable channel on the tonoplast. Wild type and *AtTPC1* over-expressers show a reduction in stomatal aperture when treated with external $Ca^{2+}$ but the mutant stomata remain unresponsive (Peiter et al., 2005). The reason for this difference in sensitivity is not known. It is possible that the transgenic plants already had more *At*TPC1 proteins and absorb the $Ca^{2+}$ in guard cells, thus making the stomatal closing less sensitive to external $Ca^{2+}$ concentration. These findings suggest that *At*TPC1 plays an essential role in voltage gated $Ca^{2+}$ transport in stomatal closure. Taken together, these findings suggest that *At*TPC1 is an important slow voltage $Ca^{2+}$ channel ubiquitously expressed with roles in stomatal closure and $Ca^{2+}$ homeostasis.

*At*TPC1 is the first characterized voltage gated $Ca^{2+}$ channel from plants and more detailed analysis will lead to a better understanding of this type of channel in plant cell signaling. Though, some of the phenotypes observed *in planta* might be related to general cation homeostasis and turgor regulation (Peiter et al., 2005). This could be due to *At*TPC1 being relatively non-selective among the mono and divalent cations. In order to determine the specific function of *At*TPC1, many factors need to be addressed. Changes in pH, as well as difference in $Ca^{2+}$ concentrations, need to be studied to see how they affect the transport properties of *At*TPC1. Over-expression of *AtTPC1* leads to ABA hypersensitivity in germinating seeds in Arabidopsis but no differences to exogenous $Ca^{2+}$ have yet to be observed in germinating seeds. The effect of $Ca^{2+}$ on ABA sensitivity of over-expressing AtTPC1 lines seeds needs to be studied given that ABA is involved in hyperpolarization of guard cells and pollen tubes (Wang et al., 2004; Prokić et al., 2005).

Another recently characterized voltage gated $Ca^{2+}$ channel, rice *Os*TPC1, functions as a PM $Ca^{2+}$ permeable channel (Kurusu et al., 2004; Kurusu et al., 2005). *Os*TPC1 is the only TPC to be PM localized which is intriguing because it may suggest some interesting differences in $Ca^{2+}$ signaling in rice. This observation begs the question as to what encodes SV channel activity in rice. Yeast has been used to characterize other SV channels from plants. To determine $Os$TPC1 function, complementation of the yeast *cch1* mutant was partially restored by expression of *OsTPC1* under $Ca^{2+}$ limited conditions (Kurusu et al., 2004). The suppression of the yeast phenotype is similar to $At$TPC1, suggesting that $Os$TPC1 functions as a $Ca^{2+}$ channel in rice. However, over-expression of *OsTPC1* in rice showed hypersensitivity to excess $Ca^{2+}$ but in limiting $Ca^{2+}$ conditions plants had a higher growth rate. This result suggests that $Os$TPC1 has a different function than $At$TPC1 *in planta*. This also suggests that when $Ca^{2+}$ is abundant, over-expression of $Os$TPC1 leads to $Ca^{2+}$ toxicity, where $At$TPC1 does not cause any $Ca^{2+}$ toxicity. Prolonged increases in $Ca^{2+}$ are harmful because it can lead to activation of apoptosis (Kass and Orrenius, 1999). Conversely, cultured cells with an insertional knockout of *OsTPC1* displayed less sensitivity to extracellular free $Ca^{2+}$ (Kurusu et al., 2004), suggesting that $Os$TPC1 has a role in $Ca^{2+}$ uptake into the cell. Similar to $At$TPC1, $Os$TPC1 functions as a $Ca^{2+}$-permeable channel with a possible role in the transduction of stimuli from either the cytoplasm or the external matrix (Medvedev, 2005).

This initial research indicates the presence of $Ca^{2+}$-permeable channels in plants. Nonetheless, there are several questions that need to be answered to better understand the biological function of $Os$TPC1. The cellular localization is one important aspect that needs to be determined. The Arabidopsis homolog localizes to the vacuole, however, it appears that $Os$TPC1 could localize to the PM due to toxicity of plant over-expressing *OsTPC1*. Another future area of inquiry is to address the role of $Os$TPC1 in stomatal closure in response to external $Ca^{2+}$. *AtTPC1* over-expression in Arabidopsis reduces stomatal aperture, suggesting that $At$TPC1 has a role in stomatal closure in plants (Peiter et al., 2005). The use of a chameleon reporter system in lines altered in *AtTPC1* expression allows us to observe the effects $At$TPC1 has in regulating calcium levels in the stomates of Arabidopsis.

## 3.2. CNGCs

The plant CNGCs (Cyclic Nucleotide Gated Channel) are permeable to mono- and divalent cations and were first isolated from barley aleurone cells (Medvedev, 2005). Plant CNGCs are inwardly rectified, ligand gated, non-selective cation channels with different selectivity profiles. In all cases published to date their conductance is increased in the presence of cyclic nucleotides (this is unique to CNGCs; Leng et al., 2002; Chan et al., 2003; Hua et al., 2003). CNGCs have both cyclic nucleotide and calmodulin domains at the C-terminus, a TM core and a pore domain similar to the Shaker family (Figure 3B; Ali et al., 2006). These channels can bind cyclic nucleotides (cAMP, cGMP) and CaM, which allows for the integration of signals that arrive from different signaling event pathways (Arazi et al., 2000). These inwardly conducting channels located on the PM are activated by cAMP (Ali et al., 2006; Lemtiri-Chlieh and Berkowitz, 2004). The Arabidopsis genome contains about 20 genes encoding CNGCs (Ali et al., 2006) and they are capable of forming heterotetramers (Kaupp and Seifert, 2002).

Recently, *CNGC1* was expressed in a mutant yeast strain to try and elucidate the function of this protein. The $Ca^{2+}$-uptake yeast mutant *mid1/cch1* was used to functionally characterize *CNGC1* and *CNGC1* mutants with domain deletions which cause these transporters to localize to different membranes (Ali *et al.*, 2006). A GFP fusion of CNGC1 shows the protein localizes to the PM in yeast cells. Conversely, mutants of CNGC1 predominantly localize to the vacuolar membrane, with some fluorescent at the PM (Ali *et al.*, 2006). This suggests that deletions in the CNGC1 protein can affect the localization of the protein and can have an affect on function. This deletion could have a positive effect on $Ca^{2+}$ transport by altering the pore selectivity filter (Ali *et al.*, 2006).

These results have shown that yeast can be an efficient model for studying the function of plant *CNGCs*. As mentioned previously, it appears that there may be factors in the yeast cell which also affect the efficacy of the assay system and need to be taken into account when heterologously expressing plant *CNGCs*. Endogenous yeast CaM and perhaps cyclic nucleotides in the yeast cytosol could affect the activity of CNGCs. Given the success of using yeast to characterize other plant transporters, this biochemical assay has the potential to be used in determining biochemical functions of other CNGCs.

CNGCs are thought to play a role in programmed cell death and tissue senescence in response to pathogen related signals (Balague *et al.*, 2003). When *CNGC2* is mutated spontaneous cell death occurs resulting in constitutively active plant defenses systems. *CNGC2*-GUS expression is constitutive at rather high levels and is repressed upon infection suggesting that *CNGC2* is repressed in response to pathogenic infection. In contrast, when *CNGC4* is mutated, plants display necrotic lesions and a constitutive expression upon activation after inoculation by a pathogen (Balague *et al.*, 2003). These lesions of clustered dead cells show an accumulation of β-glucoronidase (GUS) expression in the surrounding cells. *CNGC2* expression is higher in *cngc4* plants. Recently, a mutant, which deleted a 3kb region in the Arabidopsis genome, was made from a cross of two different ecotypes of Arabidopsis, fusing *CNGC11* and *CNGC12*. This chimera *CNGC11/CNGC12* induces pathogen resistance through multiple signaling pathways, whereas the individual genes are a component of those pathways (Yoshioka *et al.*, 2006). Together these mutants reveal that CNGCs have a role in plant stress recognition and the coordinated regulation could be related to the heterotetramer nature of the proteins in the CNCGs family (Balague *et al.*, 2003).

CNGC4 may also be involved in pathogen interactions but we know little regarding biological function. *CNGC4* expression increases in response to plant pathogens at the site of inoculation (Balague *et al.*, 2003). However, *CNGC4* expression does not increase in response to $Ca^{2+}$ (Balague *et al.*, 2003). To further determine the biological function of CNGC4, this transporter was expressed in Xenopus oocytes. *CNGC4* expression in oocytes shows more efficient activation by cGMP and is permeable by both $K^+$ and $Na^+$ but not $Ca^{2+}$. This lack of $Ca^{2+}$ transport could be related to other endogenous proteins transporting $Ca^{2+}$ therefore, masking any $Ca^{2+}$ transport change resulting from expression of *CNGC4* (Balague *et al.*, 2003). These findings suggest that CNGC4 is not activated by $Ca^{2+}$ but might be a component of the plant hypersensitive response (HR).

These results show that some CNGCs function as $Ca^{2+}$ channels but there are problems heterologously expressing these proteins. One aspect to study further is the idea that CNGCs function as heterotetramers. *CNGC* mutations have been shown to induce or alter pathogen resistance and increase the HR (Yoshioka *et al.*, 2006). Also, CNGC2 was recently shown to be involved in the pathogen response signal leading to nitric oxide production (Ali *et al.*, 2007). The low level expression of some *CNGC* makes the determination of their biological function difficult but using the sensitive $Ca^{2+}$ reporters could lead to a better understanding of their role in $Ca^{2+}$ transport.

## 3.3. GLRs

There is little known about the physiological functions of GLRs (GLutamate Receptors) but they have been implicated in several plant processes from light signaling to water balance (Davenport, 2002; Meyerhoff *et al.*, 2005). The $Ca^{2+}$ antagonist $La^{3+}$ can inhibit the increase of root $Ca^{2+}$ by GLRs suggesting, that the pore of GLRs has unique properties and can be related to the unusual pore sequences (Davenport, 2002; Hetherington and Brownlee, 2004). This pore sequence may act as a conserved filter throughout the GLR family. The conservation of the two-ligand binding domains implies a role in $Ca^{2+}$ signaling in response to elevation in extracellular glutamate levels (Davenport, 2002). The N-terminal domain has homology to mammalian GLRs and could be a site for allosteric modulation by $Ca^{2+}$ or metabolites (Davenport, 2002). In Arabidopsis, approximately 20 *GLRs* were identified with homology to animal GLRs (Lacombe *et al.*, 2001). The general structure of GLR contains two glutamate binding domains and a PM signal sequence (Figure 3 panel C). There is emerging evidence that *At*GLRs may transport a variety of cations in plants but there have only been a few GLRs that have been characterized (Davenport, 2002; Li *et al.*, 2006; Meyerhoff *et al.*, 2005).

The Arabidopsis GLR family appears to have developed by multiple local gene duplications events (Chiu *et al.*, 2002). It is possible that some of these genes have no function (Davenport, 2002). The large number of similar sequences makes the naming of GLR unique. There are 3 main groups of GLRs: *GLR1*, *GLR2* and *GLR3*. Within each of these groups different loci are given different names. For example, previous names like *GLR2* and *GLR6* are now termed *AtGLR3.1* and *AtGLR3.5* (Lacombe *et al.*, 2001). This nomenclature suggests that *AtGLR3.1* and *AtGLR3.5* are almost identical in sequence but are located in different parts of the chromosome. In the case of AtGLR3.1a and AtGLR3.1b, they are classified differently because they encode different RNAs (Lacombe *et al.*, 2001). The differences among the *AtGLRs* RNAs could indicate specific functions for the different variants.

GLRs from Arabidopsis are predicted to target to the secretory pathways on the basis of their hydrophobic N-terminal sequences. The secretory pathway involves numerous organelles suggesting GLRs could be localized to numerous plant membranes. Attempts to visualize cellular localization with C-terminal GFP tags have been unsuccessful; however, GFP fusions within the N-terminus regions of GLRs did localize to the ER, revealing the possibility of internal C-terminus targeting sequences (Davenport, 2002).

Interestingly the GUS expression data demonstrate that *AtGLR3.2* is expressed in vasculature tissue of roots and shoots and could play a role in unloading $Ca^{2+}$ into the xylem (Davenport, 2002). Localization of other *AtGLR3s* was determined by RT-PCR. *AtGLR3.4* and *AtGLR3.5* are expressed in root and shoot tissue in 2-week old Arabidopsis plants. AtGLR3.7 transcripts were also present in both leaf and root tissue (Davenport, 2002). Similarly, *AtGLR1.1* is expressed in all aerial tissue and *AtGLR2.1* is expressed in rosette leaves of Arabidopsis plants (Davenport, 2002), suggesting that GLR proteins are expressed throughout the plant.

Expression of *AtGLR3.7* in Xenopus oocytes demonstrates that this GLR is a $Ca^{2+}$ permeable non-selective cation channel. Over-expression of *AtGLR3.2* in plants produces a $Ca^{2+}$ deficiency phenotype in roots and the application of exogenous $Ca^{2+}$ alleviates these symptoms (Davenport, 2002). This agrees with the oocyte data that GLR can transport $Ca^{2+}$ into the cell.

At present, a better understanding of tissue and membrane localization, ion selectivity and the related sensitivities to different ligands is needed to define the role GLRs play in $Ca^{2+}$ signaling/homeostasis in plants. Ectopic expression of *GLRs* could cause the proteins to mis-target or cause mis-folding within the membranes, thus confounding any experimental results. Some GLR subunits could function in $Ca^{2+}$ signaling and nutrient uptake at the PM. Another function may be related to amino acid regulation due to specific distribution of glutamate within the cell from synthesis in the cytosol, plastids, mitochondria, and possibly in peroxisomes in response to the photorespiratory pathway (Medvedev, 2005). Transgenic plants and mutant analysis will help to characterize both gene function and the physiological role of GLRs. Given the numerous genes and the high occurrence of gene duplications, using multiple mutants could become too cumbersome an avenue in further characterizing GLRs.

### 3.4. Other channels

Calcium entry into the cytosol can occur from either the PM or from intracellular stores. In addition to the previously mentioned channels, there are several other types of $Ca^{2+}$ channels. Such as voltage-dependant and voltage-independent $Ca^{2+}$ selective channels on the PM and ER, inositol 3 phosphate ($IP_3$), cyclic ADP ribose (cADPR) and nicotinic acid adenine dinuclotide phosphate (NAADP) gated channels (Becker *et al.*, 2004; Hetherington and Brownlee, 2004; Reddy and Reddy, 2004; Medvedev, 2005). So far none of these channels has been cloned from Arabidopsis (Hetherington and Brownlee, 2004). The only plant NAADP characterized so far is from red beet. This channel functions in $Ca^{2+}$ release from the ER, even though the gene has not yet been cloned. Despite the lack of NAADP channels, other channels have been identified in Arabidopsis. Isolated vacuoles and tonoplast vesicles from plants treated with $IP_3$ induce $Ca^{2+}$ efflux (Medvedev, 2005). This is compelling evidence for an $IP_3$ mediated channel on the tonoplast. There also appears to be a slow voltage (SV) channel in guard cells. This channel is involved in $Ca^{2+}$ release from the vacuole by CaM activation and increases activity with increases in pH (White and Broadley, 2003). Further understanding of these channels will be a challenge and much remains to be discovered with regards to the regulation of these channels.

## 3.5. Annexins

Annexins have been suggested to be $Ca^{2+}$ binding proteins which play a role in $Ca^{2+}$ transport. They may function in either a direct manner by forming multimeric aggregation $Ca^{2+}$ channels or in an indirect manner by activation of $Ca^{2+}$ channels (White and Broadley, 2003, Dabitz et al., 2005). Multiple Arabidopsis annexin genes have been identified, but they remain poorly characterized. There is some physiological evidence of annexins playing a role in $Ca^{2+}$ activity (Medvedev, 2005; Hetherington and Brownlee, 2004, Gorecka et al., 2005). In terms of their biological function, annexin 1 (*AnnAt1*) from Arabidopsis is involved in plant response to reactive oxygen species (ROS; Gorecka et al., 2005) and subsequent activation of $Ca^{2+}$ channels (Mori and Schroeder, 2004; Rentel and Knight, 2004). ROS are part of the mutli-step pathogen response pathway in plants (Gorecka et al., 2005; Ali et al., 2007). Characterization of the redox sensing domain and identification of substrates involved in annexins response to oxidative stress has not yet been defined. Although annexins appear to have a role in plant pathogen responses, like CNGCs, their specific role is unclear and it will be interesting to determine if CNGCs and annexins are coordinately regulated.

## 4. $Ca^{2+}$ PUMPS

Plant $Ca^{2+}$ pumps are responsible for regulation of $Ca^{2+}$ concentrations in endomembrane compartments as well as the cytoplasm. Pumps function by actively moving $Ca^{2+}$ across membranes using ATP hydrolysis as the energy to drive transport against a $Ca^{2+}$ concentration gradient (Marschner, 1995). This active transport in coordination with $Ca^{2+}$ channels is used to create the specific spatiotemporal coding of a $Ca^{2+}$ signal (Medvedev, 2005). As well as being involved in plant signal transduction pathways, $Ca^{2+}$ pumps are involved in adaptation to stress conditions. After a $Ca^{2+}$ spike, the removal of calcium requires active transport into endomembranes as well as efflux from the cell through the PM (Hetherington and Brownlee, 2004; Reddy and Reddy, 2004; Medvedev, 2005). Plant calcium pumps are expressed in all plant tissues from the vasculature to the guard cells (Wu et al., 2002; Bonza et al., 2004; Schiøtt et al., 2004; Schiøtt and Palmgren, 2005; Baekgaard et al., 2006). These pumps are involved in stomatal closure in response to cold stress (Schiøtt and Palmgren, 2005), pollen tube growth and fertilization (Schiøtt et al., 2004), acclimation to salt stress (Geisler et al., 2000) and adaptation to variable nutrient conditions (Wu et al., 2002). To accomplish these processes, the pumps hydrolyze ATP to transport $Ca^{2+}$ which causes a change in protein conformation that exposes the $Ca^{2+}$ binding domain (high affinity E1) to the cytosol. Once bound, a conformational change E2 (low affinity) exposes the $Ca^{2+}$ to the opposite side of the membrane and allows the $Ca^{2+}$ to dissociate from the transporter (Figure 4A; Xu et al., 2002; Sze et al., 1999, Sørensen et al., 2004). It is also likely that a proton is exchanged for the $Ca^{2+}$ ion (Luoni et al., 2000).

Plants have two types of active $Ca^{2+}$ pumps to accomplish the transport of $Ca^{2+}$ from the cytosol, both of which are members of the P-type ATPase superfamily. The ER-type $Ca^{2+}$-ATPase (ECA) and the autoinhibited $Ca^{2+}$-ATPase (ACA) make up the two classes of calcium pumps in the Arabidopsis genome (Geisler et al., 2000). Based on the structural similarities to the animal pumps, plant $Ca^{2+}$-ATPases are classified into 2 groups, type IIA (ECAs) and IIB (ACAs). The current nomenclature for the P-type ATPase subfamilies are $P_{1B}$ (metal ATPases), $P_{2A}$ (ECAs), $P_{2B}$ (ACAs), and $P_{2A}$ (H-ATPases) (Baxter et al., 2003). Type IIA pumps (ECAs) have approximately 50% protein sequence homology with animal SERCA (sacroplasmic/endoplasmic reticulum $Ca^{2+}$-ATPases) and are not activated by

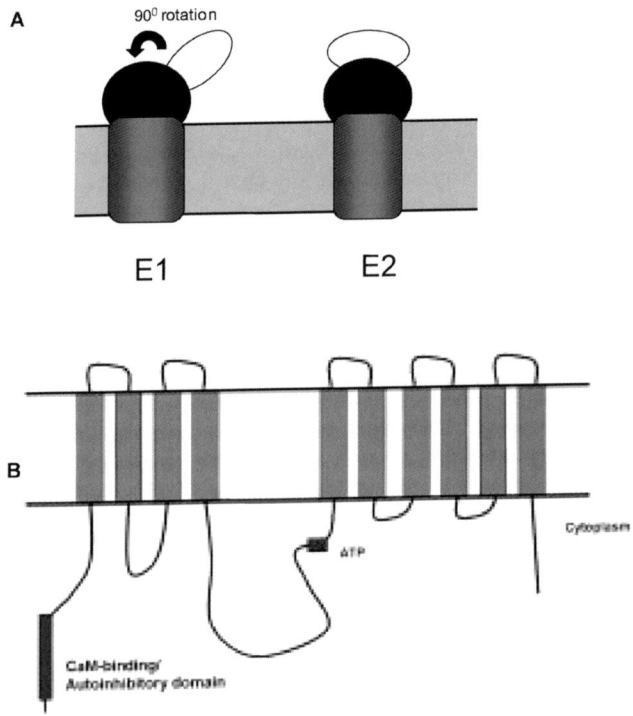

**Figure 4:** Protein conformations and topology of a plant calcium pump A) Conformation change of a calcium pump from E1 to E2 forms after activation by ATP hydrolysis (Adapted from Sørensen et al., 2004). B) Topology of the $Ca^{2+}$-ATPases, the CaM/autoinhibitory domain is only present on type IIB $Ca^{2+}$-ATPases (Adapted from Sanders et al., 2002).

calmodulin. Type IIB pumps (ACAs) have approximately 50% protein to homology with animal PMCA (PM located $Ca^{2+}$-ATPases). ACAs are activated by calmodulin but the calmodulin and autoinhibitory domains are located on the N-terminal region instead of the C-terminus, as seen in mammalian PMCAs (Geisler et al., 2000). Both of the pumps have a similar general structure (Figure 4B). Only 20% of the protein is inside the membrane and they have between 8-10 TM domains. The largest part of the protein is exposed to the cytoplasm and contains two loops with the phosphorylation and ATP binding sites (Sze et al., 1999; Sørensen et al., 2004). Only about 10% of the protein faces the non-cytoplasmic side of the membrane. Animals have three $Ca^{2+}$-ATPase types: SERCA, PMCA and a third distinct class SPCA (secretory pathway Ca-ATPases) which are Golgi localized and found in yeast (PMR1) and animals but not plants. The ECAs have some SPCA characteristics such as the ability to transport other cations including $Mn^{2+}$ (Dode et al., 2005).

**4.1. ECAs**

*ECA1* (ER $Ca^{2+}$-ATPase) appears to be the most abundantly expressed of the four Arabidopsis *ECAs* (Liang and Sze, 1998; Axelsen and Palmgren, 2001; Wu et al., 2002). ECA1 has been shown to suppress a yeast mutant defective in $Ca^{2+}$-ATPase activity by alleviating the sensitivity to $Ca^{2+}$ in the growth medium (Liang and Sze, 1998). More recently, ECA1 was shown to increase tolerance of a yeast mutant to toxic levels of $Mn^{2+}$ (1mM) and $Zn^{2+}$ (3mM; Wu et al., 2002). This result suggests that ECA1 can transport multiple cations from the cytoplasm into the ER to facilitate growth of a yeast mutant strain.

The function of ECA1 was determined using standard reverse genetic analysis. The T-DNA was inserted into the last TM domain but it did not alter growth of *eca1-1* compared to wild type plants under normal growth conditions (Wu et al., 2002). Disruption of *ECA1* does not affect normal plant growth, possibly due to functional compensation by other $Ca^{2+}$ transporters (Wu et al., 2002). When *eca1-1* lines are grown in media containing high $Mn^{2+}$ (0.5mM) these mutants have a severely reduced growth rate (66% reduction; Wu et al., 2002). These lines also have a 22% reduction in vesicular $Ca^{2+}$ pump activity. However, the plants display more tolerance to severe $Ca^{2+}$ deficiency conditions when compared to controls (Wu et al., 2002). Root hair formation in the *eca1-1* plants is disrupted in media supplemented with 0.5mM $Mn^{2+}$. Disruption of ECA1 expression alters the root hair formation by slowing hair growth but not root tip growth (Wu et al., 2002). Potentially, the plants could be slowing root hair growth to try and decrease the uptake of $Mn^{2+}$ to lessen the toxicity created in the aerial tissues. The expression of the wild type *ECA1* gene in the *eca1-1* background was sufficient to rescue the mutant. Thus the sensitivity to $Mn^{2+}$ stress was due to a disruption in ECA1 expression.

Little is known about the other three *ECAs* from Arabidopsis. This might be related to abundant expression of *ECA1* in Arabidopsis or functional redundancy. In the case of *ECA4*, which is almost 97% identical to *ECA1*, there is speculation that it could be a pseudo gene (Kabala and Klobus, 2005). Numerous other *ECAs* have been cloned and/or characterized from other plants. Type IIA pumps have been cloned from tomato (*LCA1*, Wimmers *et al.*, 1992) and rice (*OCA1*, Chen *et al.*, 1997). Ectopic expression of Arabidopsis *ECAs* in these other plant species could provide insight about their function due to their divergent genetic background. Another way to possibly characterize the functions of these other Arabidopsis *ECAs* is to make double mutants of these transporters and look for changes in calcium transport or the location of calcium pools using the chameleon reporter system. Another ECA-like pump CAP1 from maize has also identified a calmodulin binding site at the C-terminus (Subbaiah and Sachs, 2000). This is interesting because little is known about regulation of plant ECAs, which is in contrast to the well understood regulation of animal SERCA pumps.

## 4.2. ACAs

The primary function of autoinhibited $Ca^{2+}$-ATPases (ACAs) is to remove $Ca^{2+}$ from the cytosol into endomembrane compartments or into the extracellular space (Bonza, *et al*, 2004; Schiøtt *et al*, 2004; Medvedev, 2005). *ACAs* main distinction from *ECAs* is their activation by $Ca^{2+}$-dependent protein calmodulin and $Ca^{2+}$-dependent protein kinases (CDPKs) (Hwang *et al.*, 2000; Medvedev, 2005). The first ACA that was shown to be stimulated by calmodulin was the vacuolar localized $Ca^{2+}$-ATPase BCA1 from cauliflower (Malmstrom *et al.*, 2997). Unlike the C-terminus calmodulin-binding domains of animal $Ca^{2+}$-ATPase, the BCA1 binding domain is localized to the N-terminus region (Malmstrom *et al.*, 1997).

In Arabidopsis, ACAs function in pollen tube growth (Schiøtt *et al.*, 2004) and in guard cell responses to cold stress (Schiøtt and Palmgren, 2005) among other physiological functions. There are 10 *ACAs* in Arabidopsis that are divided into 4 groups: Group 1 contains *ACA1,2,7*; Group 2 contains *ACA4,11*; Group 3 contains *ACA12,13* and Group 4 contains *ACA8,9,10* (Baxter *et al.*, 2003). Many group 3 ACAs contain intronless genes, whereas group 4 genes can have as many as 33 introns (Baxter *et al.*, 2003). *ACA2* was cloned and characterized by functional expression in a yeast $Ca^{2+}$-ATPase mutant (Harper *et al.*, 1998). Truncation of the N-terminal domain of ACA2 increases transport activity 4 to 10 fold more than ACA2. This protein was also unresponsive to further stimulation by calmodulin (Hwang *et al.*, 2000). This demonstrates that ACA2 contained a regulatory domain that is activated when $Ca^{2+}$ induces calmodulin to bind to the N-terminal domain, thus suspending the inhibition. Since the characterization in yeast of ACA2, other plant $Ca^{2+}$ pumps have subsequently been shown in yeast to have an N-terminus autoinhibitory domain, e.g. ACA4, ACA8, ACA9 and soybean SCA1 (Chung *et al.*, 2000; Bonza *et al.*, 2004; Geisler *et al.*, 2000). These results show evidence that numerous ACAs contain an N-terminus autoinhibitory domain.

Within a plant cell, the ACAs are located on different membranes, (Sze *et al.*, 2000; Schiøtt *et al.*, 2004; Bonza *et al.*, 2004) unlike the animal PMCAs that are only PM pumps. This allows ACAs to tightly regulate $Ca^{2+}$ at both the PM and other endomembranes. $Ca^{2+}$ cytosolic fluctuation requires tight regulation of both plasma and endomembrane bound transporters. Two of the ACAs, ACA8 and ACA9, localize to the PM (Bonza et al., 2000; Schiøtt et al., 2004). These enzymes function to remove $Ca^{2+}$ from the cell to help control the duration and intensity of the $Ca^{2+}$ signal. Another family member, ACA2, localizes to the ER and may have a unique role in modulating the $Ca^{2+}$ efflux into endomembrane compartments (Hwang, 2000). It has been previously reported that ACA4 localizes to small vacuole membranes (Geisler et al., 2000) and ACA1 to the inner plastid envelope (Huang et al., 1993). Despite the unknown cellular localization of other ACA family members (ACA7, 10,11,12 and 13) the differing localizations implicate a diverse role of the ACAs in maintaining the spatial and temporal relationship of cellular $Ca^{2+}$ signals.

Along with the diverse cellular localization, the ACA family members also have unique tissue localization. For instance, the PM localized *ACA9* expression is primarily in pollen tubes (Schiøtt et al., 2004). Calcium efflux by ACA9 recycles calcium ions entering through the PM channels. Gene disruption of *ACA9* results in a semi-sterile phenotype in Arabidopsis plants. Three independent lines display reduction in pollen tubes and the tubes that do reach ovulation have a high frequency of aborted fertilization (Schiøtt et al., 2004). Transformation of these mutants with *ACA9* suppresses the male sterile phenotype. Complementation of these mutants is also done by expressing *ACA8* which suggests that *ACA8* functions downstream of *ACA9* in pollen tube formation. Tissue specific localization of ACA9 was shown by GUS promoter fusions. Expression of *ACA9* promoter GUS expression was in flower anthers. Conformation of *ACA9* expression by RT-PCR shows a greater than 500 fold increase in expression in stamen tissues (Schiøtt et al., 2004). This demonstrates that ACA9 function is critical for pollen tube formation and subsequent fertilization of the ovule.

Although *ACA9* is specifically expressed in pollen, other *ACAs* are expressed in different plant tissues (Schiøtt and Palmgren, 2005). Two closely related *ACAs*, *ACA8* and *ACA10*, expression is found in Arabidopsis guard cells. Expression of a promoter::GUS fusion of *ACA8* is restricted to guard cells and vasculature tissue of both roots and shoots while *ACA10* promoter::GUS expression was more ubiquitous (Schiøtt and Palmgren, 2005). Both of these tissue expression patterns were confirmed by RT-PCR.

Even though *ACA8* and *10* are expressed in similar tissues they appear to have differing functions in response to cold stress. RT-PCR analysis of plant samples taken at 2, 8, 24 and 48 hrs treatment of 5°C demonstrates that *ACA8* and *10* expression is altered by this treatment (Schiøtt and Palmgren, 2005). After 2 hrs, *ACA8* expression drops to almost zero, and then has a 5 fold increase at 8 hrs only to return to normal levels after 48 hrs. In contrast, *ACA10* expression decreases steadily all they way to a 3 fold decrease at 48 hrs (Schiøtt and Palmgren, 2005). Interestingly, mutations in either *ACA8* or *ACA10* do not cause a cold sensitive phenotype. These alterations in response to cold stress display that ACA8 and ACA10 may have a role in efflux of $Ca^{2+}$ from the cytosol after a cold induced signal.

ACAs are located in several tissues and numerous membranes. Initial findings suggest ACAs have roles in cold accumulation and pollen tube growth. Throughout this review it has been mentioned that genetic and functional redundancy creates difficulties using standard genetic approaches. Heterologous expression systems help to alleviate some of these limitations. As we discussed in our yeast section, valuable insights in *S. cerevisiae* $Ca^{2+}$ signaling were made when both $Ca^{2+}$-ATPase and $Ca^{2+}/H^+$ exchangers were mutated. Potentially these types of mutations *in planta* will be just as beneficial in further delineating biological functions of $Ca^{2+}$ pumps. Specifically, it will be interesting to determine if alterations in pollen tube growth or guard cell movement occurs in these mutants.

## 5. $Ca^{2+}$ EXCHANGERS

The vacuole is an important storage organelle for many ions, including $Ca^{2+}$. The vacuole allows for large concentrations of ions inside the luminal membrane. This helps to maintain a low concentration of $Ca^{2+}$ in the cytosol. To accomplish this, $Ca^{2+}/H^+$ exchangers use a proton gradient, created by differences in pH between the cytosol (pH ~7.5) and the vacuolar lumen (pH 3-6). This is physiologically different from the previously mentioned $Ca^{2+}$ pumps. The vacuolar $Ca^{2+}$-ATPase has a high affinity (Km = 0.1-2 μM) and low capacity for $Ca^{2+}$ (Sze *et al.*, 2000), while $Ca^{2+}/H^+$ exchangers have a low affinity (Km = 10-15 μM) and a high capacity for $Ca^{2+}$ (Hirschi *et al.*, 1996). This high capacity is thought to play a role in removing $Ca^{2+}$ from the cytosol after a signaling event, while the $Ca^{2+}$-ATPases may be involved in fine tuning a $Ca^{2+}$ oscillation (Pittman and Hirschi, 2003). The $Ca^{2+}/H^+$ exchangers may function as a component of the "off" mechanism of plant signaling events.

The $Ca^{2+}/H^+$ exchangers were identified by using a yeast suppression screen and are termed CAX1 and CAX2 (CAtion eXchangers; Hirschi *et al.*, 1996). The membrane topology model for CAXs shows 11 TM with an acidic motif between TM6 and TM7 (Figure 5A; Shigaki and Hirschi, 2006). Also indicated on the model are two distinct cation domains and an N-terminal autoregulatory domain (Shigaki *et al.*, 2001; Pittman *et al.*, 2002). The calcium domain was identified using chimeric constructs of CAX1 and CAX3 (Shigaki *et al.*, 2001). This 9 amino acid domain from CAX1, when inserted into CAX3, allows CAX3 to now confer $Ca^{2+}$ tolerance to a mutant yeast strain sensitive to high $Ca^{2+}$ (Shigaki *et al.*, 2001). This chimeric construct strategy was also useful in identifying other cation domains. CAX2 can also transport numerous other cations besides $Ca^{2+}$. Chimeric constructs were made between CAX1 and CAX2 to identify the $Mn^{2+}$ domain critical for $Mn^{2+}$ transport by CAX2 (Shigaki *et al.*, 2003). Also, similar experiments using mutagenesis of CAXs in rice (OsCAXs) have identified other domains involved in substrate specificity (Kamiya and Maeshima, 2004). Numerous reviews have mentioned how the first CAXs were characterized, but we will focus this review on recent research of the CAX family of transporters (Scrase-Field and Knight, 2003; Harper *et al.*, 2004; Reddy and Reddy, 2004; Bothwell and Ng, 2005; Hepler, 2005; Medvedev, 2005; Shigaki and Hirschi, 2006).

### 5.1. Phylogeny of CAX family

The plant CAX family has numerous members. Phylogenetic analysis of this family reveals an interesting relationship of the members. *CAX1-6* appear to be closely related to each other while *CAX7-11* are closely related to each other (Shigaki *et al.*, 2006), simply based on sequence homology. Recently, *CAX7-11* were reclassified as *CCX1-5* due to higher homology to mammalian $K^+$ dependant $Na^+/Ca^{2+}$ antiporters (Shigaki *et al.*, 2006). These phylogenetic relationships suggest that *CCXs* have a different function compared to the previously characterized *CAXs*.

One difference that might exist between CAXs and CCXs is the existence of the N-terminal autoinhibitory domain (Figure 5A and 5B). The first functional clones, later named sCAX1 and sCAX2 (short-CAX; Hirschi *et al.*, 1996; Pittman and Hirschi, 2003), were truncated cDNAs without the first 36 and 42 amino acids, respectively. The full length genes are unable to suppress the calcium sensitive phenotype of a yeast mutant when grown on high $Ca^{2+}$ medium (Pittman and Hirschi, 2003; Pittman *et al.*, 2004b). When N-terminal truncations were made in other CAXs in this CAX1-6 subfamily, these "short" forms were able to suppress the calcium sensitive yeast phenotype. In CAX1, specific residues within this N-terminal domain have been implicated in this regulation, including a putative phosphorylated serine residue and other residues that may be involved in protein interaction with another region of the CAX1 protein (Pittman *et al.*, 2002). Although this N-terminal regulatory domain is seen with the ACA $Ca^{2+}$-ATPases, this CAX

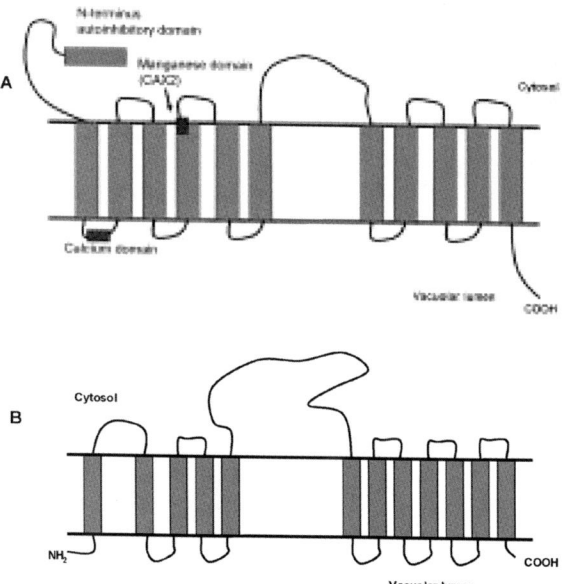

**Figure 5 :** Topology of plant cations exchangers. (A) The calcium and manganese domains are indicted on the transporter. The N-terminal autoinhibitory domain is also indicated on the transporter (Adapted from Sanders *et al.*, 2002). (B) Topology of the new classified CCXs. This model is based on the TMHMM algorithm.

N-terminal domain does not have any sequence similarity with the ACA autoinhibitory domains and does not appear to bind CaM (Pittman and Hirschi, 2003), but can interact with other proteins (Cheng and Hirschi, 2003; Cheng et al., 2004a, 2004b). CCX1-5 have very little sequence homology in the N-terminus region with CAX1-6 (Figure 5B). Rather, their N-terminal domains are highly homologous to the N–terminal domain of mammalian NCKXs (Shigaki et al., 2006). Potentially CCX transporters have different biochemical and regulatory properties than CAXs.

## 5.2. Physiological function of CAXs

As stated throughout this review, yeast is a useful tool to study plant $Ca^{2+}$ transporters. However, *in planta* gene function is still critical. Combining the heterologous yeast expression data with gene transcript expression data from microarray analysis, northern analysis and promoter-reporter fusions, can be used to reveal the role CAXs have in plants. Phylogeny suggests that the CAXs have two closely related groups. The first group consists of *CAX1*, *3*, and *4* and the second group contains *CAX2*, *5* and *6* (Shigaki et al., 2006). To characterize the *CAXs* in these two groups, *CAX* promoter::GUS constructs have been used to discern their expression patterns in different Arabidopsis tissues. *CAX1* is highly expressed in shoots but only modestly expressed in the roots (Cheng et al., 2005). Conversely, *CAX3* expression is seen in roots and modestly in the leaves (Cheng et al., 2005). RT-PCR analysis of *CAX4* shows expression throughout the plant with the highest expression in root tissues (Cheng et al., 2002). *CAX2* promoter GUS shows expression in the vasculature tissue of 10 day and 30 day old plants (Pittman et al., 2004a). Also *CAX2::GUS* expression is present in floral tissues and hydathodes. Many of the *CAX* genes are also expressed highly in inflorescence tissues (Cheng et al., 2005). CAXs are found in numerous other plants, yet only a few have been characterized. RT-PCR analysis of *OsCAXs*, from rice, shows expression in root, shoot, leaf and floral tissues (Kamiya et al., 2005).

Gene expression in response to various stimuli can be useful in determining biological function. *CAX1* and *CAX3* transcripts have been shown to be strongly induced by exogenous $Ca^{2+}$. However, *CAX4* is increased by $Ni^{2+}$ and $Mn^{2+}$ and not $Ca^{2+}$ (Cheng et al., 2002). Induction by $Na^+$ has been seen with *CAX1*, *3* and *4* and implicates a role for these *CAXs* in salt stress response (Hirschi et al., 1996; Shigaki and Hirschi, 2006; Cheng et al., 2004a). Along with induction by different cations, transcripts of *CAX1* accumulate in leaves after 24 hrs cold treatment at 4°C (Catalá et al., 2003). CAX genes from other plant species, such as rice, show induction by stress conditions. $Ca^{2+}$ and $Mn^{2+}$ can induce expression of *OsCAXs* (Kamiya et al., 2005). These results suggest that *CAXs* play a role in numerous plant physiological processes from salt stress to acclimation to cold.

To better understand the functions of CAXs in plants, analysis of specific CAX mutants was conducted. For instance, deletion of *CAX1* causes a 40% reduction in vacuolar $H^+$-ATPase (V-ATPase) activity, along with a 50% reduction in vacuolar $Ca^{2+}/H^+$ activity, despite increased expression of *CAX3* and *CAX4* (Cheng et al., 2003). This shows that CAX1 is a major $Ca^{2+}$ exchanger on the vacuole membrane.

These plants also displayed a 36% increase in vacuolar $Ca^{2+}$-ATPase activity, revealing some mode of compensation by the plants to attenuate the decrease in vacuolar $Ca^{2+}$ transport (Cheng et al., 2003). Similar reductions in V-ATPase were shown in cax3 and cax2 mutants (Cheng et al., 2005; Pittman et al., 2004b). This reduction in V-ATPase activity could mean that the deletion of CAX could disrupt the normal synthesis or assembly of the V-ATPase (Shigaki and Hirschi, 2006) or simply mean that $H^+$ pump turnover is switched down as less $H^+$ is being transported out into the cytosol. The exact nature of this alteration has yet to be understood.

Mutations of *CAX* transporters have revealed some interesting phenotypes. Mutations of *cax1* produce a gain-of-function tolerance to high $Mg^{2+}$ concentrations (Cheng et al., 2003). Additionally this same phenotype was identified in a mutant screen for tolerance to serpentine soils (low Ca:Mg ratio; Bradshaw, 2005). When mutations in multiple *CAX* genes were made, dramatic effects on the plant phenotype were observed. A double mutation of *cax1/cax3* shows plants with a severely stunted growth even though these plants still have some $Ca^{2+}/H^+$ transport activity (Cheng et al., 2005). Individually, the two mutations do not have any dramatic phenotype; however, the double mutant alludes to a functional association of CAX transporters. This could be due to some sort of heterodimer formation which could confer different biological properties (Cheng et al., 2005).

In the future, it will be paramount to determine further functions of the remaining *CAX* genes. Using yeast has been beneficial in determining substrate specificities and regulatory domains of CAX1-4 however, little is still known about *CAX5* and *CAX6*. Although single mutants in *CAX* genes show subtle phenotype, double mutants of *cax1/cax3* show dramatic phenotypes (Cheng et al., 2005). These double mutants will be good genetic backgrounds in which to study alterations in $Ca^{2+}$ partitioning within the cell using $Ca^{2+}$ reporters. Also using a comprehensive approach like micro array analysis can be useful to identify more specific characteristics of *CAXs*. Although very little is known about the newly classified *CCXs*, recently published research on *CCX5* showed increases in $Ca^{2+}$, $Cd^{2+}$, and $Zn^{2+}$ transport when ectopically expressed in tobacco (Koren'kov et al., 2007). The use of ectopic expression has been useful in characterizing *CAXs* and appears to create an avenue in which to elucidate the functions of *CCXs*.

## 6. IMPACTS IN NUTRITION

In this section we will briefly discuss some of the applied research that is taking place with $Ca^{2+}$ transporters in plants. A judicious use of transgenic plants will be an important option to improve human nutrition. Osteoporosis is a disease caused by reduced bone density and affects females to a greater extent than males. Adult humans should consume 1000 to 1200 mg/d of calcium, however Bryant and others (1999) point out that many adults consume less than this recommended daily intake. One way to overcome this problem, which has $13.8 billion in health care related costs each year, is to increase calcium levels in foods. Weaver and Plawecki (1994) state that vegetables make up the second largest source of calcium in the diet behind dairy products. Vegetables tend to be low in calcium but consumption of a wide variety of vegetables can have an impact on dietary calcium levels, including calcium-rich spinach and kale (Weaver and Plawecki, 1994). By creating vegetables or other agronomically important crops that contain higher calcium, a richer source of calcium would exist for people to use in maintaining a healthy diet.

The creation of genetically modified plants with increased nutritional benefits is an expanding field. The term "nutritional genomics" has been used to describe various studies which implement some form of plant biochemistry, genomics or human nutrition. Transgenic plants are frequently analyzed for changes in plant metabolism and this is often where the experimentation ends. Ideally, these genetically modified plants need to be labeled and used in controlled animal and human feeding studies to assess nutritional impacts. This added nutritional benefit is not without issue. This increased $Ca^{2+}$ needs to be targeted to the edible portions of the plant and will ensure that excess $Ca^{2+}$ is not sequestered into nutritionally unimportant plant tissues. Another area that needs to be investigated is accumulation of toxic metals in these plant tissues. Toxic metal might not have deleterious effects on the plants but can be detrimental to human health. When plants are grown in soils which could possibly contain heavy metals, uptake of these metals could occur due to their similar ionic radii with $Ca^{2+}$ (Shigaki and Hirschi, 2006). Accumulation of these metals in the plant's tissue needs to be assessed to determine if these foods are safe for consumption. Bioavailability of this increased $Ca^{2+}$ also needs to be determined. To date, this type of analysis to assess the nutrient value of transgenic foods is minimal.

Aside from nutritional benefits, the use of genetic engineering to increase $Ca^{2+}$ levels could improve plant productivity and extend product shelf life. Calcium has long been used to combat many post harvest issues (Kabak et al., 2006). Apples are immersed in a calcium solution to maintain firmness in shipping and prolong shelf life. Application of calcium solutions to plums (Alcaraz-Lopez et al., 2003) and pear fruits is used to increase firmness (Klein and Ferguson, 1987). Calcium is also added to soil to reduce the incidences of pathogen attack on potato tubers (Kabak et al., 2006) and to combat heat stress (Kabak et al., 2006). All of these preventative measures require time and man power to apply these calcium containing solutions to the soil or fruits. Advances in plant biotechnology can provide novel solutions to meet the benefits exogenous $Ca^{2+}$ creates for plants. One example of an alternative to exogenous applications of $Ca^{2+}$, is the expression of sCAX1 in tomatoes to increase fruit firmness and prolonged shelf life (Park et al., 2005). Another interesting aspect to increased calcium accumulation is in long term space flight. In space, providing continuous light can increase plant productivity but calcium deficiencies in plants become an issue that cannot be solved by foliar $Ca^{2+}$ applications (Bugbee, 1999). Using genetic engineering to increase calcium levels in a wide variety of plants could positively impact plant productivity, while simultaneously decreasing labor costs.

## 7. CONCLUSIONS AND FUTURE PROSPECTS

Calcium transport impacts growth, development and stress responses. We focused this overview on the different types of transporters and the methodologies used to determine their biological functions. *In planta* functional redundancy has hindered identifying specific roles to transporters. Future research promises to visualize alterations in calcium levels in lines with altered transport function. We concluded this review by detailing how the manipulation of calcium transporters can be used to alter plant productivity and benefit human nutrition.

In the future, detailing the precise biological functions of plant calcium transporters will be essential to plant biology. Using yeast as a tool to elucidate function of plant transporters is practical but the use of other heterologous systems is required. With the abundance of genomic tools available, further inferences regarding function of $Ca^{2+}$ transporters can be made. Once working hypotheses are formulated, multiple experimental approaches will aid in defining biological functions. After we better understand the biological functions of these calcium transporters, we can apply this knowledge towards improving plant productivity and human nutrition.

## 8. ACKNOWLEDGMENTS

We are grateful to Dr. Jon Pittman and Dr. Toshiro Shigaki for critical reading of the manuscript. We are also thankful to Tim Hotze, Ashley Hester and Hui Mei for their help in proof reading of the chapter. This work was supported by the National Institutes of Health grant no. IR01 DK 062366, USDA CSREES#2005-34402-16401 Designing Foods for Health, National Science Foundation Grant 0344350 and Agriculture Research Service under Cooperative Agreement 58-6250-6-001.

## 9. LITERATURE CITED

Aiello, D.P., Fu, L., Miseta, A., Sipos, K., Bedwell, D.M. (2004) The $Ca^{2+}$ homeostasis defects in a pgm2Δ strain of *Saccharomyces cerevisiae* are caused by excessive vacuolar $Ca^{2+}$ uptake mediated by the $Ca^{2+}$ -ATPase Pmc1p. *J. Biol. Chem.*, **279**: 38495-38502.

Alcaraz-Lopez, C., Botia, M., Alcaraz, C.F., Riquelme, F. (2003) Effects of foliar sprays containing calcium, magnesium and titanium on plum (*Prunus domestica* L.) fruit quality. *J. Plant Physiol.*, **160**: 1441-1446.

Ali, R., Zielinski, R.E., Berkowitz, G.A. (2006) Expression of plant cyclic nucleotide-gated cation channels in yeast. *J. Exp. Bot.*, **57**: 125-138.

Ali, R., Ma, W., Lemtiri-Chlieh, F., Tsaltas, D., Leng, O., von Bodman, S., Berkowitz, G.A. (2007) Death don't have no mercy and neither does calcium: Arabidopsis CYCLIC NUCLEOTIDE GATED CHANNEL2 and Innate Immunity. *Plant Cell*, **19**: 1081-1095.

Allen, D.G., Blinks, J.R., Prendergast, F.G. (1977) Aequorin luminescence: relation of light emission to calcium concentration—a calcium-independent component. *Science*, **195**: 996-998.

Allen, G.J., Kwak, J.M., Chu, S.P., Llopis, J., Tsein, R.Y., Harper, J.F., Schroeder, J.I. (1999) Cameleon calcium indicator reports cytoplasmic calcium dynamics in Arabidopsis guard cells. *Plant J.,* **19**: 735-747.

Arazi, T., Kaplan, B., Sunkar, R., Fromm, H. (2000) Cyclic-nucleotide- and $Ca^{2+}$/calmodulin-regulated channels in plants: targets for manipulating heavy metal tolerance, and possible physiological roles. *Biochem. Doc. Trans.,* **28**: 471-475.

Axelsen, K.B., Palmgren, M.G. (2001) Inventory of the superfamily of P-type ion pumps in Arabidopsis. *Plant Physiol.,* **126**: 696-706.

Baekgaard, L., Luoni, L., De Michelis, M.I., Palmgren, M.G. (2006) The plant plasma membrane $Ca^{2+}$ pump ACA8 contains overlapping as well as physically separated autoinhibitory and calmodulin-binding domains. *J Biol. Chem.,* **281**: 1058-1065.

Balagué, C., Lin, B., Alcon, C., Flottes, G., Malmström, S., Köhler, C., Neuhaus, G., Pelletier, G., Gaymard, F., Roby, D. (2003) HLM1, an essential signaling component in the hypersensitive response, is a member of the cyclic nucleotide-gated channel ion channel family. *Plant Cell,* **15**: 365-379.

Baxter, I., Tchieu, J., Sussman, M.R., Boutry, M., Plamgren, M.G., Gribskov, M., Harper, J.F., Axelsen, K.B. (2003) Genomic comparison of P-type ATPase ion pumps in Arabidopsis and rice. *Plant Physiol.,* **132**: 618-628.

Becker, D., Geiger, D., Dunkel, M., Roller, A., Bertl, A., Latz, A., Carpaneto, A., Dietrich, P., Roelfsema, M.R.G., Voelker, C., Schmidt, D., Mueller-Roeber, B., Czempinski, K., Hedrich, R. (2004) *At*TPK4, an Arabidopsis tandem-pore K+ channel, poised to control the pollen membrane voltage in a pH- and $Ca^{2+}$-dependent manner. *Proc. Natl. Acad. Sci.,* **101**: 15621-15626.

Berridge, M.J., Lipp, P., Bootman, M.D. (2000) The versatility and universality of calcium signaling. *Nat. Rev. Mol. Cell Biol.,* **1**: 11-21.

Bonza, M.C., Luoni, L., DeMichelis, M.I. (2004) Functional expression in yeast of an N-deleted form of *At*-ACA8, a plasma membrane $Ca^{2+}$-ATPase of *Arabidopsis thaliana*, and characterization of a hyperactive mutant. *Planta,* **218**: 814-823.

Bothwell, J.H.F., Ng, C.K.-Y. (2005) The evolution of $Ca^{2+}$ signaling in photosynthetic eukaryotes. *New Phytol.,* **16**: 21-38.

Braam, J. (2005) In touch: Plant responses to mechanical stimuli. *New Phytol.,* **165**: 373.

Bradshaw, H.D. (2005) Mutations in CAX1 produce phenotypes characteristic of plants tolerant to serpentine soils. *New Phytol.,* **167**: 81-88.

Bryant, R.J., Cadogan, J., Weaver, C.M. (1999) The new dietary reference intakes for calcium: implications for osteoporosis. *J Am. Col. Nutr.,* **8**: 406S-412S.

Bugbee, B. (1999) Engineering plants for spaceflight environments. *Gravit. Space Biol. Bull.,* **12**: 67-74.

Cai, X., Lytton, J. (2004) The cation/$Ca^{2+}$ exchanger superfamily: Phylogenetic analysis and structural implications. *Mol. Biol. Evol.,* **21**: 1692-1703.

Catalá, R., Santos, E., Alonso, J.M., Ecker, J.R., Martínez-Zapater, J.M., Salinas, J. (2003) Mutations in the $Ca^{2+}/H^+$ transporter CAX1 increase CBF/DREB1 expression and the cold-acclimation response in Arabidopsis. *Plant Cell,* **15**: 2940-2951.

Chan, C.W.M., Schorrak, L.M., Smith Jr., R.K., Bent, A.F., Sussman, M.R. (2003) A cyclic nucleotide-gated ion channel, CNGC2, is crucial for plant development and adaptation to calcium stress. *Plant Physiol.,* **132**: 728-731.

Chen, X., Chang, M., Wang, W., Wu, B. (1997) Cloning of a $Ca^{2+}$-ATPase gene and the role of cytosolic $Ca^{2+}$ in the gibberellin-dependant signaling pathway in aleurone cells. *Plant J.,* **11**: 363-371.

Cheng, N-H., Pittman, J.K., Shigaki, T., Hirschi, K.D. (2002) Characterization of CAX4, an Arabidopsis H+/cation antiporter. *Plant Physiol.,* **128**: 1245-1254.

## Calcium transporters

Cheng, N-H., Hirschi, K.D. (2003) Cloning and characterization of CXIP1, a novel PICOT domain-containing Arabidopsis protein that associates with CAX1. *J. Biol. Chem.,* **278**: 6503-6509.
Cheng, N-H, Pittman, J.K., Barkla, B., Shigaki, T., Hirschi, K.D. (2003) The Arabidopsis cax1 mutant inhibits impaired ion homeostasis, development, and hormonal responses and reveals interplay among vacuolar transports. *Plant Cell,* **15**: 347-364.
Cheng, N-H. Pittman, J.K., Zhu, J.K., Hirschi, K.D. (2004a) The protein kinase SOS2 activates the Arabidopsis $H^+/Ca^{2+}$ antiporter CAX1 to integrate calcium transport and salt tolerance. *J. Biol. Chem.,* **279**: 2922-2926.
Cheng, N-H., Liu, J.Z., Nelson, R.S., Hirschi, K.D. (2004b) Characterization of CXIP4, a novel Arabidopsis protein that activates the $H^+/Ca^{2+}$ antiporter, CAX1. *FEBS Lett.,* **559**: 99-106.
Cheng, N-H., Pittman, J.K., Shigaki, T., Lachmansingh, J., LeClere, S., Lahner, B., Salt, D.E., Hirschi, K.D. (2005) Functional association of Arabidopsis CAX1 and CAX3 is required for normal growth and ion homeostasis. *Plant Physiol.,* **138**: 2048-2060.
Chigri, F., Hormann, F., Stamp, A., Stammers, D.K., Bolter, B., Soll, J., Vothknecht, U.C. (2006) Calcium regulation of chloroplast protein translocation is mediated by calmodulin binding to Tic32. *Proc. Natl. Acad. Sci.,* **103**: 16051-16056.
Chiu, J.C., Brenner, E.D., DeSalle, R., Nitbach, M.N., Colmes, T.C., Coruzzi, G.M. (2002) Phylogentic and expresión análisis of the glutamate receptor-like gene family in *Arabidopsis thaliana. Mol. Biol. Evol.,* **19**: 1066-1082.
Chung, W.S., Lee, S.H., Kim, J.C., Heo, W.D., Kim, M.C., Park, C.Y., Park, H.C., Lim, C.O., Kim, W.B., Harper, J.F., Cho, M.J. (2000) Identification of calmodulin-regulated soybean $Ca^{2+}$-ATPase (SCA1) that is located on the plasma membrane. *Plant Cell,* **12**: 1393-1407.
Cronin, S.R., Rao, R., Hampton, R.Y. (2002) Cod1/Spf1p is a P-type ATPase involved in ER function and $Ca^{2+}$ homeostasis. *J. Cell Boil.,* **157**: 1017-1028.
Dabitz, N., Hu, N.J., Yusof, A.M., Tranter, N., Winter, A., Daley, M., Zschörnig, O., Brisson, A., Hofmann, A. (2005) Structural determinants for plant annexin-membrane interactions. *Biochem.,* **13**: 16292-16300.
Darley, C.P., van Wuytswinkel, O.C., van der Woode, K., Mager, W.H., de Boer, A.H. (2000) *Arabidopsis thaliana* and *Saccharomyces cerevisiae* NHX1 genes encode amiloride sensitive electroneutral $Na^+/H^+$ exchangers. *Biochem. J.,* **351**: 241-249.
Davenport, R., (2002) Glutamate receptors in plants. *Ann. Bot.,* **90**: 549-557.
Demidchik, V., Shabala, S.N., Coutts, K.B., Tester, M.A., Davies, J.M. (2003) Free oxygen radicals regulate plasma membrane $Ca^{2+}$- and K+-permeable channels in plant root cells. *J. Cell Sci.,* **116**: 81-88.
Dode, L., Andersen, J.P., Raeymaekers, L., Missiaen, L., Vilsen, B., Wuytack, F. (2005) Functional comparison between secretory pathway $Ca^{2+}/Mn^{2+}$-ATPase (SPCA) 1 and sarcoplasmic reticulum $Ca^{2+}$-ATPase (SERCA) 1 isoforms by steady-state and transient kinetic analyses. *J. Biol. Chem.,* **280**: 39124-39134.
Furuichi, T., Cunningham, K.W., Muto, S. (2001) A putative two pore channel AtTPC1 mediates Ca2+ flux in Arabidopsis leaf cells. *Plant Cell Physiol.,* **42**: 900-905.
Geisler, M., Frangne, N., Gomes, E., Martinoia, E., Palmgren, M.G. (2000) The ACA4 gene of Arabidopsis encodes a vacuolar membrane calcium pump that improves salt tolerance in yeast. *Plant Physiol.,* **124**:1814-1827.
Golovkin, M., Reddy, A.S.N. (2003) A calmodulin-binding protein from Arabidopsis has an essential role in pollen germination. *Proc. Natl. Acad. Sci.,* **100**: 10558-10563.

Gorecka, K.M., Konopka-Postupolska, D., Hennig, J., Buchet, R., Pikula, S. (2005) Peroxidase activity of annexin 1 from *Arabidopsis thaliana*. *Biochem.Biophys. Res. Commun.* **336**: 868-75.

Haro, R., Banuelos, M.A., Senn, M.E., Barrero-Gil, J., Rodriquez-Navarro, A. (2005) HKT1 mediates sodium uniport in roots. Pitfalls in the expresión of HKT1 in yeast. *Plant Physiol.*, **139**:1495-1506.

Harper, J.F., Hong, B., Hwang, I., Guo, H.Q., Stoddard, R., Huang, J.F., Palmgren, M.G., Sze, H. (1998) A novel calmodulin-regulated $Ca^{2+}$-ATPase (ACA2) from Arabidopsis with an N-terminal autoinhibitory domain. *J. Biol. Chem.*, **273**:1099-1106.

Harper, J.F., Breton, G., Harmon, A. (2004) Decoding $Ca^{2+}$ signals through plant protein kinases. *Annu. Rev. Plant Biol.*, **55**:263-288.

Hepler, P.K. (2005) Calcium: A central regulator of plant growth and development. *Plant Cell*, **17**:2142-2155.

Hetherington, A.M., Brownlee, C. (2004) The generation of $Ca^{2+}$ signals in plants. *Annu. Rev. Plant Biol.*, **55**:401-427.

Hirschi, K.D., Zhen, R.G., Cunningham, K.W., Rea, P.A., Fink, G.R. (1996) CAX1, an $H^+$/$Ca^{2+}$ antiporter from Arabidopsis. *Proc. Natl. Acad. Sci.*, **6**:8782-8786.

Hirschi, K.D., Korenkov, V.D., Wilganowski, N.L., Wagner, G.J. (2000) Expression of Arabidopsis CAX2 in tobacco. Altered metal accumulation and increased manganese tolerance. *Plant Physiol.*, **124**:125-133.

Hirschi, K.D. (2004) The calcium conundrum. Both versatile nutrient and specific signal. *Plant Physiol.*, **136**:2438-2442.

Hua, B.-G., Mercier, R.W., Leng, Q., Berkowitz, G.A. (2003) Plants do it differently. A new basis for potassium/sodium selectivity in the pore of an ion channel. *Plant Physiol.*, **132**:1353-1361.

Huang, L., Berkelman, T., Franklin, A.E., Hoffman, N.E. (1993) Characterization of a gene encoding a $Ca^{2+}$-ATPase-like protein in the plastid envelope. *Proc. Natl. Acad. Sci.*, **90**:10066-10070.

Hwang, I., Sze, H., Harper, J.F. (2000) A calcium-dependant protein kinase can inhibit a calmodulin-stimulated $Ca^{2+}$ pump (ACA2) located in the endoplasmic reticulum of Arabidopsis. *Proc. Natl. Acad. Sci.*, **23**:6224-6229.

Kabak, B., Dobson, A.D., Var, I., (2006) Strategies to prevent mycotoxin contamination of food and animal feed: a review. *Crit. Rev. Food Sci. Nutr.*, **46**:593-619.

Kabala, K., Klobus, G. (2005) Plant $Ca^{2+}$-ATPases. *Acta Physiol. Plant.*, **27**:559-574.

Kamiya, T., Maeshima, M. (2004) Residues in internal repeats of the rice cation/$H^+$ exchanger are involved in the transport and selection of cations. *J. Biol. Chem.*, **279**:812-819.

Kamiya, T., Akahori, T., Maeshima, M. (2005) Expression profile of the genes for rice cation/$H^+$ exchanger family and functional analysis in yeast. *Plant and Cell Physiol.*, **46**:1735-1740.

Kass, G.E., Orrenius, S. (1999) Calcium signaling and cytotoxicity. *Environ. Health Perspect.*, **107**:25-35.

Kauffman, K.J., Pridgen, E.M., Doyle, F.J., Dhurjati, P.S., Robinson, A.S. (2002) Decreased protein expression and intermittent recoveries in BiP levels results from cellular stress during heterologous protein expression in *Saccharomyces cerevisiae*. *Biotechnol. Prog.*, **18**:942-950.

Kaupp, U.B., Seifert, R. (2002) Cyclic nucleotide-gated ion channels. *Physiol. Rev.*, **82**:769-824.

Calcium transporters

Kellermayer, R., Aiello, D.P., Miseta, A., Bedwell, D.M. (2003) Extracellular $Ca^{2+}$ sensing contributes to excess $Ca^{2+}$ accumulation and vacuolar fragmentation in a pmr1Δ mutant of *S. cerevisiae*. *J. Cell Sci.*, **116**:1637-1646.
Klein, J.D., Ferguson, I.B. (1987) Effect of high temperatures on calcium uptake by suspension cultured pear fruits cells. *Plant Physiol.*, **84**:153-156.
Koren'kov, V., Park, S., Cheng, N-H, Sreevidya, C., Lachmansingh, J., Morris, J., Hirschii, K., Wagner, G.J. (2007) Enhanced $Cd^{2+}$-selective root-tonoplast-transport in tobaccos expressing Arabidopsis cation exchangers. *Planta*, **225**:403-411.
Kurusu, T., Sakurai, Y., Miyao, A., Hirochika, H., Kuchitsu, K. (2004) Identification of a putative voltage-gated $Ca^{2+}$ -permeable channel (OsTPC1) involved in $Ca^{2+}$ influx and regulation of growth and development in rice. *Plant Cell Physiol.*, **45**:693-702.
Kurusu, T., Yagala, T., Miyao, A., Hirochika, H., Kuchitsu, K. (2005) Indentification of a putative voltage-gated $Ca^{2+}$ channel as a key regulator of elicitor-induced hypersensitive cell death and mitogen-activated protein kinase activation in rice. *Plant J.*, **42**:798-809.
Lacombe, B., Becker, D., Hedrich, R., DeSalle, R., Hollman, M., Kwak, J.M., Schroeder, J.I., Le Novere, N., Nam, H.G., Spalding, E.P., Tester, M., Turano, F.J., Chiu, J., Coruzzi, G. (2001) The identity of plant glutamate receptors. *Science*, **25**:1486-1487.
Lemtiri-Chlieh, F., MacRobbie, E.A.C., Webb, A.A.R., Manison, N.F., Brownlee, C., et al. (2003) Inositol hexakisphosphate mobilizes an endomembrane store of calcium in guard cells. *Proc. Natl. Acad. Sci.*, **100**:10091-10095.
Lemtiri-Chlieh, F., Berkowitz, G.A. (2004) Cyclic adenosine monophosphate regulates calcium channels in the plasma membrane of Arabidopsis leaf guard and mesophyll cells. *J. Biol. Chem.*, **279**:35306-35312.
Leng, Q., Mercier, R.W., Hua, B.-G., Fromm, H., Berkowitz, G.A. (2002) Electrophysiological analysis of cloned cyclic nucleotide-gated ion channels. *Plant Physiol.*, **128**:400-410.
Levchenko, V., Konrad, K.R., Dietrich, P., Roelfsema, M.R.G., Hedrich, R. (2005) Cytosolic abscisic acid activates guard cell anion channels without preceding $Ca^{2+}$ signals. *Proc. Natl. Acad. Sci.*, **102**:4203-4208.
Li, J., Zhu. S., Song, X., Shen, Y., Chen, H., Yu, J., Yi, K., Liu, Y., Karplus, V.J., Wu, P., Deng, X.W. (2006) A rice glutamate receptor-like gene is critical for the division and survival of individual cells in the root apical meristem. *Plant Cell*, **18**:340-349.
Liang, F., Sze, H. (1998) A high-affinity $Ca^{2+}$ pump, ECA1, from the endoplasmic reticulum is inhibited by cyclopiazonic acid but not by thapsigargin. *Plant Physiol.*, **118**:817-825.
Luoni, L., Bonza, M.C., De Michelis, M.I. (2000) $H^+/Ca^{2+}$ exchange driven by the plasma membrane $Ca^{2+}$-ATPase of *Arabidopsis thaliana* reconstituted in proteoliposomes after calmodulin-affinity purification. *FEBS Lett.*, **482**:225-230.
Malmstrom, S., Askerlund, P., Palmgren, M.G. (1997) A calmodulin-stimulated $Ca^{2+}$-ATPase from plant vacuolar membranes with putative regulatory domains at its N-terminus. *FEBS Lett.*, **6**:324-328.
Marschner, H. (1995) Mineral Nutrition of Higher Plants, Academic Press, London.
Mäser, P., Thomine, S., Schroeder, J.I., Ward, J.M., Hirschi, K., Sze, H., Talke, I.N., Amtmann, A., Maathuis, F.J., Sanders, D., Harper, J.F., Tchieu, J., Gribskov, M., Persans, M.W., Salt, D.E., Kim, S.A., Guerinot, M.L. (2001) Phylogenetic relationships within cation transporter families of Arabidopsis. *Plant Physiol.*, **126**:1646-1667.
Medvedev, S.S. (2005) Calcium signaling system in plants. *Russian J. Plant Physiol.*, **52**:249-270.

Meyerhoff, O., Muller, K., Roelfsema, M.R., Latz, A., Lacombe, B., Hedrich, R., Dietrich, P., Becker, D. (2005) AtGLR3.4, a glutamate receptor channel-like gene is sensitive to touch and cold. *Planta*, **222**:418-427.

Miedema, H., de Boer, A.H., Pantoja, O. (2003) The gating kinetics of the slow vacuolar channel. A novel mechanism for SV channel functioning. *J. Membrane Biol.*, **194**: 11-20.

Miseta, A., Kellermayer, R., Aiello, D.P., Fu, L., Bedwell, D.M. (1999) The vacuolar $Ca^{2+}/H^+$ exchanger Vcx1p/Hum1p tightly controls cytosolic $Ca^{2+}$ levels in *S.cerevisiae*. *FEBS Lett.*, **21**:132-136.

Mori, I.C., Schroeder, J.I. (2004) Reactive oxygen species activation of plant $Ca^{2+}$ channels. A signaling mechanism in polar growth, hormone transduction, stress signaling, and hypothetically mechanotransduction. *Plant Physiol.*, **135**:702-708.

Nakamura, K., Niimi, M., Niimi, K., Colmes, A.R., Yates, J.E., Decottiginies, A., Monk, B.C., Goffeau, A., Cannon, R.D. (2001) Functional expression of *Candida albicans* drug efflux pump Cdr1p in a *Saccharomyces cerevisiae* strain deficient in membrane transport. *Antimicrob. Agents Chemother.*, **45**:3366-3374.

Nayyar, H. (2003) Calcium as environmental sensor in plants. *Curr. Sci.*, **84**:893.

Ng, C.K.-Y., McAinsh, M.R., Gray, J.E., Hunt, L., Leckie, C.P., Mills, L., Hetherington, A.M. (2001) Calcium-based signaling systems in guard cells. *New Phytol.*, **151**: 109.

Ng, C.K.-Y., McAinsh, M.R. (2003) Encoding specificity in plant calcium signaling: Hot-spotting the ups and downs and waves. *Ann. Bot.*, **92**:477-485.

Oldroyd, G.E., Downie, J.A. (2006) Nuclear calcium changes at the core of symbiosis signaling. *Curr. Opin. Plant Biol.*, **9**:351-357.

Park, S., Kim, C., Pike, L., Smith, R., Hirschi, K. (2004) Increased calcium in carrots by expression of an Arabidopsis $H^+/Ca^{2+}$ transporter. *Molecular Breeding*, **14**: 275-282.

Park, S., Cheng, N-H., Pittman, J.K., Yoo, K.S., Park, J., Smith, R.H., Hirschi, K.D. (2005) Increased Calcium levels and prolonged shelf life in tomatoes expressing Arabidopsis $H^+/Ca^{2+}$ transporters. *Plant Phys.*, **139**:1194-1206.

Peiter, E., Maathuis, F.J.M., Mills, L.N., Knight, H., Pelloux, J., Hetherington, A.M., Sanders, D. (2005) The vacuolar $Ca^{2+}$ -activated channel TPC1 regulates germination and stomatal movement. *Nature*, **434**:404-408.

Pittman, J.K., Mills, R.F., O'Connor, C.D., Williams, L.E. (1999) Two additional type IIA $CA^{2+}$-ATPases are expressed in *Arabidopsis thaliana*: evidence that type IIA sub-groups exist. *Gene*, **236**:137-147.

Pittman, J.K., Shigaki, T., Cheng, N-H., Hirschi, K.D. (2002) Mechanism of N-terminal autoinhibition in the Arabidopsis $Ca^{2+}/H^+$ antiporter CAX1. *J. Biol. Chem.*, **277**: 26452-26459.

Pittman, J.K., Hirschi, K.D. (2003) Don't shoot the second messenger: endomembrane transporters and binding proteins modulate cytosolic $Ca^{2+}$ levels. *Curr. Opin. Plant Biol.*, **6**:257-262.

Pittman, J.K., Shigaki, T., Marshall, J.L., Morris, J.M., Cheng, N-H., Hirschi, K.D. (2004a) Functional and regulatory analysis of the *Arabidopsis thaliana* CAX2 cation transporter. *Plant Mol. Biol.*, **56**:959-971.

Pittman, J.K., Cheng, N-H., Shigaki, T., Kunta, M., Hirschi, K.D. (2004b) Functional dependence on calcineurin by variants of the *Saccharomyces cerevisiae* vacuolar $Ca^{2+}/H^+$ exchanger Vcx1p. *Mol. Microbiol.*, **24**:1104-1116.

Plieth, C. (2005) Calcium: just another regulator in the machinery of life? *Ann. Bot.*, **96**: 1-8.

## Calcium transporters

Prokić, L., Jovanović, Z., Stikić, R., Vuæinić, •. (2005) The mutual effect of extracellular $Ca^{2+}$, abscisic acid, and pH on the rate of stomatal closure. *Ann. N.Y. Acad. Sci.,* **1048**:513-516.

Qu, H.Y., Shang, Z.L., Zhang, S.L., Liu L.M., Qu, J.Y. (2007) Identification of hyperpolarization-activated calcium channels in apical pollen tubes of *Pyrus pyrifolia. New Phytol.,* **174**:524-536.

Reddy, V.S., Reddy, A.S.N. (2004) Proteomics of calcium-signaling components in plants. *Phytochemistry,* **65**:1745-1776.

Rentel, M.C., Knight, M.R. (2004) Oxidative stress-induced calcium signaling in Arabidopsis. *Plant Physiol.,* **135**:1471-1479.

Samaj, J., Muller, J., Beck, M., Bohm, N., Menzel, D, (2006) Vesicular trafficking, cytoskeleton and signaling in root hairs and pollen tubes. *Trends Plant Sci.,* **11**:594-600.

Sanders, D., Pelloux, J., Brownlee, C., Harper, J.F. (2002) Calcium at the crossroads of signaling. *Plant Cell,* **14**:S401-S417.

Schiøtt, M., Romanowsky, S.M., Bækgaard, L., Jakobsen, M.K., Palmgren, M.G., Harper, J.F. (2004) A plant plasma membrane $Ca^{2+}$ pump is required for normal pollen tube growth and fertilization. *Proc. Natl. Acad. Sci.,* **101**:9502-9507.

Schiøtt, M., Palmgren, M.G. (2005) Two plant $Ca^{2+}$ pumps expressed in stomatal guard cells show opposite expression patterns during cold stress. *Physiol. Plant.,* **124**:278.

Scrase-Field, S.A.M.G., Knight, M.R. (2003) Calcium: just a chemical switch? *Curr. Opin. Plant Biol.* **6**:500-506.

Shang, Z.-I., Ma, L., Zhang, H., He, R., Wang, X., Cui, S., Sun, D. (2005) $Ca^{2+}$ influx into lily pollen grains through a hyperpolarization-activated $Ca^{2+}$ -permeable channel which can be regulated by extracellular CaM. *Plant and Cell Physiol.,* **46**:598-608.

Shigaki, T., Cheng, N-H., Pittman, J.K., Hirschi, K.D. (2001) Structural determinants of $Ca^{2+}$ transports in the Arabidopsis $H^+/Ca^{2+}$ antiporter CAX1. *J. Biol. Chem.* **276**:43152-43159.

Shigaki, T., Pittman, J.K., Hirschi, K.D. (2003) Manganese specificity determinants in the Arabidopsis metal/$H^+$ antiporter CAX2. *J. Biol. Chem.,* **278**:6610-6617.

Shigaki, T., Rees, I., Nakhleh, L., Hirschi, K.D. (2006) Identifcation of three distanct Phylogenetic groups of CAX cation/proton antiporters. *J. Mol. Evol.,* **63**: 815-825.

Shigaki, T., Hirschi, K.D. (2006) Diverse functions and molecular properties emerging for CAX cation/$H^+$ exchangers in plants. *Plant Biol.,* **8**:419-429.

Sonnhammer, E.L.L., von Heijne, G., Krogh, A. (1998) A hidden Markov model for predicting transmembrane helices in protein sequences. *Proc. ISMB,* **6**:175-182.

Sørensen, T.L-M., Møller, J.V., Nissen, P. (2004) Phosphoryl Transfer and Calcium Ion Occlusion in the Calcium Pump. *Science,* **304**:1672-1675.

Subbaiah, C.C., Sachs, M.M. (2000) Maize *cap1* encodes a novel serca-type calcium-ATPase with a calmodulin-binding domain. *J. Biol. Chem.,* **28**:21678-21687.

Sze, H., Li, X., Palmgren, M.G. (1999) Energization of plant cell membranes by $H^+$-pumping ATPases. Regulation and biosynthesis. *Plant Cell,* **11**:677-690.

Sze, H., Liang, F., Hwang, I., Curran, A.C., Harper, J.F. (2000) Diversity and regulation of plant $Ca^{2+}$ pumps; insights from expression in yeast. *Annu. Rev. Plant Physiol. Plant Mol. Biol.,* **51**:433-462.

Sze, H., Padmanaban, S., Cellier, F., Honys, D., Cheng, N-H., Bock, K.W., Conéjero, G., Li, X., Twell, D., Ward, J.M., Hirschi, K.D. (2004) Expression Patterns of a Novel *AtCHX* Gene family highlight potential roles in osmotic adjustment and $K^+$ homeostasis in pollen development. *Plant Physiol.,* **136**:2532-2547.

Tanouye, M.A., Ferrus, A., Fujita, S.C., (1981) Abnormal action potentials associated with the *Shaker* complex locus of *Drosophilia*. *Proc. Natl. Acad. Sci.,* **78**:6548-6552.

Thuleau, P., Leclerc, C., Xiong, T.C., Mazars, C., Leclerc, C., Moreau, M. (2003) Luminous plant and animals or the expression of aequorin and "chameleon" probes: a new light in calcium signaling. *J. Soc. Biol.,* **197**:291-300.

Ton, V.-K., Rao, R. (2004) Functional expression of heterologous proteins in yeast: Insights into $Ca^{2+}$ signaling and $Ca^{2+}$-transporting ATPAses. *Am. J. Physiol. Cell Physiol.,* **287**:C580-C589.

Vanoevelen, J., Dode, L., Van Baelen, K., Fairclough, R.J., Missiaen, L., Raeymaekers, L., Wuytack. (2005) The secretory pathway $Na^+/Ca^{2+}$-ATPase 2 is a golgi-localized pump with high affinity for $Ca^{2+}$ ions. *J. Biol. Chem.,* **280**:22800-22808.

Wang, Y.-F., Fan, L.-M., Zhang, W.-Z., Zhang, W., Wu, W.-H. (2004) $Ca^{2+}$-permeable channels in the plasma membrane of Arabidopsis pollen are regulated by actin microfilaments. *Plant Physiol.,* **136**:3892-3904.

Wang, Y.-J., Yu, J.-N., Chen, T., Zhang, Z.-G., Hao, Y.-J., Zhang, J.-S., Chen, S.-Y. (2005) functional analysis of a putative $Ca^{2+}$ channel gene *TaTPC1* from wheat. *J. Exp. Biol.,* **56**:3051-3060.

Weaver, C.M., Proulx, W.R., Heaney, R. (1990) Choices for achieving adequate dietary calcium with a vegetarian diet. *Am. J. Clin. Nutr.,* **70**:543S-548S.

Weaver, C.M., Plawecki, K.L. (1994) Dietary calcium: Adequacy of a vegetarian diet. *Am. J. Clin. Nutr.,* **59**:1238S-1241S.

White, P.J., Pineros, M., Testr, M., Ridout, M.S. (2000) Cation permeability and selectivity of a root plasma membrane calcium channel. *J. Membr. Biol.,* **174**:71-83.

White, P.J., Broadley, M.R. (2003) Calcium in plants. *Ann. Bot.,* **92**:487-511.

Wimmers, L.E., Ewing, N.N., Bennett, A.B. (1992) Higher plant $Ca^{2+}$-ATPase: primary structure and regulation of mRNA abundance by salt. *Proc. Natl. Acad. Sci.,* **89**:9205-9209.

Wu, Z., Liang, F., Hong, B., Young, J.C., Sussman, M.R., Harper, J.F., Sze, H. (2002) An endoplasmic reticulum-bound $Ca^{2+}/Mn^{2+}$ pump, ECA1, supports plant growth and confers tolerance to $Mn^{2+}$ stress. *Plant Physiol.,* **130**:128-137.

Xiong, T.C., Borque, S., Lecourieux, D., Amelot, N., Grat, S., Briere, C., Mazars, C., Puqin, A., Ranjeva, R. (2006) Calcium signaling in plant cell organelles delimited by a double membrane. *Biochem. Biophys. Acta,* **1763**:1209-1215.

Xu, C., Rice W.J., He, W., Stokes, D.L. (2002) A structural model for the catalytic cycle of $Ca^{2+}$-ATPase. *J. Mol. Biol.,* **8**:201-211.

Yang, T., Poovaiah, B.W. (2003) Calcium/calmodulin-mediated signal network in plants. *Trends Plant Science,* **8**:505-512.

Yoshioka, K., Moeder, W., Kang, H.G., Kachroo, P., Masmoudi, K., Berkowitz, G., Klessig, D.F. (2006) The chimeric Arabidopsis CYCLIC NUCLEOTIDE-GATED ION CHANNEL 11/12 activates multiple pathogen resistance responses. *Plant Cell,* **18**:747-763.

# Chapter 3

## NITRATE AND AMMONIUM TRANSPORTERS IN PLANTS

RANA P. SINGH[1], MANISH SAINGER[1], D.P. SINGH[1] AND PAWAN K. JAIWAL[2]

[1]Department of Environmental Science, Baba Saheb Bhimrao Ambedkar (Central) University, Lucknow - 226 025, India
[2]Advanced Centre for Biotechnology, M.D. University, Rohtak - 124 001, India
E-mail: rana_psingh@rediffmail.com

**Abstract**

Many specific low affinity and high affinity membrane bound transport proteins have been reported to be involved in uptake, transport and remobilization of $NO_3^-$ and $NH_4^+$-N, which are taken up by the plants from their rhizosphere. AtNRT1.1 and AtNRT2.1 are nitrate transporters which facilitate nitrate uptake and transport in the plants. Ammonium is transported by AMT family transporters which have been reported in Arabidopsis, rice and tomato etc. Though the genes for these transporters have been identified and some studies on their regulations have been performed, more insights are needed to manipulate these N-transporters in favor of enhanced NUE in plants.

**Keywords:** ammonia assimilation, high affinity ammonium transporters, high affinity nitrate transporters, low affinity ammonium transporters, low affinity nitrate transporters, nitrate assimilation, nitrogen use efficiency, nutrient uptake, transporter genes, transport proteins.

## 1. INTRODUCTION

Nitrate and ammonium nutrient nitrogen forms are taken up by plants through the membrane bound transport proteins, located primarily in roots, which are present ubiquitously in the plant kingdom (Aslam *et al.*, 1993; Glass and Siddiqi, 1995; Wang *et al.*, 1998; Britto and Kronzucker, 2001; Ludewig *et al.*, 2002; Glass, 2003; Suenaga *et al.*, 2003; Sonoda *et al.*, 2003; Muños *et al.*, 2004; Souza and Fernandes, 2006; Jaiwal and Singh, 2006; Loque *et al.*, 2005, 2007, Babourina *et al.*, 2007; Li *et al.*, 2007).

The plants have evolved unique capabilities to convert these inorganic N forms into amino acids, which subsequently get incorporated into proteins, nucleic acids, pigments, and secondary metabolites etc. (Singh, 1995; Miflin and Habash, 2002;

---

© CAB International 2008. *Plant Membrane and Vacuolar Transporters* (eds P.K. Jaiwal, R.P. Singh and O.P. Dhankher).

Masclaux-Daubresse et al., 2006). Both inorganic and organic forms of nitrogen can be utilized in the active plant metabolism or can be stored in the vacuoles as reserve nitrogen. A significant portion of soil N is lost by leaching or volatilization. (Wang et al., 1993; Schiltz et al., 2004; Loque et al., 2005; Singh, R.P. and coworkers, unpublished results).

Though ammonia is librated initially from the degradation of organic matter in soil, due to natural soil and microbial activities, it is oxidized to nitrate, thus, nitrate is considered to be the most abundant N source available to the plants through rhizosphere uptake in cultivated agricultural lands. Ammonium is the predominant N source in waterlogged or other anerobic soil ecosystems (Fernandes and Rossiello, 1995; Souza and Fernandes, 2006). Nitrogen use efficiency (NUE) in plants depends on many factors e.g. soil and environmental factors regulating leaching, run-off and volatilization losses, uptake and transport systems, activation of assimilation processes, and storage capabilities etc. The uptake of nitrate and ammonia is largely driven by activation and efficiency of uptake, transport and assimilation/ storage related genes and proteins which may get differentially regulated in different plants under the different agroclimatic conditions. It is also affected by the root morphology and root biomass of a plant, plant growth rates and level of the available N in the rhizosphere. The N uptake efficiency is the sum of the total N (TN) taken up by the plants until maturity, to the fertilizer N (FN), and the ratio between grain weight (GW) and total N (TN) (Finnemann and Schjoerring, 2000; Souza and Fernandes, 2006).

Thus NUE is TN/FN + GW/TN = GW/FN.

Looking at the significance of nitrogen uptake and transport in enhancing plant productivity and grain quality based on an improved NUE, the understanding of various proteins and genes involved in the uptake and transport of $NO_3^-$ and $NH_4^+$ is very important. This chapter is precisely focused on certain recent findings which can be very useful in designing NUE in plants.

## 2. BIOAVAILABILITY OF INORGANIC NITROGEN FORMS

The major reservoir of nitrogen is lithosphere where it is formed in primary igneous rocks (Williams and Haynes, 1995). Only a small portion of this nitrogen is present in the soil. The nitrogen is made available to plants in the form of $NO_3^-$ and $NH_4^+$. The process of weathering of the primary rocks to release nitrogen is very slow; consequently little of the lithosphere nitrogen is supplied to the biosphere. The major source of biosphere nitrogen is the atmosphere. Nitrogen is a very mobile element, thus readily moves between the spheres from one chemical form to another form as result of biological, chemical and physical processes. Thus the availability of nitrogen to plants is dependent upon the amount of nitrogen present in the soil as well as on the rate at which the nitrogen cycle occurs through the plant-soil system. Thus, the gain or loss of nitrogen from the soil is considered an important factor in nitrogen fertilization of plants.

A fertile mineral soil contains approximately 7000 kg N ha$^{-1}$ in the plant rooting depth (Mengel and Kirkby, 1987). At least 90% of this nitrogen is organically bound mainly as $NH_2$ in highly complex forms. The organically bound form of nitrogen is released as $NO_3^-$ and $NH_4^+$ by microbial decomposition. It can also lead to incorporation of nitrogen into microbial tissues or into humic substances which are relatively resistant to microbial attack. These two processes of mineralization and immobilization occur simultaneously and form links in the nitrogen cycle of most natural and agricultural ecosystems. Mineral nitrogen in the rhizosphere exists as $NO_3^-$, $NH_4^+$, and sometimes as $NO_2^-$ form in the soil solution as well as exchange sites of soil particles or as $NH_4^+$ held by the clay minerals. Some gaseous nitrogen may get dissolved in soil solution and remains in the soil atmosphere. $NH_3$ is rapidly hydrolyzed to $NH_4^+$ and later on is oxidized to $NO_2^-$ and $NO_3^-$ by *Nitrosomonas* and a limited number of other organisms such as *Nitrosolubus or Nitrospora*. The process of ammonification can be carried out under both the aerobic and anaerobic conditions. But nitrification requires only the aerobic state. Thus, in situations where oxygen supply is restricted, there is a predominance of $NH_4^+$ in the rhizosphere, whereas soils with sufficient oxygen i.e. aerobic soils possess a predominance of $NO_3^-$ in their rhizosphere.

Nitrate is usually the most abundant source of N and in this anionic form it is readily dissolved in soil water and thus becomes very mobile in the soil profile (Miller *et al.*, 2007). Nitrate is generated in aerable soil by the microbial conversion of other soil N forms through $NH_4^+$ and $NO_2^-$, which remain in very low concentration in the soil profile. $NO_3^-$ is lost from the soil by microbial conversion to $N_2$ gas or leaching of soil water carrying dissolved $NO_3^-$. As inhibition of generation of $NO_3^-$ from $NH_4^+$, the conversion of $NO_3^-$ to gaseous $N_2$ also occurs only when oxygen is depleted in the soil and when bacteria can use $NO_3^-$ rather than oxygen for respiration. Therefore, several factors can affect the availability of $NO_3^-$ and $NH_4^+$ in the rhizosphere e.g. soil water, soil aeration, soil temperature, soil microbial populations, soil pH and so on and the form and level of the available rhizosphere N may vary at different places even in the same area. A.J. Miller and coworkers (see Miller *et al.*, 2007) have recently demonstrated that in a transect across four arable fields with 4 m sampling intervals, soil pH varied by 2 pH units and $NO_3^-$ concentration by almost 100 fold. Ammonium concentrations were much lower (10:1 to 20:1) and less variable than $NO_3^-$ with values differing by 10 fold.

## 3. STRATEGIES OF PLANTS FOR NITRATE UPTAKE AND UTILIZATION

Nitrate uptake is an active process which is facilitated through various high affinity and low affinity nitrate transport systems (HATS and LATS) (Glass *et al.*, 1995, 2001, 2002; Zhou *et al.*, 1999; Tischner, 2000; Touraine *et al.*, 2001; Muños *et al.*, 2004; Forde, 2002; Souza and Fernandes, 2006; Remans *et al.*, 2006a,b; Miller *et al.*, 2007; Table 1; Fig. 1).

Both constitutive (cHATs) and inducible (iHATs) high affinity nitrate transport systems have been reported, whereas the low affinity nitrate transport system is considered to be constitutive only. Being an active, regulated and multiphasic transport system, nitrate uptake and transport can be manipulated for an enhanced NUE, if the various external and internal regulatory factors could be understood

**Table 1.** Some characteristics of nitrate transporters

| Transporter | Family | Type | Gene | Source | Possible Functions |
|---|---|---|---|---|---|
| AtNRT 1.1 (CHL 1) | NRT 1 | LATS HATS | AtNRT 1.1 (chl 1) | Arabidopsis thaliana | Expressed and play significant role in young tissues |
| AtNRT 1.2 | NRT 1 | cLATs | AtNRT 1.2 | A. thaliana | Constitutely expressed in root epidermal cells |
| AtNRT 1.3 | NRT 1 | iLATS | AtNRT 1.3 | A. thaliana | $NO_3^-$ induces its expression in leaf but repressed in the roots |
| AtNRT 1.4 | NRT 1 | LATS | AtNRT 1.4 | A. thaliana | May be leaf specific |
| AtNRT 2.1 | NRT 2 | cHATS iHATS | AtNRT 2.1 | A. thaliana | Expresses in leaf petiole; may be involved in $NO_3^-$ accumulation in petiole/leaf |
| NpNRT 2.1 | NRT 2 | LATS | NpNRT 2.1 | Nicotiana tabaccum | $NO_3^-$ transport, considered major contributor to HATS system |
| AtNRT 2.2 | NRT 2 | iHATs | AtNRT 2.2 | A. thaliana Oryza sativa Chlamydomonas | $NO_3^-$ transport |
| AtNRT 2.7 | NRT 2 | iHATs | AtNRT 2.7 | Xenopus laevis oocytes A. thaliana | Controls nitrate content in seeds |
| AtNRT 3.1 | NRT 3 | iHATs | AtNRT 3.1 | A. thaliana | Nitrate transport in roots |
| AtCLCa | CLC | iHATs | AtCLCa | A. thaliana | Transport of nitrate in vacuoles |

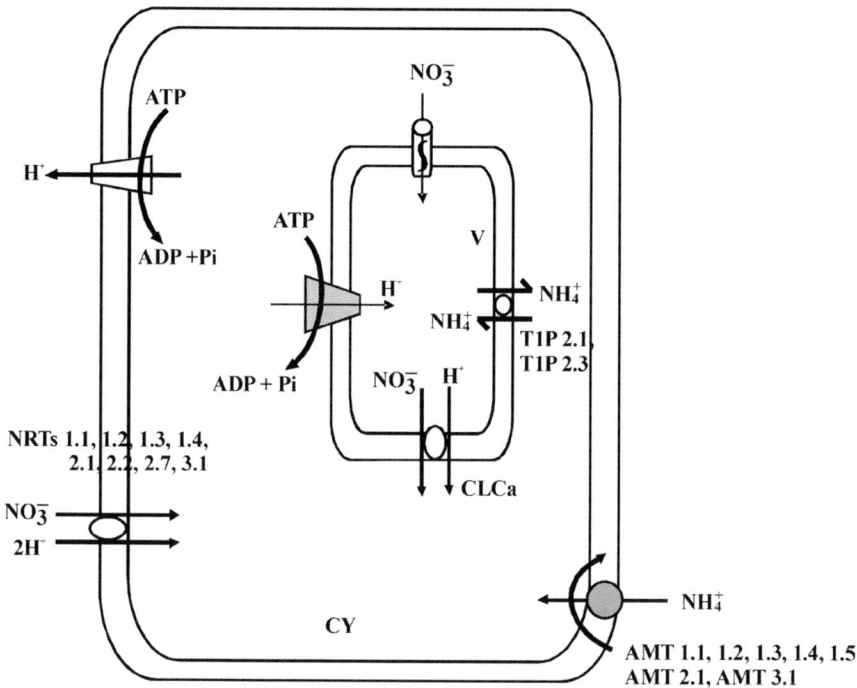

**Figure 1:** Possible localization and site of action of different nitrate and ammonium transporters. CY = cytosol, V = vacuole, NRTs 1.1-3.1, AMT 1.1-3.1, TIP 2.1, 2.3 and CLCa are various nitrate and ammonium transporters (see text for details).

adequately. Like other metabolic processes nitrate transport is also a well co-ordinated plant process, and though understanding of physical, physiological, biochemical and molecular factors involved in its regulation is significant, the successful genetic manipulation for better NUE in plants is very much dependent on the understanding of coordinated responses in plants which get initiated with nitrate signaling pathways involved not only in nitrate uptake and utilization but also in regulating the other coordinated processes e.g. N and C metabolism (see Raghuram et al., 2006; Miller et al., 2007). Occurrence of two kinds of nitrate transporters i.e. HATS and LATS are indicative of the significance of this process in NUE which assures that plants have evolved the $NO_3^-$ uptake system to cope with the varying external $NO_3^-$ concentrations (Aslam et al., 1993; Glass et al., 1995; Forde, 2002; Muños et al., 2004). Nitrate influx is driven by the $H^+$ gradient and can work against an electrochemical potential gradient (Vidmar et al., 2000). It indicates that plants have evolved genetic potential over the two constitutive nitrate transport systems i.e. cHATS and LATS to aquire soil nitrate, even if it is available in very low amounts, through the iHATS.

### 3.1. Two nitrate transport systems

The $K_m$ values of iHATS, cHATS and LATS for nitrate are in the ranges of 13-79 µM, 6-20 µM and >1 mM respectively (see Raghuram *et al.*, 2006). Nitrate acts as a regulator of nitrate uptake. The uptake of $NO_3^-$ increases after a lag phase of 30-90 minutes which follow a steady state kinetics and being biphasic it reaches a new steady state after 4-6 hours. Hole *et al.* (1990) and Aslam *et al.* (1993) have used RNA and protein synthesis inhibitors to indicate that induction of the iHATS involves gene expression and the synthesis of new transporter proteins. Studies on nitrate reductase (NR) deficient mutants of Arabidopsis and *Nicotiana plumbaginifolia* have demonstrated that NR activity is not required for nitrate transport in higher plants, unlike in lower organisms, and it is independent of nitrate reduction (Unkles *et al.*, 2004). Similarly, evidence has been provided that the inducer of iHATS is indeed nitrate ion and not its downstream metabolites (Krapp *et al.*, 1998; Lejay *et al.*, 1999). Nitrate transporters are often regulated by feedback inhibition by the downstream metabolites, e.g. $NH_4^+$ and amino acids, possibly through the transcriptional and post translational controls (Muller and Touraine, 1992; Fraisier *et al.*, 2000; Vidmar *et al.*, 2000; Glass *et al.*, 2002; Krapp *et al.*, 2002; Orsel *et al.*, 2002). Light is also known to regulate $NO_3^-$ uptake along with diurnal changes. The involvement of light may be related to the photosynthesis and the availability of energy and sugars required for the NUE (Le Bot and Kirkby, 1992; Delhon *et al.*, 1995; Sehtiya and Goyal, 2000).

An up-regulation of *AtNRT1.1* gene expression by auxin (Guo *et al.*, 2002) suggests involvement of PGRs in nitrate transport. The HATS and LATS of nitrate have also been found induced by deficiencies of other nutrients e.g. phosphate, potassium and iron (Wang *et al.*, 2001, 2002). The genome wide analyses of nitrate responsive genes indicate that NUE is an extensively coordinated process in which several other physiological processes are involved at whole plant level.

Structurally the iHATS is a multicomponent system encoded by genes of the *NRT2* family of transporters. Liu and Tsay (2003) have demonstrated a very interesting feature of the nitrate transporters. They have shown a dual affinity transporter in Arabidopsis, i.e. AtKuP1 and CHL1 or AtNRT1.1. The AtNRT1.1 function as a high affinity nitrate transporter after phosphorylation at threonine residue, whereas it acts as a low affinity transporter on dephosphorylation. Such a flexibility in mode of function of N-transporters indicates that the plant has evolved a strategy to utilize every bit of nutrient N available in its rhizosphere. In addition the AtNRT1.1 is known to belong to the PTR and POT family of $H^+$/oligopeptide transporters which can co-transport oligopeptides, amino acids, $NO_3^-$ and chlorate. Muños *et al.* (2004) have demonstrated that one nitrate transporter (NRT1.1) is involved in the regulation of another nitrate transporter (NRT2.1) in Arabidopsis. It has been shown that $NO_3^-$ transport is often facilitated by more than one transport system which work in coordination with each other as co-transporters. Low affinity transport systems get activated when external $NO_3^-$ concentration is >0.5 mM. $NO_3^-$ transport in the environment of lower external $NO_3^-$ concentrations is facilitated by the HATS

(Glass and Siddiqi, 1995). It has been demonstrated that the HATS saturates with an increase in $NO_3^-$ content at low external nitrate concentration, exhibiting Michaelis Menten Kinetics, where as in the case of LATS [($NO_3^-$) outside >0.5 mM] the $NO_3^-$ uptake velocity increases linearly with an increase in external nitrate concentration, and it does not show saturation kinetics. However, LATS mediated $NO_3^-$ transport is not a passive transport as more than 100 times cytosolic $NO_3^-$ concentration will be required to facilitate a passive $NO_3^-$ uptake at 2 mM external $NO_3^-$ concentration and -110 mV membrane potential (Crawford, 1995).

Since HATS has a constitutive component and an inducible component which can be expressed simultaneously, it appears that they are synthesized by different genes and the regulatory gene(s) may also work independently. For example, in barley, iHATS activity has been shown to increase up to 30x more than cHATS as a response to increase in external $NO_3^-$ supply (see Souza and Fernandes, 2006).

The rate of $NO_3^-$ uptake varies from species to species and is also influenced by the rhizosphere microenvironment especially availability of external $NO_3^-$, soil temperature, soil pH and other factors. An uptake rate of 4-700 µmol/g/h has been demonstrated for the LATS in Arabidopsis. Nitrate uptake seems to be a thermodynamically active process in which a co-transport of $2H^+/NO_3^-$ occurs. A depolarization followed by repolarization of the membrane potential ($\Delta\Psi$) also occurs as a result of the nitrate uptake (see Souza and Fernandes, 2006). In some situations a hyperpolarization of membrane potential can also occur, which is due to the stimulus that a depolarization has on the $H^+$-ATPase activity. It appears that a proton motive force acts on the $NO_3^-$ transport across the membranes which is also evident from an increase in $NO_3^-$ transport rates as a decreased external pH. It is considered that the first step necessary for $NO_3^-$ uptake is an active extrusion of $H^+$ through the $H^+$-ATPases, which creates a proton gradient ($\Delta\mu H^+$) across the plasma membrane. In the case of the $1H^+$:1ATP relationship, 2 moles of ATP are consumed for each $NO_3^-$ taken up. However, the concentration of anions inside and outside the membrane may influence this calculation.

### 3.2. Genetic regulation of nitrate transport systems in plants

The genes encoding $NO_3^-$ transporters have been found in two families; namely, *NRT1* and *NRT2* (Forde, 2000; Miller *et al.*, 2007). The *NRT2* family of $NO_3^-$ transporter genes in *Arabidopsis thaliana* includes seven genes (Orsel *et al.*, 2002), but the *NRT1* family has been reported to have four genes only, which belong to the large *PTR* family of transporters, with 51 members (Okamoto *et al.*, 2003; Muños *et al.*, 2004). Both families transport $NO_3^-$ together with a proton ($H^+$) in a symport mechanism that is driven by the pH gradient across membranes. Although other membrane proteins that can transport nitrate have been identified, these two families are those which are best characterized so far. There is no evidence as yet to exclude or include that $NO_3^-$ transporters are also encoded by other *PTR* genes than the four *NRT1* investigated initially. Out of the 58 putative transporters (seven *NRT2* and 51 *PTR*), only three *NRT 1.1*, *NRT 1.2* and *NRT 2.1*) have been functionally characterized *in planta* and shown to take part in the $NO_3^-$ uptake from the external medium. NRT 1.1 (also called CHL1) was the first $NO_3^-$ transporter identified in plants (Tsay et al., 1993), which was initially believed to be a $NO_3^-$ inducible low affinity transporter (Tsay et al., 1993; Touraine and Glass, 1997; Huang et al., 1999). Though the first eukaryotic $NO_3^-$ transporter gene was isolated over 15 years ago from the fungus, *Aspergillus nidulans* (Unkles et al., 1991). NRT 1.2 was also characterized as a low affinity $NO_3^-$ transporter with a constitutive expression independent to the external $NO_3^-$ (Huang et al., 1999). Thus both NRT 1.1 and NRT 1.2 were initially considered to belong to the $NO_3^-$ LATS in *A. thaliana* (Crawford and Glass, 1998). The studies on these two transporters of the same family further indicated that these transporters i.e. NRT 1.1 and NRT 1.2 do not perform the same function and get regulated differentially. The NRT 1.1 mutants appear to be strongly defective in LATS activity only when plants are supplied with $NO_3^- + NH_4^+$ as a combined N-source (Touraine and Glass, 1997; Crawford and Glass, 1998; Muños et al., 2004), whereas antisense lines of *NRT 1.2* also display a significant reduction in LATS activity, if $NO_3^-$ is supplied as sole N source (Huang et al., 1999). The reports, however, indicate that NRT 1.1 is a dual affinity transporter, belonging to both HATS and LATS (Wang et al., 1998; Liu et al., 1999). The phosphorylation of NRT 1.1 triggered by limited external $NO_3^-$ availability, is responsible for its shift from low to high affinity transporter, adapting the functional property of a transporter to the resource level in the root environment (Liu and Tsay, 2003). The *NRT 1.1* is strongly expressed in nascent organs of both root and shoot and seems to play a crucial role in the early phase of development of these young tissues (Guo et al., 2001). Mutant studies show that *NRT 1.1* mutants display altered root architecture i.e. reduced growth of primary and secondary roots, in the absence of added $NO_3^-$ in the external medium. This suggests an alternative function of *NRT1.1*, independent of $NO_3^-$ transport (Guo et al., 2001). It has also been shown that the mutation of *NRT1.1* may lead to a lower sensitivity to drought (Guo et al., 2003).

Remans et al. (2006b) have recently reported that in Arabidopsis, *NRT 1.1*, which is localized at the forefront of soil exploration by roots, is a key component of the nitrate sensing system that enables the plant to detect and exploit $NO_3^-$ rich plates. They have demonstrated that the inability of *NRT 1.1* deficient mutants to promote increased lateral root proliferation in the $NO_3^-$ rich zone impairs the efficient acquisition of $NO_3^-$ and leads to slower plant growth.

The *NRT2.1* has been shown to encode a major component of the $NO_3^-$ HATS (Lejay et al., 1999, 2003; Filleur et al., 2001; Nazoa et al., 2003). Its expression is induced by $NO_3^-$, repressed by reduced nitrogen metabolites, and stimulated by photosynthesis. Though there may be more $NO_3^-$ transporter gene(s) and protein(s) to mediate all the $NO_3^-$ transport steps in plants, these three transporters have central importance in the $NO_3^-$ uptake in *A. thaliana*. A few studies on genetic manipulation of $NO_3^-$ transporters are available which indicate that uptake and transport can be enhanced through the gene technologies. For example, Liu et al. (1999) have over-expressed a dual affinity $NO_3^-$ transporter gene *CHL1* in *Chl* mutant which resulted in recovery of $NO_3^-$ uptake for the constitutive phase of uptake. Fraisier et al. (2000) have also over-expressed a high affinity $NO_3^-$ transporter

## Nitrate and ammonium transporters in plants

*NpNRT 2.1* in tobacco and reported that it was associated with increased nitrate influx under low nitrate condition. However, the total $NO_3^-$ content was similar in the transgenic and non-transgenic tubers. It appears that though the specific $NO_3^-$ transporters can be over-expressed or silenced by tools and techniques of genetic engineering, it may not be sufficient to enhance overall NUE, which operates in co-ordination with various other processes of nitrogen and carbon metabolism and is regulated by many environmental, soil and plant factors. There may be some common regulatory switches in the form of regulatory genes in plants, which may provide more genuine insight on the functional genomics of plant nitrogen acquisition phenomena and can pave a way for more effective genetic improvement in the traits like NUE in plants. At NRT 1.2 is constitutively expressed in root epidermal cells and has a $K_m$ for $NO_3^-$ around 6 mM in oocytes (Huang *et al.*, 1999). Arabidopsis antisense lines of *At NRT 1.2* have shown 50-70% decrease in LATS, whereas that of At NRT 1.1 caused a 45% decrease in LATS. *At NRT 1.3* has been induced in the leaf, but repressed in the root and does not seem to be a significant contributor to LATS (Okamoto *et al.*, 2003). *At NRT 1.4* has a very specific pattern of expression in the leaf petiole where it has a role in $NO_3^-$ accumulation within these tissues (Chiu *et al.*, 2004).

Impaired expression of *AtNRT 2.1* and *At NRT 2.2* genes in Arabidopsis mutants has been found defective in HATS activity (Filleur *et al.*, 2001; Orsel *et al.*, 2004). These genes lie end to end in the Arabidopsis genome and encode proteins 90% identical (Orsel *et al.*, 2002). Similar proteins have also been reported from rice and *Chlamydomonas* (Quesada *et al.*, 2004; Araki and Hasegawa, 2006). At NRT 2.1 has been reported to be major contributor to iHATS and cHATs (Li *et al.*, 2007). In addition, At NRT 3.1 has recently been reported as HATS in root of Arabidopsis (Okamoto *et al.*, 2006).

Unlike the fungal transporters nrt A/B, At NRT 2.1 shows similarity in requirement for a second protein to mediate $NO_3^-$ transport (Orsel *et al.*, 2006, 2007; Okamoto *et al.*, 2006; Miller *et al.*, 2007). The regulation of NRT 2.1 expression has been thoroughly investigated at mRNA level. NRT 2.1 transcript accumulation mainly occurs in epidermis and cortex of the mature root regions (Nazoa *et al.*, 2003), and is affected by various environmental factors e.g. $NO_3^-$, $NH_4^+$, amino acids, light and sugars (Zhou *et al.*, 1999; Nazoa *et al.*, 2003; Lejay *et al.*, 2003). Both At NRT 2.1 and At NAR 2.1 have been shown to be down regulated by $NO_3^-$ itself, through a mechanism independent of the feedback repression exerted by N-metabolites, but specifically triggered by the dual-affinity transporter NRT 1.1 (Muños *et al.*, 2004; Krouk *et al.*, 2006; Miller *et al.*, 2007). A strong positive correlation exists between changes in NRT 2.1 transcript level and $NO_3^-$ HATS activity, suggesting that the transcriptional regulation of NRT 2.1 expression plays a major role in governing root high-affinity $NO_3^-$-uptake. Fraisier *et al.* (2000) over-expressed *NpNRT 2.1* in *Nicotiana plumbaginifolia*, but the plants did not accumulate more $NO_3^-$ than wild type and showed no increased biomass; possibly they lacked a necessary NAR 2-type component and over-expression of both the components of NRT 2.1 may result in a better NUE.

The post translational regulation of $NO_3^-$ transport has been demonstrated recently. Navarro *et al.* (2006) have shown that degradation of a NRT 2 type protein (YNT1) in yeast *Hansenula polymorpha* in the vacuole was associated with the removal of this transporter from the plasma membrane. The YNT1 protein is ubiquitinylated when glutamine is supplied to the cells and is transfered to the vacuole where it is rapidly degraded by a specific proteinase A (Navarro *et al.*, 2006; Miller *et al.*, 2007). A similar mechanism for Arabidopsis At NRT 2.1 degradation has also been suggested (unpublished results of J. Wirth *et al.*, as cited by Miller *et al.*, 2007). The presence of protein kinase C recognition motifs in N- and C-terminal domains of HvNRT 2.1 may suggest that phosphorylation events are involved in regulating At NRT 1.1 and At NRT 2.1 activity in response to environmental stimuli (Forde, 2000; Liu and Tsay, 2003; Miller *et al.*, 2007). Some possible 14-3-3 regulatory sites which are also considered to play a significant role in key N-assimilatory enzymes have been identified in sequence analysis of the NRT 2s (see Miller *et al.*, 2007 for details). Some cellular factors e.g. pH, membrane voltage, membrane potentials and cellular energy levels may also be involved in regulating nitrate transport which is one relatively unexplored area of investigation.

### 3.3. Nitrate efflux from cells and nitrate storage and remobilization from the vacuole

Nitrate efflux is nitrate induced, protein-mediated, passive, saturable and selective for nitrate (Aslam *et al.*, 1996). Aquaporins or anion channels are the possible route for $NO_3^-$ efflux and there may be proton-cotransport and passive mechanism both operating simultaneously (Zhou *et al.*, 1998, 2000a,b). $NO_3^-$ is assimilated in roots, however, a very high (10-30 mM) nitrate concentration has been found in xylem sap in certain conditions which indicates a rapid transport of unassimilated $NO_3^-$ to the aerial tissues (Miller and Smith, 1996).

Nitrate storage in the vacuole is important for osmotic balance and as N reserve. This ability is found in land plants but not in the aquatic plants (see Miller *et al.*, 2007; Fan *et al.*, 2007). At CLCa transporters have been reported to mediate accumulation of nitrate in the vacuole and behave as a $2NO_3^-/H^+$ antiporter (De Angeli *et al.*, 2006). Increased $NO_3^-$ storage in vacuoles may help the plants to assimilate soil $NO_3^-$ in atypical environments. Under salinity stress high $NO_3^-$ storage plants may be helpful in scavenging chloride in the vacuole with the help of some $NO_3^-$ transporters (Miller *et al.*, 2007).

As a signal molecule in addition to a nutrient, $NO_3^-$ may modulate plant metabolism and development (Crawford and Glass, 1998; Raghuram *et al.*, 2006; Miller *et al.*, 2007). At NRT 2.1 and At NRT 1.1 have been proposed as signal transducers or sensors for $NO_3^-$ availability which may cause enhanced lateral root development, root colonization, breaking of seed dormancy, enhanced NUE and co-ordinated growth and productivity (Little *et al.*, 2005; Alboresi *et al.*, 2005; Remans *et al.*, 2006a,b). Chopin *et al.* (2007) have recently demonstrated that an Arabidopsis At NRT 2.7 nitrate transporter controls nitrate content in seeds.

## 4. STRATEGIES FOR AMMONIUM UPTAKE AND UTILIZATION

Being another available source of nitrogen in the plant's rhizosphere especially during the low $O_2$ condition, $NH_4^+$ is also taken up by plants, transported through the membrane bound specific transporters and is assimilated into the primary amino acids (Suenaga et al., 2003; Sonoda et al., 2003; Muños et al., 2004; Loque et al., 2005; Souza and Fernandes, 2006; Jaiwal and Singh, 2006; Yuan et al., 2007; Babourina et al., 2007, Fig. 1, Tables 2 and 3). Both high affinity transport system (HATs) as well as low affinity transport system (LATs) have been reported to be involved in the transport of $NH_4^+$ in plants across membranes (Glass and Siddiqi, 1995; Suenaga et al., 2003; Muños, et al., 2004). Like *NRT1* and *NRT 2* families of nitrate transporters, *AMT 1* and *AMT 2* families of genes have been reported for $NH_4^+$ transporters (von-Wiren et al., 2000). After Ninnemann et al. (1994) identified the gene encoding a high affinity ammonium transporter *At AMT 1,1* from *A. thaliana* many *AMT1* homologous have been reported from Arabidopsis, rice and tomato, which indicate that the *AMT1* gene family in plants consists of at least 3 to 5 members (Suenaga et al., 2003). Another type of ammonium transporter AMT 2;1 was isolated, subsequently from *A. thaliana* by Sohlenkamp and coworkers in 2000 (Sohlenkamp et al., 2000). Suenaga et al. (2003) reported OsAMT 3;1 as a $NH_4^+$ transporter in rice which is homologous to AMT 2,1. They subsequently reported a series of functional ammonium transporter genes in rice roots which were designated as *Os AMT 1;1, OsAMT 1;2* and *OsAMT 1;3* (Sonoda et al., 2003). Though plasma-membrane transporter mediated ammonium uptake has been studied quite extensively, little is known about the remobilization and compartmentalization and sub-cellular export of ammonium. Loque et al. (2005, 2007) have recently isolated and characterized two Arabidopsis genes (*AtTIP2;1* and *AtTIP2;3*) which encode aquaporins of tonoplast intrinsic protein sub-family. They have demonstrated in transgenic plants that *AtTIPs* can mediate extra cytosolic transport of methyl ammonium or ammonium across the tonoplast membrane and thus participate in vacuolar ammonium compartmentation. These genes have shown tolerance to the toxic $NH_4^+$ analog, methyl ammonium in yeast and may be involved in detoxification of $NH_4^+$ accumulation in leaves under the consistent supply or release of $NH_4^+$ in the plant tissues.

Ammonium uptake is also shown to be a biphasic phenomenon. A high affinity transport system for $NH_4^+$, which gets induced at low $NH_4^+$ in rhizosphere shows saturation kinetics between 0.1 to 1.0 mM in maize, rye and barley (Fried et al., 1965; Souza and Fernandes, 2006). The uptake of $NH_4^+$ has been shown to be faster than that of $NO_3^-$ over a wide range of environmental conditions (Fernandes and Rosiello, 1995). The availability of $NH_4^+$ uptake in the medium and that of other competitive ions have shown to modulate the Km and $V_{max}$ for the $NH_4^+$ uptake in rice, wheat and barley (Baptista et al., 2000; Souza and Fernandes, 2006).

## 5. THE AMT SYSTEM, $NH_4^+$ UPTAKE/TRANSPORT AND REGULATION OF $NH_4^+$ TRANSPORTER GENES

Many workers have reported the AMT family transporters in Arabidopsis and their

**Table 2.** Some characteristics and possible functions of plasma membrane ammonium transporters

| Transporter | Family | Type | Gene | Source | Characters and Possible Functions |
|---|---|---|---|---|---|
| AtAMT 1:1 | AMT 1 | HATS | AtAMT 1:1 | *Arabidopsis thaliana* | • Rapid transcription in N-depleted environment |
| | | | | | • Involved in $NH_4^+$ uptake and transport across the plasma membrane |
| | | | | | • Glutamine down regulates its transcription |
| | | | | | • Active in all plant tissues |
| AtAMT 1:2 | AMT 1 | HATS(?) | AtAMT 1:2 | *A. thaliana* | Transcribe only in roots |
| AtAMT 1:3 | AMT 1 | HATS | AtAMT 1:3 | *A. thaliana* | Maintained constant in N-depletion |
| AtAMT 1:4 | AMT 1 | HATS | AtAMT 1:4 | *A. thaliana* | |
| AtAMT 1:5 | AMT 1 | HATS | AtAMT 1:5 | *A. thaliana* | |
| OsAMT 1:1 | AMT 1 | HATS | OsAMT 1:1 | *Oryza sativa* | Known to be involved in acquisition and transport of $NH_4^+$ in plasma membrane in various conditions |
| OsAMT 1:2 | AMT 1 | HATS | OsAMT 1:2 | *O. sativa* | |
| OsAMT 1:3 | AMT 1 | HATS | OsAMT 1:3 | *O. sativa* | |
| LeAMT 1:1 | AMT 1 | HATS | LeAMT 1:1 | *Lycopersicon esculentum* | |
| LeAMT 1:2 | AMT 1 | HATS | LeAMT 1:2 | *L. esculentum* | |
| LeAMT 1:3 | AMT 1 | HATS | LeAMT 1:3 | *L. esculentum* | |
| AtAMT 2:1 | AMT 2 | HATS | AtAMT 2:1 | *A. thaliana* | |
| OsAMT 3:1 | AMT 3 | ? | OsAMT 3:1 | *O. sativa* | |

Nitrate and ammonium transporters in plants

Table 3. Some characteristics and possible functions of vacuolar ammonium transporters

| Transporter | Family | Type | Gene | Source | Characters and Possible Functions |
|---|---|---|---|---|---|
| AtTIP 2:1 | TIP 2 | ? | AtTIP 2:1 | Arabidopsis thaliana | Aquaporins of tonoplast intrinsic protein, transport $NH_4^+$ across the tonoplast membrane |
| AtTIP 2:3 | TIP 2 | ? | AtTIP 2:3 | A. thaliana | |

analogs in other plants, which have been involved in the uptake and transport of $NH_4^+$ through the plasma membrane of roots. As many as five AMT transporter genes, i.e. *AtAMT1:1, 1:2, 1:3, 1:4* and *1:5* have been reported in Arabidopsis. *AtAMT 2:1* has been identified as another gene involved in the $NH_4^+$ transport in *A. thaliana* (Souza and Fernandes, 2006). The homologous genes in rice are *OsAMT 1:1, 1:2, 1:3* ( Sonoda *et al.*, 2003) and in the tomato; *LeAMT 1:1, 1:2*, and *1:3* (Glass *et al.*, 2001).

It has been demonstrated that this AMT system of transporters is specific for $NH_4^+$, and $K^+$; $Rb^+$ or $Cs^+$ can't be transported by these proteins. It appears that AMT1 family transporters are high affinity transporters (HATs) as in Arabidopsis; the *AtAMT1* gene gets a rapid transcription in a nitrogen depleted environment which decreases once the N-level increases in the rhizosphere (Gazzarrini *et al.*, 1999). A decrease in the mRNA of *AtAMT1* genes has been shown in the presence of glutamine. *AtAMT 1:1* was shown to be active in all plant tissues whereas *AtAMT 1:2* and *AtAMT 1:3* are transcribed only in roots. The other members of the AMT family of transporters (excluding AtAMT 1:1) may be active in N transport in various phases of N-metabolism in plants, as AtAMT 1:1 activity increases maximally under N-depleted condtions, whereas AtAMT1:2 and AtAMT 1:3 are maintained constant. The existence of several systems of $NH_4^+$ transport which are controlled by various genes indicate that the $NH_4^+$ transport like $NO_3^-$ transport is also a very important process for the survival of plants. Plants seem to utilize even a small amount of $NH_4^+$ available in the soil solution due to mineralization in synergy with the process of $NO_3^-$ assimilation (Kronzucker *et al.*, 1999).

Membrane proteins of the AMT1 and AMT2 sub-families are believed to represent the major pathways for high-affinity ammonium transport in plants (Loque and von Wiren, 2004; Yuan *et al.*, 2007). These proteins mediate uptake of ammonium and the substrate analog methyl ammonium when expressed in yeast (Gazzarrini *et al.*, 1999; Shelden *et al.*, 2001). When expressed in oocytes, AMT1 proteins mediate electrogenic uniport of both the above mentioned substrates in ionic form (Ludewig *et al.*, 2002, 2003; Mayer *et al.*, 2006). Located on the plasma membrane, the AMT transporters are considered to be responsible for cellular ammonium acquisition and ammonium retrieval required due to passive leakage of ammonia across the membrane (Britto et al., 2001; Sohlenkamp et al., 2000; Ludewig et al., 2003; Simon-Rosin et al., 2003; Loque et al., 2005, 2006). Ammonium influx studies revealed that At AMT 1.1 confers approximately 30% of total $NH_4^+$ uptake capacity in Arabidopsis roots, whereas At AMT 1.3 could carry another 30% $NH_4^+$ influx in an additive manner in N-deficient environments (Kaiser et al., 2002; Loque et al., 2006; Yuan et al., 2007).

It has been reported that transcriptional control in response to the N and carbon nutritional status is a major regulatory mechanism for AMTs in plants (Gazzarrini et al., 1999; Lejay et al., 2003; Loque et al., 2006). On resupply of $NH_4^+$ to N-deficient Arabidopsis plants, ammonium influx into roots showed a faster time-dependent repression relative to At AMT 1.1 mRNA levels in roots (Rawat et al., 1999). During an increased ammonium efflux, such rapid decrease in ammonium uptake capacities in root might be required to avoid cellular ammonium toxicity (Kronzucker et al., 2001; Britto and Kronzucker, 2002). Whether this rapid decrease in ammonium uptake capacity is brought about by post-transcriptional on post-translational control remains an open question. A few studies are available regarding the post transcriptional control of plant nutrient transporters by the substrates and downstream metabolites. Yuan et al. (2007) have recently demonstrated that post transcriptional regulation of *At AMT 1* mRNA levels is regulated in a N- and organ-dependent manner and suggest mRNA turnover as an additional mechanism for the regulation of At AMT 1.1 in response to the N nutritional status of plants. Babourina et al. (2007) have suggested that nitrate supply affects ammonium transport in canola roots in which it primarily affects the $NH_4^+$ low-affinity influx system and $NH_4^+$ transport is inversely linked to $Ca^{+2}$ net flux. Accumulation of $NH_4^+$ in Norway spruce seedlings is a storage mechanism of $NH_4^+$ in vacuoles with low pH possibly to protect the seedling against the toxic effects of $NH_4^+$ or $NH_3$ (Aarnes et al., 2007).

## 6. CONCLUSIONS AND FUTURE PROSPECTS

$NO_3^-$ and $NH_4^+$ are two major inorganic N forms in soil which are essentially needed for the synthesis of amino acids, proteins, nucleic acid, pigments and so many other vital metabolites necessary for the survival of all the life forms. Many factors affect the availability of $NO_3^-$ and $NH_4^+$ in soil, though $NO_3^-$ is the predominantly available N form in most of the aerable soil. Nitrogen use efficiency (NUE) is directly related to plant growth and productivity and thus regulation of acquisition of $NO_3^-$ and $NH_4^+$, their transport, storage and remobilization play a major role in determining NUE and plant productivity. Many transporter proteins and their genes, which are not only specific for nitrate or ammonium but also for LATS and HATS, organs and age of plants and targeted to plasma membrane and tonoplast membranes for specific roles e.g. uptake, transport, storage, and remobilization etc. have been characterized and studied. Like other metabolic processes, this essential event in a plant's life is also a complex regulation and more investigations are needed to understand transcriptional and post-transcriptional regulations of these transporters to enhance NUE in plants targeted at high productivity and yield.

## 7. LITERATURE CITED

Aarnes, H., Eriksen, A.B., Petersen, D. and Rise, F. (2007) Accumulation of ammonium in Norway Spruce (*Picea abies*) seedlings measured by *in vivo* $^{14}$N-NMR. *J. Exp. Bot.,* **58**: 929-934.

Alboresi, A., Gestin, C., Leydecker, M.T., Bedu, M., Meyer, C. and Truong, H.-N. (2005) Nitrate, a signal relieving seed dormancy in Arabidopsis. *Plant Cell & Environ.,* **28**: 500-512.

Araki, R. and Hasegawa, H. (2006) Expression of rice (*Oryza sativa* L.) genes involved in high affinity nitrate transport during the period of nitrate induction. *Breeding Sci.,* **56**: 295-302.

Aslam, M., Travis, R.L. and Huffaker, R.C. (1993) Comparative induction of nitrate and nitrite uptake and reduction systems by ambient nitrate and nitrite in intact roots of barley (*Hordeum vulgare* I.) seedlings. *Plant Physiol.,* **102**: 811-819.

Aslam, M., Travis, R.L. and Rains, D.W. (1996) Evidence for substrate induction of a nitrate efflux system in barley roots. *Plant Physiol.,* **112**: 1167-1175.

Babourina, O., Voltchanskii, K., McGann, B., Newman, I. and Rengel, Z. (2007) Nitrate supply affects ammonium transport in canola roots. *J. Exp. Bot.,* **58**: 651-658.

Baptista, J., Fernandes, M.S. and Souza, S.R. (2000) Cinetica de absorcao de amonio e crescimento radicular das cultivares de arroz Aghul;a e Bico Ganga. *Pesq. Agropec. Bras.,* **35**: 1325-1330.

Britto, D.T. and Kronzucker, H.J. (2001) Can unidirectional influx be measured in higher plants? A mathematical approach using parameters from efflux analysis. *New Phytol.,* **150**: 37-47.

Britto, D.T. and Kronzucker, H.J. (2002) $NH_4^+$ toxicity in higher plants. A critical review. *J. Plant Physiol.,* **159**: 567-584.

Chiu, C.-C., Lin, C.-S., Hsia, A.-P., Su, R.-C., Lin, H.-L. and Tsay, Y.-F. (2004) Mutation of a nitrate transporter, At NRT 1.4, results in a reduced petiole nitrate content and altered leaf development. *Plant Cell Physiol.,* **45**: 1139-1148.

Chopin, F., Orsel, M., Dorbe, M.-F., Chardon, F., Truong, H.-N., Miller, A.J., Krapp, A. and Daniel-Vedele, F. (2007) The Arabidopsis ATNRT2.7 Nitrate transporter controls nitrate content in seeds. *Plant Cell,* **19**: 1590-1602.

Crawford, N.M. (1995) Nitrate: nutrient and signal for plant growth. *Plant Cell,* **7**: 859-868.

Crawford, N.M. and Glass, A.D.M. (1998) Molecular and physiological aspects of nitrate uptake in plants. *Trends Plant Sci.,* **3**: 389-395.

De Angeli, A., Monachello, D., Ephritkhine, G., Frachisse, J.M., Gambale, F. and Barbier-Brygoo, H. (2006) The nitrate/proton antiporter At CLCa mediates nitrate accumulation in plant vacuoles. *Nature,* **442**: 939-942.

Delhon, P., Gojon, A., Tillard, P. and Passama, l. (1995) Diurnal regulation of $NO_3^-$ uptake in Soybean plants. I. Changes in $NO_3^-$ influx, efflux and N-utilization in the plant during the day/night cycle. *J. Exp. Bot.,* **46**: 1585-1594.

Fan, X., Jia, L., Li, Y., Smith, S.J., Miller, A.J. and Shen, Q. (2007) Compairing nitrate storage and remobilization in two rice cultivars that differ in their nitrogen use efficiency. *J. Exp. Bot.,* **58**: 1729-1740.

Fernandes, M.S. and Rossiello, R.O.P. (1995) Mineral nitrogen in plant physiology and plant nutrition. *Crit. Rev. Plant Sci.,* **14**: 111-148.

Filleur, S., Dorbe, M.F., Cerezo, M., Orsel, M., Granier, F., Gojon, A. and Daniel-Vedele, F. (2001) An Arabidopsis *T-DNA* mutant affected in NRT2 genes is impaired in nitrate uptake. *FEBS Lett.*, **489**: 220-224.

Finnemann, J. and Schjoerring, J.K. (2000) Post-translational regulation of cytosolic glutamine synthetase by reversible phosphorylation and 14-3-3 protein interaction. *Plant J.*, **24**: 171-181.

Forde, B.G. (2000) Nitrate transporters in plants: Structure and function and regulation. *Biochem. Biophys. Acta*, **1465**: 219-235.

Forde, B.G. (2002) Local and long-range signaling pathways regulating plant responses to nitrate. *Annu. Rev. Plant Biol.*, **53**: 203-224.

Fraisier, V., Gojon, A., Tillard, P. and Daniel-Vedele, F. (2000) Constitutive expression of putative high affinity nitrate transporter in *Nicotiana plumbaginifolia*: Evidence for post-transcriptional regulation by a reduced nitrogen source. *Plant J.*, **23**: 489-496.

Fried, M., Zsoldos, F., Vose, P.B. and Shatokalin, I.L. (1965) Characterizing the $NO_3^-$ and $NH_4^+$ uptake process of rice plant roots by use of $^{15}N$ labeled $NH_4NO_3$. *Physiol. Plant.*, **18**: 313-330.

Gazzarrini, S., Lejay, L., Gojon, A., Ninnemann, O., Frommer, W.B. and Von-Woren, N. (1999) Three functional transporters for constitutive, diurnally regulated and starvation-induced uptake of ammonium into Arabidopsis roots. *Plant Cell*, **11**: 937-948.

Glass, A.D.M. (2003) Nitrogen use efficiency of crop plants : Physiological constrains upon nitrogen absorption. *Crit. Rev. Plant Sci.*, **22**: 453-470.

Glass, A.D.M. and Siddiqi, M.Y. (1995) Nitrogen absorption by plant roots. In: *Nitrogen Nutrition in Higher Plants*. (Eds. Srivastava, H.S. and Singh, R.P.), Associated Pub. Co., New Delhi, India. pp 21-56.

Glass, A.D.M., Brito, D.T., Kaiser, B.N.. Tronzucker, H.J., Kumar, A., Okamoto, M., Raivat, S.R., Siddiqi, M.Y., Silim, M.Y., Vidmar, J.J. and Zhuo, D. (2001) Nitrogen transport in plants with an emphasis on the regulation of fluxes to match plant demand. *J. Plant Nutri. Soil Sci.*, **164**: 199-207.

Glass, A.D.M., Britto, D.T., Kaiser, B.N., Kinghorn, J.R., Kronzucker, H.J., Kumar, A., Okamoto, M., Rawat, S., Siddiqi, M.Y., Unkles, S.E. and Vidmar, J.J. (2002) The regulation of nitrate and ammonium transport systems in plants. *J. Exp. Bot.*, **53**: 855-864.

Guo, F.Q., Wang, R., Chen, M. and Crawford, N.M. (2001) The Arabidopsis dual affinity nitrate transporter gene *AtNRT 1.1(CHL1)* is activated and functions in nascent organ development during vegetative and reproductive growth. *Plant Cell*, **13**: 1761-1777.

Guo, F.Q., Wang, R. and Crawford, N.M. (2002) The Arabidopsis dual affinity transporter *AtNRT 1.1(CHL1)* is regulated by auxin in both shoots and roots. *J. Exp. Bot.*, **53**: 835-844.

Guo, F.Q., Young, J. and Crawford, N.M. (2003) The nitrate transporter *AtNRT 1.1(CHL1)* functions in stomatal opening and contributes to drought stress susceptibility in Arabidopsis. *Plant Cell*, **15**: 107-117.

Hole, D.J., Emran, A.M., Fares, Y. and Drew, M.C. (1990) Induction of nitrate transport in maize roots, and kinetics of influx measured with nitrogen ($N^{13}$). *Plant Physiol.*, **93**: 642-647.

Huang, N.C., Liu, K.H., Lo, H.J.and Tsay, Y.F. (1999) Cloning and functional characterization of an Arabidopsis nitrate transporter gene that encodes a constitutive component of low-affinity uptake. *Plant Cell*, **11**: 1381-1392.

Jaiwal, P.K. and Singh, R.P. (2006) Genetic manipulation of nitrogen assimilation to improve nitrogen use efficiency and yield of plants. In: *Biotechnological Approaches to Improve Nitrogen Use Efficiency in Plants* (Eds, Singh, R.P. and Jaiwal, P.K.), Studium Press, LLC, Houston (USA). pp 257-284.

Kaiser, B.N., Rawat, S.R., Siddiqi, M.Y., Masle, J. and Glass, A.D.M. (2002) Functional analysis of an Arabidosis T-DNA "Knockout" of the high-affinity transporter At AMT 1.1. *Plant Physiol.*, 130: 1263-1275.

Krapp, A., Ferrario-Mery, S. and Touraine, B. (2002) Nitrogen and signaling. In: *Photosynthetic Nitrogen Assimilation and Associated Carbon and Respiratory Metabolism* (Eds. Foyer, C.H. and Noctor, G.) Kluwer Acad. Publishers, Netherlands. pp 205-225.

Krapp, A., Fraiser, V., Scheible, W.R., Quesada, A., Gojon, A., Stitt, M., Caboche, M. and Daniel-Vedele, F. (1998) Expression studies of *nert2:1np*, a putative high affinity nitrate transporter. Evidence for its role in nitrate uptake. *Plant J.*, 14: 723-731.

Kronzucker, H.J., Siddiqi, M.Y., Glass, A.D.M. and Kirk, J.D. (1999) Nitrate-ammonium synergism in rice. A sub-cellular flux angles. *Plant Physiol.*, 119: 1041-1046.

Kronzucker, H.J., Britto, D.T., Davenport, R.J. and Tester, M. (2001) Ammonium toxicity and the real cost of transport. *Trends Plant Sci.,* 6: 335-337.

Krouk, G., Tillard, P. and Gojon, A. (2006) Regulation of the high affinity $NO_3^-$ uptake system by NRT1.1 mediated $NO_3^-$ demand signaling in Arabidopsis. *Plant Physiol.*, 142: 1075-1086.

Le Bot, J. and Kirkby, E.A. (1992) Diurnal uptake of nitrate and potassium during vegetative growth of tomato plants. *J. Plant Nutrition*, 15: 247-264.

Lejay, L., Gansel, X., Cerozo, M., Tillard, P., Muller, C., Krapp, A., von-Wiren, N., Daniel-Vedele, F. and Gojon, A. (2003) Regulation of root ion transporters by photosynthesis: Functional importance and relation with hexokinase. *Plant Cell,* 15: 2218-2232.

Lejay, L., Tillard, P., Lepetit, M., Olive, F., Filleur, S., Daniel-Vedele, F. and Gojon, A. (1999) Molecular and functional regulation of two $NO_3^-$ uptake systems by N and C status of Arabidopsis plants. *Plant J.*, 18: 509-519.

Li, W., Wang, Y., Okamoto, M., Crawford, N.M., Siddiqi, M.Y. and Glass, A.D.M. (2007) Dissection of the *AtNRT2:1: AtNRT2.2* inducible high affinity nitrate transporter gene cluster. *Plant Physiol.*, 143: 425-433.

Little, D.Y., Rao, H., Oliva, S., Daniel-Vedele, F., Krapp, A. and Malamy, J.E. (2005) The putative high-affinity nitrate transporter NRT2.1 represses lateral root initiation in response to nutritional cues. *Proc. Natl. Acad. Sci. (USA),* 102: 13693-13698.

Liu, K.H. and Tsay, Y.F. (2003) Switching between the two action modes of the dual affinity nitrate transporter CHL1 by phosphorylation. *EMBO J.*, 22: 1-9.

Liu, K.H., Huang, C.Y. and Tsay, Y.F. (1999) CHL1 is a dual affinity nitrate transporter of Arabidopsis involved in multiple phases of nitrate uptake. *Plant Cell,* 11: 865-874.

Loque, D. and von Wiren, N. (2004) Regulatory levels for the transport of ammonium in plant roots. *J. Exp. Bot.,* 55: 1293-1305.

Loque, D., Yuan, L., Kojima, S., Gojon, A., Wirth, J., Gazzarrini, S., Ishiyama, K., Kahashi, H. and von Wiren, N. (2006) Additive contribution of AtAMT1.1 and AtAMT1.3 to high affinity ammonium uptake across the plasma membrane of nitrogen deficient Arabidopsis roots. *Plant J.*, 48: 522-534.

Loque, D., Ludewig, U., Yuan, L. and von-Wiren, N. (2005) Tonoplast intrinsic proteins AtT1P2;1 and AtT1P2;3 facilitate $NH_3$ transport into the vacuole. *Plant Physiol.*, 137: 671-680.

Loque, D.S., Lalonde, S., Looger, L.L., von-Wiren, N. and Frommer, W.B. (2007) A cytosolic trans-activation domain essential for ammonium uptake. *Nature*, **446**: 195-198.

Ludewig, U., Von-Wiren, N. and Frommer, W.B. (2002) Uniport of $NH_4^+$ by root hair plasma-membrane ammonium transporter LeAMT1:1. *J. Biol. Chem.*, **277**: 13548-13555.

Ludewig, U., Wilken, S., Wu, B., Jost, W., Dlordlik, P., El Bakkourg, M., Marini, A.M., Andre, B., Hamacher, T., Boles, E. *et al.* (2003) Homo and hetero oligomerization of ammonium transporter 1, $NH_4^+$ uniporter. *J. Biol. Chem.*, **278**: 45603-45610.

Masclaux-Daubresse, C., Reisdorf-Cren, M., Pageau, K., Lelandais, M., Grandjean, O., Kronenberger, J., Valadier, M.H., Feraud, M., Jouglent, T. and Suzuki, A. (2006) Glutamine synthetase-glutamate synthetase pathway and glutamate dehydrogenase play distinct roles in the sink-source nitrogen cycle in tobacco. *Plant Physiol.*, **140**: 444-456.

Mayer, M., Schaaf, G., Mouro, L., Lopez, C., Colin, Y., Neumann, P., Cartron, J.P. and Ludewig, U. (2006) Different transport mechanisms in plant and human AMT/Rh-type ammonium transporters. *J. Gen. Physiol.*, **127**: 133-144.

Mengel, K. and Kirkby, E.A. (1987) *Principles of Plant Nutrition.* International Potash Inst., Berne.

Miflin, B.J. and Habash, D.Z. (2002) The role of glutamine synthetase and glutamate dehydrogenase in nitrogen assimilation and possibilities for improvement in nitrogen utilization of crops. *J. Exp. Bot.*, **53**: 979-987.

Miller, A.J. and Smith, S.J. (1996) Nitrate transport and compartmentation. *J. Exp. Bot.*, **47**: 843-854.

Miller, A.J., Fan, X., Orsel, M., Smith, S.J. and Wells, D.M. (2007) Nitrate transport and signalling. *J. Exp. Bot.*, **58**: 2297-2306.

Muller, B. and Touraine, B. (1992) Inhibiton of $NO_3^-$-uptake by various phloem translocated amino acids in Soybean seedlings. *J. Exp. Bot.*, **43**: 617-623.

Muños, S., Cazettes, C., Fizames, C., Gaymard, F., Tillard, P., Lepetit, M., Lejay, L. and Gojon, A. (2004) Transcript profiling in the *Chl;1-5* mutant of Arabidopsis reveals a role of nitrate transporter *NRT 1.1* in the regulation of another nitrate transporter *NRT 2.1*. *Plant Cell,* **16**: 2433-2447.

Navarro, F.J., Machin, F., Martin, Y. and Siverio, J.M. (2006) Down regulation of eukaryotic nitrate transporter by nitrogen-dependent ubiquitinylation. *J. Biol. Chem.*, **281**: 13268-13274.

Nazoa, P., Zhuo, D., Glass, A.D.M. and Touraine, B. (2003) Regulation of the nitrate transporter gene *AtNRT 2.1* in *Arabidopsis thaliana*: Response of nitrate, amino acids and development stage. *Plant Mol. Biol.*, **52**: 689-703.

Ninnemann, O., Jauniaux, J.C. and Frommer, W.B. (1994) Identification of a high affinity $NH_4^+$ transporter from plants. *EMBO J.*, **13**: 3464-3471.

Okamoto, M., Vidmar, J.J. and Glass, A.D.M. (2003) Regulation of *NRT1* and *NRT2* gene families of *Arabidopsis thaliana*: Response to nitrate provision. *Plant Cell Physiol.*, **44**: 304-317.

Okamoto, M., Kumar, A., Li, W., Wang, Y., Siddiqi, M.Y., Crawford, N.M. and Glass, A.D.M. (2006) High affinity nitrate transport in roots of Arabidopsis depends on expression of the NAR-2 like gene AtNRT 3.1. *Plant Physiol.*, **140**: 1036-1046.

Orsel, M., Krapp, A. and Daniel-Vedele, F. (2002) Analysis of *NRt2* nitrate transporter family in *Arabidopsis*: Structure and gene expression. *Plant Physiol.*, **129**: 886-896.

Orsel, M., Eulenburg, K., Krapp, A. and Daniel-Vedele, F. (2004) Disruption of the nitrate transporter gene *AtNRT 2.1* and *AtNRT 2.2* restricts growth at low external nitrate concentration. *Planta*, **219**: 714-721.

Orsel, M., Chopin, F., Leleu, O., Smith, S.J., Krapp, A., Daniel-Vedela, F. and Miller, A.J. (2006) Characterization of a two component high affinity nitrate uptake system in Arabidopsis, physiology and protein-protein interaction. *Plant Physiol.*, **142**: 1304-1317.

Orsel, M., Chopin, F., Leleu, O., Smith, S.J., Krapp, A., Daniel-Vedele, F. and Miller, A.J. (2007) Nitrate signaling and the two component high affinity uptake system in Arabidopsis. *Plant Signaling and Behaviour*, **2**: 4.

Quesada, A., Galván, A. and Fernández, E. (1994) Identification of nitrate transporter genes in *Chlamydomonas reinhardtii*. *The Plant J.*, **5**: 407-419.

Raghuram, N., Pathak, R.R. and Sharma, P. (2006) Signaling and molecular aspects of N-use efficiency in higher plants. In: *Biotechnological Approaches to Improve Nitrogen Use Efficiency in Plants* (Eds. Singh, R.P. and Jaiwal, P.K.) Studium Press LLC, Houston, USA, pp 19-40.

Rawat, S.R., Silim, S.N., Kronzucker, H.J., Siddiqi, M.Y. and Glass, A.D.M. (1999) At AMT 1 gene expression and $NH_4^+$ uptake in roots of *Arabidopsis thaliana* evidence for regulation by root glutamine levels. *Plant J.*, **19**: 143-152.

Remans, T., Nacry, P., Pervent, M., Girin, T., Tillard, P., Lepetit, M. and Gojon, A. (2006a) A central role for the nitrate transporter NRT 2.1 in the integrated morphological and physiological responses of the root system to nitrogen limitation in Arabidopsis. *Plant Physiol.*, **140**: 909-921.

Remans, T., Nacry, P., Pervent, M., Filleur, S., Diatloff, E., Mpunier, E., Tillard, P., Forde, B.G. and Gojon, A. (2006b) The Arabidopsis NRT1.1 transporter participants in the signaling pathway triggering root colonization of nitrate-rich patches. *Proc. Natl. Acad. Sci.* (USA), **103**: 19206-19211.

Schiltz, S., Gallardo, K., Huart, M., Negroni, L., SommererN. And Burstin, J. (2004) Proteome reference maps of vegetative tissues in pea. An investigation of nitrogen metabolization from leaves during seed filling. *Plant Physiol.*, **135**: 2241-2260.

Sehtiya, H.L. and Goyal, S.S. (2000) Comparative uptake of nitrate by intact seedlings of C-3 (barley) and C-4 (Corn) plants: Effects of light and exogenously supplied glucose. *Plant and Soil*, **227**: 185-190.

Shelden, M.C., deBruxelles, G.L., Whelan, J., Ryan, P.R., Howitt, S. and Udvardi, M. (2001) Arabidopsis ammonium transporter, AtAMT 1.1 and AtAMT 1.2 have different biochemical properties and functional roles. *Plant Soil*, **231**: 151-160.

Simon-Rosin, U., Wood, C.C. and Udvardi, M. (2003) Molecular and cellular characterization of Lj AMT 2.1 ammonium transporter from model legume *Lotus japonius*. *Plant Mol. Biol.*, **51**: 99-108.

Singh, R.P. (1995) Ammonia assimilation. In: *Nitrogen Nutrition in Higher Plants* (Eds. Srivastava, H.S. and Singh, R.P.), Associated Pub. Co., New Delhi, India. pp 189-203.

Sohlenkamp, C., Shelden, M., Howitt, S. and Udwardi, M. (2000) Characterization of AtAMT2, a novel ammonium transporter in plants. *FEBS Lett.*, **467**: 273-278.

Sonoda, Y., Ikeda, A., Satomi, S., von-Wiren, N., Yamaya, T. and Yamaguchi, J. (2003) Distinct expression and function of three ammonium transporter genes (*Os.AMT1; 1-1; 3*) in rice. *Plant Cell Phsyiol.*, **44**: 726-734.

Souza, S.R. and Fernandes, M.S. (2006) Nitrogen-acquisition by plants in a sustainable environment. In: *Biotechnological Approaches to Improve Nitrogen Use Efficiency in Plants* (Eds. Singh, R.P. and Jaiwal, P.K.), Studium Press, LLC, Houston (USA). pp 41-62.

Suenaga, A., Moriya, K., Sonoda, Y., Ikeda, A., von-Wiren, N., Hayakawa, T., Yamaguchi, J. and Yamaya, T. (2003) Constitutive expression of a novel type ammonium transporter *OsAMT2* in rice plants. *Plant Cell Physiol.*, **44**: 206-211.

Tischner, R. (2000) Nitrate uptake and reduction in higher and lower plants. *Plant Cell Environment*, **23**: 1005-1024.

Touraine, B. and Glass, A.D.M. (1997) $NO_3^-$ and $ClO_3^-$ fluxes in *CHL1-5* mutant of *Arabidopsis thaliana*. Does CHL 1-5 gene encode a low affinity $NO_3^-$ transporter. *Plant Physiol.*, **114**: 137-144.

Touraine, B., Daniele-Vedele, F. and Forde, B. (2001) Nitrate uptake and its regulation. In: *Plant Nitrogen* (Eds. Lea, P.J. and Mrot-Gaudry, J.F.) Berlin, Springer-Verlag. pp. 1-36.

Tsay, Y.F., Schroeder, J.I., Feldmann, K.A. and Crawford, N.M. (1993) The herbicide sensitivity gene *CHL1* of Arabidopsis encodes nitrate-inducible nitrate transporter. *Cell*, **72**: 705-713.

Unkles, S., Hawker, K., Grieve, C., Campbell, E., Montague, P. and Kinghorn, J. (1991) *crnA* encodes a nitrate transporter in *Aspergillus nidulans*. *Proc. Natl. Acad. Sci. (USA)* **88**: 204-208.

Unkles, S.E., Wang, R., Wang, Y., Glass, A.D.M., Crawford, N.M. and Kinghorn, J.R. (2004) Nitrate reductase activity is required for nitrate uptake into fungal but not in plant cells. *J. Biol. Chem.*, **279**: 28182-28186.

Vidmar, J.J., Zhuo, D., Siddiqi, M.Y., Schjoerring, J.K., Touraine, B. and Glass, A.D.M. (2000) Regulation of high affinity nitrate transporter genes and high affinity nitrate influx by nitrogen pools in roots of barley. *Plant Physiol.*, **123**: 307-318.

von-Wiren, N., Gazzarini, S., Gojon, A. and Frommer, W.B. (2000) The molecular physiology of ammonium uptake and retrieval. *Curr. Opini. Plant Biol.*, **3**: 254-261.

Wang, M.Y., Siddiqi, M.Y., Ruth, T.J. and Glass, A.D.M. (1993) Ammonium uptake by rice roots. II. Kinetics of $^{13}NH_4^+$ influx across the plasmalemma. *Plant Physiol.*, **103**: 1259-1267.

Wang, R., Liu, D. and Crawford, N. (1998) The Arabidopsis CHL1 protein plays a major role in high affinity nitrate uptake. *Proc. Natl. Acad. Sci. (USA)*, **95**: 15134-15139.

Wang, X., Wu, P., Xia, M., Wu, Z., Chen, Q. and Liu, F. (2002) Identification of genes enriched in rice roots of local nitrate treatment and their expression patterns in split-root treatment. *Gene*, **297**: 93-102.

Wang, Y.H., Garvin, D.F. and Kochian, L.V. (2001) Nitrate induced genes in tomato roots: Array analysis reveals novel genes that may play a role in nitrogen nutrition. *Plant Physiol.*, **127**: 345-359.

Williams, P.H. and Haynes, R.J. (1995) Nitrogen in plant environment. In: *Nitrogen Nutrition in Higher Plants* (Eds. Srivastava, H.S. and Singh, R.P.), Associated Pub. Co., New Delhi, India. pp 1-20.

Yuan, L., Loque, D., Ye, F., Frommer, W.B. and Von-Wiren, N. (2007) Nitrogen-dependent post-translational regulation of the ammonium transporter AtAMT1:1. *Plant Physiol.*, **143**: 732-744.

Zhou, D., Okamoto, M., Vidmar, J.J. and Glass, A.D.M. (1999) Regulation of a putative high affinity nitrate transporter (nrt2,1 At) in roots of *Arabidopsis thaliana*. *Plant J.*, **17**: 563-568.

Zhou, J.-J., Theodoulou, F.L., Muldin, I., Ingemarsson, B. and Miller, A.J. (1998) Cloning and functional characterization of a *Brassica napus* transporter which is able to transport nitrate and histidine. *J. Biol. Chem.*, **273**: 12017-12033.

Zhou, J.-J., Fernández, E., Galván, A. and Miller, A.J. (2000a) A high affinity nitrate transport system from *Chlamydomonas* requires two gene products. *FEBS Lett.*, **466**: 225-227.

Zhou, J.-J., Trueman, L., Boorer, K.J., Theodoulous, F.L., Forde, B.G. and Miller, A.J. (2000b) A high affinity fungal nitrate carrier with two transport mechanisms. *J. Biol. Chem.*, **275**: 39894-39899.

# Chapter 4

## PLANT SULFATE TRANSPORTERS

**PETER BUCHNER**
*Plant Science Department, Rothamsted Research, Harpenden AL5 2JQ, UK*
E-mail: peter.buchner@bbsrc.ac.uk

**Abstract**

Sulfur represents one of the essential macronutrients for plant nutrition. The main source of sulfur is as sulfate taken up from the soil by the roots. Cellular influx across the plasma membrane as well as efflux from the vacuoles is mediated by members of a single gene family. This plant sulfate transporter gene family is classified into 5 groups. Groups as well as subtypes differ in kinetics of transport and in patterns of expression indicating different functions in the process of sulfate uptake and whole plant transport and distribution. High affinity transport is responsible for the initial uptake by the root epidermis and cortex. Low affinity transport is involved in vascular transport. Vacuolar efflux plays a role in controlling cytoplasmic sulfate. The expression of 3 sulfate transporter groups is nutritionally regulated. In addition to a basic constitutive-like transport, sulfate uptake, vascular transport and vacuole efflux is up-regulated when sulfate is limiting. Beside this a non-responsive regulatory pathway exists. Sulfate transport is also embedded in the overall regulation of sulfur, nitrogen and carbon assimilation and metabolism.

**Keywords:** sulfate, transporter, membrane, regulation, plant nutrition

## 1. INTRODUCTION

Sulfur (S) is taken up by the roots as sulfate, and the assimilation into cysteine is considered the key entry point of the natural sulfur cycle. Although sulfur is not the most abundant macronutrient, with only 3% to 5% compared to nitrogen, as part of the essential amino acids cysteine and methionine, sulfur is essential for human and animal nutrition. The cysteine and methionine containing proteins predominantly determines the insoluble content of sulfur in plant tissues.

In addition, sulfur is found in bio-molecules like vitamins (vitamin A, B1), cofactors (biotin, thiamine, CoA and S-adenosyl-methionine) and oligopeptides (glutathione and phytochelatins), and a variety of secondary products (glucosinolates in Cruciferae and allyl cysteine sulfoxides in *Allium*; Wittstock and

---

© CAB International 2008. *Plant Membrane and Vacuolar Transporters* (eds P.K. Jaiwal, R.P. Singh and O.P. Dhankher)

Halkier, 2002; Jones *et al.*, 2004). An important function involves the thiol (sulfhydryl) group of cysteine by forming disulfide bonds to maintain protein structure. The thiol of cysteine and glutathione are often involved in redox reactions necessary for redox control and mitigation against oxidative stress in nearly all aerobic organisms including plants (Leustek and Saito, 1999; Mendoza-Cozatl *et al.*, 2005). In some plants heavy metal detoxification in mediated by phytochelatins and by cysteine-rich proteins called metallothionins. Sulfur containing secondary products act as defence compounds against herbivores and pathogenic organisms and also as signalling molecules (Zhao *et al.*, 1998; Brader *et al.*, 2001; Matsubayashi *et al.*, 2002).

Plants are also able to absorb and assimilate atmospheric $H_2S$ or sulfur dioxide ($SO_2$) via leaves but natural atmospheric concentrations do not provide sufficient sulfur for plant growth. Since minerals differ in their solubility, the concentrations of ions of different mineral elements vary widely in the soil solution. For the uptake and assimilation of inorganic nutrients, higher plants have developed multiple plasma membrane-bound transporters that are responsible for the delivery of inorganic ions to the metabolic pathways. In general they are regulated for adaptation to fluctuations of nutrient availability. Under normal conditions the rate of uptake and assimilation of sulfate will depend on the requirement for growth, which can be defined as the rate of S uptake and assimilation required per gram plant biomass produced with time (De Kok *et al.*, 2002). The sulfate requirement will fluctuate during plant development and may vary between species differing in their sulfur needs for growth and the potential sink capacity of secondary sulfur compounds.

After entry into the plant, sulfate is the also major form of transporter as well as stored sulfur. Beside of the initial uptake intracellular movement of sulfate into plastids and the vacuole as well as through consecutive cell layers for long distance transport requires a concerted action of several specific transporters. To adapt to nutrient stress, transporters with different kinetic properties are needed and the ability to control expression as well as transport activity in response to sulfate availabilities. Consequently, sulfate transport is a complex phenomenon consisting of multiple compounds.

## 2. THE PLANT SULFATE TRANSPORTER GENE FAMILY

The mechanism for plasma membrane sulfate transport is proton coupled co-transport. Studies with membrane vesicles and yeast complementation confirmed a pH dependent transport with a probable $3H^+/$ sulfate stoichiometry (Lass and Ullrich-Eberius, 1984; Hawkesford *et al.*, 1993; Smith *et al.*, 1995a). Since the first reported identification of a plant sulfate transporter in *Stylosanthes hamata* (Smith *et al.*, 1995b), many genes encoding sulfate transporters from a variety of plant species have been isolated and characterised (Bolchi *et al.*, 1999; Buchner *et al.*, 2004a,c; Howarth *et al.*, 2003; Smith *et al.*, 1995b, 1997; Vidmar *et al.*, 1999, 2000; Shibagaki *et al.*, 2002; Takahashi *et al.*, 1996, 1997, 1999a, 1999b, 2000; Yoshimoto *et al.*, 2002, 2003; Kataoka *et al.*, 2004a, b; Krusell *et al.*, 2005). With the subsequent analysis of whole plant genomes (The Arabidopsis Initiative, 2000; Feng *et al.*,

Plant sulfate transporters

2002; Goff *et al.*, 2002; Sasaki *et al.*, 2002; Yu *et al.*, 2002) it is clear that sulfate transport in plants is carried out by a complex system of transporters encoded by a large gene family. Alignment and phylogenetic analysis of the 14 Arabidopsis and rice proteins (Fig. 1a) subdivides the plant sulfate transporter family. Due to their membrane localisation and function as a transmembrane transporter, the proteins are highly hydrophobic. The Group 1-4 transporters contain about 12 transmembrane domains. The best conserved region is located in the second predicted transmembrane region. This conserved region is used as a signature pattern, the "Sulfate Transporter Signature" and is found in a number of plants, mammalian, fungal and bacterial proteins, the majority of which are known to be involved in the transport of sulfate across a membrane, as well as some as yet uncharacterised proteins. These proteins form an anion transporter super family (Sandal and Marcker, 1994; Smith *et al.*, 1995a), probably arising from a common ancestor. Substituting the proline residues in the sulfate transporter signature leads to a reduction or loss of transport activity (Shelden *et al.* 2001). A further characteristic sequence, the STAS domain, is found in the C-terminal cytoplasm part of sulfate transporters from eukaryotes and many bacteria, as well as in the bacterial anti-sigma-factor antagonists (ASAs). It was named STAS after sulfate transporters and anti-sigma-factor antagonist (Aravind and Kooni, 2000; Fig. 2b). The group 5 with two smaller proteins with only about 10 transmembrane regions and the lack of the STAS domain is more diverged but clearly related (Hawkesford, 2003).

To date sulfate transport activity has been demonstrated only for all Group 1 and 2 type transporters (Takahashi *et al.*, 2000, Yoshimoto *et al.*, 2002, 2003) and one Group 3 transporter (Krusell *et al.*, 2005). However due to the high homology between the members of the family, it is reasonable to expect that most are involved in sulfate transport.

Apart from the primary plasma membrane influx of sulfate from the soil via root epidermis and cortex, there are further requirements for transmembrane transport of sulfate. Intracellular transport of sulfate is necessary for transport into the vacuole as the main storage pool of the sulfate as well as transport into plastids where sulfate reduction takes place. Internal distribution through the vasculature and cell-to-cell symplastic movement is required to facilitate supply to all organs, tissue and cell types. Different kinetic affinities for sulfate as well as spatial expression patterns and differential expression in response to nutritional status indicate different functions of the individual groups, and also of individual transporters representing functional subtypes (Hawkesford, 2000).

## 3. THE PRIMARY SULFATE UPTAKE

Epstein and co-workers (Leggett and Epstein, 1956; Epstein, 1966) described sulfate uptake which was resolved into a saturable high affinity phase and a non-saturable low affinity phase. Yeast complementation of the first identified and isolated plant sulfate transporter cDNAs from *Stylosanthes hamata* distinguished two different sulfate transporter isoforms, which may be responsible for the dual pattern of high

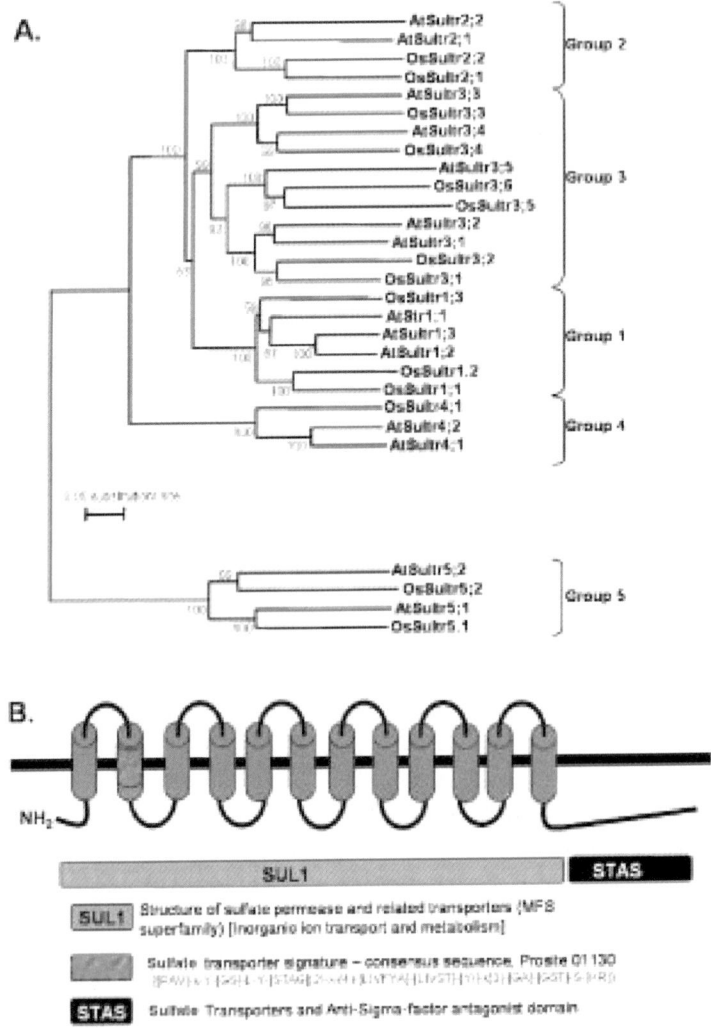

**Figure 1:** (a) Phylogenetic analysis: Neighbour-joining tree (MEGA V. 2.1, Kumar *et al.*, 2001) from the multiple alignment (ClustalX V.1.81, Thompson *et al.*, 1997) of the coding cDNAs of the *Arabidopsis* (AB018695, AB042322, AB049624, AB003591, D85416, D89631, AB004060, AB023423, AB054645, AB061739, AB008782, AB052775, AC018848, AC006053) and *Oryza sativa* sulfate transporter family (from genomic sequences - Feng *et al.*, 2002; Goff *et al.*, 2002; Sasaki *et al.*, 2002; Yu *et al.* 2002, accession and protein ID AF493790, AAN59764.1, BAC98594, AAN59769, AAN59770, NP_921514, AAN06871, AK104831, AK067270, NM_192602, NM_191791, AF493791, BAC05530, BAB03554). (b) Domain structure of plant proton/*sulfate* co-transporter family based on the Pfam (Bateman *et al.*, 2004), CCD (Marchler-Bauer *et al.*, 2003) and PROSITE (Sigrist *et al.*, 2002) databases of protein families and domains.

and low affinity transport (Smith et al., 1995b). Finally, mutant and more detailed expression analysis has confirmed the high affinity component as the initial primary step for the entry of sulfate into the plant root (Shibagaki et al., 2002; Yoshimoto et al., 2002; Maruyama-Nakashita et al., 2003).

The high affinity sulfate transporters ($K_m$s in the range of 1.5-10 µM) belong to the Group 1 sulfate transporter (Smith et al., 1995b, 1997; Shibagaki et al., 2002; Yoshimoto et al., 2002; Howarth et al., 2003). The primary uptake of sulfate by the root is mediated by two different Group 1 sulfate transporter isoforms. Expression of the these Group 1 types is predominantly in the roots and expression of these two high affinity transporters was found in the root tip and root epidermis, including root hairs and in the cortical cells of the mature root (Takahashi et al., 2000; Shibagaki et al., 2002; Yoshimoto et al., 2002; Howarth et al., 2003; Rae and Smith,

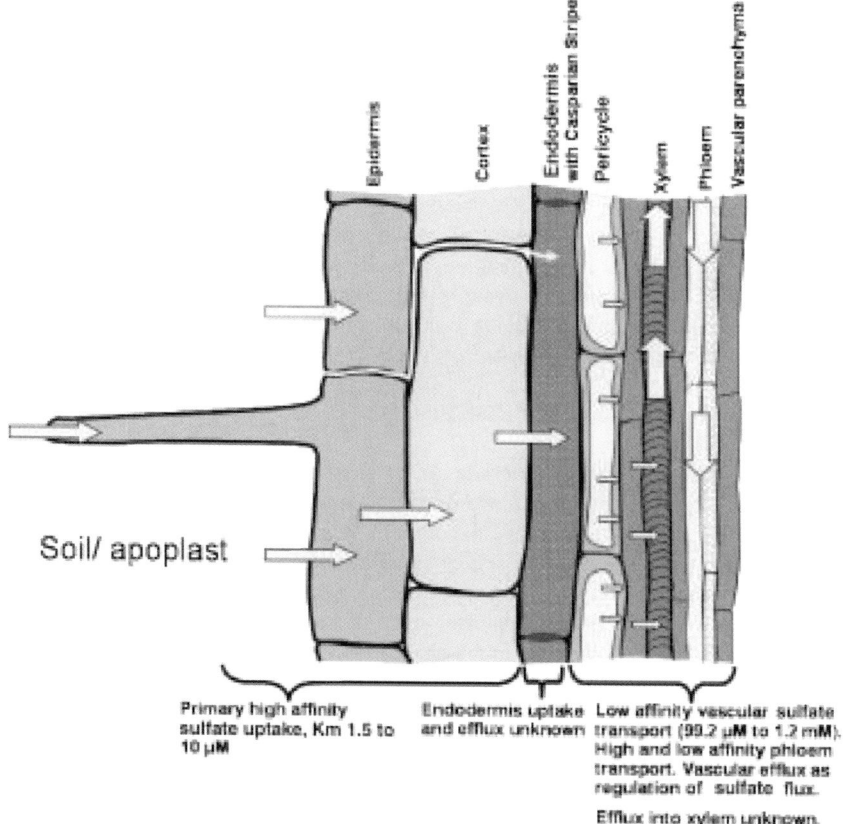

**Figure 2:** Longitudinal model illustration of *sulfate* uptake, movement and transport in different cell layers of Arabidopsis roots. Arrows indicate uptake and potential movement of *sulfate*.

2002). This suggests that all these tissues have the capacity for the primary high-affinity sulfate influx across the plasma membrane into the symplast (Fig. 2 for overview).

The induction of the high affinity Group 1 sulfate transporters has been correlated with increased capacity for uptake in roots in response to limiting S-availability. This response maximises the ability of the plant to capture sulfate from the pedosphere and may be associated with root proliferation. Many studies have shown that the changes in the capacity were paralleled by changes in the steady state contents of mRNAs and protein of the Group 1 sulfate transporters (Clarkson *et al.*, 1992; Smith *et al.*, 1995b, 1997; Takahashi *et al.*, 2000; Hawkesford and Wray, 2000; Shibagaki *et al.*, 2002; Yoshimoto *et al.*, 2002; Howarth *et al.*, 2003). The two distinct Group 1 sulfate transporters differ in their inducibilities in relation to the nutritional status of the plant. In Arabidopsis or *Brassica oleracea* the Sultr1;2 mediates uptake of sulfate under both sulfur replete and sulfur deficient conditions, and expression is relatively insensitive to external sulfate concentrations. The second transporter Sultr1;1 is highly inducible under sulfate limitation but almost absent in non sulfur-stressed plants (Yoshimoto *et al.*, 2002; Buchner *et al.*, 2004a). This dual inducible uptake system was verified by the identification of selenate resistant *(sel)* mutants of Arabidopsis (Shibagaki *et al.*, 2002). The mutation related lesion in the *AtSultr1;2* sulfate transporter isoform restricted the uptake of both, sulfate and its toxic analogue, selenate. No mutations of the inducible isoform were found in the screen. Analysis of another mutation of *AtSultr1;2*, *(sel1-10)*, indicated that AtSultr1;2 serves as a major facilitator for the acquisition of sulfate. Although *AtSultr1;1* expression was up-regulated in the *sel1-10* mutant, reduced growth indicated that the AtSultr1;1 was not able to compensate for the missing AtSultr1;2 (Maruyama-Nakashita *et al.*, 2003).

## 4. SUBCELLULAR SULFATE DISTRIBUTION

Because the plastid is the major site of the assimilatory reductive pathway an effective delivery of sulfate is of fundamental importance to plant sulfur assimilation. Excess sulfate transported into cells accumulates mainly in the vacuoles for cytosolic ion homeostasis, which serves as an internal nutritional storage reservoir (Fig. 3 for overview).

The nature of the plastid transporter has not been unequivocally identified to date, but has been the subject of much speculation (Clarkson *et al.*, 1993; Leustek *et al.*, 2000). One idea is that sulfate and phosphate influxes into the stroma of the plastid are linked (Hampp and Ziegler, 1977; Mourioux and Douce, 1979).

Alternatively existence of a putative ABC-type chloroplast envelope-localised sulfate transporter was described in the model unicellular green alga *Chlamydomonas reinhardtii* (Melis and Chen, 2005). Two of the four nuclear genes encoding this sulfate permease holocomplex, are coding for chloroplast envelope-targeted transmembrane proteins (SulP and SulP2). The sulfate permease holocomplex is postulated to consist of a SulP-SulP2 chloroplast envelope transmembrane heterodimer, flanked by the Sabc and the Sbp proteins on the stroma side and the

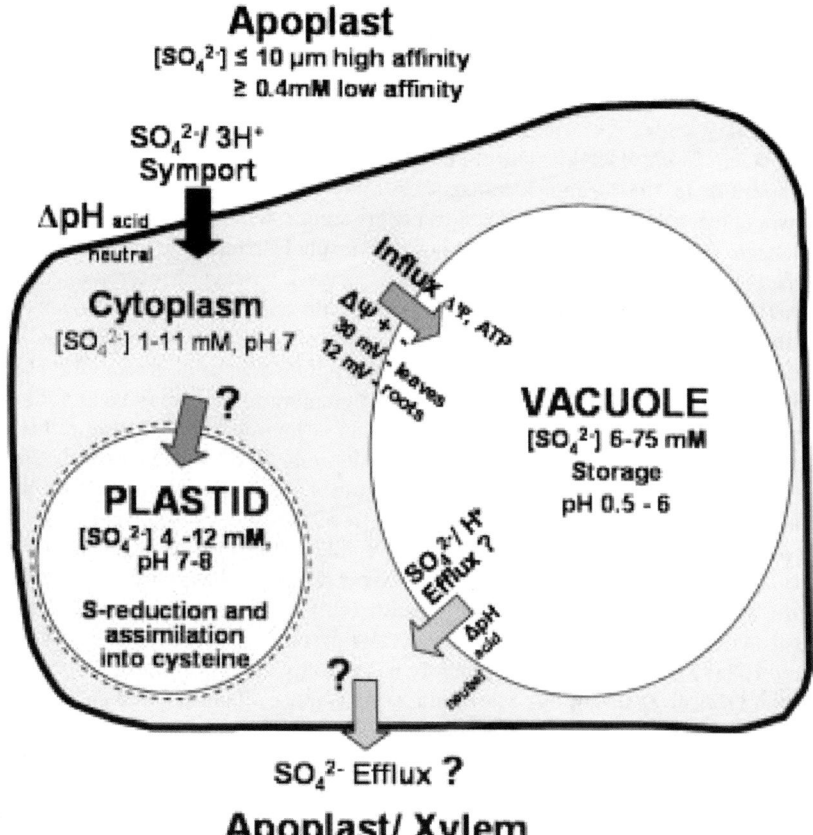

**Figure 3:** Transmembrane *sulfate* movement, driving forces and potential mechanisms adapted and modified from Hawkesford and Wray, 2000: Cytoplasmic, vacuole, plastid *sulfate* concentrations, pH and tonoplast driving forces based on Kaiser *et al.*, 1989 and Miller *et al.*, 2001. Plasma membrane uptake mechanism based on Lass and Ullrich-Eberius, 1984 and Hawkesford *et al.*, 1993. Arrows indicate transmembrane fluxes of *sulfate*.

cytosolic side of the inner envelope, respectively. The mature SulP and SulP2 proteins contain seven transmembrane domains and one or two large hydrophilic loops, which are oriented toward the cytosol. The function of the SulP protein was analysed in antisense transformants of *Chlamydomonas reinhardtii*. Results showed that cellular sulfate uptake capacity was lowered as a consequence of attenuated SulP gene expression in the cell. These putative sulfate transporters are highly homologous to the *CysT* gene product of the cyanobacterial sulfate transport system (Kertesz, 2001), however SulP and SulP2 genes or any of its homologues, have not been identified in vascular plants, e.g. *Arabidopsis thaliana*, although they are encountered in the chloroplast genome of the liverwort *Marchantia polymorpha* (Ohyma *et al.*, 1986).

A transient expression assay of the Arabidopsis *Sultr4;1* in which the putative transit peptide region (N-terminal 99 amino acids) or the truncated transporter (1-672 amino acids) was fused to GFP indicated localisation in the chloroplast (Takahashi *et al.*, 1999a). The high sequence similarity of the Group 4 sulfate transporters to the Group 1 transporters may support a proton-coupled sulfate transport mechanism. Proton linked transport across the chloroplast inner envelope has been reported for phosphoglycerate/ phosphate exchange as well as nitrite transport (Flügge *et al.*, 1983; Shingles *et al.*, 1996), however Mourioux and Douce (1979) found no stimulation of sulfate import into chloroplast by acidic solutions. Further studies of the Arabidopsis Sultr4;1 sulfate transporter indicated that fusion of the whole cDNA encoding the complete 685 amino acid protein to a GFP reporter gene localised the transporter in the tonoplast membrane. This may be an indication that the region encompassing residues 673-685 is involved in targeting rather than the putative plastidal transit peptide (Takahashi *et al.*, 2003; Kataoka *et al.*, 2004b).

It is assumed that the sulfate concentrations in the cytoplasm and in the chloroplast are quite similar (Kaiser *et al.*, 1989). Photosynthesis is known to be sensitive to sulfate: sulfate is a competitive inhibitor of ribulose-1,5-biphosphate carboxylase and inhibits photophosphorylation (Kaiser *et al.*, 1986; Jagendorf and Ryrie, 1971). Clearly cytosolic and plastidic sulfate homeostasis is important. To avoid toxification excess sulfate accumulates as nitrate and phosphate (Martinoia *et al.*, 1981; Mimura *et al.*, 1990) mainly in vacuoles and constitutes a large internal reserve (Kaiser *et al.*, 1989; Martinoia *et al.*, 2000). Transport of sulfate across the tonoplast has been investigated in detail only in yeast and barley mesophyll vacuoles (Hirata *et al.*, 2002; Kaiser *et al.*, 1989). For both, sulfate uptake was stimulated by MgATP and driven by the membrane potential ($\Delta\Psi$) but not by a pH gradient. The uptake of sulfate into barley mesophyll vacuoles follows a biphasic kinetic with a saturable component below 1 mM and a concentration-dependent component above 1 mM (Kaiser *et al.*, 1989; Dietz *et al.*, 1992). These observations suggest a transporter specific for import of sulfate into the vacuoles requiring $\Delta\Psi$ as the main driving force. To date, genes encoding vacuole sulfate influx transporters have not been identified. Preliminary data suggest a tonoplast localisation of one of the Group 5 sulfate transporters (P. Buchner, unpublished). The sequence divergence of the Group 5 compared to the Group 1-4 sulfate transporters suggests that these may be functionally distinct. Whether the Group 5 transporters are involved in vacuole sulfate influx remains to be verified.

For the vacuole to serve as an internal reservoir of sulfate there is a need for a specific efflux system. The inside-acidic pH gradient favours the action of proton/ sulfate cotransporter for efflux. In addition to the verification of the subcellular localisation of the Arabidopsis Group 4 sulfate transporters, the measurement of an increased accumulation of sulfate in isolated vacuoles and decrease of cysteine and glutathione contents of T-DNA mutant plants grown on low sulfate (Takahashi *et al.*, 2003; Kataoka *et al.*, 2004b) favours the group 4 sulfate transporter as vacuole sulfate efflux transporter. The expression of *Sult4;1*-GFP construct in the double knock out mutant is able to complement this phenotype back to wild-type level. The drastic reduction of root sulfate concentrations under sulfate deficiency is accompanied by an upregulation of *Sultr4;1* and *Sultr4;2* in root and leaves of *Brassica napus* and *Brassica oleracea* (Hawkesford et al., 2003; Buchner et al., 2004a). Increased expression of this transporter would maximise vacuolar efflux of stored sulfate under these conditions.

In Arabidopsis *Sultr4;1* and *Sultr4;2* are predominantly expressed in pericycle and xylem parenchyma cells of roots and hypocotyls. Because of the specific potential function for unloading of inorganic nutrients to the xylem stream in the vascular root tissue the functions of the Sultr4;1 and Sultr4;2 vacuolar sulfate transporter may be the control of the flux of sulfate at the root xylem parenchyma cells (Kataoka et al., 2004b). The expression in the vasculature of the hypocotyls would additionally maintain the continuous flow of root-to-shoot transfer of sulfate. Increased transcript of *Sultr4;1* and *Sultr4;2* is also found in leaves under sulfur deficiency (Kataoka et al., 2004b; Buchner et al., 2004a). Besides controlling the flux of sulfate to the xylem in roots, the efflux from vacuoles in leaves would allow a remobilisation of sulfate to overcome S-limited growth conditions.

## 5. LONG-DISTANCE SULFATE TRANSPORT

Translocation of sulfate to other tissues and organs requires several distinct steps. Once inside the symplast, radial transport across the root to the central stele and subsequently unloading into the xylem are necessary for translocation to the shoot. Discharging the sulfate from the xylem vessels into the apoplastic continuum and uptake into the symplast will bring the sulfate to the cells of the target organs and tissues for reduction in the plastids or storage in the vacuoles (Figs 2 and 4 for overview).

Plasmodesmata may allow the radial transport from the epidermis through the cortex and endodermis without traversing plasma membranes. From the stele, sulfate has to be loaded into the xylem. The nature of the efflux system for sulfate from plant cells is unknown. There is no evidence that Group 1 and 2 sulfate transporters are able to act in the reverse direction. Frachisse et al. (1999) reported a voltage-dependent anion channel in Arabidopsis hypocotyl cells, which is activated by sulfate and deactivated by nucleotides. Such a channel may be involved in the delivery to the vascular system, especially to the xylem from vascular parenchyma cells or contribute to homeostasis. The Arabidopsis Group 2 sulfate transporters expression has been localised in the vascular tissues (Takahashi et al., 1997; Takahashi et al., 2000). The Group 2 sulfate transporters are responsible for low affinity sulfate transport with $K_m$s between 99.2 µM and 1.2 mM (Smith et al., 1995b; Takahashi et al., 2000). Differences in the kinetic and expression pattern of the two Arabidopsis Group 2 transporters, AtSultr2;1 ($K_m$ 0.41 mM) and AtSultr2;2 ($K_m$ 1.2 mM), suggested specific functions in the process of vascular movement of sulfate. In leaves, *AtSultr2;1* is expressed in the xylem parenchyma and phloem cells, but in the root in xylem parenchyma and pericycle cells. In contrast, AtSultr2;2 is localised specifically in the phloem of roots and in vascular bundle sheath cells of leaves (Fig. 2; Takahashi *et al.*, 2000). The expression of *AtSultr2;1* in the root pericycle and xylem parenchyma cells may indicate an efflux of sulfate from the endodermal cells leading to a high concentration of sulfate in the apoplast of the vascular tissue. AtSultr2;1 would reabsorb this sulfate and will optimise the amount of sulfate transferred to the shoots during sulfate deficiency. The observed upregulation of *AtSultr2;1* in roots during sulfate starvation, and the increase of the mRNA level of *AtSultr2;1* under selenate treatment, may be an indication of this function (Takahashi *et al.*, 2000). The leaf phloem expression suggests a role in phloem loading for sulfate transport to other organs. The leaf xylem parenchyma localisation of AtSultr2;1 might indicate absorption of sulfate from the xylem vessels or re-absorption for further xylem transport. The localisation of AtSultr2;2 in the root indicates a role in sulfate transport via the phloem. In leaves however, the expression in the bundle sheet cells surrounding the vascular veins suggests uptake of sulfate released from xylem vessels at millimolar concentrations for transfer to the primary sites of assimilation in leaf palisade and mesophyll cells.

The expression pattern of both Group 2 transporters suggests that the two transporters are involved in balancing the vascular movement of sulfate in relation to the sulfate status of the different tissues (Takahashi *et al.*, 2000).

The expression of the Arabidopsis Group 3 *Sultr3;5* is co-localised with the *Sultr2;1* expression in parenchyma and pericycle cells in roots. The Arabidopsis Sultr3;5 alone was not able to complement yeast; however, cells co-expressing both, *Sultr3;5* and *Sultr2.1*, showed considerably higher uptake activity than with *Sultr2;1* expression alone. Further a restriction of root-to-shoot transport of sulfate was found in the *Sultr3;5* knock-out mutant, under conditions of high *Sultr2;1* expression in sulfur starved roots. Kataoka *et al.* (2004a) suggested that *Sultr3;5* is constitutively expressed in the root stele, but its function to reinforce the capacity of the Sultr2;1 low-affinity transporter is only essential when *Sultr2.1* expression is induced by sulfur deprivation. This provides a maximum capacity of sulfate transport activity to facilitate retrieval of apoplastic sulfate to the xylem parenchyma cell for subsequent transport of sulfate to the shoot.

A specialised function was found for the third Group 1 high affinity sulfate transporter. The Arabidopsis AtSultr1;3 is localised in the sieve elements-companion cell complexes of the phloem and seems to be involved in the source-to-sink translocation of sulfate through the plants. Analysis of a *AtSultr1;3* T-DNA insertion mutant provided direct evidence for this function by restricting movement of labelled sulfate from the cotyledon to the other organs (Yoshimoto *et al.*, 2003). The expression of high affinity Group 1 sulfate transporters in the vasculature is not restricted to Sultr1;3. In sulfate-deprived barley roots, *HvSultr1;1* was expressed within the stele (Rae and Smith, 2002). This was not observed for the homologous Arabidopsis *AtSultr1;1* and *1;2*. Rae and Smith (2002), speculated that in barley, a single transporter is responsible for functions carried out by more than one transporter in *Arabidopsis*. In tomato, *LeSultr1;1* is expressed under sulfate deficient conditions in the pericycle (Howarth *et al.*, 2003). This may indicate that under sulfate stress, plants are able to induce an additional high affinity sulfate transport to maintain vascular movement of sulfate under low sulfate concentrations.

## 6. THE WHOLE PLANT DEMAND FOR SULFATE TRANSPORT

There is constitutive demand for sulfur to synthesise protein, sulfolipid and other essential S-containing molecules for plant growth. Different growth stages including different tissues and organs differ in their demands for sulfur, which in turn may depend also on function. Responding to sulfur availability, a demand driven regulation, also influenced by environmental factors, is believed to control uptake and subsequent distribution of sulfate (Fig. 4 for overview). The specific expression of low or high affinity transporters in fast growing tissues like the root tip as well as in axillary buds, indicates the importance of an adequate sulfate supply (Takahashi et al., 1997; 2000; Rae and Smith, 2002). In addition, in root tips, high levels of expression are likely to be of functional value to facilitate 'foraging'. Young

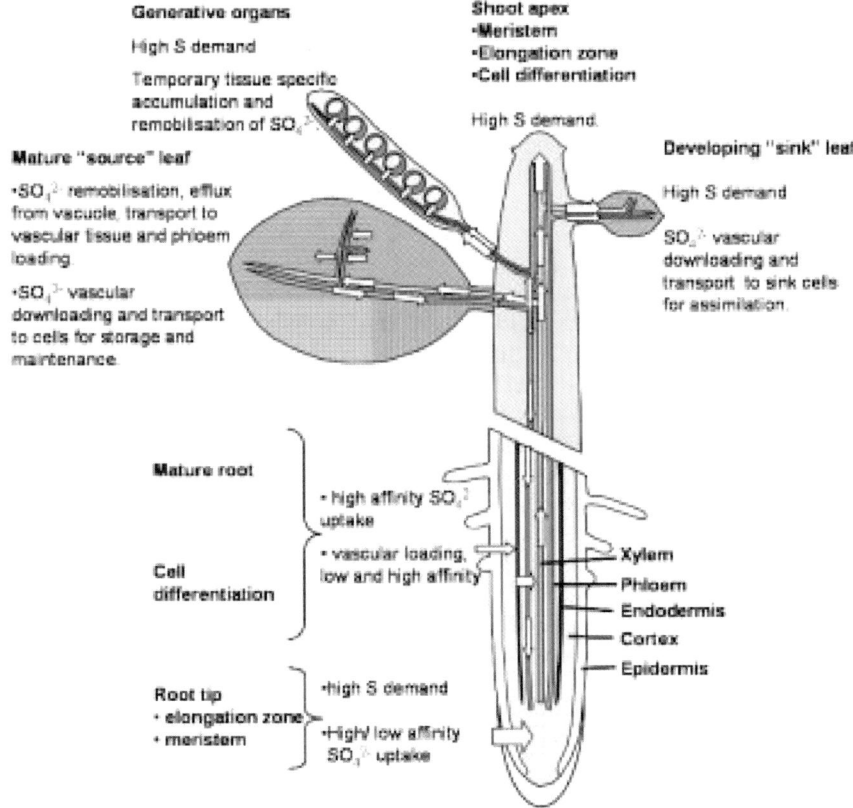

**Figure 4:** Model of sulfate uptake and movement on the whole plant level. Illustration of the proposed uptake, distribution, import, remobilisation and export of sulfate based on the function and tissue as well as developmental specific demand of sulfate. Arrows indicate uptake and potential movement of sulfate.

developing leaves are strong S-sinks, but may show a net loss of S after full expansion (Sunarpi and Anderson, 1996). Different studies have shown sulfate fluxes between organs and tissues. Feeding experiments in poplar trees indicated a transport of sulfate of $^{35}$S mainly into developing tissues and an allocation of $^{35}$S from mature leaves to developing leaves and storage organs including an exchange of sulfate between xylem and phloem (Hartmann et al., 2000). In oilseed rape, a clear difference in the distribution of S between leaves of different ages was found. At sufficient S-supply more than 50% of the total S in the young leaves was assimilated into insoluble S, glutathione (2%) and glucosinolates (6%) and 42% accumulated as sulfate. In contrast, in the middle and oldest leaves 70-90% of S accumulated as sulfate. Under S-starvation the sulfate pool was diminished in older leaves, possibly to supply young developing leaves under sulfur deficiency (Blake-Kalff et al., 1998). In shoots of Arabidopsis Sultr2;1 is suggested to mediate the transport of excess sulfate back into xylem and phloem parenchyma cells for efficient transfer of sulfate to other organs. The upregulation of Arabidopsis *AtSultr2;2* and *1;3* in leaves under sulfate starvation implicated also a participation of both transporters in the vascular allocation of sulfate from leaves to other tissues (Takahashi et al., 2000; Yoshimoto et al., 2003). On the other side this expression pattern was not found for the *Sultr1;3* and *Sultr2;2* homologues genes in *Brassica oleracea* or oilseed rape (Buchner et al., 2004a; Buchner personal communication) although in oilseed rape sulfate is allocated from the mature to the younger leaves (Blake-Kalff et al., 1998). This may implicate that in different plant species different transporter are responsible for the same action. The tomato high affinity sulfate transporter, *LeSultr1;2*, was expressed in the stem and in the leaves and was upregulated under S-stress (Howarth et al., 2003). Arabidopsis and *Brassica Sultr1;1* and *1;2* expression are induced in leaves under long term S-starvation. *AtSultr1;2* promoter activity was found in the guard cells under normal growth conditions (Buchner et al., 2004a; Yoshimoto et al., 2002). Apart from this example, spatial localisation and function of high affinity sulfate transporters in the allocation of sulfate in leaves is unknown. An induction of the tomato *Sultr1;2* expression was also found in the vascular tissue of the *Verticillium*-resistant tomato line, GCR 218, after infection by *Verticillium dahliae*. This may suggest an involvement of the sulfate transporter activity in the mechanism of *Verticillium* resistance involving elemental sulfur formation. The up-regulation of the Arabidopsis Group 4 transporter, *AtSultr4;1*, under S-deficiency in roots as well as in leaves (Takahashi et al., 2000), indicated the importance of vacuolar efflux of sulfate regulated by the sulfur demand.

Generative tissues have a high demand for S. Developing seeds require sulfur amino acids to synthesise storage proteins, utilised for germination and early growth of the next generation. Although the process of sulfate reduction and cysteine synthesis predominantly takes place in the plastids of green tissues, where the reducing power from photosynthesis can be utilised (Leustek et al., 2000; Saito, 2004), all enzymes of the sulfur assimilation pathway are in multiple compartments (Lunn et al., 1990; Ruffet et al., 1995; Noji et al., 1998) as well as also in non-photosynthetic organs (Rolland et al., 1992). In addition, Sexton and

Plant sulfate transporters

Shibles (1999) reported that seed tissue is the dominant site of ATP sulfurylase activity during reproductive growth of soybean. In wheat endosperm, an accumulation of sulfate was found in early development (Fitzgerald et al., 2001). In soybean, sulfate accumulated in pods and decreased with the onset of grain enlargement. The data are consistent with a model in which, under S-limiting conditions, the pods act as sinks for sulfate and grain growth initiates the assimilation of sulfate into hGSH in the pods, and then into developing grains, where it is incorporated into grain proteins. (Sunarpi and Anderson, 1996). After incubation of developing lupin seed with $^{35}$S-sulfate incorporation of $^{35}$S into the seed storage proteins was found (Tabe and Droux 2001; 2002).

Expression of sulfate transporter genes in generative tissues has been reported. In wheat, the high affinity sulfate transporter homologue to the AtSultr1;1 was detected in aleurone cells of wheat grains by immuno-localisation (Mills, 2003). Antisense inhibition of the Arabidopsis AtSultr2;1 lowered the sulfate to nitrate ratio in mature seeds, indicating dysfunction of low affinity transport in delivery of sulfate into seeds (Awazuhara et al., 2001). Localisation of the AtSultr2;1 in the funiculus of siliques was indicative of its role in controlling flux of sulfate in the conductive tissue to the developing seed (Awazuhara et al., 2005) In chickpeas, expression of the low affinity *CaSultr2;1* and group 3 was not restricted to vegetative tissues, but was also found in pods and in the testa of developing embryos (Tabe et al., 2003). Additional *CaSultr3.1* was also detected in developing chickpea embryos. The Expression of *CaSultr3;3* was detectable in almost all organs with high abundance in the developing chickpea embryo, suggesting participation in sulfate transport in many cell types (Tabe et al., 2003).

In Arabidopsis, *AtSultr3;1, 3;2* and *3;3* expression seems to be restricted to the leaves (Takahashi et al., 2000). In *Brassica oleracea* transcripts of all group 3 transporters are detectable in roots and only *BolSultr3.1* and *BolSultr3.3* are expressed in leaves (Buchner et al., 2004a). The spatial and subcellular localisation of these transporters in the leaf tissue is unknown, and the specific function in the process of sulfate transport remains to be verified.

## 7. REGULATION OF SULFATE TRANSPORT

External changes of the soil sulfate concentration as well as the internal demand of sulfate to fulfil the requirement for plant growth and development need an effective control system of root sulfate uptake and transport to be able to sense the changes.

In general, high concentrations of cysteine and glutathione repress sulfate uptake/ transport activities and assimilation, while sulfur starvation or high demand for sulfur metabolites results in increased activities of key enzymes in the sulfate transport and assimilatory pathway. Gene and protein expression studies have confirmed that increases in transporter activity occurs predominantly at the transcriptional level (Hawkesford and Wray, 2000; Hawkesford et al., 1993; Smith et al., 1995b, 1997; Takahashi et al., 2000; Yoshimoto et al., 2002; Kataoka et al., 2004b; Marayuma-Nakashita et al., 2004a). Apart from the differences in spatial expression patterns, sulfate transport consists of both, constitutive and sulfur-nutrition regulated transporters. Decreased intracellular content of sulfate, cysteine and glutathione is concomitant with increasing transporter activity (Smith et al., 1997).

The expression of the Group 3 sulfate transporters is not influenced by the S-nutritional status of the plant (Takahashi et al., 2000; Buchner et al., 2004a; Kataoka et al., 2004a). The expression pattern of the genes of the Group 1, 2 and 4 sulfate transporters reflects a dual pattern of regulation: semi-constitutive and inducible. Under adequate sulfur-nutrition and the saturable uptake kinetics of the high and low affinity transporters a semi-constitutive expression of one high affinity Group 1, one low affinity Group 2 together with undefined Group 3 transporters, and one vacuolar sulfate efflux Group 4 transporter is responsible/sufficient for root sulfate uptake and transport (Smith et al., 1997; Takahashi et al., 2000; Buchner et al., 2004b). Increasing excess external sulfate supply is not further influencing this constitutive sulfate transport systems and the root sulfate uptake capacity and therefore will not prevent excess toxic internal sulfate accumulation if supply continues to exceed demand, although this is an unlikely scenario.

The inducible transporters are normally repressed under excess sulfur nutrition. When internal demand for sulfate/ sulfur is increasing or external sulfate supply is decreasing the repression of sulfate transport is being modified.

The control of expression of genes for sulfate transporters may be mediated by feedback loops involving key metabolites of Cys biosynthesis. The nature of the signal(s) that are sensed under conditions of sulfate deficiency, and which are transduced to the transporter genes and promoters are unclear.

A regulatory model has been proposed that suggests that S starvation leads to an accumulation of *O*-acetylserine (*O*AS) and that this accumulation positively regulates ST gene expression (Hawkesford and Smith, 1997; Smith et al., 1997). Treatment with external *O*AS increased sulfate transporter gene expression consistent with a model of transcriptional induction by *O*AS (Smith et al., 1997; Hopkins et al., 2005). Increased cysteine and glutathione pools, particularly in the potato leaves, over-expressing the *Escherichia coli* gene *cysE* (serine acetyltransferase), which synthesises *O*AS, to modify internal metabolite pools, had increased sulfate transporter expression in the roots. However, the small increases in the *O*AS pools in the root tissue were not supportive of the hypothesis that this molecule is the signal of sulfur (S) nutritional status. In potato as well as *Brassica oleracea* the content of S-containing compounds decreased as a result of S starvation (consistent with de-repression as a mechanism of regulation) but the *O*AS pools increased only following extended starvation after the observed increased sulfate transporter mRNA abundance in the roots, probably as a consequence of the S starvation (Buchner et al., 2004a; Hopkins et al., 2005). Sulfate uptake and transport to the shoots of tobacco (*Nicotiana tabacum*) plants was repressed by glutathione and cysteine (Herschbach and Rennenberg, 1994). Cysteine or glutathione (Bolchi et al., 1999; Lappartient et al., 1999; Maruyama-Nakashita et al., 2004a) have been suggested to repress expression of ATP-sulfurylase and sulfate uptake. However, results of the *cysE* over-expressing potato plants excluded that this signal could not be reduced levels of Cys or glutathione, as root Cys and glutathione pools were elevated in the transgenic line, either from elevated biosynthesis in the root or from biosynthesis in the shoot followed by translocation to the root (Hopkins *et al.*, 2005).

The change of gene expression is very rapid when sulfur starts to be limiting. Increased transcript amounts of high and low affinity transporters as well the vacuolar efflux transporter are found after 4 hours sulfate starvation (Maruyama-Nakashita *et al.*, 2005) suggesting the capability of the root to sense external sulfate concentrations leading to the de-repression of the transporter expression. This sulfate starvation-induced, coordinated de-repression leads to complete activation of the whole root sulfate transport machinery by higher expression of the constitutive-like transporters and also induction of expression of genes of the other high and low affinity transporters including those involved in vacuolar efflux (Buchner *et al.*, 2004a, b), thus maximising sulfate uptake and vascular transport capacity including release of sulfate from vacuoles.

The regulation of sulfate transport is also embedded in the overall regulation of sulfur, nitrogen and carbon assimilation and metabolism. The up-regulation of the transporters is much reduced in response to sulfur deficiency when nitrogen is limiting (Clarkson *et al.*, 1989, 1999) and is stimulated by nitrate application (Vidmar *et al.*, 1999; Maruyama-Nakashita *et al.*, 2004a). The regulation of sulfate assimilation is not only aimed at coordination with nitrate metabolism, but also with carbon assimilation; the expression of root sulfate transporters is also induced by light. The light induction of sulfate transporter expression could be mimicked by addition of sucrose to the nutrition medium (Lejay *et al.*, 2003; Maruyama-Nakashita *et al.*, 2004a). As sucrose also induced nitrate and ammonium transporters (Lejay *et al.*, 2003) the assimilation of sulfate and nitrate must be interconnected. *O*-acetylserine links assimilatory sulfate reduction with carbohydrate and nitrogen metabolism and has been proposed as a signalling molecule coordinating these three pathways (Brunold, 1993). Feeding experiments indicated that the sugar regulation of sulfate transporter and APS reductase expression is independent from *O*-acetylserine regulation (Hesse *et al.*, 2003, Lejay *et al.*, 2003) which may indicate that signals from nitrogen and carbon metabolism act synergistically and that positive signalling from sugars can override negative signalling from nitrate assimilation (Hesse *et al.*, 2003). In contrast to the *OAS* positive regulatory effect, a reduced sulfur compound such as cysteine or glutathione may act as a negative regulator (Vauclare *et al.*, 2002; Maruyama-Nakashita *et al.*, 2004a). Glutathione has been suggested to be a phloem-translocated signal molecule that represses genes of sulfur assimilation (Lappartient *et al.*, 1999). However observations on sulfate uptake in *Brassica* species as well as transgenic potatoes over-expressing *cysE* suggest that shoot to root signalling of the regulation is unlikely to be mediated by the size of the thiol pool and that the concentration of sulfate itself may be the determining influence as a direct or an indirect signal (Buchner *et al.*, 2004a; Hopkins *et al.*, 2005).

Peter Buchner

The nature of the input nutrient signal and the mechanisms of signal transduction finally regulating sulfate transporter expression are still unknown at the molecular level. Sulfur responsive promoters have been examined to determine the regulatory factors involved in transcriptional activation or repression. Studies on the $\beta$-subunit of $\beta$-conglycinin (Awazuhara et al., 2001) and *NIT3* nitrilase (Kutz et al., 2002) provided initial data indicating the presence of regulatory regions in the 5'-promoter sequences of these sulfur-responsive genes. Within the -2777/-2761 promoter region of the Arabidopsis *Sultr1;1* a sulfur responsive *cis*-acting element, (SURE = **su**lfur **r**esponsive **e**lement) was identified which was essential for the regulation of *Sultr1;1* expression under –S condition. The SURE core sequence (GGAGACA) was also found in the promoter region of the sulfate transporters *Sultr2;1* and *Sultr4;2* and the *APS reductase 3* gene of Arabidopsis (Maruyama-Nakashita et al., 2005). In contrast no SURE element was present in the promoter region of the other genes –S regulated sulfate transport and assimilation genes indicating a different regulatory mechanism involved in sulfur nutrition regulation of different isoforms.

The SURE dependent expression was down-regulated by cysteine and glutathione suggesting that the internal sulfur status represented by the thiol content in plants controls expression (Maruyama-Nakashita et al., 2005). Alternatively, addition of thiols may disturb turnover of sulfate, affecting conversion of downstream assimilatory pathways. In contrast to cysteine and glutathione, *O*AS, which has been reported as a positive effector for the regulation of sulfur responsive genes (Smith et al., 1997; Kim et al., 1999; Koprivova et al., 2000; Hirai et al., 2003), did not show any effect on modulating the SURE dependent expression. From these results it is suggested that S deprivation regulated gene expression is differently regulated by sulfur and *O*AS. The S-inducible gene expression of the Arabidopsis *SULTR1;1* required a protein phosphatase and an upstream regulatory factor (Marayuma-Nakashita et al., 2004a). In *Chlamydomonas*, a Snf1-like Ser/Thr kinase, Sac3, is involved in the regulation of sulfate uptake and arylsulfatase activities (Davies et al., 1999) indicating that phosphorylation and dephosphorylation are important in the regulation of S-responsive genes in plants and algae.

Some plant hormones are influencing gene expression related to sulfate transport. Several studies on Arabidopsis transcriptome analysis suggest that auxin and methyl jasmonate are involved in the sulfur-deficiency stress response (Hirai et al., 2003; Maruyama-Nakashita et al., 2003; Nikiforova et al., 2003). Cytokinin has been shown to down-regulate high-affinity sulfate transporter gene expression in Arabidopsis roots (Maruyama-Nakashita et al., 2004b). The regulatory pathway involved a two-component phosphor-relay system initiated by CRE1/WOL/AHK4 cytokinin receptor histidine kinase. The *cre1-1* mutant of Arabidopsis was unable to regulate the expression of high affinity sulfate transporter in response to cytokinin (Maruyama-Nakashita et al., 2004b). The cellular levels of cytokinin do not significantly change in response to the sulfur status (Ohkama et al., 2002) which may suggest a non sulfur-related regulatory pathway. A model was suggested that sulfur- and cytokinin-derived signals modulate uptake of sulfate as two independent regulatory events. By using promoter-reporter systems for *in planta*

## Plant sulfate transporters

screening of regulatory elements in Arabidopsis a sulfur response-less mutant could be identified, which was unable to induce sulfur responsive genes including the high affinity sulfate transporters. This **s**ulfur **lim**itation response 1 (*SLIM1*) gene codes for a transcription factor that generally regulates sulfur responsive genes in the Arabidopsis root (Maruyama-Nakashita and Takahashi, 2005; Takahashi 2005; Maruyama-Nakashita *et al.*, 2006). Recent studies have shown a novel function of microRNAs (miRNAs) in regulating plant adaptive responses to nutrient stresses. The microRNAs miR395 is up-regulated during sulfate deficiency and seems to participate in sulfate assimilation and allocation via adjusting the expression of the Arabidopsis sulfate transporter AtSultr2;1 and ATP-sulfurylase (Chiou, 2007).

Sulfate transporters in animals and plants are structurally conserved. The carboxyl-terminal region has significant similarity with the *Bacillus* sp. anti-anti-sigma protein SpoIIAA, and is referred to as the STAS domain (**s**ulfate **t**ransporter and **a**nti-**s**igma antagonist) (Fig. 1b). Although the exact function of the STAS has not been elucidated, mutations in the STAS domain of human sulfate transporters result in serious diseases, including diastrophic dysplasia and congenital chloride diarrhea. These findings suggest that the STAS domain contributes to the catalytic, biosynthetic, or regulatory aspects of anion transporters (Everett and Green, 1999).

Studies of STAS truncated Arabidopsis high affinity transporters and chimeric transporter constructs found that the function of the STAS domain is more concerned with plasma membrane localisation and stability, than with regulation of transporter activity. The Arabidopsis sulfate transporter, Sultr1;2, with a deleted STAS domain was unable to complement the yeast sulfate transporter mutant strain CP154-7B. Fusing the STAS domain from other sulfate transporters to the STAS-deleted Sultr1;2 restored function and plasma membrane localisation, however the kinetics of sulfate uptake in the transformants were dependent on the origin of the STAS domain probably due to significant differences among the STAS domains of the different *A. thaliana* sulfate transporters. These results suggest that the STAS domain is essential, either directly or indirectly, for facilitating localisation of the transporters to the plasma membrane, but it also appears to influence the kinetic properties of the catalytic domain of transporters (Shibagaki and Grossman, 2004). Substitutions made of the putative Thr-587 phosphorylation site resulted in a complete loss of the sulfate transport function of Arabidopsis Sultr1.2. The reduction or suppression of sulfate transport of the Sultr1.2 mutants in yeast was not due to an incorrect targeting to the plasma membrane. Both three-dimensional modelling and mutational analyses strengthen the hypothesis that the STAS domain is involved in protein-protein interactions. Such interaction could depend on the phosphorylation status of the STAS domain and could control sulfate transport activity (Rouached *et al.*, 2005). The S-starvation inducible gene expression of the Arabidopsis *Sultr1;1* required a protein phosphatase and an upstream regulatory factor (Marayuma-Nakashita *et al.*, 2004c). The maximal activity of the Arabidopsis sulfate transporter AtSultr2;1 requires co-expression of the *AtSultr3;5*, raising the possibility of posttranslational protein-protein interaction as regulatory mechanisms (Kataoka *et al.*, 2004b). A lack of correlation between sulfate transporter mRNA level and protein level as compared to measurable sulfate uptake capacity was particularly evident in transgenic potatoes over-expressing the *E. Coli cysE* gene, suggesting additional post-translational control (Hopkins *et al.*, 2005).

## 8. CONCLUSIONS AND FUTURE PROSPECTS

The identification and current state of analysis of the plant sulfate transporter family provides a model of the role of these transporters in plant S-nutrition. The differences in the regulation by nutrition including different transport kinetics as well as cellular expression patterns including the differences in the sub-cellular localisation of the different Groups and subtypes of the plant sulfate transporter family indicate the complexity of root sulfate uptake and transport. After entry of sulfate into the xylem sulfate is also the major form of transported and stored sulfur. At the whole plant level, in relation to growth and development, sulfate transport is potentially even more complex. Although there are slight differences between the Arabidopsis and rice genomes, the occurrence of the multigene sulfate transporter family, with the specific clustering of subtypes, seems to be ubiquitous amongst plant species.

Not all cellular and sub-cellular sulfate transport mechanisms in plants are understood. The nature of the mechanism of sulfate transport into plant plastids and the transporter responsible for sulfate influx into the vacuole and efflux from cells are completely unknown. Besides of the Arabidopsis Sultr3;5 that, the role of the Group 3 transporters is not known. The visible sub-clustering of the Group 3 transporters (Fig. 1) may change the subdivision of the family after more detailed information on the different Group 3 subtypes becomes available.

The functional analysis of the whole sulfate transporter family and the identification of the missing transport mechanisms remain as challenges for the future. These could provide information for transgenic approaches or for breeders to optimise plant sulfate uptake and transport. Several approaches to increase cysteine and/or methionine content in seed by over-expression of foreign sulfur rich proteins indicated a limitation of uptake of sulfate into the cotyledon (Tabe and Droux, 2002; Chiaiese *et al.*, 2004). Optimising not only sulfate uptake but also delivery to sink tissues/organs may be an option to support approaches to increase the content of the essential amino acids cysteine and methionine in crop plants. Because sulfate is also the main storage form, a coordinated remobilisation from the vacuoles of source tissues for transport and delivery to developing sink especially during seed growth/ seed filling may improve the sulfur utilisation efficiency.

A relation of pathogen resistance and sulfate transport was found in tomato infected by *Verticillium dahliae*, a fungal vascular pathogen. Increased expression of a high affinity transporter was found in vascular tissue of infected resistant plants suggesting involvement of sulfate transport in the resistance mechanism (Howarth *et al.*, 2003). Different field surveys have shown that sulfur (S) fertilisation can increase the resistance of agricultural crops against fungal pathogens (Bloem *et al.*, 2004). Enhancing root sulfate uptake would be an option for pathogen control even in crop plants without natural resistances.

The overall picture of the regulation of sulfate transport indicates an adaptation of sulfate transport to changing environmental conditions and to the availability of other nutrients. This adaptation can differ between plant species, which may reflect species-specific adaptations and demands in sulfur metabolism. Transcriptional control is of major importance; however there are also roles for post transcriptional control in fine tuning the network of sulfur distribution and utilisation in the context of specific metabolic pathways and the whole plant repression or de-repression of expression is controlled by the balance of sulfate supply and sink demand. The analysis of the sulfur responsive *cis*-acting SURE element and the cytokinin-derived signals indicated that sulfate transport is regulated by S-responsive and non-responsive pathways which are not fully understood yet. The identification the *trans*-acting *SLIM1* transcription factor and miRNAs looks very promising to get into these regulatory pathways. The understanding of the identity of the signal metabolites including the cascade of regulation will be the next major breakthrough, arising either from transcriptomics, metabolomics or specifically targeted approaches. The upcoming progresses may be the basis for the genetic engineering of sulfate transport and assimilation in plants to improve sulfur qualities of crop plant species.

## 9. ACKNOWLEDGEMENTS

Rothamsted Research receives grant-aided support from the Biotechnology and Biological Science Research Council (BBSRC) of the UK.

## 10. LITERATURE CITED

Aravind, L. and Kooni, E.V. (2000) The STAS domain – a link between anion transporters and antisigma-factor antagonists. *Curr. Biol.,* **10**: 53-55.

Awazuhara, M., Takahashi, H., Watanabe-Takahashi, A., Hayashi, H., Fujiwara, T. and Saito, K. (2001) Function of sulfate transporter Sultr2;1 in seeds of *Arabidopsis thaliana*. In: *Plant Nutrition – Food Security and Sustainability of Agro-ecosystems.* (Eds. Horst, W.J., Schenk, M.K. and Burkert, A.), Kluwer Academic Publishers, Dordrecht, The Netherlands, pp 38-39.

Awazuhara, M., Fujiwara, T., Hayashi, H., Watanabe-Takahashi, A., Takahashi, H. and Saito, K. (2005) The function of SULTR2;1 sulfate transporter during seed development in *Arabidopsis thaliana*. *Physiol Plant.,* **125**: 95-105.

Bateman, A., Coin, L., Durbin, R., Finn, R.D., Hollich, V., Griffiths-Jones, S., Khanna, A., Marshall, M., Moxon, S., Sonnhammer, E.L., Studholme, D.J., Yeats, C. and Eddy, S.R. (2004) The Pfam protein families database. *Nucleic Acids Res.,* **32**: 138-141.

Blake-Kalff, M.M., Harrison, K.R., Hawkesford, M.J., Zhao, F.J. and McGrath, S.P. (1998) Distribution of sulfur within oilseed rape leaves in response to sulfur deficiency during vegetative growth. *Plant Physiol.,* **118**: 1337-1344.

Bloem, E., Riemenschneider, A., Volker, J., Papenbrock, J., Schmidt, A., Salac, I., Haneklaus, S. and Schnug, E. (2004) Sulphur supply and infection with *Pyrenopeziza brassicae* influence L-cysteine desulphydrase activity in *Brassica napus* L. *J. Exp. Bot.,* **55**: 2305-2312.

Bolchi, A., Petrucco, S., Tenca, P.L., Foroni, C. and Ottonello, S. (1999) Coordinate modulation of maize sulfate permease and ATP sulfurylase mRNAs in response to variations in sulfur nutritional status: stereospecific downregulation by L-cysteine. *Plant Mol. Biol.,* **39**: 527-537.

Brader, G., Tas, E. and Palva, E.T. (2001) Jasmonate-dependent induction of indole glucosinolates in Arabidopsis by culture filtrates of the nonspecific pathogen *Erwinia carotovora*. *Plant Physiol.,* **126**: 849-860.

Brunold, C. (1993) Regulatory Interactions between Sulfate and Nitrate Assimilation. In: *Sulfur Nutrition and Sulphur Assimilation in Higher Plants* (Eds. De Kok, L.J., Stulen, I., Rennenberg, H., Brunold, C. and Rauser, W.E.) SPB Academic Publishing. The Hague, The Netherlands, pp. 61-75.

Buchner, P., Stuiver, C.E.E., Westermann, S., Wirtz, M., Hell, R., Hawkesford, M.J. and Kok, L.J. (2004a) Regulation of sulfate uptake and expression of sulfate transporter genes in *Brassica oleracea* L. as affected by atmospheric $H_2S$ and pedospheric sulfate nutrition. *Plant Physiol.,* **136**: 3396-3408.

Buchner, P., Takahashi, H. and Hawkesford, MJ. (2004b) Plant sulphate transporters – coordination of uptake, intracellular and long distance transport. *J. Exp. Bot.,* **55**: 1765-1773.

Buchner, P., Prosser, I. and Hawkesford, M.J. (2004c) Phylogeny and expression of paralogous and orthologous sulphate transporter genes in diploid and hexaploid wheats. *Genome,* **47**: 526-534.

Chiaiese, P., Ohkama-Ohtsu, N., Molvig, L., Godfree, R., Dove, H., Hocart, C., Fujiwara, T., Higgins, T.J. and Tabe, L.M. (2004) Sulphur and nitrogen nutrition influence the response of chickpea seeds to an added, transgenic sink for organic sulphur. *J. Exp. Bot.,* **55**: 1889-1901.

Chiou, J.J. (2007) The role of micro RNAs in sensing nutrient stress. *Plant Cell Environ.,* **30**: 323-332.

Clarkson, D.T., Saker, L.R. and Purves, J.V. (1989) Depression of nitrate and ammonium transport in barley plants with diminished sulfate status. Evidence for co-regulation of nitrogen and sulfate uptake. *J. Exp. Bot.,* **40**: 953-963.

Clarkson, D.T., Diogo, E. and Amancio, S. (1999) Uptake and assimilation of sulfate by sulfur deficient *Zea mays* cells: The role of O-acetyl-L-serine in the interaction between nitrogen and sulfur assimilatory pathways. *Plant Physiol. and Biochem.,* **37**: 283-290.

Clarkson, D.T., Hawkesford, M.J., Davidian, J.-C. and Gignon, C. (1992) Contrasting responses of sulphate and phosphate-transport in barley (*Hordeum vulgare* L.) roots to protein-modifying reagents and inhibition of protein synthesis. *Planta,* **187**: 306-314.

Clarkson, D.T., Hawkesford, M.J. and Davidian, J.-C. (1993) Membrane and long-distance transport of sulfate. In: Sulfur nutrition and assimilation in higher plants (Eds. Kok L.J., Stulen I., Rennenberg H., Brunold C. and Rauser W.E.), SPB Academic Publishing, The Hague, The Netherlands, pp 3-19.

Davies, J.P., Yildiz, F.H. and Grossman, A.R. (1999) Sac3, an Snf1-like serine threonine kinase that positively and negatively regulates the responses of *Chlamydomonas* to sulfur limitation. *Plant Cell,* **11**: 1179-1190.

De Kok, L.J., Stuiver, C.E.E., Westerman, S. and Stulen, I. (2002) Elevated levels of hydrogen sulfide in the plant environment: nutrient or toxin. In: *Air Pollution and Biotechnology in Plants* (Eds. Omasa, K., Saji, H., Youssefian, S. and Kondo, N.), Springer-Verlag, Tokyo, pp. 201-213.

Plant sulfate transporters

Dietz, K-J., Brune, B. and Pfanz, H. (1992) Trans-tonoplast transport of the sulfur containing compounds sulfate, methionine, cysteine and glutathione. *Phyton,* 32: 37-40.
Epstein, E. (1966) Dual pattern of ion absorption by plant cells and by plants. *Nature,* 212: 1324-1327.
Everett, L.A. and Green, E.D.A (1999) A family of mammalian anion transporters and their involvement in human genetic diseases. *Hum. Mol. Genet.,* 8: 1883-1891.
Feng, Q., Zhang, Y., Hao, P., Wang, S., Fu, G., Huang, Y., Li, Y., Zhu, J., Liu, Y., Hu, X. et al. (2002) Sequence and analysis of rice chromosome 4. *Nature,* 420: 316-320.
Fitzgerald, M.A., Ugalde, T.D. and Anderson, J.W. (2001) Sulfur nutrition affects delivery and metabolism of S in developing endosperm of wheat. *J. Exp. Bot.,* 52: 1519-1526.
Flügge, U.I., Gerber, J. and Heldt, H.W. (1983) Regulation of the reconstituted chloroplast phosphate translocator by a $H^+$ gradient. *Biochim Biophys Acta.,* 725: 229-237.
Frachisse, J.-M., Thomine, S., Colombet, J., Guern, J. and Barbier-Brygoo, H. (1999) Sulfate is both a substrate and an activator of voltage-dependent anion channel of Arabidopsis hypocotyls cells. *Plant Physiol.,* 121: 253-261.
Goff, S.A., Ricke, D., Lan, T.H., Presting, G., Wang, R., Dunn, M., Glazebrook, J., Sessions, A., Oeller, P., Varma, H. et al. (2002) A draft sequence of the rice genome (*Oryza sativa* L. ssp. japonica). *Science,* 296: 92-100.
Hampp, R. and Ziegler, I. (1977) Sulfate and sulfite translocation via the phosphate translocator of the inner envelope membrane of chloroplasts. *Planta,* 137: 309-312.
Hartmann, T., Mult, S., Suter, M., Rennenberg, H. and Herschbach, C. (2000) Leaf age-dependent differences in sulphur assimilation and allocation in poplar (*Populus tremula* × *P. alba*) leaves. *J. Exp. Bot.,* 51: 1077-1088.
Hawkesford, M.J., Davidian, J.-C. and Grignon, C. (1993) Sulfate proton cotransport in plasma vesicles isolated from roots of *Brassica napus* L. Increased transport in membranes isolated from sulphur deprived plants. *Planta,* 190: 297-304.
Hawkesford, M.J. and Smith, F.W. (1997) Molecular biology of higher plant sulphate transporters. In: *Sulphur Metabolism in Higher Plants.* (Eds. Cram, W.J., Kok, L.J., Stulen, I., Brunold, C. and Rennenberg, H.), Backhuys Publishers, Leiden, The Netherlands, pp 13–25.
Hawkesford, M.J. (2000) Plant responses to sulphur deficiency and genetic manipulation of sulphate transporters to improve S-utilisation efficiency. *J. Exp. Bot.,* 51: 131-138.
Hawkesford, M.J. and Wray, J.L. (2000) Molecular genetics of sulphur assimilation. *Adv Bot Res.,* 33: 159-223.
Hawkesford, M.J. (2003) Transporter gene families in plants: the sulphate transporter gene family – redundancy or specialization? *Physiol. Plant.,* 117: 115-163.
Hawkesford, M.J., Buchner, P., Hopkins, L. and Howarth, J.R. (2003) The plant sulphate transporter family: specialized functions and integration with whole plant nutrition. In: *Sulfur Transport and Assimilation in Plants* (Eds. Davidiasn, J.-C., Grill, D., Kok, L.J., Stulen, I., Hawkesford, M.J., Schnug, E. and Rennenberg, H.), Backhuys Publishers, Leiden, The Netherlands pp 1-10.
Herschbach, C. and Rennenberg, H. (1994) Influence of glutathione (GSH) on net uptake of sulphate and sulphate transport in tobacco plants. *J Exp Bot.,* 45: 1069-1076.
Hesse, H., Trachsel, N., Suter, M., Kopriva, S., von Ballmoos, P., Rennenberg, H. and Brunold, C. (2003) Effect of glucose on assimilatory sulfate reduction in *Arabidopsis thaliana* roots. *J. Exp. Bot.,* 54: 1701-1709.
Hirai, M.Y., Fujiwara, T., Awazuhara, M., Kimura, T., Noji, M. and Saito, K. (2003) Global expression profiling of sulfur-starved Arabidopsis by DNA macroarray reveals the role of O-acetyl-l-serine as a general regulator of gene expression in response to sulfur nutrition. *Plant J.,* 33: 651-663.

Hirata, T., Wada, Y. and Futai, M. (2002) Sodium and sulfate ion transport in yeast vacuoles. *J Biochem.*, **131**: 261-265.

Hopkins, L., Parmar, S., Blaszczyk, A., Hesse, H., Hoefgen, R. and Hawkesford, M.J. (2005) O-acetylserine and the regulation of expression of genes encoding components for sulfate uptake and assimilation in potato. *Plant Physiol.*, **138**: 433-440.

Howarth, J., Fourcroy, P., Davidian, J.-C., Smith, F.W. and Hawkesford, M.J. (2003) Cloning of two contrasting high-affinity sulphate transporters from tomato induced by low sulphate and infection by the vascular pathogen *Verticillium dahlia*. *Planta*, **218**: 58-64.

Jagendorf, A.T. and Ryrie, I.J. (1971) Inhibition of photophosphorylation in chloroplasts by inorganic sulfate. *J. Biol. Chem.*, **246**: 582-588.

Jones, M.G., Hughes, J., Tregova, A., Milne, J., Tomsett, A.B. and Collin, H.A. (2004) Biosynthesis of the flavour precursors of onion and garlic. *J. Exp. Bot.*, **55**: 1903-1919.

Kaiser, G., Martinoia, E., Schroppel-Maier, G. and Heber, U. (1989) Active transport of sulfate into the vacuole of plant cells provides halotolerance and can detoxify $SO_2$. *J. Plant Physiol.*, **133**: 756-763.

Kaiser, W.M., Schroppel-Meier, G. and Wirth, E. (1986) Enzyme activities in an artificial stroma medium. An experimental model for studying effects of dehydration on photosynthesis. *Planta*, **161**: 292-299.

Kataoka, T., Hayashi, N., Yamaya, T. and Takahashi, H. (2004a) Root-to-shoot transport of sulfate in Arabidopsis. Evidence for the role of SULTR3;5 as a component of low-affinity sulfate transport system in the root vasculature. *Plant Physiol.*, **136**: 4198-4204.

Kataoka, T., Watanabe-Takahashi, A., Hayashi, N., Ohnishi, M., Mimura, T., Buchner, P., Hawkesford, M.J., Yamaya, T. and Takahashi, H. (2004b) Vacuolar sulfate transporters are essential determinants controlling internal distribution of sulfate in Arabidopsis. *Plant Cell*, **16**: 2693-2704.

Kertesz, M.A. (2001) Bacterial transporters for sulfate and organosulfur compounds. *Res. Microbiol.*, **152**: 279-290.

Kim, H., Hirai, M.Y., Hayashi, H., Chino, M., Naito, S. and Fujiwara, T. (1999) Role of O-acetyl-L-serine in the coordinated regulation of the expression of a soybean seed storage-protein gene by sulfur and nitrogen nutrition. *Planta*, **209**: 282-289.

Koprivova, A., Suter, M., Op den Camp, R., Brunold, C. and Kopriva, S. (2000) Regulation of sulfate assimilation by nitrogen in Arabidopsis. *Plant Physiol.*, **122**: 737-746.

Krusell, L., Krause, N., Ott, T., Desbrosses, G., Kramer, U., Sato, S., Nakamura, Y., Tabata, S., James, E.K., Sandal, N., Stougaard, J., Kawaguchi, M., Miyamoto, A., Suganuma, N. and Udvardi, M.K. (2005) The sulphate transporter SST1 is crucial for symbiotic nitrogen fixation in *Lotus japanicus* root nodules. *Plant Cell*, **17**: 1625-1636.

Kumar, S., Tamura, K., Jakobsen, I.B. and Nei, M. (2001) MEGA2: molecular evolutionary genetics analysis software. *Bioinformatics*, **17**: 1244-1245.

Kutz, A., Müller, A., Hennig, P., Kaiser, W.M., Piotrowski, M. and Weiler, E.W. (2002) A role for nitrilase 3 in the regulation of root morphology in sulphur-starving *Arabidopsis thaliana*. *Plant J.*, **30**: 95-106.

Lappartient, A.G., Vidmar, J.J., Leustek, T., Glass, A.D.M. and Touraine, B. (1999) Inter-organ signaling in plants: regulation of ATP sulfurylase and sulfate transporter genes expression in roots mediated by phloem-translocated compound. *Plant J.*, **18**: 89-95.

Lass, B. and Ullrich-Eberius, C.I. (1984) Evidence for proton/ sulfate cotransport and its kinetic in *Lemna gibba* G1. *Planta*, **161**: 53-60.

Leggett, J.E. and Epstein, E. (1956) Kinetics of sulfate absorption by barley roots. *Plant Physiol.*, **31**: 222-226.
Lejay, L., Gansel, X., Cerezo, M., Tillard, P., Müller, C., Krapp, A., Von Wiren, N., Danniel-Vedele, F. and Gojon, A. (2003) Regulation of root ion transporters by photosynthesis: Functional importance and relation with hexokinase. *Plant Cell*, **18**: 2218-2232.
Leustek, T. and Saito, K. (1999) Sulfate transport and assimilation in plants. *Plant Physiol.*, **120**: 637-643.
Leustek, T., Martin, M.N., Bick, J.-A. and Davies, J.P. (2000) Pathways and regulation of sulphur metabolism revealed through molecular and genetic studies. *Ann. Rev. Plant Physiol. Plant Mol. Biol.*, **51**: 141-165.
Lunn, J.E., Droux, M., Martin, J. and Douce, R. (1990) Localization of ATP sulfurylase and O-acetylserin (thiol) lyase in spinach leaves. *Plant Physiol.*, **94**: 1345-1352.
Marchler-Bauer, A., Anderson, J.B., DeWeese-Scott, C., Fedorova, N.D., Geer, L.Y., He, S., Hurwitz, D.I., Jackson, J.D., Jacobs, A.R., Lanczycki, C.J., Liebert, C.A., Liu, C., Madej, T., Marchler, G.H., Mazumder, R., Nikolskaya, A.N., Panchenko, A.R., Rao, B.S., Shoemaker, B.A., Simonyan, V., Song, J.S., Thiessen, P.A., Vasudevan, S., Wang, Y., Yamashita, R.A., Yin, J.J. and Bryant, S.H. (2003) CDD: a Conserved Domain Database for protein classification. *Nucleic Acids Res.*, **31**: 383-387.
Martinoia, E., Heck, U. and Wiemken, A. (1981) Vacuoles as storage compartments for nitrate in barley leaves. *Nature*, **289**: 292-294.
Martinoia, E., Massonneau, A. and Frangne, N. (2000) Transport processes of solutes across the vacuolar membrane of higher plants. *Plant Cell Physiol.*, **51**: 1175-1186.
Maruyama-Nakashita, A., Inoue, E., Watanabe-Takahashi, A., Yamaya, T. and Takahashi, H. (2003) Transcriptome profiling of sulfur-responsive genes in Arabidopsis reveals global effects of sulfur nutrition on multiple metabolic pathways. *Plant Physiol.*, **132**: 597-605.
Maruyama-Nakashita, A., Nakamura, Y., Tohge, T., Saito, K. and Takahashi, H. (2006) Arabidopsis SLIM1 is a central transcriptional regulator of plant sulfur response and metabolism. *Plant Cell*, **18**: 3235-3251.
Maruyama-Nakashita, A., Nakamura, Y., Yamaya, T. and Takahashi, H. (2004a) Regulation of high-affinity sulphate transporters in plants: towards systematic analysis of sulphur signalling and regulation. *J. Exp. Bot.*, **55**: 1843-1849.
Maruyama-Nakashita, A., Nakamura, Y., Yamaya, T. and Takahashi, H. (2004b) A novel regulatory pathway of sulfate uptake in Arabidopsis roots: implication of CRE1/WOL/AHK4-mediated cytokinin-dependent regulation. *Plant J.*, **38**: 779-789.
Maruyama-Nakashita, A., Nakamura, Y., Watanabe-Takahashi, A., Yamaya, T. and Takahashi, H. (2004c) Induction of SULTR1;1 sulfate transporter in Arabidopsis roots involves protein phosphorylation/dephosphorylation circuit for transcriptional regulation. *Plant Cell Physiol.*, **45**: 340-345.
Maruyama-Nakashita, A., Nakamura, Y., Watanabe-Takahashi, A., Inoue, E., Yamaya, T. and Takahashi, H. (2005) Identification of a novel cis-acting element conferring sulfur deficiency response in Arabidopsis roots. *Plant J.*, **42**: 305-14.
Maruyama-Nakashita, A. and Takahashi, H. (2005) Transcriptional regulation of Sultr1;1 and Sultr1;2 in Arabidopsis roots. In: *Sulfur Transport and Assimilation in Plants in the Post Genomic Era* (Eds. Saito, K., De Kok, L.J., Stulen, I., Hawkesford, M.J., Schnug, E., Sirko, A. and Rennenberg, H.), Backhuys Publishers, Leiden, The Netherlands, pp 43-44.

Matsubayashi, Y., Ogawa, M., Morita, A. and Sakagami, Y. (2002) An LRR receptor kinase involved in perception of a peptide plant hormone, phytosulfokine. *Science,* **296**: 1470-1472.

Melis, A. and Chen, H.C. (2005) Chloroplast sulphate transport in green algae – gene, proteins and effects. *Photosynth Res.,* **86**: 299-307.

Mendoza-Cozatl, D., Loza-Tavera, H., Hernandez-Navarro, A. and Moreno-Sanchez, R. (2005) Sulfur assimilation and glutathione metabolism under cadmium stress in yeast, protists and plants. *FEMS Microbiol Rev.,* **29**: 653-671.

Miller, A.J., Cookson, S.J., Smith, S.J. and Wells, D.M. (2001) The use of microelectrodes to investigate compartmentation and the transport of metabolized inorganic ions in plants. *J. Exp. Bot.,* **52**: 541-549.

Mills, T. (2003) Localisation of sulphate transporters in wheat. Oxford Brookes University PhD Thesis, Oxford, UK.

Mimura, T., Dietz, K.J., Kaiser, W., Schramm, M.J., Kaiser, G. and Heber, U. (1990) Phosphate transport macros biomembranes and cytosolic phosphate homeostasis in barley leaves. *Planta,* **180**: 139-146.

Mourioux, D. and Douce, R. (1979) Transport du sulphate à travers la double membrane limitante, ou envelope, des chloroplasts d'epinard. *Biochimie,* **61**: 1283-1292.

Nikiforova, V., Freitag, J., Kempa, S., Adamik, M., Hesse, H. and Hoefgen, R. (2003) Transcriptome analysis of sulfur depletion in *Arabidopsis thaliana*: interlacing of biosynthetic pathways provides response specificity. *Plant J.,* **33**: 633-50.

Noji, M., Inoue, K., Kimura, N., Gouda, A. and Saito, K. (1998) Isoform-dependent differences in feedback regulation and subcellular localization of serine acetyltransferase involved in cysteine biosynthesis from *Arabidopsis thaliana*. *J. Biol. Chem.,* **273**: 32739-32745.

Ohkama, N., Takei, K., Sakakibara, H., Hayashi, H., Yoneyama, T. and Fujiwara, T. (2002) Regulation of sulfur-responsive gene expression by exogenously applied cytokinins in *Arabidopsis thaliana*. *Plant Cell Physiol.,* **43**: 1493-1501.

Ohyma, K., Fukuzawa, H., Kohchi, T., Shirai, H., Sano, T., et al. (1986) Chloroplast gene organization deduced from complete sequence of liverwort *Marchantia polymorpha* chloroplast DNA. *Nature,* **322**: 572-574.

Rae, A.L. and Smith, F.W. (2002) Localisation of expression of a high-affinity sulfate transporter in barley roots. *Planta,* **215**: 565-568.

Rolland, N., Droux, M. and Douce, R. (1992) Subcellular distribution of O-acetylserine (thiol) lyase in cauliflower (*Brassica oleracea* L.) inflorescence. *Plant Physiol.,* **98**: 927-935.

Rouached, H., Berthomieu, P., El Kassis, E., Cathala, N., Catherinot, V., Labesse, G., Davidian, J.C and Fourcroy, P. (2005) Structural and functional analysis of the C-terminal STAS (sulfate transporter and anti-sigma antagonist) domain of the *Arabidopsis thaliana* sulfate transporter SULTR1.2. *J. Biol. Chem.,* **280**: 15976-15983.

Ruffet, M.L., Lebrun, M., Droux, M. and Douce, R. (1995) Subcellular distribution of serine acetyltransferase from *Pisum sativum* and characterization of an *Arabidopsis thaliana* putative cytosolic isoform. *Eur. J. Biochem.,* **227**: 500-509.

Saito, K. (2004) Sulfur assimilation metabolism. The long and smelling road. *Plant Physiol.,* **136**: 2443-2450.

Sandal, N.N. and Marcker, K.A. (1994) Similarities between a soybean nodulin, Neurospora crassa sulphate permease II and a putative human tumour suppressor. *Trends Biochem. Sci.,* **19**: 19.

Sasaki, T., Matsumoto, T., Yamamoto, K., Sakata, K., Baba, T., Katayose, Y., Wu, J., Niimura, Y., Cheng, Z., Nagamura, Y. et al. (2002) The genome sequence and structure of rice chromosome 1. *Nature,* **420**: 312-316.

Sexton, P.J. and Shibles, R.M. (1999) Activity of ATP sulphurylase in reproductive soybean. *Crop Sci.,* **39**: 131-135.

Shelden, M.C., Loughlin, P., Tierney, M.L. and Howitt, S.M. (2001) Proline residues in two tightly coupled helices of the sulphate transporter, SHST1, are important for sulphate transport. *Biochem. J.,* **356**: 589-594.

Shibagaki, N., Rose, A., McDermott, J.P., Fujiwara, T., Hayashi, H., Yoneyama, T. and Davies, J.P. (2002) Selenate-resistant mutants of *Arabidopsis thaliana* identify Sultr1;2, a sulfate transporter required for efficient transport of sulfate into roots. *Plant J.,* **29**: 475-486.

Shibagaki, N. and Grossman, A.R. (2004) Probing the function of STAS domains of the Arabidopsis sulfate transporters. *J. Biol. Chem.,* **279**: 30791-30799.

Shingles, R., Roh, M.H. and McCarty, R.E. (1996) Nitrite transport in chloroplast inner envelope vesicles. I. Direct measurement of proton-linked transport. *Plant Physiol.,* **112**: 1375-1381.

Sigrist, C.J, Cerutti, L., Hulo, N., Gattiker, A., Falquet, L., Pagni, M., Bairoch, A. and Bucher, P. (2002) PROSITE: a documented database using patterns and profiles as motif descriptors. *Brief Bioinform.,* **3**: 265-274.

Smith, F.W., Hawkesford, M.J., Prosser, I.M. and Clarkson, D.T. (1995a) Isolation of a cDNA from *Saccharomyces cerevisiae* that encodes a high affinity sulphate transporter at the plasma membrane. *Mol. Gen. Genet.,* **247**: 709-715.

Smith, F.W., Ealing, P.M., Hawkesford, M.J. and Clarkson, D.T. (1995b) Plant members of a family of sulfate transporters reveal functional subtypes. *Proc. Natl. Acad. Sci. USA,* **92**: 9373-9377.

Smith, F.W., Hawkesford, M.J., Ealing, P.M., Clarkson, D.T., van den Berg, P.J., Belcher, A.R. and Warrilow, A.G.S. (1997) Regulation of expression of a cDNA from barley roots encoding a high affinity sulphate transporter. *Plant J.,* **12**: 875-884.

Sunarpi and Anderson, J.W. (1996) Effect of sulfur nutrition on the redistribution of sulfur in vegetative soybean plants. *Plant Physiol.,* **112**: 623-631.

Tabe, L.M. and Droux, M. (2001) Sulfur assimilation in developing lupin cotyledons could contribute significantly to the accumulation of organic sulfur reserves in the seed. *Plant Physiol.,* **126**: 176-187.

Tabe, L.M. and Droux, M. (2002) Limits to sulfur accumulation in transgenic lupin seeds expressing a foreign sulfur-rich protein. *Plant Physiol.,* **128**: 1137-1148.

Tabe, L.M., Venables, I., Grootemaat, A. and Lewis, D. (2003) Sulfur transport and assimilation in developing embryos of chickpea (*Cicer arietinum*). In: *Sulfur Transport and Assimilation in Plants.* (Eds. Davidian, J.-C., Grill, D., Kok, L.J., Stulen, I., Hawkesford, M.J., Schnug, E. and Rennenberg, H.), Backhuys Publishers, Leiden, The Netherlands, pp 335-337.

Takahashi, H., Sasakura, N., Noji, M. and Saito, K. (1996) Isolation and characterization of a cDNA encoding a sulfate transporter from *Arabidopsis thaliana*. *FEBS Lett.,* **392**: 95-99.

Takahashi, H., Yamazaki, M., Sasakura, N., Watanabe, A., Leustek, T., de Almeida-Engler, J., Engler, G., van Montagu, M. and Saito, K. (1997) Regulation of sulfur assimilation in higher plants: a sulfate transporter induced in sulphate starved roots plays a central role in *Arabidopsis thaliana*. *Proc. Natl. Acad. Sci. USA,* **94**: 11102-11107.

Takahashi, H., Asanuma, W. and Saito, K. (1999a) Cloning of an Arabidosis cDNA encoding a chloroplast localizing sulfate transporter isoform. *J. Exp. Bot.,* **50**: 1713-1714.

Takahashi, H., Sasakura, N., Kimura, A., Watanabe, A. and Saito, K. (1999b) Identification of two leaf-specific sulfate transporter in *Arabidopsis thaliana* (accession no. AB012048 and AB004060) (PGR99-154). *Plant Physiol.,* **121**: 686.

Takahashi, H., Watanabe-Takahashi, A., Smith, F.W., Blake-Kalf, M., Hawkesford, M.J. and Saito, K. (2000) The roles of three functional sulfate transporters involved in uptake and translocation of sulfate in *Arabidopsis thaliana. Plant J.,* **23**: 171-182.

Takahashi, H., Watanabe-Takahashi, A. and Yamaya, T. (2003) T-DNA insertion mutagenesis of sulfate transporters in Arabidopsis. In: *Sulfur Transport and Assimilation in Plants.* (Eds. Davidian, J.C., Grill, D., Kok, L.J., Stulen, I., Hawkesford, M.J., Schnug E., Rennenberg H.). Backhus Publishers, Leiden, The Netherlands, pp. 339-340.

Takahashi, H. (2005) Functions and regulation of plant sulfate transporters. In: *Sulfur Transport and Assimilation in Plants in the Post Genomic Era* (Eds. Saito, K., De Kok, L.J., Stulen, I., Hawkesford, M.J., Schnug, E., Sirko, A. and Rennenberg, H.), Backhuys Publishers, Leiden, The Netherlands, pp 13-21.

The Arabidopsis Initiative (2000) Analysis of the genome sequence of the flowering plant *Arabidopsis thaliana. Nature,* **408**: 796-815.

Thompson, J.D., Gibson, T.J., Plewniak, F., Jeanmougin, F. and Higgins, D.G. (1997) The ClustalX windows interface: flexible strategies for multiple sequence alignment aided by quality analysis tools. *Nucleic Acids Res.,* **24**: 4876-4882.

Vauclare, P., Kopriva, S., Fell, D., Suter, M., Sticher, L., von Ballmoos, P., Kahnenbuhl, U., Op den Camp, R. and Brunold, C. (2002) Flux control of sulphate assimilation in *Arabidopsis thaliana*: adenosine 5'-phosphosulphate reductase is more susceptible than ATP sulphurylase to negative control by thiols. *Plant J.,* **31**: 729-740.

Vidmar, J.J., Schjoerring, J.K., Touraine, B. and Glass, A.D.M. (1999) Regulation of the *hvst1* gene encoding a high-affinity sulfate transporter from *Hordeum vulgare. Plant Mol. Biol.,* **40**: 883-892.

Vidmar, J.J., Tagmount, A., Cathala, N., Touraine, B. and Davidian, J.-C. (2000) Cloning and characterization of a root specific high-affinity sulfate transporter from *Arabidopsis thaliana. FEBS Lett.,* **475**: 65-69.

Wittstock, U. and Halkier, B.A. (2002) Glucosinolate research in the Arabidopsis era. *Trends Plant Sci.,* **7**: 263-270.

Yoshimoto, N., Takahashi, H., Smith, F.W., Yamaya, T. and Saito, K. (2002) Two distinct high-affinity sulfate transporters with different inducibilities mediate uptake of sulfate in Arabidopsis root. *Plant J.,* **29**: 465-473.

Yoshimoto, N., Inoue, E., Saito, K., Yamaya, T. and Takahashi, H. (2003) Phloem-localizing sulfate transporter, Sultr1;3, mediates re-distribution of sulfur from source to sink organs in Arabidopsis. *Plant Physiol.,* **131**: 1511-1517.

Yu, J., Hu, S., Wang, J., Wong, G.K.S., Li, S., Liu, B., Deng, Y., Dai, L., Zhou, Y., Zhang, X. et al. (2002) A draft sequence of the rice genome (*Oryza sativa* L. ssp. indica). *Science,* **296**: 79-92.

Zhao, J.M., Williams, C.C. and Last, R.L. (1998) Induction of Arabidopsis tryptophan pathway enzymes and camalexin by amico acid starvation, oxidative stress, and an abiotic elicitor. *Plant Cell,* **10**: 359-370.

# Chapter 5

## PHOSPHATE UPTAKE AND TRANSPORT TO PLANT CELLS

### TOSHIO SANO[1] AND TOSHIYUKI NAGATA[2]

[1]*Graduate School of Frontier Sciences, The University of Tokyo, Tokyo, Japan*
[2]*Graduate School of Science, The University of Tokyo, Tokyo, Japan*
E-mail: tsano@k.u-tokyo.ac.jp

**Abstract**

Phosphorus is one of the 17 plant essential elements. Phosphate (Pi), the available form of phosphorus for plants and other living organisms, however, exists only at a very low concentration in soil. As plants are usually exposed to the Pi limiting condition, we have to add sufficient amounts of Pi fertilizer to increase crop yield at reasonable levels. This handling causes two problems. One is the shortage of the phosphate rocks for the Pi fertilizer, while the other is the water pollution caused by the excess Pi remains in soil. In this review, to circumvent these Pi problems, we first provide an overview of physiological characteristics of Pi uptake into plants and molecular characterization of plant Pi transport proteins. Secondly, we will focus on plant responses to the Pi limiting condition together with the regulation of Pi-starvation inducible (PSI) gene expression including Pi transporter genes. Thirdly, plant cell growth and responses to Pi uptake will be described. Finally, possible ways to improve Pi acquisition by plants will be discussed in considering the views of economical Pi recycling.

**Keywords:** phosphate, transporters, fertilizer, phosphate-starvation

## 1. INTRODUCTION

Phosphorus (P) is one of the 17 essential elements required for plant growth and development. This molecular species taken up into the plant cells is used as building blocks for macromolecules such as nucleic acids, sugar-phosphate intermediates and phospholipids (Taiz and Zeiger, 1998). P is also used in high energetic transfer processes such as ATP and in signal transduction pathways in which phosphorylation and de-phosphorylation play molecular switches of biological reactions. Phosphorus has a role in buffering action in cytoplasm. Thus, the P uptake is one of prerequisites for plant cell growth as well as basic biological processes.

---

© CAB International 2008. *Plant Membrane and Vacuolar Transporters* (eds P.K. Jaiwal, R.P. Singh and O.P. Dhankher)

In the rhizosphere, P may be abundant; however, 20 to 80% of P is found in organic forms or insoluble complexes with cations, which are unavailable for plants and other living organisms (Schachtman et al., 1998; Vance et al., 2003). Plants are the source of phosphorus for other organisms. The available form of phosphorus for plants and other living organisms is phosphate (Pi), which exists only in very low concentrations on earth. In soil, the available Pi seldom exceeds 10 µM that is about 1000 times lower than that exists in plants (Bieleski, 1973; Raghothama, 1999). This low concentration of Pi tends to easily result in Pi deficient conditions of plants, along with its low diffusion rate in soil.

To increase crop yields to reasonable levels, we have to add a large amount of Pi fertilizers as well as nitrogen and potassium ones. Regarding this application of Pi fertilizers, two points should carefully be considered. One point is that the sources for the fertilizer mostly of phosphate rocks, are quite limited and are said to be exhausted within 100 years (Runge-Metzger, 1995). The other point is, because of the low efficiency of Pi uptake ability in plants and low mobility of Pi in soil, the excess Pi fertilizers flow into watercourses and often result in environmental problems, such as water pollution and eutrophication. To circumvent these environmental problems and to possibly reduce the consumption of the fertilizer, one strategy may be to cultivate crops requiring less Pi fertilizers. In this process, Pi transporters should potentially play key roles, as such plants could possibly circumvent above-mentioned problems. Therefore, this review first aims to provide an overview of the physiological nature of Pi uptake into plant cells and the molecular characteristics of Pi transporters mediating this process. Secondly, it will be considered to review strategies for plants to adapt to Pi limiting conditions and the mode of cell growth under Pi deficiency and recovery of growth upon addition of Pi. Finally, possible ways to improve Pi acquisition by plants will be discussed in considering the views of economical Pi recycling. More detailed discussions are referred to in previous reviews on the physiological nature of Pi uptake and translocation (Bieleski, 1973; Mimura, 1999), on responses of plants to Pi starvation (Raghothama, 1999; Vance et al., 2003) and on the molecular mechanism of Pi transport (Rausch and Bucher, 2002).

## 2. PHYSIOLOGICAL NATURE OF PHOSPHATE UPTAKE INTO PLANT CELLS

Pi transport across plasma membrane has been measured in various plant species, usually by using radioisotopic tracers ($^{32}P$ and $^{33}P$). Calculation of the kinetic parameters of $V_{max}$ (the maximum reaction rate at infinite substrate concentration) and $K_m$ (the substrate concentration required at $V_{max}/2$ that shows affinity to the substrate) indicates the presence of multiple Pi uptake systems in plants (Furihata et al., 1992; Mimura, 1999). One is a high affinity system that could uptake very low concentrations (below 10 µM) of Pi, while the other is a low affinity system that could transport relatively high concentration of Pi. Dependent on the Pi nutritional status or the level of Pi supply, plants utilize such systems as shown below.

The $K_m$ value of the high affinity Pi transport system (usually several micro molars) is fit for the Pi concentration in soil. Because the Pi concentrations in plant cells are calculated to be about 10 mM which is 1000 to 10000 fold higher than that in soil, Pi has to be transported against both electrical and concentration gradients. Thus Pi is believed to be transported into the cells via cotransport with $H^+$ by using an electro-chemical $H^+$ gradient produced by plasma membrane $H^+$-ATPases (Mimura, 1999). The mode of this transport mechanism has been demonstrated by measuring the change in the extracellular and cytoplasmic pH upon Pi uptake. The former can be measured with a glass pH electrode, while the latter by a fluorescent pH indicator, 2',7'-bis (2-carboxyethyl)-5(6)-carboxyfluorescein (BCECF). In *Catharanthus roseus* cultured cells, upon the Pi uptake, extracellular pH is increased, while the cytoplasmic pH is decreased (Sakano, 1990; Sakano *et al.*, 1992). Similar cytoplasmic acidification by Pi uptake is observed in barley leaves (Mimura *et al.*, 1992). After the exhaustion of Pi in the medium, both the extracellular and cytoplasmic pH recovered to the original values by the $H^+$ export activity of the plasma membrane $H^+$-ATPase (Sakano *et al.*, 1992). In the low affinity Pi-transport system, mediation by $H^+$/Pi cotransporters has also been suggested from inhibition of the activity by $H^+$ uncouplers (Schmidt *et al.*, 1992). The $H^+$/Pi cotransport in the low affinity system was confirmed by the pH dependent activity of PHT2;1, an Arabidopsis low affinity Pi transport protein, in yeast complementation (Daram *et al.*, 1999).

## 3. PLASMA MEMBRANE AND ORGANELLE PHOSPHATE TRANSPORTERS

Recent molecular biological studies have identified many plant Pi transporter genes that mediate Pi transport activities mentioned above. The plant Pi transporter genes were first identified in Arabidopsis and potato by searching for homologs of yeast PHO84 Pi transporter (Muchhal *et al.*, 1996; Mitsukawa *et al.*, 1997a; Smith *et al.*, 1997; Leggewie *et al.*, 1997). The Arabidopsis Pi transporters were first named by various names such as AtPT, APT and PHT, but after a proposal to systematically categorize plant Pi transporters, these proteins are now called ARAth;Pht1;x and belong to a Pht1 family (Table 1, Rausch and Bucher, 2002). The canonical structure of the Pht1 family proteins contains 12 transmembrane domains (TMs), in which a large hydrophilic loop was observed between TM6 and TM7, resulting in a repeat of six TM configurations. In the Arabidopsis genome, there have been found nine Pht1 family transporters and until now homologs were identified in other higher plant organisms such as tomato (Daram *et al.*, 1998) and alfalfa (Liu *et al.*, 1998b), while the predicted gene products are registered from grains and industrial crops such as rice, soybean, wheat, maize and tobacco. The Pi transport activities of ARAth;Pht1;1 and ;4 and potato StPT1 and 2 proteins were determined by complementation of yeast *pho84* mutant and are suggested to mediate high-affinity Pi transport system (Muchhal *et al.*, 1996; Leggewie *et al.*, 1997). Many Pht1 family genes are induced in roots especially under Pi depleted condition, implying their role in Pi uptake into roots (Table 1).

**Table 1.** List of representative members of plants and fungi phosphate transporter/translocators. Gene IDs of the Arabidopsis Pi transporters are from 'The Arabidopsis Information Resource' (TAIR, http://www.arabidopsis.org/).

| Species | Gene name | Gene ID | Published name(s) | Acc. No. | Tissue | (-) Pi induced |
|---|---|---|---|---|---|---|
| Arabidopsis | ARAth;Pht1;1 | At5g43350 | PHT1[a] | D86608 | Root[b,c] | Yes[b,c,q] |
| | | | APT2[b] | Y07682 | | |
| | | | AtPT1[c] | U62330 | | |
| | ARAth;Pht1;2 | At5g43370 | PHT2[d] | AB000094 | Root[b] | Yes[b,q] |
| | | | APT1[b] | Y07681 | | |
| | ARAth;Pht1;3 | At5g43360 | PHT3[d] | AB000094 | Root[q] | Yes[q] |
| | | | AtPT4[c] | U97546 | | |
| | ARAth;Pht1;4 | At2g38940 | PHT4[f] | AB016166 | Root[c,q], Leaf[q] | Yes[c,q] |
| | | | AtPT2[c] | U62331 | | |
| | ARAth;Pht1;5 | At2g32830 | PHT5[f] | AB000093 | Root, Leaf[q] | Yes[q] |
| | ARAth;Pht1;6 | At5g43340 | PHT6[f] | AB005746 | Flower[q] | No[q] |
| | ARAth;Pht1;7 | At3g54700 | | | Flower[q] | Yes[q] |
| | ARAth;Pht1;8 | At1g20860 | | | Root[q] | Yes[q] |
| | ARAth;Pht1;9 | At1g76430 | | | Root[q] | Yes[q] |
| | ARAth;Pht2;1 | At3g26570 | Pht2;1[g] | AJ302645 | Leaf[g], Chloroplast[r] | No[g,r] |
| | ARAth;Pht3;1 | At5g14040 | | AB016066 | | n.d. |
| | ARAth;Pht3;2 | At3g48850 | | | | |
| | ARAth;Pht3;3 | At2g17270 | | | | |
| | AtTPT | At5g46110 | | AY050811 | | n.d. |
| | AtPPT | At5g33320 | | AY080788 | | n.d. |
| | AtGPT1 | At5g54800 | | AY057679 | | n.d. |

**Table 1.** Continued

| Species | Gene name | Gene ID | Published name(s) | Acc. No. | Tissue | (-) Pi induced |
|---|---|---|---|---|---|---|
| | *AtGPT2* | At1g61800 | | AY081479 | | n.d. |
| | *AtXPT* | At5g17640 | *XPT*[h] | AF209211 | Flower, Leaf, Stem, Root[h] | n.d. |
| | *PHO1* | At3g23430 | *PHO1*[i] | AF474076 | Root stelar cells[i] | Yes[i] |
| Potato | | | *StPT1*[j] | X98890 | Root, Leaf[j] | Yes[j] |
| | | | *StPT2*[j] | X98891 | Root[j] | Yes[j] |
| | | | *StPT3*[k] | AJ318822 | Arbuscule-containing root[k] | No[k] |
| Tomato | | | *LePT1*[l] | AF022973 | Root, Leaf[l] | Yes[l] |
| | | | *LePT2*[l] | AF022874 | Root[l] | Yes[l] |
| *Medicago truncatula* | | | *MtPT1*[m] | AF000354 | Root, Root hair[s] | Yes[s] |
| *Saccharomyces cerevisiae* | | | *PHO84*[n] | D90346 | | Yes[n] |
| *Neurospora crassa* | | | *pho-5*[+o] | L36127 | | n.d. |
| *Glomus versiforme* | | | *GvPT*[p] | U38650 | | n.d. |

[a]Mitsukawa *et al.*, 1997a; [b]Smith *et al.*, 1997; [c]Muchhal *et al.*, 1996; [d]Mitsukawa *et al.*, 1997b; [e]Lu *et al.*, 1997; [f]Okumura *et al.*, 1998; [g]Daram *et al.*, 1999; [h]Eicks *et al.*, 2002; [i]Hamburger *et al.*, 2002; [j]Leggewie *et al.*, 1997; [k]Rausch *et al.*, 2001; [l]Liu *et al.*, 1998a; [m]Liu *et al.*, 1998b; [n]Bun-ya *et al.*, 1991; [o]Versaw, 1995; [p]Harrison and Van Buuren, 1995; [q]Mudge *et al.*, 2002; [r]Versaw *et al.*, 2002; [s]Chiou *et al.*, 2001; n.d.: not determined

The second family of the Pi transporter genes is *Pht2* and one gene of *ARAth;Pht2;1* has been identified in Arabidopsis (Daram *et al.*, 1999). The predicted protein structure showed 12 TMs that have similarity to the proteins in the Pht1 family. However, the Pht2;1 shows no homology with the plant high affinity $H^+$-Pi cotransporter family. Yeast complementation with the Pht2;1 protein showed a Pi transport activity of low affinity (Daram *et al.*, 1999). Reduction of both the accumulation of Pi in leaves and the allocation of Pi throughout the plant in the *pht2;1-1* null mutant suggests its role in the Pi allocation within plants. Versaw and Harrison (2002), however, showed a localization of a Pht2;1-GFP fusion protein in the chloroplast inner envelope and suggested its role in Pi translocation into chloroplasts. Thus far, the mechanisms by which the Pht2;1 affects these processes is still the subject of debate for future investigations.

Arabidopsis *pho1* and *pho2* mutants have been isolated in a screening for plants with altered Pi levels in tissues. The *pho1* mutant showed a defect in Pi transport into shoots, while the *pho2* mutant accumulated excessive Pi in shoots (Poirier *et al.*, 1991; Delhaize and Randall, 1995). PHO1 promoter-GUS analysis revealed a predominant expression in the stelar cells of roots, suggesting its role in Pi efflux out of roots into xylem (Hamburger *et al.*, 2002). Although molecular cloning showed membrane-spanning domains in the predicted protein, PHO1 structure appears quite distinct from other Pi transporters. In addition, as no Pi transport activity of this protein has been demonstrated in heterologous systems thus far (Hamburger *et al.*, 2002), it is supposed that PHO1 may not be involved directly in Pi transport, but rather it could influence the activity of Pi loading to the xylem. The PHO2 protein is suggested to play a part in Pi transport in the phloem between shoots and roots from the mutant analysis (Dong *et al.*, 1998). Recently, the *PHO2* gene was identified and found to encode an E2 conjugase.

After transport into the cells, the fate of Pi is separated into at least three pathways. One pathway is in cytoplasm and plays a role in cytoplasmic Pi homeostasis. The second one is translocation into mitochondria and plastids and used in metabolic pathways such as organic Pi components. The third one is the storage in the vacuole.

As for the plastids, Pi is translocated by proteins that belong to the Pi translocator (PT) family. These proteins antiport Pi with phosphorylated C3 compounds as a counter substrate (Flügge, 1999). Thus far, four groups of PT proteins (TPT, PPT, GPT and XPT) have been identified, which antiport triose phosphate, phosphoenolpyruvate, glucose 6-phosphate and xylulose 5-phosphate, respectively, as counter substrates. PT proteins have 6 TMs, being thought to function as dimers after integration into inner envelope membranes (Eicks *et al.*, 2002).

As for the mitochondria, the majority of Pi is known to be translocated by Pi carriers (Pic) that catalyze the activity of Pi/$OH^-$ antiport in animal systems and yeast. The first plant putative mitochondrial Pi transporter (MPT) gene was identified as an ozone inducible gene in birch (Kiiskinen *et al.*, 1997). Further search revealed the presence of plant MPT genes in herbaceous plants of soybean, maize, rice and Arabidopsis (Takabatake et al., 1999). The predicted MPT proteins have 6 TMs and are proposed to belong to the third family of the Pi transporter (Pht3, Rausch and Bucher, 2002). The recombinant soybean MPT proteins showed Pi transport activity in liposomes (Takabatake et al., 1999); however, the significance of the role of the plant MPTs in relation to mitochondrial metabolism remains to be clarified for further study.

Vacuoles play a role in the cytoplasmic Pi homeostasis as a pool of Pi (Mimura, 1995). In sycamore suspension-cultured cells, *in vivo* measurement of Pi using $^{31}$P-NMR showed a decrease of the vacuolar Pi level upon Pi depletion, while that in cytoplasm was almost constant. When Pi was applied, the cytoplasmic Pi level first increased and subsequently that of the vacuole followed (Rebeille et al., 1983). Measurement of Pi concentrations in isolated protoplasts and vacuoles from barley leaves also showed the above-mentioned tendency (Mimura et al., 1990). The Pi transport activity into isolated vacuoles of cultured *Catharanthus roseus* cells indicated that the building-up of an electrochemical gradient by the H$^+$ pumps was essential for Pi uptake (Massonneau et al., 2000). Recently, proteomic analysis of isolated vacuoles from Arabidopsis has been reported (Carter et al., 2004; Shimaoka et al., 2004); however, molecular identification of the vacuolar Pi transporters has not yet been done. Further identification and characterization of organelle Pi transporters will explain Pi utilization and metabolism after taken-up into cells.

## 4. TRANSCRIPTIONAL CONTROL IN THE RESPONSE TO PHOSPHATE STARVATION

Under Pi limiting conditions, various Pi starvation inducible (PSI) genes have been identified, that include acid phosphatase and RNase together with many Pht1 transporter genes described above (Raghothama, 1999). In yeast, extensive studies revealed a regulatory network in gene expression by Pi depletion called the PHO-regulon (Oshima et al., 1996). In this regulon, induction of the Pi transpoter gene, *PHO84*, in addition to those of acid and alkaline phosphatases (*PHO5* and *PHO8*) upon Pi depleted condition is mediated by a PHO4 transcription factor (Bun-ya et al., 1991). In the high-Pi medium, the PHO4 is negatively regulated by export from nucleus upon phosphorylation by a complex of PHO80 cyclin and PHO85 cyclin-dependent kinase (CDK, O'Neill et al., 1996; Kaffman et al., 1998). Upon Pi-depleted condition, a CDK inhibitor, PHO81, inhibits the function of the PHO80-PHO85 complex, allowing the PHO4 proteins to remain in the nucleus (Schneider et al., 1994). A similar regulon has been suggested to exist in another fungus *Neurospora crassa* and in *E. coli* (Wanner, 1993; Peleg et al., 1996). In plants, the reported examples of many PSI genes suggest the presence of a kind of Pi starvation regulon as well; however, the respective regulon has not yet been identified as plant genes. A recent screening of Pi starvation response mutants has identified genes of *PHR1* (Phosphate Starvation Response1, Rubio et al., 2001) and *PHF1* (Phosphate Transporter Traffic Facilitator1, González et al., 2005). The PHR1 protein contains a putative MYB domain with a predicted coiled-coil domain for protein-protein interaction, suggesting an involvement of transcriptional regulation in Pi starvation, while the PHF1 protein is related to the SEC12 proteins, suggesting a role for the exits of phosphate transporter proteins from ER. On the other hand, a negative regulation of gene expression upon Pi starvation has been suggested by dissociation of the nuclear protein factors under the Pi starved condition from a promoter region of the PSI gene, *ARAth;Pht1;4* (the former name was AtPT2, Mukatira *et al.*, 2001). In addition, repression of many PSI gene expression including the Pi transporter gene of *ARAth;Pht1;1* by cytokinins and impairment of this repression in a cytokinin receptor mutant of *cre1* (Martín *et al.*, 2000; Franco-Zorrilla *et al.*, 2002; Franco-Zorrilla *et al.*, 2004) suggests an involvement of phytohormonal signaling in the Pi-sensing pathway.

Investigation of the regulatory mechanism of Pi-starvation response revealed an induction of a micro RNA of miR399 expression and down-regulation of the predicted target gene of *UBC24* (Chiou *et al.*, 2006). Since a genomic fragment of *UBC24* in *pho2* had a single nucleotide mutation from G to A and introduction of *UBC24* into *pho2* complemented the Pi overaccumulating phenotype, *PHO2* was concluded to be *UBC24* that encoded an E2 conjugase (Aung *et al.*, 2006; Bari *et al.*, 2006). Over-expression of miR399 over-accumulated Pi in shoots which was similar to the *pho2* mutant phenotype and *UBC24* was targeted by the miR399 (Aung *et al.*, 2006; Chiou *et al.*, 2006). These observations imply that plant Pi response includes post-transcriptional and post-translational regulation in addition to transcriptional regulation, which would be more complex than in bacteria and fungi.

## 5. PLANT RESPONSES TO PHOSPHATE LIMITING CONDITIONS

In addition to the induction of Pi transporter genes, plants can adapt to the Pi limiting condition by increasing the amounts of available Pi by several ways. One way is altering plant root architecture (Lynch, 1995). Upon Pi deficiency, it is known that the root-shoot ratio increases and root hair production is enhanced. Their responses are considered to enhance the total root surface area available for soil exploitation. The second way to increase the available Pi is to synthesize and exude organic compounds such as malate and citrate from roots. These organic compounds allow the chelation of $Al^{3+}$, $Fe^{3+}$ and $Ca^{2+}$ from the Pi bound insoluble salts and as a consequence increase the free Pi concentration in soil (Vance *et al.*, 2003). In white lupin (*Lupinus albus*), specially-developed bottlebrush-like clusters of rootlets called 'proteoid roots' are formed, from which more than 10 fold of citrate and malate are synthesized and exuded into soil upon Pi deficiency (Neumann *et al.*, 1999). Secretion of acid phosphatase enzymes (APases) from roots to rhizosphere is also observed, which could release Pi from the soil organic P-esters (Goldstein *et al.*, 1988). These activities by APases, secreted from suspension cultured cells or roots upon Pi starvation, are observed in several plant species such as tomato, barley, rose and *Brassica nigra* (Goldstein *et al.*, 1988; Lee, 1988; Lefebvre *et al.*, 1990; Duff *et al.*, 1991). Molecular cloning has identified the APase cDNAs from Arabidopsis, white lupin and tomato (del Pozo *et al.*, 1999; Wasaki *et al.*, 1999; Baldwin *et al.*, 2001); however, characterization of these proteins does not necessarily mean evidence for their secretion into soil at present. Their induction upon the Pi limited condition suggested a role of these APases in internal mobilization of Pi. The first evidence for its secretion was provided by Haran *et al.* (2000), in which they showed a secretion of a recombinant GFP with the APase signal peptide, identified in Arabidopsis, by the roots into the medium.

Plants could also acquire Pi from symbiotic fungi (Schachtman *et al.*, 1998). In more than 90% of the land plants, symbiotic associations called mycorrhizae are formed between the plant roots and fungi. In the mycorrhiza, fungal hyphae often extend between the root cells and form tuft-like branched tissue called arbuscules. The plant plasma membrane extends to surround the arbuscule and between them, nutrients being transferred bidirectionally; the host plants obtain Pi mainly from the fungus, while the fungus in turn gains carbohydrates from plants. Since the mycorrhizal fungi extend their hyphae in soil areas distant from the plant roots, the plant roots could take up Pi transported through the hyphae. In the germ tubes of the mycorrhizal fungus *Gigaspora margarita* showed two Pi uptake systems of low and high affinities (Thomson *et al.*, 1990). This aspect of Pi utilization by plants is quite important, in particular from viewpoints of ecophysiological consideration. Molecular cloning and characterization of a fungus Pi transporter GvPT from *Glomus versiforme* showed its similarity with the yeast PHO84 and the members of plant Phi1 family and its physiological nature of the high affinity Pi transport activity ($K_m$ = 18 μM) (Harrison and Van Buuren, 1995). In the subsequent transfer process from the arbuscule to the plant roots, two Pi transport steps are required; the Pi transfer from the fungus to the apoplastic space across the arbuscular membrane and subsequently into the plant root cells across the plant plasma membrane. The molecular mechanism for Pi efflux across the arbuscular membrane is not still clear, while regarding the Pi uptake into the plant cells in the mycorrhiza, a Pi transporter named StPT3 was identified from potato (Rausch *et al.*, 2001). StPT3 is similar to other Pht1 proteins and a complementation of the yeast mutant defective in the two high-affinity Pi transporter genes showed its high-affinity Pi transport activity with a $K_m$ value of 64 μM (Rausch *et al.*, 2001). Reporter gene expression by StPT3 promoter in the root sectors where mycorrhizal structures are formed suggests the mediation of StPT3 in the Pi uptake from mycorrhizae to plant cells (Rausch *et al.*, 2001). For further details of the arbuscular mycorrhizal symbiosis, readers can refer Strack *et al.* (2003), Harrison (2005) and Bucher (2007).

## 6. PHOSPHATE UPTAKE AND PLANT CELL GROWTH

Pi taken up into plant cells is used as building blocks for various macromolecules. The metabolism of Pi after the uptake into the cells, however, is not so clear. When the consumption of the mineral nutrients was measured in the culture medium with tobacco BY-2 suspension cultured cells, Pi was found to be consumed almost completely at the initial stage of the batch culture, while other components such as $NO_3^-$, $SO_4^{2-}$, $Ca^{2+}$ and $K^+$ remained in the medium (Kato *et al.*, 1977). Since the cell growth continued even after the complete consumption of Pi in the medium, the supply of Pi from vacuoles as a Pi pool was implied. On the other hand, an increase of the initial concentration of $KH_2PO_4$ in the medium and the resultant growth rate of tobacco BY-2 cells or *C. roseus* cell culture suggest the role of Pi as a key limiting factor for the growth of these cells (Kato *et al.*, 1977; Amino *et al.*, 1983). Application of $KH_2PO_4$ to the Pi-depleted *C. roseus* or tobacco BY-2 cell cultures induced a certain level of cell synchrony (ca 8% of mitotic index for *C. roseus*, while ca 30% for tobacco, Amino *et al.*, 1983; Sano *et al.*, 1999), while in tobacco BY-2 cells, Pi depletion arrested the cell cycle at the G1 phase, which was confirmed by flow cytometric analysis and measurement of DNA synthesis by BrdU incorporation (Wilson *et al.*, 1998; Sano *et al.*, 1999). These observations indicate the induction of cell cycle re-entry from G1 to S phase by the Pi application to the Pi-starved cells. Parallelism of the cell cycle re-entry upon Pi application to the Pi depleted cell culture to a rapid and transient activation of a MAP kinase activity has been reported (Wilson *et al.*, 1998), while two Pi-induced cDNAs named *phi-1* and *phi-2* were identified in BY-2 cells (Sano *et al.*, 1999; Sano and Nagata, 2002). Although a predicted Phi-1 protein showed no significant homology to proteins with known functions, Phi-2 protein had a bZIP motif found in transcriptional factors involved in ABA-signaling pathways. Upon addition of 2.7 mM $KH_2PO_4$ that induced cell cycle re-entry and resulted in high growth rate of the cell culture, the cell cycle re-entry delayed compared to that upon addition of 10 times lower concentration of $KH_2PO_4$ (Sano and Nagata, 2002). This delay of the cell cycle re-entry was accompanied by the increase of the medium pH and the induction of *phi-1* and *phi-2*. Since the medium pH increase by Pi application suggests the cytoplasmic acidification by the co-transported $H^+$ with Pi, Phi-1 and Phi-2 proteins may have roles in alleviating changes of intracellular pH; however, the mechanism of pH homeostasis by Phi-1 and Phi-2 proteins has not yet been determined.

In order to understand the fate of Pi taken up into plant cells, the metabolic pattern of Pi has been measured using biochemical methods. Using high-performance anion-exchange chromatography with pulsed amperometric detection (HPAEC-PAD) and ion chromatography coupled to electrospray ionization tandem mass spectrography (IC-ESI-MS-MS, Sekiguchi *et al.*, 2004; Sekiguchi *et al.*, 2005), 17 compounds of sugar phosphate in Arabidopsis plants could be detected. These methods allowed the detection of phosphorus compounds which were usually interfered with high concentrations of anion peaks and other metabolites produced in higher plants. Myo-inositol hexakisphosphate ($InsP_6$; phytic acid) was a major storage form of Pi and accumulated in vacuoles (Bieleski, 1973). Recently, the mode of its synthesis and vacuolar accumulation was measured in *C. roseus* cultured cells using ion chromatography method (Mitsuhashi *et al.*, 2005). These metabolomic approaches together with the proteome analysis described above and transcriptome analysis (Hammond *et al.*, 2003) will be powerful tools to understand the Pi metabolism and its regulation, which should give us a basic knowledge on Pi turnover, resulting in finding clues to understand the improvement of plant growth.

## 7. TOWARDS OBTAINING PLANTS WITH HIGH PHOSPHATE ACQUISITION

Based on the above-mentioned information, improvement of Pi acquisition by plants has been tried in several ways (Hammond *et al.*, 2004). Tobacco plants overproducing citrate by expressing a citrate synthase by genetic means enhanced Pi uptake under the Pi limiting condition and increased the plant dry weight (López-Bucio *et al.*, 2000). Similar results were obtained in carrot cultured cells and in Arabidopsis plants over-expressing citrate synthase (Koyama *et al.*, 1999; Koyama *et al.*, 2000). The exudated citrate is presumed to chelate $Al^{3+}$ from the insoluble Al-Pi complex, resulting in the increase of $Al^{3+}$ tolerance in addition to the increased Pi availability in soil (de la Fuente *et al.*, 1997; Koyama *et al.*, 1999). However, Delhaize *et al.* (2001) have reported that this method was not easily reproducible and discussed the sensitivity of the citrate synthase to environmental conditions. Arabidopsis plants secreting phytase by expressing a phytase gene from *Aspergillus niger* could grow on the medium supplemented with phytic acid as a sole P nutrient (Richardson *et al.*, 2001; Mudge *et al.*, 2003). These plants are supposed to utilize insoluble inorganic Pi as well as organic P fractions which occupy major portions of the soil P. Modification of plants to secrete more APases or enhance the symbiosis with mycorrhizae would also improve the Pi acquisition. Over-expression of the Pi transporter genes may improve the Pi acquisition. In fact, Mitsukawa *et al.* (1997a) have shown that the over-expression of *ARAth;Pht1;1* in tobacco BY-2 cell culture increased the Pi uptake activity. However, the total amount of the cell mass which was determined one week after the start of cell culture was almost the same that was obtained by the non-transformed cell culture. This will be explained by the fact that even if cells have an increased Pi uptake activity, the final growth mass could become the same, since the final Pi amounts taken up into the cell culture would not be changed much.

To reduce the consumption of the phosphate rock for the source of Pi fertilizer, processing of phosphorus recycling has been developed from sludge and industrial sewage by biotechnological means (de-Bashan and Bashan, 2004). Today, as the main civil engineering processes, chemical precipitation removes phosphorus from the waste water by the addition of alkali, in such forms as iron, alum and lime (Donnert and Salecker, 1999). Phosphorus in the sewage is taken up by biological means into bacteria, in which the Pi is stored as polyphosphate (Ohtake *et al.*, 1985; Stratful *et al.*, 1999). These processes could remove phosphorus from the waste water and thus reduce the water pollution; however, these methods do not recycle phosphorus for the fertilizers, since phosphorus is removed along with various other waste products such as heavy metals. Therefore, to recycle phosphorus, methods for the extraction are being pursued. Recently from the polyphosphate-accumulated bacteria, an efficient method of phosphorus extraction was developed by a simple heat treatment of the bacteria (Kuroda *et al.*, 2002). With this method, incubation of the bacteria at 70°C for 1 h eluted the most amount of the polyphosphate into the water phase. The isolated polyphosphate can be precipitated by the addition of calcium, resulting in forming artificial rock phosphate. The phosphorus recycling by the chemical methods may be a direct crystallization of the Pi raw materials such as struvite in the waste water (de-Bashan and Bashan, 2004). Struvite is a common name of magnesium ammonium phosphate hexahydrate ($MgNH_4PO_4 \cdot H_2O$) and spontaneously precipitates in some waste water processes (Booker et al., 1999; Stratful et al., 1999). Now controlled and cost-effective methods for the formation and collection are being searched for. Adsorption of phosphorus in the waste water using a strongly sorbing filter material will be another option. With this method, two possible filter materials of blast furnaces, slag and opoka, were proposed (Johansson and Gustafsson, 2000). The former is an industrial byproduct that derives from the separation of iron from ore, while the latter is a name of siliceous sedimentary rock produced in Poland and Russia. Using these materials, according to the absorption of P, Ca concentration in the waste water was decreased (Johansson and Gustafsson, 2000), resulting in precipitation of hydroxyapatite, which has a similar structure to fluoricapatite, a major compound of the phosphate rock. These crystallization methods may produce potential raw materials for the fertilizer industry; however, they should be improved further to a commercially profitable level.

## 8. CONCLUSION AND FUTURE PROSPECTS

Understanding of efficient Pi utilization in plants is very important as sources of Pi in nature are very limited and its loss causes a serious environmental hazard such as water pollution and eutrophication in water bodies. Many Pi transporter proteins and their respective genes have been identified and characterized in plants and other organisms, which are induced by Pi-starvation. Like other plant processes, Pi-uptake and transport is also a well coordinated process and more studies on molecular mechanism involved in its regulation will pave a way to develop more efficient Pi-utilizing plants.

## 9. LITERATURE CITED

Amino, S., Fujimura, T. and Komamine, A. (1983). Synchrony induced by double phosphate starvation in a suspension culture of *Catharanthus roseus*. *Physiol. Plant.*, **59**: 393–396.

Aung, K., Lin, S.I., Wu, C.C., Huang, Y.T., Su, C.L. and Chiou, T.J. (2006). *pho2*, a phosphate overaccumulator, is caused by a nonsense mutation in a microRNA399 target gene. *Plant Physiol.*, **141**: 1000-1011.

Baldwin, J.C., Athikkattuvalasu, S.K. and Raghothama, K.G. (2001). *LEPS2*, a phosphorus starvation-induced novel acid phosphatase from tomato. *Plant Physiol.*, **125**: 728–737.

Bari, R., Datt Pant, B., Stitt, M. and Scheible, W.R. (2006) PHO2, microRNA399 and PHR1 define a phosphate-signaling pathway in plants. *Plant Physiol.*, **141**: 988-999.

Bieleski, R.L. (1973). Phosphate pools, phosphate transport and phosphate availability. *Annu. Rev. Plant Physiol.*, **24**: 225-252.

Booker, N.A., Priestley, A.J. and Fraser, I.H. (1999). Struvite formation in wastewater treatment plants opportunities for nutrient recovery. *Environ. Technol.*, **20**: 777–782.

Bucher, M. (2007). Functional biology of plant phosphate uptake at root and mycorrhiza interfaces. *New Phytol.*, **173**: 11-26.

## Phosphate uptake to plant cells

Bun-Ya, M., Nishimura, M., Harashima, S., Oshima, Y. (1991). The *PHO84* gene of *Saccharomyces cerevisiae* encodes an inorganic phosphate transporter. *Mol. Cell Biol.,* **11**: 3229-3238.

Carter, C., Pan, S., Zouhar, J., Avila, E.L., Girke, T. and Raikhel, N.V. (2004). The vegetative vacuole proteome of *Arabidopsis thaliana* reveals predicted and unexpected proteins. *Plant Cell,* **16**: 3285-3303.

Chiou, T.J., Liu, H. and Harrison, M.J. (2001). The spatial expression patterns of a phosphate transporter (MtPT1) from *Medicago truncatula* indicate a role in phosphate transport at the root/soil interface. *Plant J.,* **25**: 281-293.

Chiou, T.J., Aung, K., Lin, S.I., Wu, C.C., Chiang, S.F. and Su, C.L. (2006) Regulation of phosphate homeostasis by microRNA in Arabidopsis. *Plant Cell,* **18**: 412-421.

Daram, P., Brunner, S., Persson, B.L., Amrhein, N. and Bucher, M. (1998). Functional analysis and cell-specific expression of a phosphate transporter from tomato. *Planta,* **206**: 225-233.

Daram, P., Brunner, S., Rausch, C., Steiner, C., Amrhein, N. and Bucher, M. (1999). *Pht2;1* encodes a low-affinity phosphate transporter from Arabidopsis. *Plant Cell,* **11**: 2153-2166.

de-Bashan, L.E. and Bashan, Y. (2004) Recent advantages in removing phosphorus from wastewater and its future use as fertilizer (1997-2003). *Water Res.,* **38**: 4222-4246.

de la Fuente, J.M., Ramírez-Rodríguez, V., Cabrera-Ponce, J.L. and Herrera-Estrella, L. (1997). Aluminum tolerance in transgenic plants by alteration of citrate synthesis. *Science,* **276**: 1566-1568.

Delhaize, E. and Randall, P.J. (1995). Characterization of a phosphate-accumulator mutant of *Arabidopsis thaliana*. *Plant Physiol.,* **107**: 207-213.

Delhaize, E., Hebb, D.M. and Ryan, P.R. (2001). Expression of a *Pseudomonas aeruginosa* citrate synthase gene in tobacco is not associated with either enhanced citrate accumulation or efflux. *Plant Physoil.,* **125**: 2059-2067.

del Pozo, J.C., Allona, I., Rubio, V., Layva, A., de la Peña, A., Aragoncillo, C. and Paz-Ares, J. (1999). A type 5 acid phosphatase gene from *Arabidopsis thaliana* is induced by phosphate starvation and by some other types of phosphate mobilizing/oxidative stress conditions. *Plant J.,* **19**: 579-589.

Dong, B., Rengel, Z. and Delhaize, E. (1998). Uptake and translocation of phosphate by *pho2* mutant and wild-type seedlings of *Arabidopsis thaliana*. Planta, **205**: 251-256.

Donnert, D. and Salecker, M. (1999). Elimination of phosphorus from waste water by crystallization. *Environ. Technol.,* **20**: 735–742.

Duff, S.M.G., Plaxton, W.C. and Lefebvre, D.D. (1991). Phosphate-starvation response in plant cells: de novo synthesis and degradation of acid phosphatase. *Proc. Natl. Acad. Sci. USA,* **88**: 9538-9542.

Eicks, M., Maurino, V., Knappe, S., Flügge, U.I. and Fischer, K. (2002). The plastidic pentose phosphate translocator represents a link between the cytosolic and the plastidic pentose phosphate pathways in plants. *Plant Physiol.,* **128**: 512-522.

Flügge, U.I. (1999). Phosphate translocators in plastids. *Annu. Rev. Plant Physiol. Plant Mol. Biol.,* **50**: 27-45.

Franco-Zorrilla, J.M., Martín, A.C., Solano, R., Rubio, V., Leyva, A. and Paz-Ares, J. (2002). Mutations at *CRE1* impair cytokinin-induced repression of phosphate starvation responses in Arabidopsis. *Plant J.,* **32**: 353–360.

Franco-Zorrilla, J.M., González, E., Bustos, R., Linhares, F., Leyva, A. and Paz-Ares, J. (2004). The transcriptional control of plant responses to phosphate limitation. *J. Exp. Bot.,* **55**: 285–293.

Furihata, T., Suzuki, M. and Sakurai, H. (1992). Kinetic characterization of two phosphate uptake systems with different affinities in suspension-cultured *Catharanthus roseus* protoplasts. *Plant Cell Physiol.,* **33**: 1151-1157.

Goldstein, A.H., Baertelein, D.A. and McDaniel, R.G. (1988). Phosphate starvation inducible metabolism in *Lycopersicon esculentum. Plant Physiol.,* **87**: 711-715.

González, E., Solano, R., Rubio, V., Leyva, A. and Paz-Ares, J. (2005). PHOSPHATE TRANSPORTER TRAFFIC FACILITATOR1 is a plant-specific SEC12-related protein that enables the endoplasmic reticulum exit of a high-affinity phosphate transporter in Arabidopsis. *Plant Cell,* **17**: 3500-3512.

Hamburger, D., Rezzonico, E., MacDonald-Comber Petétot, J., Somerville, C. and Poirier, Y. (2002). Identification and characterization of the Arabidopsis *PHO1* gene involved in phosphate loading to the xylem. *Plant Cell,* **14**: 889-902.

Hammond, J.P., Bennett, M.J., Bowen, H.C., Broadley, M.R., Eastwood, D.C., May, S.T., Rahn, C., Swarup, R. Woolaway, K.E. and White, P.J. (2003). Changes in gene expression in Arabidopsis shoots during phosphate starvation and the potential for developing smart plants. *Plant Physiol.,* **132**: 578-596.

Hammond, J.P., Broadley, M.R. and White, P.J. (2004). Genetic responses to phosphorus deficiency. *Ann. Bot.,* **94**: 323-332.

Haran, S., Longendra, S., Bratanova, M. and Raskin, I. (2000). Characterization of Arabidopsis acid phosphatase promoter and regulation of acid phosphatase expression. *Plant Physiol.,* **124**: 615-626.

Harrison, M.J. (2005). Signaling in the arbuscular mycorrhizal symbiosis. *Annu. Rev. Microbiol.,* **59**: 19-42.

Harrison, M.J. and Van Buuren, M.L. (1995). A phosphate transporter from the mycorrhizal fungus *Glomus versiforme. Nature,* **378**: 626-629.

Johansson, L. and Gustafsson, J.P. (2000). Phosphate removal using blast furnace slags and opoka-mechanisms. *Water Res.,* **34**: 259-265.

Kaffman, A., Rank, N.M., O'Neill, E.M., Huang, L.S. and O'Shea, E.K. (1998). The receptor Msn5 exports the phosphorylated transcription factor *Pho4* out of the nucleus. *Nature,* **396**: 482-486.

Kato, A., Fukasawa, A., Shimizu, Y., Soh, Y. and Nagai, S. (1977). Requirements of $PO_4^{3-}$, $NO_3^-$, $SO_4^{2-}$, $K^+$, and $Ca^{2+}$ for the growth of tobacco cells in suspension culture. *J. Ferment. Technol.,* **55**: 207–212.

Kiiskinen, M., Korhonen, M. and Kangasjarvi, J. (1997). Isolation and characterization of cDNA for a plant mitochondrial phosphate translocator (Mpt1): ozone stress induces *Mpt1* mRNA accumulation in birch (*Betula pendula* Roth). *Plant Mol. Biol.,* **35**: 271-279.

Koyama, H., Takita, E., Kawamura, A., Hara, T. and Shibata, D. (1999). Over expression of mitochondrial citrate synthase gene improves the growth of carrot cells in Al-phosphate medium. *Plant Cell Physiol.,* **40**: 482–488.

Koyama, H., Kawamura, A., Kihara, T., Hara, T., Takita, E. and Shibata, D. (2000) Over-expression of mitochondrial citrate synthase in *Arabidopsis thaliana* improved growth on a phosphorus-limited soil. *Plant Cell Physiol.,* **41**: 1030-1037.

Kuroda, A., Takiguchi, N., Gotanda, T., Nomura, K., Kato, J., Ikeda, T. and Ohtake, H. (2002). A simple method to release polyphosphate from activated sludge for phosphorus reuse and recycling. *Biotechnol. Bioeng.,* **78**: 333–338.

Lee, R.B. (1988). Phosphate influx and extracellular phosphatase activity in barley roots and rose cells. *New Phytol.,* **109**: 141-148.

Phosphate uptake to plant cells

Lefebvre, D.D., Duff, S.M., Fife, C., Julien-Inalsingh, C. and Plaxton, W.C. (1990). Response to phosphate deprivation in *Brassica nigra* suspension cells. Enhancement of intracellular, cell surface and secreted phosphatase activities compared to increases in Pi-absorption rate. *Plant Physiol.,* **93**: 504-511.
Leggewie, G., Willmitzer, L. and Riesmeier, J.W. (1997). Two cDNAs from potato are able to complement a phosphate uptake-deficient yeast mutant: identification of phosphate transporters from higher plants. *Plant Cell,* **9**: 381-392.
Liu, C., Muchhal, U.S., Uthappa, M., Kononowicz, A.K. and Raghothama, K.G. (1998a). Tomato phosphate transporter genes are differentially regulated in plant tissues by phosphorus. *Plant Physiol.,* **116**: 91-99.
Liu, H., Trieu, A.T., Blaylock, L.A. and Harrison, M.J. (1998b). Cloning and characterization of two phosphate transporters from *Medicago truncatula* roots: regulation in response to phosphate and to colonization by arbuscular mycorrhizal (AM) fungi. *Mol. Plant Microbe Interact.,* **11**: 14-22.
López-Bucio, J., de la Vega, O.M., Guevara-Garcia, A. and Herrera-Estrella, L. (2000). Enhanced phosphorus uptake in transgenic tobacco plants that overproduce citrate. *Nature Biotech.,* **18**: 450-453.
Lu, Y.-P., Zhen, R.-G. and Rea, P.A. (1997). *AtPT4*: a fourth member of the Arabidopsis phosphate transporter gene family. *Plant Physiol.,* **114**: 747.
Lynch, L. (1995). Root architecture and plant productivity. *Plant Physiol.,* **109**: 7-13.
Martín, A.C., del Pozo, J.C., Iglesias, J., Rubio, V., Solano, R., de la Peña, A., Leyva, A. and Paz-Ares, J. (2000). Influence of cytokinins on the expression of phosphate starvation responsive genes in Arabidopsis. *Plant J.,* **24**: 559-567.
Massonneau, A., Martinoia, E., Dietz, K.J. and Mimura, T. (2000). Phosphate uptake across the tonoplast of intact vacuoles isolated from suspension-cultured cells of *Catharanthus roseus* (L.) G. Don. *Planta,* **211**: 390-395.
Mimura, T. (1995). Homeostasis and transport of inorganic phosphate in plants. *Plant Cell Physiol.,* **36**: 1-7.
Mimura, T. (1999). Regulation of phosphate transport and homeostasis in plant cells. *Int. Rev. Cytol.,* **191**: 149-200.
Mimura, T., Dietz, K., Kaiser, W., Schramm, M., Kaiser, G. and Heber, U. (1990). Phosphate transport across biomembranes and cytosolic phosphate homeostasis in barley leaves. *Planta,* **180**: 139-146.
Mimura, T., Yin, Z.H., Wirth, E. and Dietz, K.J. (1992). Phosphate transport and apoplastic phosphate homeostasis in barley leaves. *Plant Cell Physiol.,* **33**: 563-568.
Mitsuhashi, N., Ohnishi, M., Sekiguchi, Y., Kwon, Y.U., Chang, Y.T., Chung, S.K., Inoue, Y., Reid, R.J., Yagisawa, H. and Mimura, T. (2005). Phytic acid synthesis and vacuolar accumulation in suspension-cultured cells of *Catharanthus roseus* induced by high concentration of Pi and cations. *Plant Physiol.,* **138**: 1607-1614.
Mitsukawa, N., Okumura, S., Shirano, Y., Sato, S., Kato, T., Harashima, S. and Shibata, D. (1997a). Overexpression of an *Arabidopsis thaliana* high-affinity phosphate transporter gene in tobacco cultured cells enhances cell growth under phosphate-limited conditions. *Proc. Natl. Acad. Sci. USA,* **94**: 7098-7102.
Mitsukawa, N., Okumura, S. and Shibata, D. (1997b). High-affinity phosphate transporter genes of *Arabidopsis thaliana*. *Soil Sci. Plant Nutr.,* **43**: 971-974.
Muchhal, U.S., Pardo, J.M. and Raghothama, K.G. (1996). Phosphate transporters from the higher plant *Arabidopsis thaliana*. *Proc. Natl. Acad. Sci. USA,* **93**: 10519-10523.
Mudge, S.R., Rae, A.L., Diatloff, E. and Smith, F.W. (2002). Expression analysis suggests novel roles for members of the Pht1 family of phosphate transporters in Arabidopsis. *Plant J.,* **31**: 341-353.

Mudge, S.R., Smith, F.W. and Richardson, A.E. (2003). Root-specific and phosphate-regulated expression of phytase under the control of a phosphate transporter promoter enables Arabidopsis to grow on phytate as a sole P source. *Plant Sci.*, **165**: 871-878.

Mukatira, U., Liu, C., Varadarajan, D.K. and Raghothama, K.G. (2001). Negative regulation of phosphate starvation-induced genes. *Plant Physiol.*, **127**: 1854-1862.

Neumann, G., Massonneau, A., Martinoia, E. and Romheld, V. (1999). Physiological adaptations to phosphorus deficiency during proteoid root development in white lupin. *Planta*, **208**: 373-382.

Ohtake, H., Takahashi, K., Tsuzuki, Y. and Toda, K. (1985). Uptake and release of phosphate by a pure culture of *Acinetobacter calcoaceticus*. *Water Res.*, **19**: 1587-1594.

Okumura, S., Mitsukawa, N., Shirano, Y. and Shibata, D. (1998). Phosphate transporter gene family of *Arabidopsis thaliana*. *DNA Res.*, **5**: 261-269.

O'Neill, E.M., Kaffman, A., Jolly, E.R. and Shea, E.K. (1996). Regulation of PHO4 nuclear localization by the PHO80-PHO85 cyclin-CDK complex. *Science*, **271**: 209-212.

Oshima, Y., Ogawa, N. and Harashima, S. (1996). Regulation of phosphatase synthesis in *Saccharomyces cerevisiae* - a review. *Gene*, **179**: 171-177.

Peleg, Y., Addison, R., Aramayo, R. and Metzenberg, R.L. (1996). Translocation of *Neurospora crassa* transcription factor NUC-1 into the nucleus in induced by phosphorous limitation. *Fungal Gent. Biol.*, **20**: 185-191.

Poirier, Y., Thoma, S., Somerville, C. and Schiefelbein, J. (1991). A mutant of Arabidopsis deficient in xylem loading of phosphate. *Plant Physiol.*, **97**: 1087-1093.

Raghothama, K. (1999). Phosphate acquisition. *Annu. Rev. Plant Physiol. Plant Mol. Biol.*, **50**: 665-693.

Rausch, C., Daram, P., Brunner, S., Jansa, J., Laloi, M., Leggewie, G., Amrhein, N. and Bucher, M. (2001). A phosphate transporter expressed in arbuscule-containing cells in potato. *Nature*, **414**: 462-470.

Rausch, C. and Bucher, M. (2002). Molecular mechanisms of phosphate transport in plants. *Planta*, **216**: 23-37.

Rebeille, F., Bligny, R., Martin, J.-B. and Douce, R. (1983). Relationship between the cytoplasm and vacuole phosphate pool in *Acer pseudoplatanus* cells. *Arch. Biochem. Biophy.*, **225**: 143-148.

Richardson, A.E., Hadobas, P.A. and Hayes, J.E. (2001). Extracellular secretion of *Aspergillus* phytase from Arabidopsis roots enables plants to obtain phosphorus from phytate. *Plant J.*, **25**: 641-649.

Rubio, V., Linhares, F., Solano, R., Martin, A.C., Iglesias, J., Leyva, A. and Paz-Ares, J. (2001). A conserved MYB transcription factor involved in phosphate starvation signaling both in vascular plants and in unicellular algae. *Genes Dev.*, **15**: 2122-2133.

Runge-Metzger, A. (1995). Closing the cycle: obstacles to efficient P management for improved global security. In: *Phosphorus in the Global Environment: Transfers, Cycles and Management* (Ed. Tiessen H.), John Wiley and Sons, New York, pp 27-42.

Sakano, K. (1990). Proton/phosphate stoichiometry in uptake of inorganic phosphate by cultured cells of *Catharanthus roseus* (L.) G.-Don. *Plant Physiol.*, **93**: 479-483.

Sakano, K., Yazaki, Y. and Mimura, T. (1992). Cytoplasmic acidification induced by inorganic phosphate uptake in suspension cultured *Catharanthus roseus* cells - Measurement with fluorescent pH indicator and P-31 nuclear-magnetic-resonance. *Plant Physiol.*, **99**: 672-680.

Sano, T., Kuraya, Y., Amino, S. and Nagata, T. (1999). Phosphate as a limiting factor for the cell division of tobacco BY-2 cells. *Plant Cell Physiol.*, **40**: 1-8.

## Phosphate uptake to plant cells

Sano, T. and Nagata, T. (2002). The possible involvement of a phosphate-induced transcription factor encoded by *Phi-2* gene from tobacco in ABA-signaling pathways. *Plant Cell Physiol.,* **43**: 12–20.
Schachtman, D.P., Reid, R.J. and Ayling, S.M. (1998). Phosphorous uptake by plants; from soil to cell. *Plant Physiol.,* **116**: 447-453.
Schmidt, M.E., Heim, S., Wylegalla, C., Helmbrecht, C. and Wagner, K. (1992). Characterization of phosphate uptake by suspension cultured *Catharanthus roseus* cells. *J. Plant Physiol.,* **140**: 179-184.
Schneider, K.R., Smith, R.L. and O'Shea, E.K. (1994). Phosphate-regulated inactivation of the kinase PHO80-PHO85 by the CDK inhibitor PHO81. *Science,* **266**: 122-126.
Sekiguchi, Y., Mitsuhashi, N., Inoue, Y., Yagisawa, H. and Mimura, T. (2004). Analysis of sugar phosphates in plants by ion chromatography on titanium dioxide column with pulsed amperometric detection. *J. Chromatogr. A,* **1039**: 71-76.
Sekiguchi, Y., Mitshashi, N., Kokaji, T., Miyakoda, H. and Mimura, T. (2005). Development of a comprehensive analytical method for phosphate metabolites in plants by ion chromatography coupled with tandem mass spectrometry. *J. Chromatogr. A,* **1085**: 131-136.
Shimaoka, T., Ohnishi, M., Sazuka, T., Mitsuhashi, N., Hara-Nishimura, I., Shimazaki, K., Maeshima, M., Yokota, A., Tomizawa, K. and Mimura, T. (2004). Isolation of intact vacuoles and proteomic analysis of tonoplast from suspension-cultured cells of *Arabidopsis thaliana. Plant Cell Physiol.,* **45**: 672-683.
Smith, F.W., Ealing, P.M., Dong, B. and Delhaize, E. (1997). The cloning of two Arabidopsis genes belonging to a phosphate transporter family. *Plant J.,* **11**:83-92.
Strack, D., Fester, T., Hause, B., Schliemann, W. and Walter, M.H. (2003). Arbuscular mycorrhiza: biological, chemical, and molecular aspects. *J. Chem. Ecol.,* **29**: 1955-1979.
Stratful, I., Brett, S., Scrimshaw, M.B. and Lester, J.N. (1999). Biological phosphorus removal, its role in phosphorus recycling. *Environ. Technol.,* **20**: 681–695.
Taiz, L. and Zeiger, E. (1998). *Plant Physiology*, second edition. Sinauar Associates, Inc., Publishers, Sunderland, Massachusetts.
Takabatake, R., Hata, S., Taniguchi, M., Kouchi, H., Sugiyama, T. and Izui, K. (1999). Isolation and characterization of cDNAs encoding mitochondrial phosphate transporters in soybean, maize, rice, and Arabidopsis. *Plant Mol. Biol.,* **40**: 479-486.
Thomson, B.D., Clarkson, D.T. and Brain, P. (1990). Kinetics of phosphorus uptake by the germ-tubes of the vesicular-arbuscular fungus *Gigaspora margarita. New Phytol.,* **116**: 647-653.
Vance, C.P., Uhde-Stone, C. and Allan, D.L. (2003). Phosphorous acquisition and use: critical adaptations by plants for securing a nonrenewable resource. *New Phytol.,* **157**: 423-447.
Versaw, W.K. (1995). A phosphate-repressible, high-affinity phosphate permease is encoded by the *pho-5+* gene of *Neurospora crassa. Gene,* **153**: 135-139.
Versaw, W.K. and Harrison, M.J. (2002). A chloroplast phosphate transporter PHT2;1 influences allocation of phosphate within the plant and phosphate-starvation responses. *Plant Cell,* **14**: 1751-1766.
Wanner, B.L. (1993). Gene regulation by phosphate in enteric bacteria. *J. Cell Biochem.,* **51**: 47-54.
Wasaki, J., Omura, M., Osaki, M., Ito, H., Matsui, H., Shinano, T. and Tadano, T. (1999). Structure of a cDNA for an acid phosphatase from phosphate-deficient lupin (*Lupinus albus* L.) roots. *Soil Sci. Plant Nutr.,* **45**: 439–449.
Wilson, C., Pfosser, M., Jonak, C., Hirt, H., Heberle-Bors, E. and Vicente, O. (1998). Evidence for the activation of a MAP kinase upon phosphate-induced cell cycle re-entry in tobacco cells. *Physiol. Plant.,* **102**: 532–538.

# Chapter 6

## IRON UPTAKE AND TRANSPORT IN PLANTS

TZVETINA BRUMBAROVA AND PETRA BAUER
*Department of Biological Sciences – Botany, Saarland University, PO Box 151150, D-66041 Saarbrücken, Germany*
E-mail: p.bauer@mx.uni-saarland.de

**Abstract**

Iron is an essential micronutrient. Low iron uptake leads to leaf chlorosis and yield loss, whereas excessive iron uptake results in leaf bronzing. Iron uptake and transport is controlled by the plant at the level of mobilization and entry from the rhizosphere into the root, distribution and availability within the plant. Best characterized are the processes for iron mobilization and iron uptake into roots. Two strategies are distinguished based either on reduction of iron and uptake of divalent iron (Strategy I) or chelation of iron by phytosiderophores (Strategy II). Here, we summarize the physiological and molecular aspects that are relevant for iron uptake and transport in plants.

**Keywords:** iron uptake, iron transport, iron reduction, Strategy I, iron chelation, Strategy II

## 1. INTRODUCTION

Iron (Fe) is an essential nutrient for every organism. As a component of many vital enzymes, it is required for a wide range of biological functions, such as electron transport in the respiratory chain (cytochromes), DNA synthesis (ribonucleotide reductase), photosynthesis (chlorophyll synthesis and chloroplast structure/ function), nitrogen fixation (symbiotic root nodules establishment/ function) and hormone synthesis (lipoxygenase, ethylene precursor) (Briat and Lobreaux, 1997).

Fe deficiency is the world's most prevalent human nutritional disorder (WHO: http://www.who.int/nut/ida.htm). In many developing countries, the use of staple crops with naturally low Fe content can cause nutritional problems, especially when vegetables are the predominant food source. This effect can be severe even in plants with high Fe content such as legumes due to the presence of antinutrients, such as oxalic acid or phytate that decrease the bioavailability of Fe.

Iron deficiency is also often a problematic factor for reduced yield in agriculture, especially in regions with calcareous and alkaline soils. Iron deficiency results in

---

© CAB International 2008. *Plant Membrane and Vacuolar Transporters* (eds P.K. Jaiwal, R.P. Singh and O.P. Dhankher)

iron leaf chlorosis and reduced growth. Since in the future agricultural areas will be extended it is likely that iron deficiency chlorosis may increase. The foliar application of Fe chelators such as Fe-EDTA or Fe-EDDHA has been recommended to cure iron chlorosis (Chen, 1997). These chemicals are, however, very expensive for extensive use. Fe deficiency of crops growing on calcareous soils can be cured to some extent with fertilizers. However, such treatments are costly and cannot be precisely targeted to the deficient parts of the plant, causing, in some cases, Fe excess followed by yield reduction.

To have significant impact on the Fe nutrition of humans, improvement strategies are under way to fortify crops with Fe. That is, to develop new varieties of major crops with increased amounts of bioavailable Fe.

In this respect, understanding the control of Fe uptake mechanisms in plants is of vital importance for efficient Fe fortification efforts.

## 2. BIOAVAILABILITY OF IRON

Despite being generally present in high quantities in soils (the fourth most abundant element in the lithosphere), Fe has a very limited bioavailability in aerobic and neutral pH environments. In aerobic soils, Fe is found predominantly in the form of $Fe^{3+}$, mainly as a constituent of oxyhydroxide polymers with extremely low solubility. Due to that, the equilibrium concentration of free $Fe^{3+}$ in such environments is limited to approximately $10^{-17}$ M. Such a value is far below that required for the optimal growth of plants and microbes – $10^{-9}$ to $10^{-4}$ and $10^{-7}$ to $10^{-5}$, respectively. The insufficient Fe availability can be particularly pronounced in plants grown on calcareous soils, which cover approximately one-third of Earth's surface. Therefore, without active mechanisms for extracting Fe from the soil, most plants would exhibit Fe-deficiency symptoms, such as chlorotic (yellowed) interveinal areas in young leaves, which leads to reduced crop yields or even complete crop failure.

In contrast, in acidic, waterlogged soils, excess $Fe^{2+}$ can be toxic for the plants. It promotes the formation of reactive oxygen-based radicals that are able to damage vital cellular components (notably membranes, by lipid peroxidation), leading to a loss of integrity and possible cell death. Plants exposed to excessive levels of Fe show bronzing (coalesced tissue necrosis), flaccidity and/ or blackening of the roots.

As a consequence of these properties of solubility and toxicity, Fe homeostasis in the whole organism, as well as in the cells, must be balanced to supply enough Fe for cell metabolism and to avoid excessive, toxic levels. In this way, plants have evolved different mechanisms to control Fe uptake.

## 3. STRATEGIES FOR IRON UPTAKE IN PLANTS

As a strategy for restricting excessive uptake of Fe, wetland species have evolved mechanisms for oxidizing ferrous Fe ($Fe^{2+}$) in the rhizosphere. Plants, living under aerobic soil conditions, have developed two phylogenetically distinct strategies to cope with the extremely low availability of soluble Fe compounds (Marschner and Römheld, 1994).

Dicots and nongraminaceous monocots employ an Fe acquisition mechanism termed Strategy I based on the reductive detachment of Fe from its ligand. Under Fe-deficient conditions, such plants exhibit enhanced proton extrusion in the rhizosphere, increased $Fe^{3+}$ reduction capacity at the root surface, followed by an uptake of $Fe^{2+}$ via a ferrous transporter on the root plasma membrane (Römheld and Marschner, 1983). As a result, plants elevate the Fe availability in the rhizosphere and enhance its uptake.

In response to Fe deficiency, graminaceous monocots release high-affinity Fe-chelating substances from the mugineic acid family, called phytosiderophores (PS). These substances solubilize the inorganic $Fe^{3+}$ compounds from the soil and the resulting $Fe^{3+}$-PS complexes are taken up by the root cells via a specific plasma membrane transport system without reduction of the ferric ion. This mechanism is termed Strategy II (Römheld and Marschner, 1986) and it might resemble the microbial siderophore strategy.

### 3.1. Strategy I

Strategy I-type plants respond to Fe deficiency with both morphological and physiological changes (Römheld, 1987), which aim at an increased root surface area for reduction and transport of Fe. The changes in morphology include formation of root hairs, swelling of root tips, enhanced lateral root development and reduced lateral root growth (Schmidt, 1999). Molecular components are most studied in Arabidopsis and tomato (Bauer et al., 2004).

*3.1.1. Rhizosphere acidification*

A main physiological response to Fe deficiency is the increased acidification of the rhizosphere due to activation of a $H^+$-ATPase, which leads to extrusion of protons from the roots and aids in rendering Fe more soluble. This process can be quite fast – within a few hours the roots may lower the pH in the soil solution to values of 3 or lower. At the same time, a pH decrease of 1 releases $10^3$ times more $Fe^{3+}$ ions into the rhizosphere (Bienfait, 1985). This not only helps to acidify the extracellular space but it also has a pivotal role in establishing the electrochemical gradient (a proton moving force) that drives the uptake of solutes through their respective carriers and channels (Sussman, 1999). The capacity of a plant to acidify the rhizosphere in response to Fe deficiency depends to some extent on the cation/anion uptake balance and the nitrogen (N) nutrition of the plant. One member of the family of P-type $H^+$-ATPases, *AHA2* ('Arabidopsis $H^+$-ATPase'), is most abundantly expressed in root hairs (Sussman, 1994) and may encode an isoform involved in the uptake of mineral nutrients (Fox and Guerinot, 1998).

In addition, small chelating agents, especially citrate, are secreted into the rhizosphere which aids in solubilizing $Fe^{3+}$.

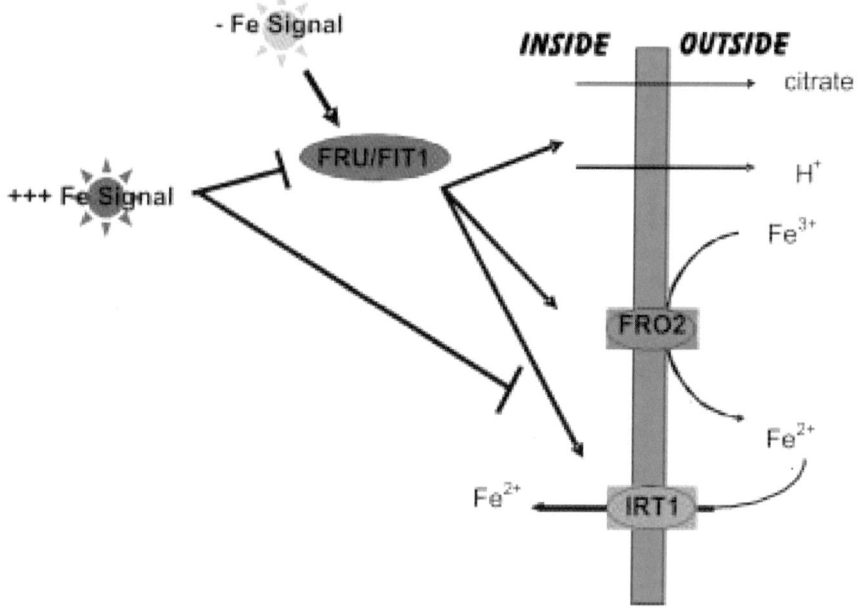

**Figure 1:** Strategy I for Fe uptake in plant roots (exemplified for Arabidopsis). The uptake of Fe in roots of dicots and nongraminaceous monocots is based on iron reduction and involves acidification of the rhizosphere through an $H^+$ ATPase activity, which solubilizes iron, followed by reduction of iron through a plasma membrane-bound iron-chelate reductase (FRO2 in Arabidopsis), and uptake of $Fe^{2+}$ into the root epidermal cells through a divalent iron transporter (IRT1 in Arabidopsis). Fe deficiency is sensed and a yet unknown Fe signal triggers the activation of the bHLH transcription factor FRU also known as FIT1 (bHLH029) in Arabidopsis, which acts as a positive regulator of the above-described Strategy I components. Upon iron resupply the system is switched off at post-transcriptional and transcriptional level. FRO2, IRT1 and FRU/FIT1 are indicated as these are three essential components for iron uptake in Arabidopsis roots. Loss of function of any of these three components leads to iron deficiency chlorosis.

## 3.1.2. $Fe^{3+}$ reduction

The second and most typical physiological Strategy I response to Fe deficiency is an enhanced $Fe^{3+}$ reduction. It is thought to be the primary factor in making Fe available for absorption (Guerinot and Yi, 1994). Ferric reduction takes place at the plasma membrane of root epidermal cells. It catalyzes transmembrane electron transport from cytosolic reduced pyridine nucleotides to extracellular Fe compounds serving as electron acceptors. This process is an obligatory prerequisite for Fe uptake in Strategy I plants and it is activated at insufficient Fe supply (Buckhout et al., 1989). Plasma membranes isolated from roots of Fe-deficient plants contained 2- to 3-fold higher specific activities for $Fe^{3+}$-chelate reductase than plasma membranes isolated from plants grown under Fe-sufficient conditions (Buckhout et al., 1989). The enzymatic activity of ferric reductase in the root epidermal cells is additionally optimized by the pH decrease on the root surface, as a result of the $H^+$-ATPase activity, since the optimum pH of the plant membrane reductase is generally around 6 (Chaney et al., 1972).

### 3.1.3. Uptake and transport of $Fe^{2+}$

Fe is transported across the root plasma membrane as free $Fe^{2+}$ via a divalent metal transporter acting downstream of the $Fe^{3+}$-chelate reductase. The capacity of uptake and translocation of Fe is greatly enhanced upon Fe starvation (Yi and Guerinot, 1996). Recently, transport of metals in general was reviewed by Colangelo and Guerinot (2006).

### 3.1.4. Genes involved in the mobilization of iron in Strategy I plants

3.1.4.1. Reductase genes

***AtFRO2*** - The characterization of three Arabidopsis mutants (*frd1-1*, *frd1-2* and *frd1-3*) which do not show induction of $Fe^{3+}$-chelate reductase under Fe-deficient conditions, confirms that Fe must be reduced prior to its transport and that $Fe^{3+}$ reduction can be uncoupled from proton release (Yi and Guerinot, 1996).

The $Fe^{3+}$-chelate reductase gene *AtFRO2* has been cloned by Robinson et al. (1999) from Fe-deficient Arabidopsis roots, based on sequence similarity with the yeast *FRE* genes, and it was shown to be mutated in *frd1-1*. When expressed in *frd1-1* mutant lines, it restores $Fe^{3+}$-reductase activity.

*AtFRO2* is upregulated in roots under Fe-deficiency conditions (Robinson et al., 1999). *AtFRO2* mRNA was detected in root epidermal cells, similarly to observations for *AtIRT1* localization (see below). Additionally, a posttranscriptional level of control on *LeFRO2* gene expression was revealed by analysis of transgenic 35S::AtFRO2 plants, which failed to induce $Fe^{3+}$-reductase activity under sufficient Fe-supply conditions regardless of the constitutive expression of the gene (Connolly et al., 2003).

AtFRO2 shares similarities with human phagocytic NADPH gp91$^{phox}$ oxidoreductase and with the yeast $Fe^{3+}$-chelate reductases, specifically in the heme- and nucleotide cofactor-binding sites. This is consistent with its function in electron transfer from cytosolic NADPH to extracellular $Fe^{3+}$. Therefore, AtFRO2 belongs to the superfamily of flavocytochromes that transport electrons across membranes (Robinson et al., 1999).

*AtFRO2* belongs to an eight-member gene family in Arabidopsis. *AtFRO3* is also strongly induced upon Fe deficiency, which suggests it has a similar function as *AtFRO2* (Wu et al., 2005). Some *FRO* family members are also expressed in shoots where they may serve iron mobilization or increased production of $Fe^{2+}$ in leaves (Feng et al., 2006; Mukherjee et al., 2005; Wu et al., 2005).

In pea, **PsFRO1** represents the reductase involved in root Fe acquisition (Waters et al., 2002), supported by the observation that *PsFRO1* mRNA levels in plants correlated with $Fe^{3+}$-chelate reductase activity. In contrast to *AtFRO2*, *PsFRO1* is expressed in both root and shoot (upregulated by Fe deficiency), suggesting an additional role in Fe distribution throughout the plant.

In tomato, the $Fe^{3+}$-chelate reductase is encoded by *LeFRO1*, which is expressed in roots, leaves, cotyledons, flowers, and young fruits. Expression of *LeFRO1* in roots is dependent on the Fe status as well as the FER protein, the transcription factor homolog of AtFRU/AtFIT1 that is required for iron uptake responses in roots. However, *LeFRO1* is constitutively expressed in leaves (Li *et al.*, 2004).

3.1.4.2. Transporter genes

*AtIRT1* - Expression of Arabidopsis cDNA clones in the yeast *fet3fet4* (*FERROUS TRANSPORTER*) double mutant strain, impaired in both low- and high-affinity Fe transport, enabled cloning of a plant $Fe^{2+}$ transporter by screening for complementation of the mutant phenotype (Eide *et al.*, 1996). It was designated *AtIRT1* (*IRON-REGULATED TRANSPORTER1*). The *AtIRT1* gene encodes the founding member of a class of eukaryotic metal ion transporters, referred to as the ZIP (ZRT, IRT-LIKE TRANSPORTERS) family (Guerinot, 2000), with related sequences in plants, yeast, animals and humans. It encodes a protein with eight transmembrane (TM) domains. Four histidine-glycine repeats constitute potential metal-binding sites between TM domains 3 and 4 (Eng *et al.*, 1998).

In addition, AtIRT1 mediates uptake of $Mn^{2+}$ and $Zn^{2+}$ in the yeast *smf1* and *zrt1zrt2* mutants, respectively defective in Mn and Zn transport, but cannot restore growth of the Cu uptake-deficient yeast mutant *ctr1*, implying that this transporter is not involved in the uptake of Cu (Korshunova *et al.*, 1999). Inhibition of Fe uptake in *AtIRT1*-expressing yeast by excess of several transition metals such as Cd, Co, Mn and Zn was observed, showing that AtIRT1 is also able to transport $Cd^{2+}$ and $Co^{2+}$. The determinants for this broad substrate specificity of AtIRT1 have been investigated by site-directed mutagenesis (Rogers *et al.*, 2000).

*AtIRT1* expression is induced in roots of plants grown under Fe-deficiency, and it is dependent on the iron uptake regulatory transcription factor FRU/FIT1 (Connolly *et al.*, 2002; Eide *et al.*, 1996; Vert *et al.*, 2002; Jakoby *et al.*, 2004; Colangelo and Guerinot, 2004), suggesting a role of AtIRT1 in Fe uptake in planta. This was confirmed by the characterization of an Arabidopsis *irt1* knock-out mutant (Vert *et al.*, 2002). The *irt1* mutant plant is chlorotic and has a severe growth defect in soil, leading to death, which can be rescued by foliar application of Fe. Additionally, roots of the *irt1* mutant are defective in Fe uptake, and do not accumulate Zn, Cd, Mn, and Co under Fe-deficient conditions. This is in agreement with the observation that Fe-deficient plants have increased levels of root-associated Mn, Zn, Cd and Co, suggesting that, in addition to Fe, AtIRT1 mediates uptake of these metals into plant cells (Korshunova *et al.*, 1999). The fact the AtIRT1 has plasma membrane localization in root epidermal cells supports a transporter function in Fe uptake from the soil. These lines of evidence for the function of AtIRT1 in planta have been confirmed by two other independently obtained *irt1* mutant lines (Henriques et al., 2002; Varotto et al., 2002). Thus, AtIRT1 is considered as the major Fe transporter at the root surface in *A. thaliana*.

AtIRT1 production is further regulated at the protein level as AtIRT1 protein accumulation is repressed by sufficient Fe and Zn. 35S::AtIRT1 transgenic plants express *AtIRT1* mRNA constitutively, but are unable, under Fe-deficient conditions, to produce AtIRT1 protein in any plant tissue except the root (Connolly et al., 2002). This additional level of control of *AtIRT1* expression provides the plant with an effective mechanism to shut off Fe-uptake activity when not needed.

AtIRT1 homologs have also been characterized in pea and tomato (Cohen et al. 1998; Eckhardt et al., 2001). The pea gene, called **PsRIT1**, is upregulated under Fe deficiency and complements both the *fet3fet4* and *zrt1zrt2* yeast mutants, thus potentially mediating high-affinity Fe and Zn uptake in plants.

In tomato, **LeIRT1** and **LeIRT2** are both expressed in roots but only *LeIRT1* appears to be upregulated in response to Fe deficiency and dependent on the regulator FER (Eckhardt et al., 2001). *LeIRT1* was constitutively expressed in roots irrespective of iron supply and FER phenotype. Both genes restore the growth defect of the *fet3fet4*, *zrt1zrt2*, and *smf1* yeast mutants (Eckhardt et al., 2001). However, the tomato genes are able to complement the Cu transport-deficient yeast strain *ctr1* whereas *AtIRT1* does not (Eide et al., 1996).

The Strategy II plant rice possesses iron deficiency-regulated **OsIRT1** and **OsIRT2** genes whose function might be relevant under flooding conditions to take up $Fe^{2+}$ (Ishimaru et al., 2006).

*AtIRT2* – AtIRT2 is a gene belonging to the ZIP family and closely related to *AtIRT1*. AtIRT2 is able to transport Fe and Zn, but, unlike AtIRT1, it cannot transport Mn and Cd, when expressed in yeast (Vert et al., 2001). *AtIRT2* is expressed only in roots, in the same territories as *AtIRT1*, and is upregulated by Fe deficiency. However, both the level of expression of *AtIRT2* and its induction by Fe deficiency are much lower compared to *AtIRT1*. A null *irt2* mutant has no apparent phenotype, and overexpression of *AtIRT2* in *irt1* mutants does not rescue the *irt1* growth defect, which raises the question of the function of AtIRT2 in planta (Varotto et al., 2002; Vert et al., 2002).

*AtNRAMPs* – The *NRAMP* (*NATURAL RESISTANCE-ASSOCIATED MACROPHAGE PROTEIN*) gene family of metal transporters in Arabidopsis has seven members (Mäser et al., 2001). Generally, *NRAMPs* are widely distributed throughout living organisms, functioning in the transport of a broad range of divalent metal cations, including Fe (Gunshin et al., 1997). Their name was derived from phagosomal NRAMP1 which functions as an efflux pump in the membrane and in this way enhances resistance against intracellular bacteria by reducing metal availability (Lafuse et al., 2000).

One of the seven Arabidopsis members, *AtEIN2*, is involved in ethylene response and its function in metal transport has not been demonstrated yet.

On the other hand, the role of AtNRAMP1, AtNRAMP3, and AtNRAMP4 in metal transport has been shown both in yeast and *in planta* (Curie et al., 2000;

Thomine et al., 2000). In yeast, expression of *AtNRAMP1*, *AtNRAMP3*, or *AtNRAMP4* complements the phenotype of strains defective in Mn or Fe uptake. In addition, heterologous expression of *AtNRAMP3* or *AtNRAMP4* increases yeast sensitivity to Cd, indicating that these genes encode metal transporters with multiple specificities (Curie et al., 2000; Thomine et al., 2000).

In Arabidopsis, *AtNRAMP1*, *AtNRAMP2*, *AtNRAMP3*, and *AtNRAMP4* are expressed in both roots and shoots, but only the accumulation of *AtNRAMP1*, *AtNRAMP3*, and *AtNRAMP4* increases in roots in response to Fe deficiency (Curie et al., 2000).

*AtNRAMP1* overexpression in plants confers increased resistance to toxic Fe levels (Curie et al., 2000). The closely related genes *AtNRAMP3* and *AtNRAMP4* share similar tissue-specific expression patterns, transcriptional regulation by Fe, and subcellular localization at the vacuolar membrane (Lanquar et al., 2005; Thomine et al., 2003). Although neither single mutant has a dramatic phenotype, the germination of *nramp3 nramp4* double mutants is arrested only under low Fe nutrition and fully rescued by high Fe supply (Lanquar et al., 2005). Additionally, mutant seeds have wild type Fe content, but fail to retrieve Fe from the vacuolar globoids. These results indicate that AtNRAMP3 and AtNRAMP4 function redundantly in the mobilization of Fe from the vacuole during early seed development.

The tomato homologs ***LeNRAMP1*** and ***LeNRAMP3*** (Bereczky et al., 2003) encode functional NRAMP metal transporters in yeast, where they were shown to be Fe regulated and localized mainly to intracellular vesicles. *LeNRAMP1*, in contrast to *LeNRAMP3*, has a root-specific expression and is strongly upregulated by Fe deficiency and dependent on FER. Additionally, *LeNRAMP1* was expressed in the vascular root parenchyma. A role for LeNRAMP1 in Fe mobilization within the plant was suggested (Bereczky et al., 2003).

*AtYSL* – A search for homologs of the maize *YS1* gene in Arabidopsis identified eight genes, named *YSL* (*YELLOW STRIPE-LIKE*) (Curie et al., 2001). YS1 is essential for Fe-phytosiderophore (PS) import into roots of graminaceous plants (see Strategy II) (Curie et al., 2001). Since nongraminaceous plants do not synthesize or secrete PS, it was suggested that YSL proteins mediate the uptake of metals that are complexed with plant-derived PS or nicotianamine (NA). NA is produced in all plants and has the ability to bind Fe (von Wiren et al., 1999). Based on sequence similarity to ZmYS1, *A. thaliana* has eight predicted AtYSL proteins. Two family members, *AtYSL1* and *AtYSL2* have recently been studied in some detail.

*AtYSL1* is a shoot-specific gene, expressed in the xylem parenchyma of leaves, whose transcript levels increase in response to Fe excess (Le Jean et al., 2005). Based on the phenotype of the *ysl1* mutant, a role of AtYSL1 in long-distance circulation of Fe and NA and their delivery to the seed was suggested.

*AtYSL2* is expressed in many cell types in both shoot and root, such as the xylem-associated cells within the vasculature of expanded leaves, and in the pericycle and endodermis of the roots (Schaaf et al., 2005). *AtYSL2* transcript accumulation increases under conditions of Fe sufficiency or Fe resupply (DiDonato *et al.*, 2004; Schaaf *et al.*, 2005), and responds also to Cu (DiDonato *et al.*, 2004) and Zn (Schaaf *et al.*, 2005). Based on its expression pattern and its apparent protein localization in lateral membranes, a major function of AtYSL2 might be in the lateral transport of metals from the vasculature (DiDonato *et al.*, 2004). The *ysl2-1* single mutant does not have an obvious phenotype which may reflect functional redundancy within the Arabidopsis *YSL* family (DiDonato *et al.*, 2004).

*AtIREG2* – Another class of transporters involved in the intracellular distribution of Fe might be represented by AtIREG2 (or AtFPN2) (Colangelo and Guerinot, 2004). AtIREG1, AtIREG2, and AtIREG3 are related by sequence to the animal IRON-REGULATED PROTEINS (IREG), also called FERROPORTINS (FPN), for which a function in Fe export has been demonstrated, as for example for the mammalian *IREG1* gene (McKie *et al.*, 2000). This iron export from duodenum cells is required to deliver iron to blood and liver cells. Further studies on the plant *IREG* genes are required.

*AtFRD3* – The Arabidosis *frd3* mutant, also isolated as *man1* (Delhaize, 1996), is unable to turn off the root iron-reductase activity at sufficient Fe supply (Yi and Guerinot, 1996). *frd3* accumulates a variety of metals, such as Fe and Mn, due to the upregulation of *AtIRT1* (Delhaize, 1996).

The *AtFRD3* gene encodes a transmembrane protein belonging to the MULTIDRUG AND TOXIN EFFLUX TRANSPORTERS (MATE) family (Rogers and Guerinot, 2002) and is therefore likely to transport small organic molecules. *AtFRD3* is expressed in the root vascular cylinder where it seems required but not in shoots. The phenotype of *frd3* mutant plants, which is consistent with a defect in either Fe-deficiency signaling or Fe distribution, indicates that AtFRD3 could be involved in delivering iron to the xylem (Green and Rogers, 2004; Rogers and Guerinot, 2002).

3.1.4.3. Nicotianamine (NA)

The tomato mutant *chloronerva* accumulates high levels of Fe (Stephan and Scholz, 1993) and behaves as if it is always experiencing Fe deficiency, even when grown under Fe-sufficient conditions. Among the typical symptoms of Fe deficiency is the characteristic intercostal chlorosis in young leaves. On the other hand, the Fe-uptake mechanisms in the mutant, including proton extrusion and reductase activity, are constitutively expressed. Molecular Fe-deficiency responses are induced (Bereczky *et al.*, 2003; Brumbarova and Bauer, 2005). As a result, it accumulates too much Fe in its shoots, leading to retarded growth and development of necrotic spots on the leaves. Grafting the *chloronerva* mutant onto wild type or vice versa normalizes the mutant phenotype, indicating that it is due to the lack of a transportable substance.

The observed abnormalities in *chloronerva* have been correlated with a deficiency in NA synthesis (Stephan and Grün, 1989) due to a mutation in the gene *LeNAS* encoding the enzyme NA synthase (NAS) that converts S-adenosyl methionine to NA (Ling *et al.*, 1999).

*Chloronerva* is an NA auxotroph – application of NA to the roots or leaves of mutant plants leads to their phenotypic recovery (Stephan and Scholz, 1993). It was thought that in *chloronerva* cells Fe is unable to react with the sensor protein without the aid of NA, leaving the repressor unsaturated and Fe uptake to continue in excess of cellular needs (Scholz *et al.*, 1992). This could explain why the NA-free mutant suffers from apparent Fe deficiency and fails to repress inducible Fe-uptake processes (Stephan and Scholz, 1993).

NA occurs in all plants and chelates metal cations, including Fe. Its role as a mediator of Fe transport in the phloem could also explain the various phenotypes of the *chloronerva* mutant (Stephan *et al.*, 1994; 1996). Evidence for this is that the concentrations of NA in the phloem correlate with those of Fe and other metals, and that the NA-free mutant *chloronerva* has a phenotype indicative of Fe deficiency (Pich and Scholz, 1996; Stephan and Scholz, 1993). It was also shown that NA chelates both $Fe^{2+}$ and $Fe^{3+}$, and a role for NA in scavenging Fe to protect the cell from oxidative damage, resulting from the Fenton reaction, was proposed (von Wiren *et al.*, 1999). Further roles of NA were reviewed by Hell and Stephan (2003).

Additional evidence for the role of NA was obtained from transgenic tobacco plants that constitutively expressed the barley *HvNAAT* gene (Takahashi *et al.*, 2003). In gramineous plants, NAAT catalyzes the amino group transfer of NA for the biosynthesis of phytosiderophores. In this way, the transgenic plants experienced NA shortage, which caused disorders in internal metal transport, leading to interveinal chlorosis of young leaves (similar to the *chloronerva* phenotype) and abnormally shaped and sterile flowers. These findings demonstrated again the essential role of NA for growth, flower development, and fertility in plants (Takahashi *et al.*, 2003).

### 3.1.4.4. Other known mutants impaired in Strategy I responses

Several plant mutants are known which exhibit increased rates of Fe uptake. The genes responsible for the mutant phenotypes are not yet identified.

***brz*** - The pea mutant *brz* (*bronze*) accumulates high levels of Fe (Kneen *et al.*, 1990; Welch and Kochian, 1992) similar to the tomato mutant *chloronerva*. The *brz* mutant develops bronze necrotic spots on its leaves probably due to the 50-fold increased leaf-Fe content compared to leaves of wild type plants. The basis for the excessive Fe accumulation appears to be the increased $Fe^{3+}$ reduction and $Fe^{2+}$ uptake regardless of plant Fe status (Grusak *et al.*, 1990). The *brz* mutant also accumulates high levels of other divalent cations (Mg, Mn, Zn). The *brz* mutation is monogenic, recessive and maps to chromosome 4 (Kneen *et al.*, 1990).

***dgl*** – A similar phenotype is observed for the *dgl* (*degenerated leaflets*) mutant in pea (Grusak and Pezeshgi, 1996). It has an increased capacity to acidify the rhizosphere. Reductase studies using plants with reciprocal shoot:root grafts demonstrated that shoot expression of the *dgl* gene leads to the generation of a transmissible signal that enhances $Fe^{3+}$-reductase activity in roots. The *dgl* gene product may alter or interfere with a normal component of a signal transduction mechanism regulating Fe homeostasis in plants (Grusak and Pezeshgi, 1996).

Iron transport

3.1.4.5. Ferritin

Due to the potential of free Fe to cause damage in the cell through radical formation, plants need to regulate iron uptake and store it in a safe and soluble form. Two important compartments involved in this function are the apoplastic space and the vacuoles (Briat *et al.*, 1995). The ferritins, a class of multimeric proteins, also act as an Fe buffer inside the cell (Harrison and Arosio, 1996). The importance of ferritin function is emphasized by its ubiquitous distribution among living species – plants, animals, fungi and bacteria.

Ferritins have a highly conserved three-dimensional structure – they are hollow spheres with a 24-subunit shell and a cavity containing an Fe core (Theil, 1987). Phytoferritin is localized in the plastids of shoots and also roots where it serves to store excess Fe to avoid oxidative stress (Deák *et al.*, 1999; Lescure *et al.*, 1991; Savino *et al.*, 1997). Exogenous treatment with ozone or ethylene, as well as impaired photosynthesis or Fe overload, induce ferritin accumulation. Developmental studies demonstrate the role of ferritin as an Fe buffer for important Fe-dependent processes, such as photosynthesis and nitrogen fixation (Ragland and Theil, 1993). Furthermore, it was shown that post-transcriptional regulatory mechanisms are involved in ferritin gene regulation, at least at protein stability level (Ragland and Theil, 1993). This suggests that both environmental and developmental signals are involved in plant ferritin gene regulation and that their abundance is controlled by precise regulatory mechanisms that reflect the cellular need for maintaining Fe homeostasis (Briat *et al.*, 1999). Transgenic ferritin expression in seeds may lead to increased iron content of seeds, however iron uptake into the roots might be required (Qu *et al.*, 2005). This suggests that iron storage and iron uptake regulation are connected.

3.1.4.6. Transcriptional regulators

In plants, the maintenance of Fe homeostasis is so far mainly described at the level of gene expression.

Expression of most of the above described Arabidopsis and tomato Fe-deficiency response genes is controlled by Fe status at the level of transcript accumulation. Their mRNAs accumulate in roots in response to Fe deficiency and are switched off by Fe resupply. For *AtIRT1* and *AtFRO2*, additional posttranscriptional control was observed, which did not allow accumulation of the protein or induction of the $Fe^{3+}$-reductase activity, respectively, in Fe-sufficient conditions and outside of the root in overexpressing 35S::AtIRT1 and 35S::AtFRO2 plants (Connolly *et al.*, 2002, 2003).

The *LeFER* gene, identified in tomato, encodes a basic helix-loop-helix transcription factor whose absence in the *fer* mutant results in a complete loss of Fe-deficiency responses of the root, such as $Fe^{3+}$-reductase activity and iron uptake (Ling *et al.*, 2002).

The Arabidopsis homolog of *LeFER*, *AtFRU* (also known as *AtFIT1* or *bHLH29*) (Colangelo and Guerinot, 2004; Jakoby *et al.*, 2004; Yuan *et al.*, 2005) was shown to control Fe-deficiency responses in the root in a similar manner, suggesting a conserved regulatory mechanism in dicots. Both, *LeFER/AtFRU/ATFIT1* are essential for up-regulation of the *FRO/IRT* system in roots. The *BHLH* genes are themselves up-regulated by low Fe supply although this up-regulation is not as strong as that of *AtIRT1*. Overexpression studies showed that LeFER and AtFRU can enhance iron uptake responses but only at low Fe supply suggesting additional posttranscriptional activation by low Fe or inactivation by high Fe (Bereczky *et al.*, 2003; Jakoby *et al.*, 2004; Brumbarova and Bauer, 2005).

### 3.2. Strategy II

Strategy II plants are characterized by the release of phytosiderophores (PS) (e.g. mugineic acid (MA) in barley and avenic acid in oat) which efficiently solubilize

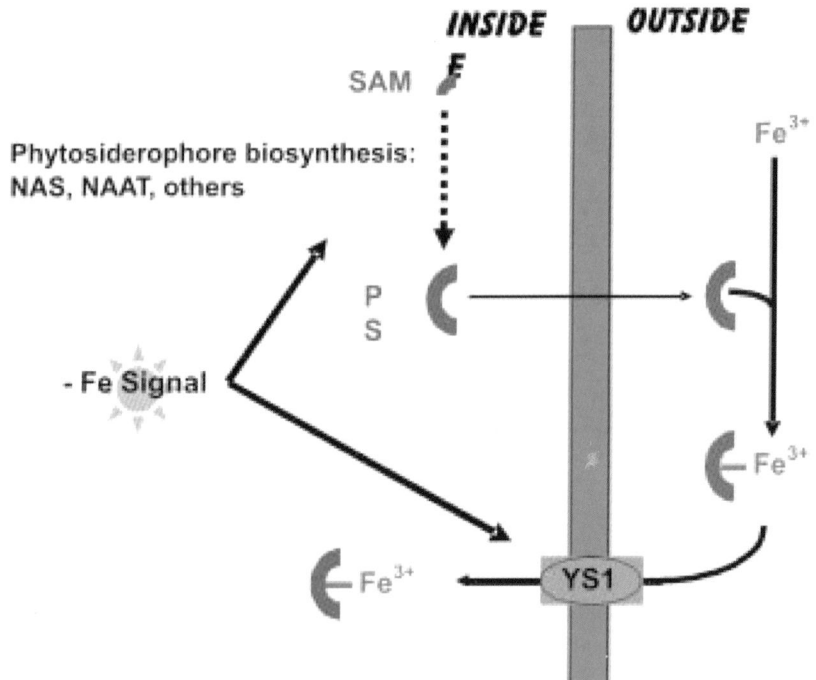

**Figure 2:** Strategy II for Fe uptake in grasses. The uptake of Fe in roots of graminaceous monocots is based on $Fe^{3+}$ chelation and involves biosynthesis of $Fe^{3+}$-chelating substances – phytosiderophores (PS) from S-adenosyl methionine (SAM) through the enzymatic activity of nicotianamine synthase (NAS), nicotianamine aminotransferase (NAAT) and other enzymes, excretion of PS into the rhizosphere, where they chelate $Fe^{3+}$ from insoluble compounds, followed by uptake of $Fe^{3+}$-PS complexes into the root epidermal cells through a specific transporter named YS1 in maize. The iron uptake process is induced by Fe deficiency.

inorganic $Fe^{3+}$ by chelation, and by the induction of a high-affinity uptake system for $Fe^{3+}$-PS complexes that transports the $Fe^{3+}$chelates as intact molecules (Römheld, 1987). Both processes are induced in response to Fe deficiency through upregulation of the underlying genes (Fig. 2).

This strategy is considered to be more efficient than Strategy I. A good illustration for this is the fact that grasses can grow on calcareous soils which do not support the growth of dicots. One reason for this may be that Strategy II is less pH dependent than Strategy I (Guerinot and Yi, 1994).

Knowledge of the Strategy II Fe-acquisition mechanism has increased considerably since the discovery of PS in washings from the roots of Fe-deficient rice and oats (Takagi, 1976; 1993). It was found that PS are structurally related to nicotianamine (NA). Later, it became evident that NA is an intermediate in the biosynthesis of the mugineic acid family of PS. The sole precursor of MAs was found to be methionine (Mori and Nishizawa, 1987). Methionine is converted to S-adenosylmethionine (SAM) by SAM synthetase (Shojima et al., 1989). HvSAM synthetase enzyme activity shows no difference in Fe-deficient and Fe-sufficient barley roots (Takizawa et al., 1996). Therefore, it was concluded that there is no specific Fe-deficiency *HvSAM* synthase gene and that constitutive HvSAM synthase activity is sufficient for MA synthesis.

Then, nicotianamine synthase (NAS) combines three molecules of SAM to form one molecule of NA (Higuchi et al., 1996; Shojima et al., 1989). Seven cDNAs encoding HvNAS were cloned from barley (Higuchi et al., 1999). Based on partial peptide sequence of purified HvNAS, *HvNAS*-cDNAs were isolated from an Fe-deficient barley root cDNA library (Higuchi et al., 1999; Herbik et al., 1999). HvNAS has an increased activity in response to Fe deficiency, which is 100-fold higher in roots than in leaves (Higuchi et al., 1999). In maize, three *ZmNAS* genes were isolated (Mizuno et al., 2003). *ZmNAS1* and *ZmNAS2* were shown to be upregulated in Fe-deficient roots, whereas *ZmNAS3* was downregulated. Based on their expression pattern and subcellular localization, the NAS proteins in maize were divided in two types (Mizuno et al., 2003). Further *NAS* genes were characterized from rice (Higuchi et al., 2001; Inoue et al., 2003).

Although NA is produced by both monocots and dicots, the subsequent steps leading to MA synthesis are specific to grasses. The critical enzyme in this specific pathway is nicotianamine-aminotransferase (NAAT) (Kanazawa et al., 1994; 1995; Ohata et al., 1993; Shojima et al., 1990) that catalyzes the transfer of an amino residue to NA, resulting in the production of 2'-deoxymugineic acid (DMA), the precursor of all other MAs (Shojima et al., 1990). Two *HvNaat* cDNAs were cloned from Fe-deficient barley roots (Takahashi et al. 1999). Northern analysis showed that Fe deficiency induces *HvNAAT-A* and *HvNAAT-B*, with *HvNAAT-B* having a different expression pattern. A barley genomic fragment containing the *HvNAAT-A* and *HvNAAT-B* genes, when introduced into rice, resulted in increased PS secretion and enhanced tolerance to low iron availability (Takahashi et al., 2001). HvNAAT activity is strongly induced by Fe deficiency in barley roots and represents a limiting step in the production of MAs.

Subsequent hydroxylation of DMA results in the formation of other members of the MA family. Two barley cDNA clones specifically expressed in Fe-deficient roots, *HvIds2* (Okumura et al., 1994) and *HvIds3* (Nakanishi et al., 1993) (*IRON DEFICIENCY SPECIFIC CLONE*), were shown to encode dioxygenases involved in the hydroxylation of DMA (Nakanishi et al., 2000). HvIDS2 is thought to catalyze the hydroxylation that converts DMA in 3-epihydroxy-2'-deoxymugineic acid (epiHDMA) and MA in 3-epihydroxy-mugineic acid (epiHMA), whereas HvIDS3 is responsible for converting DMA into MA (Nakanishi et al., 2000). Transgenic rice with introduced HvIDS3 was able to produce MA under Fe deficiency, whereas wild rice can produce only DMA (Kobayashi et al., 2001). That supports the hypothesis that HvIDS3 is an MA-synthase.

In contrast to the biosynthetic pathway of MAs, the molecular mechanisms of PS secretion remain poorly understood. It has been suggested that vesicular transport may be involved, since the appearance of swollen vesicles in Fe-deficient barley roots correlates with PS release. The production of PS is increased in response to Fe deficiency, and tolerance to Fe deficiency is correlated to the quantity and the kind of PS secreted (Negishi et al., 2002).

The extrusion of PS from Strategy II plants has a complex ecological background. PS are released with a distinct diurnal rhythm (Römheld, 1991) and in large quantities, which is supposed to increase the probability that PS will mobilize Fe before being degraded by rhizosphere microorganisms, as it was shown for MA (Jurkevitch et al., 1993).

The *yellow-stripe 3* (*ys3*) maize mutant shows interveinal chlorosis characteristic for Fe deficiency, due to a defect in MA secretion, and can be rescued by exogenous application of MAs or cocultivation with wild type plants (Basso et al., 1994).

The uptake of $Fe^{3+}$-PS complexes in Strategy II plants occurs through a specialized transporter. The gene encoding for this transporter was discovered by investigating the *yellow-stripe 1* (*ys1*) mutant of maize, which is unable to respond to Fe deficiency due to a defect in the uptake of $Fe^{3+}$-PS complexes (von Wiren et al., 1994). The *ZmYS1* gene encodes a plasma membrane protein from the OLIGOPEPTIDE TRANSPORTER (OPT) family (Yen et al., 2001). *ZmYS1* expression in the *fet3fet4* mutant strain restores growth on $Fe^{3+}$-DMA-containing medium, but not on $Fe^{3+}$-citrate-containing medium. In addition, both *ZmYS1* mRNA and protein are upregulated by Fe deficiency in roots and shoots, where ZmYS1 functions as a proton-coupled symporter to transport $Fe^{3+}$-PS and $Fe^{3+}$-NA (Curie et al., 2001; Roberts et al., 2004; Schaaf et al., 2004).

## 4. IRON TRANSPORT THROUGHOUT THE PLANT AND SYSTEMIC SIGNALING

Accumulation of a metal depends on uptake capacity and intracellular binding sites. In a multicellular organism, the situation is complicated by tissue- and cell-specific differences and also by intercellular transport. The processes that are assumed to influence metal accumulation rates in plants are the following: mobilization and uptake from the soil, compartmentalization and sequestration within the root, efficiency of xylem loading and transport, distribution between metal sinks in the aerial parts, sequestration and storage in leaf cells. At every level, concentration and affinities of chelating molecules, as well as the presence and selectivity of transporters, affect metal accumulation rates (Clemens et al., 2002).

The uptake of Fe in the plant starts in the apoplast of the root epidermal cells (Bienfait et al., 1985). Once taken up into the root symplast, Fe has to be shielded from oxygen to prevent precipitation and generation of oxygen radicals. For this purpose, it is assumed to be chelated by NA for several reasons: NA forms stable complexes with both Fe oxidation states at neutral and weakly alkaline pH (Stephan et al., 1996) and is ubiquitously present in higher plants in all tissues (Scholz et al., 1992). Fe-NA complexes are relatively poor Fenton reagents (von Wiren et al., 1999) and NA concentrations correlate with localization and levels of Fe (Pich et al., 2001). Radial transport of Fe from the root epidermis to the xylem vessels may occur as an Fe-NA complex on a symplastic route (Stephan et al., 1996). Fe is loaded in the xylem sap and translocated into the aerial parts of the plant through the transpiration stream. This process is mediated by xylem parenchyma or transfer cells and represents a separate control point for nutrient transport to the shoot. Organic acids, especially citrate, are the main metal chelators in the xylem (White et al., 1981) and it is generally agreed that Fe is oxidized when released into the xylem vessels and then transported as an $Fe^{3+}$-citrate complex (Lopez-Millan et al., 2000; Tiffin, 1966).

Further up in the shoot, Fe is unloaded from the xylem into the apoplastic spaces of leaf mesophyll cells. The uptake of Fe by leaf mesophyll appears to depend on a reduction step via a plasma membrane-bound $Fe^{3+}$-chelate reductase, which most likely releases $Fe^{2+}$ from $Fe^{3+}$-citrate (Brüggemann et al., 1993). There is also some evidence that $Fe^{3+}$ reduction *in vivo* may be aided by intermediate superoxide radical formation or by strong blue light (Brüggemann et al., 1993). In pea and tomato, *PsFRO1* and *LeFRO1*, respectively, are also expressed in leaves and induced by Fe deficiency (Li et al., 2004; Waters et al., 2002), making them good candidates for an $Fe^{3+}$ reductase in the leaf.

The distribution of Fe from the cells adjacent to the veins to the leaf lamina is probably again mediated by an Fe-NA complex, because in the NA-deficient mutant *chloronerva* most of the Fe is deposited along the veins. As a result, a characteristic interveinal chlorosis develops (Stephan and Scholz, 1993).

Fe moves to sink tissues via the phloem sap. NA is one of the potential phloem metal transporters (Stephan and Scholz, 1993). It binds preferentially $Fe^{2+}$, and not $Fe^{3+}$, in the phloem sap (von Wiren et al., 1999). There is a low but significant steady-state concentration of $Fe^{2+}$ in the phloem (Maas et al., 1988). The bulk of Fe in the phloem is in the form of $Fe^{3+}$ and can be bound to the IRON TRANSPORT PROTEIN (ITP) (belonging to the Late Embryogenesis Abundant (LEA) family) (Krueger et al., 2002). It is suggested that NA may play a role in loading and unloading Fe in the phloem by chelating $Fe^{2+}$ in the transition to and from $Fe^{3+}$-ITP. The YSL transporters are good candidates for transporting Fe-NA in and out of the phloem.

For the induction of Fe-deficiency responses, small amounts of extracellular Fe are more favorable than no Fe. For example, increased reductase activity was found in *Glycine max* roots supplied with 0.32 µM Fe relative to those supplied with 0.1 µM (Chaney *et al.*, 1972). Similar results have been reported for various species (Jolley *et al.*, 1986; Miller and Olsen, 1986; Stephan and Grün, 1989). Likewise, in mutants that are characterized by increased Fe-deficiency responses (such as *chloronerva*, for example), higher reductase activity is observed with normal amounts of Fe relative to Fe-free grown plants (Grusak and Pezeshgi, 1996; Grusak *et al.*, 1990; Stephan and Grün, 1989). A stimulation of the Fe uptake system by low, non-zero, Fe concentrations may be advantageous in ecological terms. Besides sensing changes in Fe supply in the rhizosphere, plants are able to monitor the shoot Fe status and to send a signal from the shoot to the root to activate uptake mechanisms. The identity of this signal is still unknown but it is believed to be transmitted by the phloem (Maas *et al.*, 1988).

When split-root plants were grown continuously with a localized supply of Fe, an increase in reduction rates was evident in the Fe-supplied portion of the root system. Apparently, the reductase activity of the Fe-supplied roots is controlled by the shoot Fe requirement, compensating for the decreased percentage of roots participating in Fe acquisition (Schmidt *et al.*, 1996).

The regulation of the root high-affinity Fe-uptake system by whole-plant signals was investigated in Arabidopsis, through monitoring the gene expression of *AtFRO2* and *AtIRT1* (Vert *et al.*, 2003). Split-root experiments indicated that the expression of *AtFRO2* and *AtIRT1* is controlled by a local induction from the root Fe pool and through a systemic pathway involving a shoot-borne signal, both signals being integrated to control production of the root Fe uptake proteins. Additionally, the expression of *AtFRO2* and *AtIRT1* is diurnally regulated (expressed during the day and downregulated at night) but this level of control can be overruled by Fe starvation (Vert *et al.*, 2003).

In summary, information trafficking between different plant parts appears to be of crucial importance for the regulation of appropriate internal concentrations of Fe in higher plants. Although root cells are capable of sensing intracellular and possibly external Fe concentrations and of inducing Fe deficiency responses, this control can be overruled by shoot-derived signals. This is most obvious in cases where, despite adequate external Fe concentrations, either the translocation or the uptake of Fe by leaf cells is inhibited or a high demand for shoot growth is clear (e.g. when Fe supply is locally restricted as in the case of split-root plants) (Schmidt, 1999).

## 5. CONCLUSIONS AND FUTURE PROSPECTS

Research has shown that iron uptake and transport is mediated by multiple different proteins that participate in mobilization, transport, binding and storage of iron as well as in regulation of these responses. Among angiosperms, iron homeostasis components are conserved. However, with respect to iron uptake into the root, we distinguish the Strategy I and Strategy II. By far, the list of gene and protein functions is not complete yet. A promising approach to date is to continue identifying novel genes through forward genetic approaches as well as finding gene functions through reverse genetics. Reverse genetics can be very rapid when genome sequence is available and functional genomics projects have been established. Since iron transport genes are often part of gene families, gene redundancy may retard immediate functional analysis in reverse genetic approaches. In such cases, multiple mutant combinations have to be generated and analyzed. Knowing the genes and proteins involved in iron uptake allows new insights regarding their functions: We can first of all, establish where in the plant and inside the cell the respective components are present, for example we can determine if the components are involved in vacuolar iron transport or plasmamembrane iron transport, if that is for example in the root epidermis or in the stele. Furthermore, we can study which signals regulate expression and protein stability. Are the proteins induced or repressed by iron deficiency? What happens in response to other metals? Through reverse genetics it is possible to knock out gene function, and overexpression transgenic approaches can help to study gain of function, to determine in which process the gene is involved and what the overall impact of the gene is for the respective process, for example iron uptake into roots or rather iron loading of seeds.

Iron uptake and its regulation is not an isolated process in the plant. It is clear that multiple signals interfere with regulation of iron uptake and that altered iron homeostasis affects whole plant physiology (Buckhout and Thimm, 2003; Wang *et al.*, 2002). There is a nutritional crosstalk, interference with plant hormone signaling, metabolic and developmental pathways. Future research involving whole genome approaches will undoubtedly reveal connections between these systems. Moreover, many mutants affected in the various physiological processes are available to check for iron transport phenotypes and altered iron transporter gene expression.

## 6. LITERATURE CITED

Basso, B., Bagnaresi, P., Bracale, M. and Soave, C. (1994) The yellow-stripe-1 and -3 mutants of maize: nutritional and biochemical studies. *Maydica,* **39**: 97-105.

Bauer, P., Thiel, T., Klatte, M., Brumbarova, T., Hell, R. and Grosse, I. (2004) Analysis of sequence, map position and gene expression reveals conserved essential genes for iron uptake in Arabidopsis and tomato. *Plant Physiol.,* **136**: 4169-4183.

Bereczky, Z., Wang, H.Y., Schubert, V., Ganal, M. and Bauer, P. (2003) Differential regulation of *nramp* and *irt* metal transporter genes in wild type and iron uptake mutants of tomato. *J. Biol. Chem.,* **278**: 24697-24704.

Bienfait, H.F. (1985) Regulated redox processes at the plasmalemma of plant root cells and their function in iron uptake. *J. Bioenerg. Biomembr.,* **17**: 73-83.

Bienfait, H.F., van den Briel, W. and Mesland-Mul, N.T. (1985) Free space iron pools in roots: Generation and mobilization. *Plant Physiol.,* **78**: 596-600.

Briat, J.F., Fobis-Loisy, I., Grignon, N., Lobréaux, S., Pascal, N., Savino, G., Thoiron, S., Von Wiren, N. and Wuytswinkel, O.V. (1995) Cellular and molecular aspects of iron metabolism in plants. *Biol. Cell,* **84**: 69-81.

Briat, J.F. and Lobreaux, S. (1997) Iron transport and storage in plants. *Trends Plant Sci.,* **2**: 187-193.

Briat, J.F., Lobreaux, S., Grignon, N. and Vansuyt, G. (1999) Regulation of plant ferritin synthesis: how and why. *Cell Mol. Life Sci.*, **56**: 155-166.
Brüggemann, W., Maas-Kantel, K. and Moog, P.R. (1993) Iron uptake by leaf mesophyll cells: The role of the plasma membrane-bound ferric-chelate reductase. *Planta*, **190**: 151-155.
Brumbarova, T. and Bauer, P. (2005) Iron-mediated control of the basic helix-loop-helix protein FER, a regulator of iron uptake in tomato. *Plant Physiol.*, **137**: 1018-1026.
Buckhout, T.J., Bell, P.F., Luster, D.G. and Chaney, R.L. (1989) Iron-stress induced redox activity in tomato (*Lycopersicon esculentum* Mill.) is localized on the plasma membrane. *Plant Physiol.*, **90**: 151-156.
Buckhout, T.J. and Thimm, O. (2003) Insights into metabolism obtained from microarray analysis. *Curr. Opin. Plant Biol.*, **6**: 288-296.
Chaney, R.L., Brown, J.C. and Tiffin, L.O. (1972) Obligatory reduction of ferric chelates in iron uptake by soybeans. *Plant Physiol.*, **50**: 208-213.
Chen, Y. (1997) Remedy of iron deficiency - present and future. *9th International Symposium on Iron Nutrition and Interactions in Plants*: 20-25 July, Stuttgart, p 51.
Clemens, S., Palmgren, M.G. and Kramer, U. (2002) A long way ahead: understanding and engineering plant metal accumulation. *Trends Plant Sci.*, **7**: 309-315.
Cohen, C.K., Fox, T.C., Garvin, D.F. and Kochian, L.V. (1998) The role of iron-deficiency stress responses in stimulating heavy-metal transport in plants. *Plant Physiol.*, **116**: 1063-1072.
Colangelo, E.P. and Guerinot, M.L. (2004) The essential basic helix-loop-helix protein FIT1 is required for the iron deficiency response. *Plant Cell*, **16**: 3400-3412.
Colangelo, E.P. and Guerinot, M.L. (2006) Put the metal to the petal: metal uptake and transport throughout plants. *Curr. Opin. Plant Biol.*, **9**: 322-330.
Connolly, E.L., Campbell, N.H., Grotz, N., Prichard, C.L. and Guerinot, M.L. (2003) Overexpression of the FRO2 ferric chelate reductase confers tolerance to growth on low iron and uncovers posttranscriptional control. *Plant Physiol.*, **133**: 1102-1110.
Connolly, E.L., Fett, J.P. and Guerinot, M.L. (2002) Expression of the IRT1 metal transporter is controlled by metals at the levels of transcript and protein accumulation. *Plant Cell*, **14**: 1347-1357.
Curie, C., Alonso, J.M., Le Jean, M., Ecker, J.R. and Briat, J.F. (2000) Involvement of NRAMP1 from *Arabidopsis thaliana* in iron transport. *Biochem. J.*, **347** Pt 3: 749-755.
Curie, C., Panaviene, Z., Loulergue, C., Dellaporta, S.L., Briat, J.F. and Walker, E.L. (2001) Maize *yellow stripe1* encodes a membrane protein directly involved in Fe(III) uptake. *Nature*, **409**: 346-349.
Deák, M., Horváth, G.V., Davletova, S., Török, K., Sass, L., Vass, I., Barna, B., Király, Z. and Dudits, D. (1999) Plants ectopically expressing the ironbinding protein, ferritin, are tolerant to oxidative damage and pathogens. *Nat. Biotech.*, **17**: 192-196.
Delhaize, E. (1996) A metal-accumulator mutant of *Arabidopsis thaliana*. *Plant Physiol.*, **111**: 849-855.
DiDonato, R.J. Jr., Roberts, L.A., Sanderson, T., Eisley, R.B. and Walker, E.L. (2004) Arabidopsis *Yellow Stripe-Like2* (*YSL2*): a metal-regulated gene encoding a plasma membrane transporter of nicotianamine-metal complexes. *Plant J.*, **39**: 403-414.
Eckhardt, U., Mas Marques, A. and Buckhout, T.J. (2001) Two iron-regulated cation transporters from tomato complement metal uptake-deficient yeast mutants. *Plant Mol. Biol.*, **45**: 437-448.

Iron transport

Eide, D., Broderius, M., Fett, J. and Guerinot, M.L. (1996) A novel iron-regulated metal transporter from plants identified by functional expression in yeast. *Proc. Natl. Acad. Sci. USA,* **93**: 5624-5628.
Eng, B.H., Guerinot, M.L., Eide, D. and Saier, M.H. Jr. (1998) Sequence analyses and phylogenetic characterization of the ZIP family of metal ion transport proteins. *J. Membr. Biol.,* **166**: 1-7.
Feng, H., An, F., Zhang, S., Ji, Z., Ling, H.Q. and Zuo, J. (2006) Light-regulated, tissue- and cell differentiation-specific expression of the Arabidopsis Fe(III)-chelate reductase gene AtFRO6. *Plant Physiol.,* **140**: 1345-1354.
Fox, T.C. and Guerinot, M.L. (1998) Molecular biology of cation transport in plants. *Annu. Rev. Plant Physiol. Plant Mol. Biol.,* **49**: 669-696.
Green, L.S. and Rogers, E.E. (2004) FRD3 controls iron localization in Arabidopsis. *Plant Physiol.,* **136**: 2523-2531.
Grusak, M.A. and Pezeshgi, S. (1996) Shoot-to-root signal transmission regulates root Fe(III) reductase activity in the *dgl* Mutant of Pea. *Plant Physiol.,* **110**: 329-334.
Grusak, M.A., Welch, R.M. and Kochian, L.V. (1990) Physiological characterization of a single-gene mutant of *Pisum sativum* exhibiting excess iron accumulation: I. Root iron reduction and iron uptake. *Plant Physiol.,* **93**: 976-981.
Guerinot, M.L. (2000) The ZIP family of metal transporters. *Biochim. Biophys. Acta,* **1465**: 190-198.
Guerinot, M.L. and Yi, Y. (1994) Iron: nutritious, noxious, and not readily available. *Plant Physiol.,* **104**: 815-820.
Gunshin, H., Mackenzie, B., Berger, U.V., Gunshin, Y., Romero, M.F., Boron, W.F., Nussberger, S., Gollan, J.L. and Hediger, M.A. (1997) Cloning and characterization of a mammalian proton-coupled metal-ion transporter. *Nature,* **388**: 482-488.
Harrison, P.M. and Arosio, P. (1996) The ferritins: molecular properties, iron storage function and cellular regulation. *Biochim. Biophys. Acta,* **1275**: 161-203.
Hell, R. and Stephan, U.W. (2003) Iron uptake, trafficking and homeostasis in plants. *Planta,* **216**: 541-551.
Henriques, R., Jasik, J., Klein, M., Martinoia, E., Feller, U., Schell, J., Pais, M.S. and Koncz, C. (2002) Knock-out of Arabidopsis metal transporter gene *IRT1* results in iron deficiency accompanied by cell differentiation defects. *Plant Mol. Biol.,* **50**: 587-597.
Herbik, A., Koch, G., Mock, H.P., Dushkov, D., Czihal, A., Thielmann, J., Stephan, U.W. and Baumlein, H. (1999) Isolation, characterization and cDNA cloning of nicotianamine synthase from barley. A key enzyme for iron homeostasis in plants. *Eur. J. Biochem.,* **265**: 231-239.
Higuchi, K., Kanazawa, K., Nishizawa, N.K. and Mori, S. (1996) The role of nicotianamine synthase in response to Fe nutrition status in Gramineae. *Plant Soil,* **178**: 171-177.
Higuchi, K., Suzuki, K., Nakanishi, H., Yamaguchi, H., Nishizawa, N.K. and Mori, S. (1999) Cloning of nicotianamine synthase genes, novel genes involved in the biosynthesis of phytosiderophores. *Plant Physiol.,* **119**: 471-480.
Higuchi, K., Watanabe, S., Takahashi, M., Kawasaki, S., Nakanishi, H., Nishizawa, N.K. and Mori, S. (2001) Nicotianamine synthase gene expression differs in barley and rice under Fe-deficient conditions. *Plant J.,* **25**: 159-167.
Inoue, H., Higuchi, K., Takahashi, M., Nakanishi, H., Mori, S. and Nishizawa, N.K. (2003) Three rice nicotianamine synthase genes, *OsNAS1, OsNAS2,* and *OsNAS3* are expressed in cells involved in long-distance transport of iron and differentially regulated by iron. *Plant J.,* **36**: 366-381.

Ishimaru, Y., Suzuki, M., Tsukamoto, T., Suzuki, K., Nakazono, M., Kobayashi, T., Wada, Y., Watanabe, S., Matsuhashi, S., Takahashi, M., Nakanishi, H., Mori, S. and Nishizawa, N.K. (2006) Rice plants take up iron as an $Fe^{3+}$-phytosiderophore and as $Fe^{2+}$. *Plant J.*, **45**: 335-346.

Jakoby, M., Wang, H.Y., Reidt, W., Weisshaar, B. and Bauer, P. (2004) *FRU (BHLH029)* is required for induction of iron mobilization genes in *Arabidopsis thaliana*. *FEBS Lett.*, **577**: 528-534.

Jolley, V.D., Brown, J.C., Davies, T.D. and Walser, R.H. (1986) Increased iron efficiency in soybeans (*Glycine max*) through plant breeding related to increased response to iron deficiency stress. I. Iron stress response. *J. Plant Nutr.*, **9**: 373-386.

Jurkevitch, E., Hadar, Y., Chen, Y., Chino, M. and Mori, S. (1993) Indirect utilization of the phytosiderophore mugineic acid as an iron source to rhizosphere fluorescent *Pseudomonas*. *Biometals*, **6**: 119-123.

Kanazawa, K., Higuchi, K., Nishizawa, N.K., Fushiya, S., Mitsuo, C. and Mori, S. (1994) Nicotianamine aminotransferase activities are correlated to the phytosiderophore secretions under Fe-deficient conditions in Gramineae. *J. Exp. Bot.*, **45**: 1903-1906.

Kanazawa, K., Higuchi, K., Nishizawa, N.K., Fushiya, S. and Mori, S. (1995) Detection of two distinct isozymes of nicotianamine aminotransferases in Fe-deficient barley roots. *J. Exp. Bot.*, **46**: 1241-1244.

Kneen, B.E., LaRue, T.A., Welch, R.M. and Weeden, N.F. (1990) Pleiotropic effects of *brz*: A mutation ir *Pisum sativum* (L.) cv 'sparkle' conditioning decreased nodulation and increased iron uptake and leaf necrosis. *Plant Physiol.*, **93**: 717-722.

Kobayashi, T., Nakanishi, H., Takahashi, M., Kawasaki, S., Nishizawa, N.K. and Mori, S. (2001) *In vivo* evidence that *Ids3* from *Hordeum vulgare* encodes a dioxygenase that converts 2'-deoxymugineic acid to mugineic acid in transgenic rice. *Planta*, **212**: 864-871.

Korshunova, Y.O., Eide, D., Clark, W.G., Guerinot, M.L. and Pakrasi, H.B. (1999) The IRT1 protein from *Arabidopsis thaliana* is a metal transporter with a broad substrate range. *Plant Mol. Biol.*, **40**: 37-44.

Krueger, C., Berkowitz, O., Stephan, U.W. and Hell, R. (2002) A metal-binding member of the Late Embryogenesis Abundant protein family transports iron in the phloem of *Ricinus communis* L. *J. Biol. Chem.*, **277**: 25062-25069.

Lafuse, W.P., Alvarez, G.R. and Zwilling, B.S. (2000) Regulation of *Nramp1* mRNA stability by oxidants and protein kinase C in RAW264.7 macrophages expressing *Nramp1* (Gly169). *Biochem. J.*, **351** Pt 3: 687-696.

Lanquar, V., Lelievre, F., Bolte, S., Hames, C., Alcon, C., Neumann, D., Vansuyt, G., Curie, C., Schroder, A., Kramer, U., Barbier-Brygoo, H. and Thomine, S. (2005) Mobilization of vacuolar iron by AtNRAMP3 and AtNRAMP4 is essential for seed germination on low iron. *EMBO. J.*, **24**: 4041-4051.

Le Jean, M., Schikora, A., Mari, S., Briat, J.F. and Curie, C. (2005) A loss-of-function mutation in *AtYSL1* reveals its role in iron and nicotianamine seed loading. *Plant J.*, **44**: 769-782.

Lescure, A.M., Proudhon, D., Pesey, H., Ragland, M., Theil, E.C. and Briat, J.F. (1991) Ferritin gene transcription is regulated by iron in soybean cell cultures. *Proc. Natl. Acad. Sci. USA*, **88**: 8222-8226.

Li, L., Cheng, X. and Ling, H.Q. (2004) Isolation and characterization of Fe(III)-chelate reductase gene *LeFRO1* in tomato. *Plant Mol. Biol.*, **54**: 125-136.

Ling, H.Q., Bauer, P., Bereczky, Z., Keller, B. and Ganal, M. (2002) The tomato *fer* gene encoding a bHLH protein controls iron-uptake responses in roots. *Proc. Natl. Acad. Sci. USA*, **99**: 13938-13943.

Ling, H.Q., Koch, G., Baumlein, H. and Ganal, M.W. (1999) Map-based cloning of *chloronerva*, a gene involved in iron uptake of higher plants encoding nicotianamine synthase. *Proc. Natl. Acad. Sci. USA*, **96**: 7098-7103.

Lopez-Millan, A.F., Morales, F., Abadia, A. and Abadia, J. (2000) Effects of iron deficiency on the composition of the leaf apoplastic fluid and xylem sap in sugar beet. Implications for iron and carbon transport. *Plant Physiol.*, **124**: 873-884.

Maas, F.M., van de Wetering, D.A.M., van Beusichem, M.L. and Bienfait, H.F. (1988) Characterization of phloem iron and its possible role in the regulation of Fe-efficiency reactions. *Plant Physiol.*, **87**: 167-171.

Marschner, H. and Römheld, V. (1994) Strategies of plants for acquisition of iron. *Plant Soil*, **165**: 375-388.

Mäser, P., Thomine, S., Schroeder, J.I., Ward, J.M., Hirschi, K., Sze, H., Talke, I.N., Amtmann, A., Maathuis, F.J., Sanders, D., Harper, J.F., Tchieu, J., Gribskov, M., Persans, M.W., Salt, D.E., Kim, S.A. and Guerinot, M.L. (2001) Phylogenetic relationships within cation transporter families of Arabidopsis. *Plant Physiol.*, **126**: 1646-1667.

McKie, A.T., Marciani, P., Rolfs, A., Brennan, K., Wehr, K., Barrow, D., Miret, S., Bomford, A., Peters, T.J., Farzaneh, F., Hediger, M.A., Hentze, M.W. and Simpson, R.J. (2000) A novel duodenal iron-regulated transporter, IREG1, implicated in the basolateral transfer of iron to the circulation. *Mol. Cell*, **5**: 299-309.

Miller, R.O. and Olsen, R.A. (1986) Changes in the roots of sunflower under iron stress. *J. Plant Nutr.*, **9**: 815-822.

Mizuno, D., Higuchi, K., Sakamoto, T., Nakanishi, H., Mori, S. and Nishizawa, N.K. (2003) Three nicotianamine synthase genes isolated from maize are differentially regulated by iron nutritional status. *Plant Physiol.*, **132**: 1989-1997.

Mori, S. and Nishizawa, N. (1987) Methionine as a dominant precursor of phytosiderophores in Gramineae plants. *Plant Cell Physiol.*, **28**: 1081-1092.

Mukherjee, I., Campbell, N.H., Ash, J.S. and Connolly, E.L. (2005) Expression profiling of the Arabidopsis ferric chelate reductase (*FRO*) gene family reveals differential regulation by iron and copper. *Planta*, 1-13.

Nakanishi, H., Okumura, N., Umehara, Y., Nishizawa, N.K., Chino, M. and Mori, S. (1993) Expression of a gene specific for iron deficiency (*Ids3*) in the roots of *Hordeum vulgare*. *Plant Cell Physiol.*, **34**: 401-410.

Nakanishi, H., Yamaguchi, H., Sasakuma, T., Nishizawa, N.K. and Mori, S. (2000) Two dioxygenase genes, *Ids3* and *Ids2*, from *Hordeum vulgare* are involved in the biosynthesis of mugineic acid family phytosiderophores. *Plant Mol. Biol.*, **44**: 199-207.

Negishi, T., Nakanishi, H., Yazaki, J., Kishimoto, N., Fujii, F., Shimbo, K., Yamamoto, K., Sakata, K., Sasaki, T., Kikuchi, S., Mori, S. and Nishizawa, N.K. (2002) cDNA microarray analysis of gene expression during Fe-deficiency stress in barley suggests that polar transport of vesicles is implicated in phytosiderophore secretion in Fe-deficient barley roots. *Plant J.*, **30**: 83-94.

Ohata, T., Mihashi, S., Nishizawa, N.K., Fushiya, S., Nozoe, S., Chino, M. and Mori, S. (1993) Biosynthetic pathway of phytosiderophores in iron-deficient gramineous plants. *Soil Sci. Plant Nutr.*, **39**: 745-749.

Okumura, N., Nishizawa, N.K., Umehara, Y., Ohata, T., Nakanishi, H., Yamaguchi, T., Chino, M. and Mori, S. (1994) A dioxygenase gene (*Ids2*) expressed under iron deficiency conditions in the roots of *Hordeum vulgare*. *Plant Mol. Biol.*, **25**: 705-719.

Pich, A., Manteuffel, R., Hillmer, S., Scholz, G. and Schmidt, W. (2001) Fe homeostasis in plant cells: does nicotianamine play multiple roles in the regulation of cytoplasmic Fe concentration? *Planta,* 213: 967-976.

Pich, A. and Scholz, G. (1996) Translocation of copper and other micronutrients in tomato plants (*Lycopersicon esculentum* Mill.): nicotianamine-stimulated copper transport in the xylem. *J. Exp. Bot.,* 47: 41-47.

Qu, L.Q., Yoshihara, T., Ooyama, A., Goto, F. and Takaiwa, F. (2005) Iron accumulation does not parallel the high expression level of ferritin in transgenic rice seeds. *Planta,* 222: 225-233.

Ragland, M. and Theil, E.C. (1993) Ferritin (mRNA, protein) and iron concentrations during soybean nodule development. *Plant Mol. Biol.,* 21: 555-560.

Roberts, L.A., Pierson, A.J., Panaviene, Z. and Walker, E.L. (2004) Yellow stripe1. Expanded roles for the maize iron-phytosiderophore transporter. *Plant Physiol.,* 135: 112-120.

Robinson, N.J., Procter, C.M., Connolly, E.L. and Guerinot, M.L. (1999) A ferric-chelate reductase for iron uptake from soils. *Nature,* 397: 694-697.

Rogers, E.E., Eide, D.J. and Guerinot, M.L. (2000) Altered selectivity in an Arabidopsis metal transporter. *Proc. Natl. Acad. Sci. USA,* 97: 12356-12360.

Rogers, E.E. and Guerinot, M.L. (2002) FRD3, a member of the multidrug and toxin efflux family, controls iron deficiency responses in Arabidopsis. *Plant Cell,* 14: 1787-1799.

Römheld, V. (1987) Different strategies for iron acquisition in higher plants. *Physiol. Plant.,* 70: 231-234.

Römheld, V. (1991) The role of phytosiderophores in acquisition of iron and other micronutrients in gramineaceous species: an ecological approach. In: *Iron Nutrition and Interactions in Plants.* (Eds. Chen, Y., Hadar, Y.) Kluwer Academic Publishers, Boston, MA, pp 159-166.

Römheld, V. and Marschner, H. (1983) Mechanisms of iron uptake by peanut plant. I. Fe$^{III}$ reduction, chelate splitting, and release of phenolics. *Plant Physiol.,* 71: 949-954.

Römheld, V. and Marschner, H. (1986) Evidence for a specific uptake system for iron phytosiderophore in roots of grasses. *Plant Physiol.,* 80: 175-180.

Savino, G., Briat, J.F. and Lobreaux, S. (1997) Inhibition of the iron-induced *ZmFer1* maize ferritin gene expression by antioxidants and serine/threonine phosphatase inhibitors. *J. Biol. Chem.,* 272: 33319-33326.

Schaaf, G., Ludewig, U., Erenoglu, B.E., Mori, S., Kitahara, T. and von Wiren, N. (2004) ZmYS1 functions as a proton-coupled symporter for phytosiderophore- and nicotianamine-chelated metals. *J. Biol. Chem.,* 279: 9091-9096.

Schaaf, G., Schikora, A., Haberle, J., Vert, G., Ludewig, U., Briat, J.F., Curie, C. and von Wiren, N. (2005) A putative function for the Arabidopsis Fe-phytosiderophore transporter homolog AtYSL2 in Fe and Zn homeostasis. *Plant Cell Physiol.,* 46: 762-774.

Schmidt, W. (1999) Mechanisms and regulation of reduction-based iron uptake in plants. *New Phytol.,* 141: 1-26.

Schmidt, W., Boomgaarden, B. and Ahrens, V. (1996) Reduction of root iron in *Plantago lanceolata* during recovery from Fe deficiency. *Physiol. Plant.,* 98: 587-593.

Scholz, G., Becker, R., Pich, A. and Stephan, U.W. (1992) Nicotianamine - A common constituent of Strategies I and II of iron acquisition in plants. *Rev. J. Plant Nutr.,* 15: 1649-1665.

Shojima, S., Nishizawa, N. and Mori, S. (1989) Establishment of a cell free system for the biosynthesis of nicotianamine. *Plant Cell Physiol.,* 30: 673-677.

Iron transport

Shojima, S., Nishizawa, N.K., Fushiya, S., Nozoe, S., Irifune, T. and Mori, S. (1990) Biosynthetic pathway of phytosiderophores in iron-deficient gramineous plants. *Plant Physiol.*, **93**: 1497-1503.
Stephan, U.W. and Grün, M. (1989) Physiological disorders of the nicotianamine auxotroph tomato mutant chloronerva at different levels of iron nutrition. II. Iron deficiency response and heavy metal metabolism. *Biochem. Physiol. Pflanzen.*, **185**: 189-200.
Stephan, U.W., Schmidke, I. and Pich, A. (1994) Phloem translocation of Fe, Cu, Mn, and Zn in *Ricinus* seedlings in relation to the concentrations of nicotianamine, an endogenous chelator of divalent metal ions, in different seedling parts. *Plant Soil*, **165**: 181-188.
Stephan, U.W., Schmidke, I., Stephan, V.W. and Scholz, G. (1996) The nicotianamine molecule is made-to-measure for complexation of metal micronutrients in plants. *Biometals*, **9**: 84-90.
Stephan, U.W. and Scholz, G. (1993) Nicotianamine: mediator of transport of iron and heavy metals in the phloem? *Physiol. Plant.*, **88**: 522-529.
Sussman, M.R. (1994) Molecular analysis of proteins in the plant plasma membrane. *Ann. Rev. Plant Physiol. Plant Mol. Biol.*, **45**: 211-234.
Sussman, M.R. (1999) Pumping iron. *Nat. Biotech.*, **17**: 230-231.
Takagi, S. (1976) Naturally occuring iron-chelating compounds in oat- and rice-root washing. I. *Soil Sci. Plant Nutr.*, **22**: 423-433.
Takagi, S. (1993) Production of phytosiderophores. In: *Iron chelation in plants and soil microorganisms.* (Eds. Barton, L., Hemming, B.) Academic Press, San Diego, pp 111-131.
Takahashi, M., Nakanishi, H., Kawasaki, S., Nishizawa, N.K. and Mori, S. (2001) Enhanced tolerance of rice to low iron availability in alkaline soils using barley nicotianamine aminotransferase genes. *Nat. Biotechnol.*, **19**: 466-469.
Takahashi, M., Terada, Y., Nakai, I., Nakanishi, H., Yoshimura, E., Mori, S. and Nishizawa, N.K. (2003) Role of nicotianamine in the intracellular delivery of metals and plant reproductive development. *Plant Cell*, **15**: 1263-1280.
Takahashi, M., Yamaguchi, H., Nakanishi, H., Shioiri, T., Nishizawa, N.K. and Mori, S. (1999) Cloning two genes for nicotianamine aminotransferase, a critical enzyme in iron acquisition (Strategy II) in graminaceous plants. *Plant Physiol.*, **121**: 947-956.
Takizawa, R., Nishizawa, N.K., Nakanishi, H. and Mori, S. (1996) Effect of iron deficiency on S-adenosylmethionine in barley roots. *J. Plant Nutr.*, **19**: 1189-1200.
Theil, E.C. (1987) Ferritin: structure, gene regulation, and cellular function in animals, plants, and microorganisms. *Annu. Rev. Biochem.*, **56**: 289-315.
Thomine, S., Lelievre, F., Debarbieux, E., Schroeder, J.I. and Barbier-Brygoo, H. (2003) AtNRAMP3, a multispecific vacuolar metal transporter involved in plant responses to iron deficiency. *Plant J.*, **34**: 685-695.
Thomine, S., Wang, R., Ward, J.M., Crawford, N.M. and Schroeder, J.I. (2000) Cadmium and iron transport by members of a plant metal transporter family in Arabidopsis with homology to *Nramp* genes. *Proc. Natl. Acad. Sci. USA*, **97**: 4991-4996.
Tiffin, L.O. (1966) Iron translocation I. Plant culture, exudate sampling, iron-citrate analysis. *Plant Physiol.*, **41**: 510-514.
Varotto, C., Maiwald, D., Pesaresi, P., Jahns, P., Salamini, F. and Leister, D. (2002) The metal ion transporter IRT1 is necessary for iron homeostasis and efficient photosynthesis in *Arabidopsis thaliana*. *Plant J.*, **31**: 589-599.
Vert, G., Briat, J.F. and Curie, C. (2001) Arabidopsis IRT2 gene encodes a root-periphery iron transporter. *Plant J.*, **26**: 181-189.

Vert, G., Grotz, N., Dedaldechamp, F., Gaymard, F., Guerinot, M.L., Briat, J.F. and Curie, C. (2002) IRT1, an Arabidopsis transporter essential for iron uptake from the soil and for plant growth. *Plant Cell*, **14**: 1223-1233.

Vert, G.A., Briat, J.F. and Curie, C. (2003) Dual regulation of the Arabidopsis high-affinity root iron uptake system by local and long-distance signals. *Plant Physiol.*, **132**: 796-804.

von Wiren, N., Klair, S., Bansal, S., Briat, J.F., Khodr, H., Shioiri, T., Leigh, R.A. and Hider, R.C. (1999) Nicotianamine chelates both FeIII and FeII. Implications for metal transport in plants. *Plant Physiol.*, **119**: 1107-1114.

von Wiren, N., Mori, S., Marschner, H. and Römheld, V. (1994) Iron Inefficiency in maize mutant *ys1* (*Zea mays* L. cv Yellow-Stripe) is caused by a defect in uptake of iron phytosiderophores. *Plant Physiol.*, **106**: 71-77.

Wang, Y.H., Garvin, D.F. and Kochian, L.V. (2002) Rapid induction of regulatory and transporter genes in response to phosphorus, potassium, and iron deficiencies in tomato roots. Evidence for cross talk and root/rhizosphere-mediated signals. *Plant Physiol.*, **130**: 1361-1370.

Waters, B.M., Blevins, D.G. and Eide, D.J. (2002) Characterization of FRO1, a pea ferric-chelate reductase involved in root iron acquisition. *Plant Physiol.*, **129**: 85-94.

Welch, R.M. and Kochian, L.V. (1992) Regulation of iron accumulation in food crops: studies using single gene pea mutants. In: *Biotechnology and Nutrition*. (Eds. Bills, D.D., Kung, S.-D.) Butterworth-Heinemann, Boston, MA, pp 325-344.

White, M.C., Baker, F.D., Chaney, R.L. and Decker, A.M. (1981) Metal complexation in xylem fluid. II. Theoretical equilibrium model and computational computer program. *Plant Physiol.*, **67**: 301-310.

Wu, H., Li, L., Du, J., Yuan, Y., Cheng, X. and Ling, H.Q. (2005) Molecular and biochemical characterization of the Fe(III) chelate reductase gene family in *Arabidopsis thaliana*. *Plant Cell Physiol.*, **46**: 1505-1514.

Yen, M.R., Tseng, Y.H. and Saier, M.H. Jr. (2001) Maize Yellow Stripe1, an iron-phytosiderophore uptake transporter, is a member of the oligopeptide transporter (OPT) family. *Microbiology*, **147**: 2881-2883.

Yi, Y. and Guerinot, M.L. (1996) Genetic evidence that induction of root Fe(III) chelate reductase activity is necessary for iron uptake under iron deficiency. *Plant J.*, **10**: 835-844.

Yuan, Y.X., Zhang, J., Wang, D.W. and Ling, H.Q. (2005) AtbHLH29 of *Arabidopsis thaliana* is a functional ortholog of tomato FER involved in controlling iron acquisition in strategy I plants. *Cell Res.*, **15**: 613-621.

## Chapter 7

## MECHANISMS OF MANGANESE ACCUMULATION AND TRANSPORT

**JON K. PITTMAN**
*Faculty of Life Sciences, University of Manchester, 3.614 Stopford Building, Oxford Road, Manchester, M13 9PT, U.K.*
E-mail: jon.pittman@manchester.ac.uk

**Abstract**

Manganese (Mn) is an essential mineral nutrient. Mn transport proteins have a critical role in mediating acquisition of Mn by the roots, partitioning Mn into organelles such as the vacuole and chloroplast, controlling the long-distance translocation of Mn, and regulating Mn loading into organs such as leaves, flowers and seeds. Mn transporters also play a major role in providing tolerance to toxic concentrations of Mn, by mediating sequestration into internal cellular compartments or efflux from the cell, and have been implicated in adaptive strategies to overcome Mn deficiency. Recently, members of various metal transporter families involved in Mn transport have been identified. These include members of the ZIP, Nramp, CAX, CDF, OPT/YSL and P-type ATPase transporter families. An emerging feature of many plant Mn transporters, particularly those involved in cytosolic Mn influx, is their broad substrate selectivity, often coupled with iron (Fe) transport and regulated by Fe deficiency. In contrast, some of the transporters that mediate Mn sequestration into the vacuole, such as the $Mn^{2+}/H^+$ antiporters CAX2 and ShMTP1, appear to be more specific for Mn and are not influenced by Fe nutrition. This review describes the characterisation of these putative Mn transporters, and discusses their potential roles in various physiological processes. The genetic manipulation of Mn transport and homeostasis proteins as a strategy to overcome plant Mn stress, and the identification and characterisation of Mn homeostasis mutants and Mn hyperaccumulator plants as a means to identify novel components in Mn homeostasis, is also discussed.

**Keywords:** transport, manganese, iron, manganese toxicity, manganese deficiency

## 1. INTRODUCTION

Manganese (Mn) is an essential mineral nutrient in plants that has a wide number of roles, most notably in mediating water oxidation in photosystem II, and as a co-

© CAB International 2008. *Plant Membrane and Vacuolar Transporters* (eds P.K. Jaiwal, R.P. Singh and O.P. Dhankher)

factor for various enzymes including Mn-dependent superoxide dismutase (SOD), catalase, glycosyltransferase, and pyruvate carboxylase (Marschner, 1995). Mn content can vary widely between plants, ranging from 25 to 1000 µg Mn $g^{-1}$ dry weight (DW), but can exceed 10 000 µg $g^{-1}$ DW for a few Mn hyperaccumulating species (Foy et al., 1988; Bidwell et al., 2002). This variation is due largely to the increased availability and uptake efficiency of Mn by some species. While the requirement for Mn is essential, the amount required for most species is relatively low (~20 µg $g^{-1}$ DW), but for many plants the amount accumulated is much greater than that needed (Clarkson, 1988; Marschner, 1995). This suggests that plants do not tightly regulate the accumulation of Mn. Mn toxicity can therefore be a significant problem for some plants, particularly when soil pH is acidic (below pH 5.5) causing an increase in Mn(II) abundance.

Plants are proposed to have a variety of mechanisms to overcome Mn toxicity stress, including the sequestration of Mn into internal compartments, particularly the vacuole, efflux of Mn from the cell into the apoplast, and chelation of Mn with organic acids such as malate and oxalate, and possibly the subsequent vacuolar sequestration of the Mn-acid complex (Horst and Maier, 1999; Bidwell et al., 2002; Pittman, 2005). In addition, mechanisms to overcome the production of reactive oxygen species (ROS) that can result from Mn toxicity have been observed, which include maintenance of high levels of the antioxidant ascorbate and various antioxidant enzymes such as SOD and ascorbate peroxidase (González et al., 1998; Fecht-Christoffers et al., 2003). As well as having a critical role in Mn detoxification, transporters are obviously required for the acquisition of Mn from the soil, and in mediating Mn translocation throughout the plant and partitioning to various subcellular locations where needed. Some of the physiological and biochemical characteristics of plant Mn accumulation and transfer throughout the plant have been determined over a number of years (Clarkson, 1988; Loneragan, 1988; Rengel, 2000). Recent sequencing of plant genomes has revealed the presence of many multigene families encoding putative metal transporters (Williams et al., 2000; Mäser et al., 2001; Hall and Williams, 2003). While many of these genes remain to be characterised, a number have now been identified as having roles in Mn transport.

## 2. MN TRANSPORT IN MICROORGANISMS

Mn transport and homeostasis processes have been extensively studied in microorganisms, particularly in the yeast *Saccharomyces cerevisiae* and various bacterial species. Many genes encoding Mn transporters have been cloned from these species. While the physiological characteristics of Mn transport will differ between plants and microorganisms, there is likely to be some conservation in transport processes between them. Therefore, some insights into plant Mn transport may be gained from comparative analysis with microorganisms.

### 2.1. Mn transport pathways in bacteria

Many bacteria are dependent on Mn; for example, *Bacillus subtilis* requires Mn for sporulation. In contrast *Escherichia coli* and *Salmonella enterica* can survive on very low concentrations of Mn but are highly dependent on iron (Fe); however, regulated Mn homeostasis and transport is important in these species for virulence (Kehres and Maguire, 2003). Some pathogenic bacteria no longer utilise Fe but have an essential requirement for Mn instead (Posey and Gherardini, 2000), and bacteria that accumulate high concentrations of Mn but low Fe are resistant to gamma radiation (Daly et al., 2004). It has been known for decades that bacteria possess high-affinity Mn uptake pathways, but only relatively recently have genes encoding these transport pathways been identified. The first bacterial Mn transporter to be identified was MntABC in the cyanobacterium *Synechocystis* sp. PCC 6803 (Bartsevich and Pakrasi, 1996). MntABC is a member of the ATP-binding cassette (ABC) transporter family and is a high-affinity Mn(II) transporter that mediates accumulation of Mn into the cell during Mn starvation. Subsequently, MntABC transporters, also called Mn permeases, have been identified and functionally characterised in many bacterial species (Horsburgh et al., 2002b). It is not clear how MntABC transports Mn(II), whether as the $Mn^{2+}$ ion or as a Mn(II)-complex, and while most MntABC transporters are specific for Mn, some can transport other metals, such as cadmium (Cd) and Fe. A second bacterial Mn transport pathway is the MntH transporter, a $H^+$-dependent $Mn^{2+}$ transporter which is a member of the ubiquitous natural resistance-associated macrophage protein (Nramp) family. Following their identification in animals and demonstration that they are broad specificity divalent cation transporters, they were found to be widespread in many bacteria (Kehres and Maguire, 2003). Like MntABC, MntH is a high-affinity $Mn^{2+}$ transporter. Most characterised bacterial MntH transporters are highly selective for $Mn^{2+}$, although many have the ability to transport $Cd^{2+}$, and some can also transport other cations including $Fe^{2+}$. An additional novel bacterial Mn transport mechanism has been identified in the lactic acid bacterium *Lactobacillus plantarum* which accumulates high concentrations of Mn (~ 30 mM). In addition to MntABC and MntH, this bacterium has a P-type ATPase named MntA, a high-affinity $Mn^{2+}$ transporter that can also transport $Cd^{2+}$ and which is induced by Mn starvation (Hao et al., 1999). Interestingly, sequence analysis suggests that MntA is similar to the $P_{3A}$-type $H^+$-ATPases, which are widespread in plants (Axelsen and Palmgren, 2001). As yet, no other MntA-like $Mn^{2+}$ transporter has been identified in any other organism.

Mn transport is tightly regulated at the transcriptional level in bacteria. The molecular mechanisms of Mn regulation are well understood, particularly in *Staphylococcus aureus*, *B. subtilus* and *S. enterica* (Horsburgh et al., 2002b; Kehres and Maguire, 2003; Moore and Helmann, 2005). The transcriptional regulator MntR negatively regulates *mntH* and *mntABC* by functioning as a Mn-dependent repressor (Que and Helmann, 2000; Horsburgh et al., 2002a). When intracellular Mn levels rise MntR binds to the promoter regions of these genes and represses transcription. In various species, the regulation of Mn homeostasis significantly overlaps with Fe homeostasis and oxidative stress. This is probably due in part to the diverse roles of Mn and Fe in bacterial oxidative stress. Fe can contribute to the production of ROS that lead to oxidative stress, by reacting with hydrogen peroxide to form hydroxyl radicals. Conversely, Mn contributes to detoxification of

ROS as a co-factor for Mn-SOD and Mn-dependent catalase. Furthermore, Mn itself is a potent antioxidant and can scavenge superoxide and hydrogen peroxide (Horsburgh et al., 2002b). This is exemplified in *L. plantarum* which lacks any SOD activity and instead uses its high concentrations of Mn bound to phosphate or lactate to scavenge ROS. To allow an accurate balance between Mn and Fe levels in response to oxidative stress, there is coordinated sensing and regulation of these metals, and metal-dependent sensing and regulation of ROS levels. In some species, in addition to the MntR Mn homeostasis regulator, Fe homeostasis and oxidative stress regulators also regulate Mn transport. The Fe-dependent Fe uptake regulator Fur represses Mn transporters in *E. coli* and *S. enterica* and the peroxide-sensitive Mn-dependent peroxide regulator PerR (OxyR in *S. enterica*) represses MntABC in *S. aureus* and MntH in *S. enterica* (Patzer and Hantke, 2001; Horsburgh et al., 2002a; Kehres et al., 2002). In addition, Fur and PerR regulate various oxidative stress response genes.

Mn is of particular importance in photosynthetic organisms for water oxidation, therefore the mechanisms of Mn homeostasis in photosynthetic cyanobacteria are of interest due to its potential as a comparative model for plant chloroplasts. Upon Mn starvation, *Synechocystis* 6803 transports Mn into the cell via the high-affinity MntABC transporter (Bartsevich and Pakrasi, 1996). MntABC is transcriptionally regulated, but rather than by a MntR-type regulator, it is negatively regulated by a two-component Mn sensing system comprising of a histidine kinase ManS that senses external Mn and a response regulator ManR that binds to the *mntCAB* operon that encodes MntABC to repress expression (Ogawa et al., 2002; Yamaguchi et al., 2002). A MntH transporter has not been identified in *Synechocystis* but other Mn accumulation mechanisms have been observed including light-dependent uptake (Keren et al., 2002), and a second high-affinity pathway (Bartsevich and Pakrasi, 1996), although the genes encoding these pathways have yet to be identified.

## 2.2. Mn transport pathways in yeast

As with other eukaryotic organisms, yeast requires Mn as a cofactor for many enzymes. A variety of Mn transport pathways and homeostatic mechanisms have been identified in the yeast *S. cerevisiae* (Culotta et al., 2005). Mn accumulation into the cell occurs via the high affinity $Mn^{2+}$ transporter Smf1, which is an Nramp transporter related to the MntH bacterial transporters (Supek et al., 1996). Expression studies have shown that Smf1 also has the ability to transport other metals such as copper (Cu), Fe, zinc (Zn) and Cd, but whether it transports these metals into the yeast under physiological conditions is unclear. During non-stressed, Mn replete conditions, a yeast cell contains relatively low concentrations of Mn, determined to be $\sim 1.0 \times 10^6$ Mn atoms per cell when grown in a medium containing 11 µM $MnCl_2$ (Eide et al., 2005). Under such conditions, Smf1 has a limited role in Mn accumulation and it has been suggested that additional low affinity Mn uptake transporters are present at the plasma membrane (Culotta et al., 2005). However, under Mn starvation conditions, Smf1 and another Nramp transporter Smf2 are up-regulated by post-translational mechanisms. Smf2 is also involved in

$Mn^{2+}$ transport, but unlike Smf1, is critical for Mn accumulation even under Mn-replete conditions. Smf2 has been localised to small intracellular vesicles, possibly Golgi vesicles, and is thought to be involved in the trafficking of Mn through the cell to other organelles, particularly the Golgi and mitochondria (Luk and Culotta, 2001). The P-type $Ca^{2+}/Mn^{2+}$-ATPase Pmr1 mediates $Mn^{2+}$ transport into the Golgi which is required by various glycosyltransferases (Lapinskas et al., 1995; Dürr et al., 1998). The mechanism for Mn accumulation into the mitochondria is unclear, but Mtm1, a member of the mitochondria carrier family is localised at the inner envelope and is required for delivering Mn from the unknown transporter to the mitochondrial matrix-localised Mn-SOD Sod2 (Luk et al., 2003).

During Mn excess conditions, and even relatively Mn replete conditions (> 1 µM Mn), Smf1 is down-regulated by the action of a secretory pathway protein Bsd2, which targets Smf1 to the vacuole for degradation rather than to the plasma membrane, while under Mn starvation, Smf1 is not recognised by Bsd2 and is targeted to the plasma membrane to facilitate $Mn^{2+}$ uptake (Liu et al., 1997; Liu and Culotta, 1999). However, even during Mn excess conditions, Mn can accumulate into the cell, via the high-affinity phosphate transporter Pho84 which functions as a low-affinity Mn transporter by transporting $MnHPO_4$ when the yeast is exposed to high Mn concentrations (Jensen et al., 2003). Interestingly, this mechanism does not appear to be unique to yeast, as algal phosphate transporters are up regulated by Mn starvation, indicating of a role in Mn accumulation (Allen et al., 2007). As the yeast cell cannot prevent Mn uptake, detoxification mechanisms are required to prevent cytosolic accumulation. There is no evidence of transporter-mediated Mn efflux out of the cell, but it has been suggested that Mn pumped into the Golgi by Pmr1 is then removed from the cell by the trafficking of Mn-containing secretory pathway vesicles to the cell surface (Culotta et al., 2005). Intracellular Mn is predominantly localised in the vacuole, thus this compartment has a significant role in Mn detoxification and proteins have been identified that have a role in vacuolar sequestration of Mn. The vacuolar $Ca^{2+}/H^+$ antiporter Vcx1 (also named Hum1) was previously proposed to also function as a $Mn^{2+}$ transporter due to its ability to confer modest Mn tolerance to a yeast strain lacking calcineurin (Pozos et al., 1996), although subsequent studies have shown that Vcx1 contributes little to Mn tolerance and has no significant $Mn^{2+}/H^+$ antiport activity (Pittman et al., 2004a). The Ccc1 transporter has been implicated in vacuolar Mn and Fe accumulation (Lapinskas et al., 1996; Li et al., 2001). Recently, a plant orthologue of Ccc1 called VIT1 was identified which was able to complement the Fe and Mn vacuolar transport deficiency of a ccc1 mutant (Kim et al., 2006). Despite this ability to transport both Fe and Mn in yeast, further experiments suggested that VIT1 was only involved in Fe transport in vivo. Finally, vacuolar protein Cos16 is involved in Mn sequestration, although there is no evidence that Cos16 is a transporter (Paidhungat and Garrett, 1998).

## 3. MN ACCUMULATION INTO THE PLANT

The accumulation of Mn into a plant depends on its oxidation state as plants can efficiently accumulate Mn(II), taken up as the cation $Mn^{2+}$, or in chelated forms such as Mn-EDTA, but cannot accumulate Mn(III) or Mn(IV) (Clarkson, 1988; Laurie et al., 1995). The availability of Mn(II) in the soil is therefore a major determining factor in plant Mn accumulation. When the soil solution pH is low, such as in acidic or waterlogged soils, the proportion of Mn(II) available increases to the extent that in very acidic soils Mn toxicity can be a major cause of plant yield loss. Conversely, in alkaline soils, Mn deficiency can be a significant problem. Plants may improve Mn(II) availability by $H^+$ pumping into the rhizosphere or by the exudation of organic compounds (Marschner, 1995; Rengel and Marschner, 2005). The presence of rhizosphere bacteria can have an influence on Mn(II) availability depending on whether the bacteria are Mn reducers or oxidisers (Rengel and Marschner, 2005).

Fe and Zn deficiency have been demonstrated to enhance Mn uptake (Welch and Norvell, 1993; Cohen et al., 1998). The effect of Fe deficiency may be due in part to activation of ferric reductase. The root plasma membrane ferric reductase reduces Fe(III) to Fe(II) under Fe deficiency conditions for Fe(II) accumulation by Strategy I plants. It has been suggested that the ferric reductase may also reduce Mn(III)/Mn(IV), although there is no evidence of ferric reductase up-regulation by Mn deficiency (Norvell et al., 1993; Delhaize, 1996). Metals such as Fe, calcium (Ca), Cd and Zn can inhibit Mn accumulation (Roomizadeh and Karimian, 1996; Hernandez et al., 1998), suggesting that Mn uptake may be mediated by broad specificity transporters. For example, some $Ca^{2+}$ influx channels are permeable to a range of cations including $Mn^{2+}$ (White, 1998), and various other non-specific transition metal transporters can transport Mn (see below). Inhibition of Mn uptake by Fe has been observed in many species. As in many bacterial species, there is a close interaction between plant Fe and Mn nutrition (Moraghan, 1992; Roomizadeh and Karimian, 1996; Alam et al., 2001; Izaguirre-Mayoral and Sinclair, 2005). While high Fe concentration can inhibit Mn accumulation, plants exposed to high Mn display Fe deficiency symptoms. In addition, Mn toxicity can be ameliorated by Fe application (Alam et al., 2001). The effect of Fe on Mn acquisition may be due in part to direct inhibition of Mn uptake into the plant, such as observed with the metal transporter IRT1 (Korshunova et al., 1999). However, some studies have found that the cause of reduced Mn content by Fe was not just due to inhibition of Mn uptake into the root, but an inhibition by Fe on Mn translocation from the root to the shoot (Ghasemi-Fasaei et al., 2003, 2005). Relatively few studies have analysed Mn uptake kinetics in intact plants grown under physiological conditions (Rengel, 2000), but it appears that in some species at least, there is both a high-affinity and low-affinity Mn uptake pathway, with the high-affinity uptake pathway functional at the low nanomolar range (Pedas et al., 2005).

## 4. MN EFFLUX FROM THE CYTOSOL AND INTERNAL COMPARTMENTATION

One means for detoxification of elevated concentrations of Mn in the cell is the sequestration of $Mn^{2+}$ or a Mn(II)-chelate complex into an internal compartment such as the vacuole. The main sink for Mn in a cell is the vacuole and to a lesser extent chloroplasts, and in addition there is some association of Mn with the cell wall (McCain and Markley, 1989; Quiquampoix et al., 1993; González and Lynch, 1999). The distribution of Mn between the vacuole and chloroplast can vary depending on species, plant age, and the Mn stress condition which the plant is exposed to. Unlike Mn uptake into a cell, Mn transport into an internal compartment or efflux from the cytosol across the plasma membrane is an energy dependent process, requiring either direct energisation by ATP or utilisation of a $H^+$ electrochemical gradient. One potential pathway for Mn transport into the vacuole is by $Mn^{2+}/H^+$ antiport, and various genes have been identified that could facilitate this pathway (see below). Mechanisms of Mn accumulation into the chloroplast remain unclear in plants. Whether ATP-dependent Mn accumulation into chloroplasts analogous to the ATP-dependent Mn uptake into photosynthetic cyanobacteria does occur remains to be determined. Biochemical studies have demonstrated transport of $Fe^{2+}$ across the pea chloroplast inner envelope membrane that was energised by a membrane potential-dependent mechanism (Shingles et al., 2002). This $Fe^{2+}$ uptake was significantly inhibited by a range of cations including $Mn^{2+}$. Various Mn-dependent enzymes are present in other plant organelles. For example, Mn-SOD has been identified in peroxisomes and mitochondria, and Mn-dependent glycosyltransferases are present in the Golgi (Marschner, 1995). The mechanisms of Mn transport into these organelles are unknown, but Mn is known to be transported into the endoplasmic reticulum (ER) by $Mn^{2+}$-ATPase activity (Wu et al., 2002). An alternative mechanism of Mn detoxification is Mn removal from the cytosol by efflux out of the cell. It has been suggested that Mn could be exported from the cell by $Mn^{2+}/H^+$ antiport (Clarkson, 1988). A recent study has indeed provided evidence that $Mn^{2+}/H^+$ antiport-mediated efflux across the plasma membrane occurs in cucumber roots. In this work, $H^+$-coupled $Mn^{2+}$ transport with a $K_m$ for $Mn^{2+}$ of 5 mM was observed (Migocka and Klobus, 2007). $H^+$-coupled $Cd^{2+}$, $Pb^{2+}$ and $Ni^{2+}$ transport was also detected, however it is unclear whether these activities are provided by a single or multiple transport proteins. A rice gene (*OsCAX3*) has also been identified that may have a role in $Mn^{2+}/H^+$ antiport at the plasma membrane (Qi et al., 2005; see below).

## 5. EMERGING MOLECULAR MECHANISMS OF MN TRANSPORT

### 5.1. ZIP transporters

Members of the phylogenetically widespread ZRT/IRT-like protein (ZIP) transporter family are often involved in Zn or Fe transport. An indication that plant ZIP transporters may have a role in Mn transport comes from analysis of the *Arabidopsis thaliana* Fe transporter IRT1. When expressed in yeast IRT1 has the ability to transport Mn, Zn and Cd (Korshunova et al., 1999; Table 1). IRT1 is up-regulated by Fe deficiency both at the RNA and protein level, and mediates Fe accumulation into the root (Connolly et al., 2002; Vert et al., 2002). Knockout of *IRT1* has a very severe morphological phenotype that can be rescued by application of exogenous Fe, but not any other metal, indicating that this phenotype is due to loss of Fe acquisition (Vert et al., 2002). Mn cannot rescue the *irt1* phenotype, yet the *irt1* plants have a complete loss of Mn content in the roots, in addition to a loss of Fe, Zn and cobalt (Co) content, demonstrating that IRT1 is required for the uptake of these metals into the plant. This indicates that not only is IRT1 the major pathway for Mn accumulation into the Arabidopsis root, but because *IRT1* is up-regulated during Fe deficiency, Mn accumulation only occurs by this pathway during Fe deficiency. Kinetic studies show that IRT1 has a $K_m$ for $Mn^{2+}$ of $9 \pm 1$ µM (Korshunova et al., 1999) which is equivalent to some estimates of $Mn^{2+}$ uptake kinetics into maize roots (Landi and Fagioli, 1983). There are 15 ZIP genes in Arabidopsis, many of which remain to be characterised, but IRT1 is the only one that has to date been shown to transport Mn (Mäser et al., 2001). $Zn^{2+}$ transport by Arabidopsis ZIP3 is slightly inhibited by Mn (Grotz et al., 1998), but whether this indicates $Mn^{2+}$ transport needs confirmation. ZIP transporters from tomato, pea, *Medicago truncatula*, and *Thlaspi japonicum* may also transport Mn in addition to other metals, as determined by the ability of these proteins to rescue the growth of Mn uptake-deficient yeast mutants (Eckhardt et al., 2001; Cohen et al., 2004; López-Millán et al., 2004; Mizuno et al., 2005; Table 1). Whether these proteins have an *in planta* role in Mn transport is not known. It is not clear from ZIP sequence analysis what determines the ability to transport Mn. A mutagenesis study of IRT1 indicated a role of Asp-100 or Asp-136 in determining selectivity for Fe and Mn (Rogers et al., 2000). However, these residues are conserved in ZIP transporters that do not transport Mn. Moreover, residues have not been identified that are a specific determinant for Mn transport alone.

## 5.2. Nramp transporters

Nramp transporters are critical for Mn homeostasis in yeast and bacteria (see above). They are also conserved throughout plants (Mäser et al., 2001; Thomine and Schroeder, 2004); therefore one may ask whether plant Nramps have a role in Mn accumulation and transport. All of the Nramps from various plant species that have been tested appear to have the ability to transport Mn and are broad specificity metal transporters (Thomine et al., 2000; Bereczky et al., 2003; Kaiser et al., 2003; Table 1). Mn transport activity is inferred by the ability to complement the Mn uptake-deficient phenotype of the *smf1* mutant yeast. Likewise, transport of other metals is inferred by complementation of other yeast mutant phenotypes. In only a few studies has direct $^{55}$Fe transport activity been confirmed (Thomine et al., 2000; Kaiser et al., 2003). However, unlike the yeast Nramp transporters, there is no clear evidence that plant Nramps have a role in Mn homeostasis under physiological or metal stress conditions. Yeast Smf1 has the ability to transport a wide range of metals as determined by heterologous expression in oocytes or in yeast (Culotta et al., 2005). Yet *in vivo* it has a specific role in Mn accumulation because it is up-regulated by Mn starvation. Many plant Nramp transporters are up-regulated by Fe deficiency rather than Mn deficiency, indicating a role in Fe homeostasis (Thomine et al., 2000; Bereczky et al., 2003). For example, AtNramp3 and AtNramp4 are vacuolar-localised transporters and are critical for Fe mobilisation from the vacuole of seeds (Thomine et al., 2003; Lanquar et al., 2005). It was previously observed that AtNramp3 can transport multiple metals in the plant. During Fe deficiency conditions, knockout of *AtNramp3* leads to an enhancement of Mn and Zn content in seedlings, while overexpression of *AtNramp3* down-regulates Mn and Zn accumulation (Thomine et al., 2003). However, AtNramp3 and AtNramp4 are specific for Fe mobilisation from the seed, as the phenotype of an *nramp3 nramp4* double knockout cannot be rescued by Mn or Zn, and the sites of metal storage in Arabidopsis seeds, the vacuolar globoid crystals, do not contain Mn (Lanquar et al., 2005). It remains to be determined whether any plant Nramp transporter is specifically required for Mn homeostasis or involved in cellular Mn accumulation analogous to Smf1, particularly as no plant Nramp has yet been localised to the plasma membrane.

One photosynthetic eukaryotic organism in which Nramp transporters may have a clear role in Mn homeostasis is the unicellular green alga, *Chlamydomonas reinhardtii*. A Chlamydomonas Nramp transporter called Nramp1 (also named DMT1) has been shown to complement the Mn uptake deficiency of the *smf1* yeast mutant (Rosakis and Köster, 2005). Furthermore, *Nramp1* is significantly up-regulated upon Mn starvation (Allen et al., 2007). However, the subcellular location of Nramp1 is unknown.

### 5.3. CAX transporters

Biochemical studies have shown that one potential pathway for the accumulation of Mn into the vacuole is by $Mn^{2+}/H^+$ antiport activity (González et al., 1999). $Mn^{2+}/H^+$ antiporters can be encoded by members of the cation exchanger (CAX) gene family. Arabidopsis has a multigene family of CAX genes which are homologous to yeast *VCX1* (Mäser et al., 2001; Shigaki and Hirschi, 2006). The CAX genes cloned and characterised to date, including *CAX1* and *CAX2*, all encode vacuolar membrane-localised proteins. CAX transporters were originally identified as Ca transporters, and are proposed to have a critical function in endomembrane Ca partitioning and signalling (Pittman and Hirschi, 2003). Expression studies in tobacco and yeast have shown that CAX2 has the ability to transport a range of transition metals including Cd and Mn in addition to Ca (Hirschi et al., 2000; Schaaf et al., 2002; Shigaki et al., 2003; Table 1). Heterologous expression of CAX2 in a Mn-sensitive yeast mutant was able to suppress this sensitivity due to vacuolar $Mn^{2+}/H^+$ antiport activity (Shigaki et al., 2003). Although yeast expression experiments have shown that CAX2 can transport Ca and Mn, whether transport of both of these metals is physiologically relevant in the plant is not fully clear. However, knockout plants lacking *CAX2* have significantly reduced vacuolar $Mn^{2+}/H^+$ antiport activity while $Ca^{2+}/H^+$ antiport activity is unchanged compared to wild-type, indicating that CAX2 is indeed an important component in Arabidopsis vacuolar Mn sequestration (Pittman et al., 2004b). In a recent yeast Mn tolerance screen in which an Arabidopsis cDNA library was expressed in yeast and grown on high Mn media, the only cDNA identified out of over 100,000 transformants that could mediate Mn tolerance was *CAX2* (Schaaf et al., 2002). Despite this result, CAX2

Jon K. Pittman

is not the only Arabidopsis transporter that can provide Mn tolerance (see below), particularly as vacuolar Mn transport was significantly reduced but not completely abolished in *cax2* plants (Pittman et al, 2004b) indicating that other transporters contribute to this pathway. A *CAX2*-like gene has been identified in the Zn hyperaccumulator *Arabidopsis halleri*. Expression of *AhCAX2* is up-regulated by low Zn, but it is unable to provide tolerance to Cd or Zn stress when expressed in yeast (Becher *et al.*, 2004). The possible role of AhCAX2 in Mn homeostasis was not tested.

CAX transporters from rice have also been shown to transport Mn into the vacuole. Vacuolar-localised OsCAX1a can suppress the Mn sensitive phenotype of a yeast strain, indicating the ability to transport Mn (Kamiya and Maeshima, 2004). Like CAX2, OsCAX1a can also transport Ca and is up-regulated in rice by Ca but not Mn (Kamiya *et al.*, 2006), so it is possible that physiologically OsCAX1a may be more important in Ca transport rather than Mn. Interestingly, OsCAX1a is more similar to the Arabidopsis CAX1 transporter, which can transport Ca but not Mn, than CAX2; but another rice CAX, OsCAX3, which has greater similarity to CAX2, can also transport Ca and Mn (Kamiya *et al.*, 2005). There is evidence that OsCAX3 is localised at the plasma membrane rather than the vacuole (Qi *et al.*, 2005), suggesting that this transporter could be important for metal detoxification by efflux from the cell rather than sequestration into the vacuole. Although there is no direct Mn transport measurements for any rice CAX, biochemical analysis of Arabidopsis CAX2 shows that it is a low-affinity $Mn^{2+}$ transporter (Shigaki *et al.*, 2003), indicating that Mn transport by CAX transporters is important during Mn toxicity conditions, when cytosolic Mn becomes elevated.

### 5.4. CDF transporters

The cation diffusion facilitator (CDF) family of metal transporters are, like Nramp and ZIP transporters, found in many organisms from bacteria to man. The CDF transporters characterised to date are involved in either the transfer of various transition metals out of the cell or into internal organelles (Williams *et al.*, 2000; Mäser *et al.*, 2001). In several plant species, many CDF family members have been identified and characterised, and shown to be involved in the transport of metals, particularly Zn (Hall and Williams, 2003; Mäser *et al.*, 2001). These plant transporters, named metal tolerance proteins (MTP), include the Arabidopsis vacuolar Zn transporter AtMTP1 (previously called ZAT) (Kobae *et al.*, 2004) and the *Thlaspi goesingense* plasma membrane Zn transporter TgMTP1 (Kim *et al.*, 2004). CDF transporters are thought to be $H^+$ dependent (Hall and Williams, 2003), therefore CDF transporter-mediated metal efflux will be by cation/$H^+$ antiport, in a manner similar to the CAX transporters. A CDF transporter has been identified that can transport $Mn^{2+}$ into the vacuole (Table 1). In *Stylosanthes hamata*, a Mn-tolerant legume, ShMTP1 is suggested to mediate vacuolar Mn accumulation (Delhaize *et al.*, 2003). This assumption was based on expression of ShMTP1 in Arabidopsis and yeast. This transporter could specifically provide tolerance to Mn in both organisms, but not tolerance to other metals, and ShMTP1 was shown to localise to the tonoplast. Three related proteins, ShMTP2, ShMTP3 and ShMTP4 can also confer Mn tolerance in yeast.

Recently, an Arabidopsis orthologue of *ShMTP1* (now renamed *ShMTP8* to maintain consistency with Arabidopsis MTP gene nomenclature) called *AtMTP11* has been cloned and shown to provide tolerance to Mn when expressed in yeast (Delhaize *et al.*, 2007; Peiter *et al.*, 2007). A knockout mutant of *AtMTP11* has increased sensitivity to Mn but not other metals; furthermore, direct transport measurements of AtMTP11 and ShMTP8 demonstrate that they can transport $Mn^{2+}$ in a mechanism which is $H^+$ dependent, but other metals cannot be transported (Delhaize *et al.*, 2007). This suggests that like ShMTP8, AtMTP11 is a $Mn^{2+}/H^+$ antiporter and is specific for Mn. However, unlike ShMTP8, AtMTP11 does not appear to localise to the vacuolar membrane. Characterisation of AtMTP11 by two independent studies has come to contrasting conclusions as to the membrane localisation of this $Mn^{2+}$ transporter. Delhaize *et al.* (2007) provided evidence to suggest that AtMTP11 is localised at pre-vacuolar compartments. Peiter *et al.* (2007) infer from their data that AtMTP11 is localised at Golgi membranes and suggest that Golgi-based Mn accumulation may result in Mn tolerance through a vesicular trafficking and exocytosis mechanisms, analogous to the pathway of Mn export via Pmr1 in yeast (Culotta *et al.*, 2005). Further experiments are therefore required to ascertain the function and localisation of AtMTP11 in more detail.

Phylogenetic characterisation of CDF transporter genes identified from genomes across kingdoms suggests that substrate specificity can be inferred from sequence information, and indicates that a wide range of specific Mn transporting CDF transporters are present in many organisms (Montanini *et al.*, 2007). Transporters from this sub-group of CDF transporters may therefore be suitable candidates to identify proteins involved in Mn-specific homeostasis. It is also interesting that mammals and prokaryotes do not appear to possess Mn-specific CDF transporters, according to this analysis.

### 5.5. P-type ATPase transporters

*S. cerevisiae* possesses a Golgi-localised P-type ATPase, Pmr1 that can transport Ca and Mn (Lapinskas *et al.*, 1995; Dürr *et al.*, 1998). Pmr1 was initially identified as a $Ca^{2+}$-ATPase with homology to mammalian PMCA- and SERCA-type $Ca^{2+}$ pumps. Subsequently it has become evident that Pmr1 belongs to a third distinct class of P-type $Ca^{2+}$-ATPase called the secretory pathway $Ca^{2+}$-ATPase (SPCA), which has members in yeast and animals, and most of which can transport Mn as efficiently as Ca (Van Baelen *et al.*, 2004). However, plants are an exception and do not contain Pmr1-like ATPases (Axelsen and Palmgren, 2001). Plants have two multi-gene families of $Ca^{2+}$-ATPase referred to as type $P_{2A}$ (similar to SERCA pumps) and type $P_{2B}$ (similar to PMCA pumps), with the $P_{2A}$-type $Ca^{2+}$-ATPases being most closely related to *PMR1* (Pittman *et al.*, 1999). The Arabidopsis $P_{2A}$ $Ca^{2+}$-ATPase ECA1, localised at the ER, can also transport Mn (Wu *et al.*, 2002; Table 1). Ca transport activity mediated by ECA1 can be blocked by Mn, expression of ECA1 in yeast is able to suppress the Mn hypersensitivity of a yeast mutant,

**Table I.** Transporters with a putative role in Mn homeostasis in plants

| Transporter family/Name (species) | Potential metal substrate | Regulation | Membrane location (direction of flow) | Tissue expression |
|---|---|---|---|---|
| **ZIP transporter** | | | | |
| IRT1 (Arabidopsis) | Cd, Co?, Fe, Mn, Zn | Fe deficiency | Plasma membrane (influx) | Roots, flowers – anther filament |
| LeIRT1 (Tomato) | Cd, Cu, Fe, Mn, Ni, Zn | Fe deficiency | nt* | Roots, flowers |
| LeIRT2 | Cd, Cu, Fe, Mn, Ni, Zn | No Fe regulation | nt | Roots |
| MtZIP4 (*Medicago*) | Fe?, Mn | Metal deficiency | nt | Leaves, roots (Zn deficient) |
| MtZIP7 | Mn | None detected | nt | Leaves |
| TjZNT1 (*Thlaspi*) | Cd, Mn, Ni, Zn | nt | nt | nt |
| TjZNT2 | Cd, Mn, Ni | nt | nt | nt |
| RIT1 (Pea) | Cd, Fe, Mn, Zn | Fe deficiency | Plasma membrane? | Roots |
| **Nramp transporter** | | | | |
| AtNramp1 (Arabidopsis) | Cd, Fe, Mn | Fe deficiency | Internal? | Roots, shoots |
| AtNramp3 | Cd, Fe, Mn | Fe deficiency | Vacuole (release) | Roots, shoots – vascular |
| AtNramp4 | Cd, Fe, Mn, Zn | Fe deficiency | Vacuole (release) | Roots, shoots – vascular |
| AtNramp5 | Cd, Fe, Mn, Zn | nt | nt | nt |

**Table I.** Continued

| Transporter family/Name (species) | Potential metal substrate | Regulation | Membrane location (direction of flow) | Tissue expression |
|---|---|---|---|---|
| GmDMT1 (Soybean) | Cu, Fe, Mn, Zn | nt | Peribacteroid membrane | Root nodule |
| LeNramp1 (Tomato) | Fe, Mn? | Fe deficiency | Internal | Root vascular |
| LeNramp3 | Fe, Mn | Fe deficiency | Internal | Roots, leaves |
| **CAX transporter** | | | | |
| CAX2 (Arabidopsis) | Ca, Cd, Mn, Zn? | Post-translational? | Vacuole (sequestration) | Roots, leaves, flowers – vascular |
| OsCAX1a (Rice) | Ca, Mn | Ca levels, post-translational? | Vacuole (sequestration) | Roots, leaves – both vascular, flowers, grain |
| OsCAX3 | Ca, Mn | nt | Plasma membrane? (efflux) | nt |
| **CDF transporter** | | | | |
| ShMTP1 (*Stylosanthes*) | Mn | nt | Vacuole (sequestration) | nt |
| AtMTP11 (Arabidopsis) | Mn | nt | PVC, Golgi? | Roots, Shoots |
| **P-type ATPase** | | | | |
| ECA1 (Arabidopsis) | Ca, Mn, Zn? | nt | ER (sequestration) | Roots, leaves |

**Table I.** Continued

| Transporter family/Name (species) | Potential metal substrate | Regulation | Membrane location (direction of flow) | Tissue expression |
|---|---|---|---|---|
| **YSL/OPT transporter** | | | | |
| ZmYS1 (Maize) | Cd?-, Cu-, Fe-, Mn?-, Ni-, Zn-PS | Fe deficiency | nt | Roots, shoots |
| OsYSL2 (Rice) | Fe-NA, Mn-NA | Fe deficiency | Plasma membrane (influx?) | Leaves – vascular, grain |
| AtOPT3 (Arabidopsis) | Cu, Fe, Mn | Mn, Fe?, Cu? deficiency | nt | Roots, leaves, flowers – vascular, seed |

*nt – not tested

and an *eca1* knockout shows significant Mn sensitivity compared to wild-type when grown on high Mn media, including a reduction in fresh weight, inhibition of root elongation and leaf growth, and loss of root hair tip growth. $Mn^{2+}$ transport activity has not been demonstrated for any other plant ATPase but it is conceivable that other ECA1-like transporters have similar $Ca^{2+}$ and $Mn^{2+}$ transport properties.

## 5.6. YSL/OPT transporters

Recent analysis of a maize mutant *yellow stripe 1* (*ys1*), which is deficient in Fe(III)-phytosiderophore (PS) complex accumulation, demonstrated that the mutated *ZmYS1* gene encodes a Fe(III)-PS transporter localised at the plasma membrane which is up-regulated by Fe deficiency (Curie *et al.*, 2001). Various Yellow Stripe-like (YSL) genes have subsequently been identified in various plant species and shown to be involved in metal-complex transport. PS are mugineic acid compounds that are synthesised by grass species and secreted from the root into the rhizosphere where they can chelate metals such as Fe(III) and promote their uptake into the root. Non-grass species such as Arabidopsis do not produce PS but do utilise other compounds with metal binding properties including nicotianamine (NA), which is a precursor in the biosynthetic pathway of mugineic acid. Metal-NA complexes are therefore substrates of YSL transporters in non-grass species (Le Jean *et al.*, 2005). The emerging data suggest that many YSL transporters can transport Fe(II) or Fe(III) complexes and are regulated by Fe nutrition. For example, ZmYS1 is up-regulated by Fe deficiency and can transport Fe(III)-PS (Roberts *et al.*, 2004), and AtYSL1 is up-regulated by Fe excess and involved in Fe loading into the seed (Le Jean *et al.*, 2005). Seed from the *ysl1* mutant had increased Mn levels, possibly an indirect consequence of altered Fe nutrition. There is some evidence that YSL transporters can transport other metal-complexes (Table 1). ZmYS1 can also transport Zn(II)-PS, Cu(II)-PS and Ni(II)-PS, and to a lesser extent Mn(II)-PS and Cd(II)-PS (Schaaf *et al.*, 2004), while rice OsYSL2 can transport Fe(II)-NA and Mn(II)-NA with equivalent efficiency (Koike *et al.*, 2004). The energisation of metal-complex transport by the YSL transporters is by $H^+$ symport (Schaaf *et al.*, 2004).

Phylogenetic analysis has found that the YS and YSL transporters are members of the oligopeptide transporter (OPT) family (Yen *et al.*, 2001), which are present in bacteria, yeast and plants, and shown to be $H^+$-dependent tetra- and pentapeptide transporters. Arabidopsis has 9 OPT genes most of which have been functionally characterised and shown to transport peptides (Stacey *et al.*, 2006). Interestingly, some OPT transporters may have a role in metal transport. There is evidence that AtOPT6 can transport a Cd-glutathione complex (Cagnac *et al.*, 2004). *AtOPT2* is up-regulated by Zn and Fe deficiency, while *AtOPT3* is significantly up-regulated by Mn deficiency and to a lesser extent by Fe and Cu deficiency (Wintz *et al.*, 2003; Stacey *et al.*, 2006). The up-regulation of *AtOPT3* by Mn deficiency is noteworthy as this is the first plant transporter shown to be significantly regulated by this metal. AtOPT3 can also restore growth defects of Mn, Fe and Cu uptake-defective yeast mutants, indicating that it can transport these metals (Wintz *et al.*, 2003). The PS and NA metal chelator substrates of YSL transporters are peptide-like compounds, therefore it is conceivable that some peptide derivatives transported by OPT transporters may be conjugated with a metal substrate. Further work is needed to determine the exact metal-peptide substrates of the OPT transporters.

## 6. REGULATION OF MN TRANSPORT

Unlike for metal transport pathways in bacteria and yeast, our understanding of the regulatory mechanisms of transition metal transport in plants, including Mn, is poor. It is apparent that many metal transporters that can transport Mn are broad specificity transporters. This is a characteristic that is similar to many yeast and bacteria transporters such as Smf1 and MntH, yet these transporters are specifically regulated in response to alterations in Mn levels, indicating a physiological role in Mn homeostasis. We do not yet know which of the candidate plant Mn transporters are important for Mn nutrition *in planta*, but elucidation of their regulatory properties will be critical in determining this. Various Fe transporters can also transport Mn and are regulated by Fe status, such as IRT1, AtNramp3 and OsYSL2. Some Fe transporters are regulated in response to Fe deficiency by a basic helix-loop-helix (bHLH) transcription factor called FER in tomato and FIT1 (or AtbHLH29) in Arabidopsis (Ling *et al.*, 2002; Colangelo and Guerinot, 2004; Yuan *et al.*, 2005). The tomato *fer* mutant has a reduction in leaf Fe content and a very dramatic reduction in Mn content (Yuan *et al.*, 2005), probably due in part to altered regulation of a transporter. Indeed, the Fe/Mn transporters AtIRT1 and

LeIRT1 have been shown to be downregulated in Arabidopsis *fit1* and tomato *fer* mutants, respectively (Ling *et al.*, 2002; Colangelo and Guerinot, 2004). The effect of altered Mn status on metal transport has been examined in very few cases, one being the significant up-regulation of *AtOPT3* by Mn deficiency (Wintz *et al.*, 2003). The $Mn^{2+}/H^+$ antiporter CAX2 may be regulated by a post-translational mechanism involving the N-terminal region of the protein. It has previously been shown that the $Ca^{2+}/H^+$ antiporter CAX1 is regulated by modulation of an N-terminal domain to activate or repress transport activity (Pittman and Hirschi, 2003; Shigaki and Hirschi, 2006). Heterologous expression of CAX2 in yeast is unable to mediate Mn transport activity unless the N-terminus is truncated (Schaaf *et al.*, 2002; Pittman *et al.*, 2004b). This suggests that CAX2 may also be regulated in a manner similar to CAX1, possibly due to autoinhibition by the N-terminus disrupting transport activity. This transport inhibition is not specific for Mn as full-length CAX2 expressed in yeast is also disrupted for Ca transport (Pittman *et al.*, 2004b). However, we have yet to determine the mechanism of transport activation of CAX2, and this could be mediated in a Mn-specific manner. Mn and Ca transport by OsCAX1a is also dependent on N-terminal truncation, but mutagenesis of the N-terminus of OsCAX1a allowed a full-length protein to transport Ca but Mn transport was lost, indicating that this region may have a role in metal selectivity in addition to regulation (Kamiya *et al.*, 2005).

## 7. WHOLE PLANT PARTITIONING OF MN

### 7.1. Long distance translocation of Mn

Following the accumulation of Mn into the root, some Mn will be sequestered within root cells, but to allow the transfer of Mn to aerial parts of the plant it must enter the vascular system. Mn can enter the xylem and be transported up the plant in the transpiration stream. The loading of Mn into the xylem is tightly controlled as all metals can only enter the xylem by the symplastic route because the casparian strip surrounding the endodermal cell layer prevents movement into the stele apoplast. Therefore transporters regulate the translocation of Mn into the root stele and loading into the xylem. Mn has been detected in the xylem of many species, with concentrations of up to 1 mM measured (Loneragan, 1988; Rengel, 2000). It has been estimated that approximately 60% of the xylem Mn is in the form of the divalent cation $Mn^{2+}$, while the remainder is present as a complex, possibly with citrate or malate (White *et al.*, 1981). The possible complexation of Mn with the chelator NA was not considered by White *et al.* (1981), but subsequent analysis suggests that NA is not involved in Mn translocation in the xylem, at least in tomato (Pich and Scholz, 1996). After the translocation of Mn in the xylem it is then unloaded in various aerial parts of the plant including leaves, flower organs and seeds (see below). Once Mn is transferred to the leaves, remobilisation can only occur via the phloem. In addition, metals could transfer from xylem to phloem throughout the root and stem. Mn is often regarded as being relatively phloem immobile and in most plant species the phloem sap Mn concentration is low (< 20 μM) (Loneragan, 1988; Rengel, 2000). However, Mn has been detected at relatively high concentrations in the phloem of some species, such as lupin and castor bean (Hocking *et al.*, 1977; Schmidke and Stephan, 1995). Studies with castor bean indicate that phloem-localised Mn may be chelated to NA (Schmidke and Stephan, 1995; Stephan *et al.*, 1996). There are conflicting reports as to whether Mn can be recirculated in the plant via the phloem following unloading into leaves, and it appears that this ability may be species-specific, age-specific or dependent on the Mn stress conditions that the plant is experiencing (Rengel, 2000). For example, some phloem remobilisation of Mn from leaves to roots in Douglas fir has been reported (Ducic *et al.*, 2006) while no Mn remobilisation in the phloem was observed in young wheat plants (Pearson and Rengel, 1994; Page and Feller, 2005).

Various putative Mn transporters have vascular expression. CAX2 is expressed in vascular tissue throughout the plant, particularly in the root, petiole, mature leaf, and very strongly in the flower (Pittman *et al.*, 2004b). As CAX2 is a tonoplast protein, rather than mediating translocation into the vascular pathway, it may be involved in sequestering Mn in vascular cells, such as phloem companion cells immediately following unloading to allow later remobilisation of the Mn. Many OPT and YSL transporters are expressed in vascular tissue, including AtOPT3 in vascular tissue of seedlings and mature plants (Stacey *et al.*, 2002, 2006), AtOPT6 in cambial cells (Cagnac *et al.*, 2004), OsYSL2 in phloem companion cells (Koike *et al.*, 2004), and AtYSL2 in leaf xylem (Le Jean *et al.*, 2005). This is indicative of a role of these transporters in long-distance metal-chelate transport throughout the plant.

## 7.2. Seed loading and unloading of Mn

Mineral content in the seed is critical to allow the seedling to sustain growth until establishment of the root system. Furthermore, the concentration of seed mineral reserves can determine the efficiency of germination and seedling development during soil mineral deficiency. Seeds are also an important mineral nutrient source for a balanced human diet. Many cereals have significant Mn content in the grain embryo and aleurone layer rather than the endosperm (Loneragan, 1988), although maize has exceptionally high endosperm Mn content (Bityutskii *et al.*, 2002). In many plants, Mn in the embryo is concentrated in globoid crystals and associated with phytate (Loneragan, 1988). The mineral content in the developing Arabidopsis seed has been studied in detail (Otegui *et al.*, 2002). In a young Arabidopsis seed, the developing embryo globoid crystals contain various minerals but no Mn; however, high levels of Mn associated with phytate are present in the endosperm and specifically in the ER. The pathway for Mn transport into the endosperm ER is unknown, but the $Mn^{2+}$-ATPase ECA1 could be a candidate. During later seed development, Mn is translocated from the endosperm to the embryo, and this translocation coincides with increased abundance of MnSOD and PSII proteins, and chlorophyll content in the embryo (Otegui *et al.*, 2002).

The mechanism of Mn loading into seeds has been well studied in the wheat grain system. Mn taken up from the roots of young wheat plants is rapidly transferred to the shoot rather than leaves and may be stored in the shoot before later transfer via the xylem to the developing grains (Pearson and Rengel, 1995). Storage of Mn in shoot tissue allows easy remobilisation into the xylem which cannot occur if the Mn is stored in leaves. Large amounts of Mn accumulate in spikelet structures in the wheat ear via the xylem (Pearson *et al.*, 1995). Some remaining Mn is transferred to the grain stalk and taken up by transfer cells, then loaded into the grain phloem (Pearson *et al.*, 1995, 1996). This step will require the action of transport proteins. Mn then enters the crease and pericarp tissues of the grain via the crease phloem, and during grain development, some Mn is retranslocated to the embyro and the endosperm, although the rate of Mn accumulation is much faster into the embryo (Pearson *et al.*, 1996, 1998). Less is known regarding Mn unloading from the seed during early seedling development. In wheat, Mn is remobilised into the root and shoot, mostly from the seed coat during a 3 to 12 day period after imbibition (Moussavi-Nik *et al.*, 1997). However, the proportion of Mn remobilised from the seed coat is small. In maize seed, Mn is mainly stored in the endosperm, and some in the scutellum, and the efficiency of translocation from these stores to the germinating seedling is much greater than in wheat (Bityutskii *et al.*, 2002).

Loading of Mn into the grain phloem, translocation into the embryo and seed Mn remobilisation into the developing root and shoot are all steps that require transport proteins, as yet unidentified. Recently, AtYSL1 has been shown to be involved in loading of Fe into the Arabidopsis seed (Le Jean *et al.*, 2005), while AtNramp3 and AtNramp4 are required for mobilisation of Fe from the seed during germination (Lanquar *et al.*, 2005). However, Mn is not loaded or unloaded by these transporters. Rice *OsYSL2* expression is observed in the vascular bundle of spikelets before anthesis and this increases after fertilisation (Koike *et al.*, 2004). In addition, *OsYSL2* expression increases during seed development with strong expression in the embryo and outer endosperm layer, indicating a role of this transporter in Fe and Mn loading into the rice grain. *AtOPT3* is expressed in vascular tissue in various flower parts including the seed funiculus, and in the endosperm and embryo of the developing Arabidopsis seed (Stacey *et al.*, 2002). Furthermore, a knockout of *AtOPT3* gives an embryo lethal phenotype. Since AtOPT3 may have a role in Mn, Fe and Cu transport (Wintz *et al.*, 2003), it may be essential in translocation of minerals into the developing embryo.

## 8. NATURAL VARIATION IN PLANT MN ACCUMULATION AND HOMEOSTASIS

### 8.1. Plants with increased tolerance to low Mn availability

Mn deficiency can be a serious problem for growth yield, particularly on alkaline soils in which oxidation of Mn to the Mn(III)/Mn(IV) species is preferential, therefore the availability of Mn(II) to the plant is reduced. It has been long observed that there is large variability between species and genotypes in their ability to tolerate and grow on soils with low Mn availability, but little is known about the physiological, biochemical or genetic mechanisms involved. The efficiency of plants in releasing Mn-chelating and Mn-reducing exudates to improve Mn(II) availability, and the efficiency of Mn uptake either due to altered uptake kinetics or the size of the root system, have been proposed to be among the factors involved in providing tolerance to Mn deficiency, although variation in the plant's metabolic requirement for Mn could also be a factor (Graham, 1988). In addition, increased Mn concentration in the seed can allow a plant to have increased yield on Mn deficient soils compared to seed with low Mn concentration (Longnecker et al., 1991). In a recent study, two barley genotypes were compared, one which is tolerant to low Mn availability, referred to as Mn-efficient and one which is Mn-inefficient, and their Mn transport characteristics were analysed (Pedas et al., 2005). High- and low-affinity Mn transport pathways were identified in both genotypes, and it was found that the kinetics of the high-affinity pathway varied between the two, with the Mn efficient genotype having a significantly higher $Mn^{2+}$ uptake capacity ($V_{max}$) than the inefficient genotype, while $K_m$ values were equivalent between the two. This increased capacity of high-affinity $Mn^{2+}$ uptake mediated increased Mn accumulation, by up to approximately 75% in the shoots of the Mn efficient genotype. Another study analysed Mn efficient and inefficient wheat genotypes and found that the Mn inefficient genotypes had greater root growth inhibition during low Mn supply in the soil, but no significant differences in $Mn^{2+}$ uptake kinetics were observed (Sadana et al., 2005).

## 8.2. Plants with increased tolerance to Mn toxicity including Mn hyperaccumulators

There is wide variation between plant species and genotypes within species in tolerance to Mn toxicity (Foy et al., 1988). For example, some rice varieties are very tolerant to high Mn concentrations, and this tolerance has been shown to be associated with significant accumulation of Mn in shoot tissue and storage in the chloroplast (Lidon, 2001). Another example is the significant variation in Mn toxicity observed between two varieties of Douglas fir. This variation is due to 3-fold less total Mn uptake in the tolerant variety compared to the sensitive variety, and nearly all of the Mn is retained in the roots in the tolerant variety, whereas the sensitive variety translocates over 60% of the accumulated Mn into the shoot (Ducic et al., 2006). Mn tolerance phenotypes are particularly dramatic in Mn hyperaccumulating species that have the ability to accumulate and tolerate Mn concentrations greater than 10000 µg $g^{-1}$ DW (Reeves and Baker, 2000). To date 12 Mn hyperaccumulators have been identified out of the few hundred metal hyperaccumulating plants known, indicating that Mn hyperaccumulation is a rare trait (Bidwell et al., 2002; Xue et al., 2004). These plants are mostly from subtropical regions. The most recently identified Mn hyperaccumulator is *Phytolacca acinosa* which was identified growing on Mn contaminated soils in Southern China, and could accumulate almost 20000 µg Mn $g^{-1}$ DW in leaves (Xue et al., 2004). Further analysis of these plants indicates that Mn is mostly accumulated in intracellular compartments, probably the vacuole of mature leaf epidermal cells (Xu et al., 2006). Another recently identified Mn hyperaccumulator is the Australian tree species *Gossia bidwillii* (previously named *Austromyrtus bidwillii*) that can accumulate

19200 µg Mn g$^{-1}$ DW in leaves and 26500 µg Mn g$^{-1}$ DW in young bark (Bidwell et al., 2002). Analysis of *G. bidwillii* demonstrated that it contained significant concentrations of organic acids in leaf extracts, particularly oxalic, succinic, malic and malonic acids, which were approximately three times the molar equivalent of total Mn. It was therefore suggested that the excess concentrations of these organic acids were a major determinant of the Mn tolerance trait (Bidwell et al., 2002). Unlike for *Phytolacca acinosa* and other previously characterised metal hyperaccumulators, in *G. bidwillii* Mn accumulates in vacuoles of photosynthetic palisade mesophyll cells, rather than epidermal cells (Fernando et al., 2006). This might be associated with high photosynthetic activity in these cells. The conclusion from these studies is therefore that this hyperaccumulator is tolerant to excess Mn due to vacuolar sequestration of Mn-organic acid compounds. Further biochemical and genetic analysis of these and other Mn hyperaccumulators should allow the identification of the various components that determine Mn tolerance, some of which may be transporters.

## 9. GENETIC MANIPULATION OF MN TRANSPORT AND ACCUMULATION IN PLANTS

### 9.1. Overexpression studies

The identification of various genes with roles in Mn transport and homeostasis offers the potential for genetic engineering strategies to overcome Mn deficiency by improvement of Mn accumulation or to provide tolerance to Mn toxicity. Furthermore, we are beginning to identify amino acid residues or domains involved in determining Mn specificity in various transporters (Pittman, 2005), allowing the possibility of engineering plants to provide enhanced transport for a single metal such as Mn. Overexpression of the cation/H$^+$ antiporter CAX2 in tobacco yielded plants which had a significant 2-fold increase in Mn content in root tissue and a slight increase in shoot tissue, yet these plants could tolerate a toxic concentration (0.5 mM) of Mn compared to control plants due to an increase in Mn accumulation into the vacuole (Hirschi et al., 2000). Overexpression of another vacuolar Mn transporter, ShMTP1 in Arabidopsis was able to confer significant tolerance to various concentrations of Mn up to 3.5 mM both in short term and long term experiments, up to 21 days, and was associated with an increase in internal Mn concentration (Delhaize et al., 2003). The ER Ca$^{2+}$ and Mn$^{2+}$-ATPase ECA1 has a role in Mn homeostasis. Overexpression of ECA1 in the *eca1* knockout background was able to rescue the Mn-sensitive knockout phenotype, but there was no difference in fresh weight of the overexpressing lines compared to wild-type when grown on 0.5 mM Mn conditions (Wu et al., 2002). These lines were not tested on higher Mn concentrations, but it was observed that despite using the 35S promoter, ECA1 protein levels were comparable to wild-type, possibly due to translational regulation or protein degradation, indicating that overexpression of ECA1 may not be a suitable strategy for providing tolerance to dramatically elevated concentrations of Mn.

Manganese transport

One potential means to improve growth on Mn deficient soils is to improve Mn accumulation efficiency, such as by manipulating the uptake transporter IRT1. As described above, an *irt1* knockout mutant had loss of accumulation of Fe, Mn, Zn and Co, and reduced sensitivity to Cd (Vert *et al.*, 2002), indicating that it is a pathway for the uptake of these metals into the plant. Overexpression of *IRT1* using the 35S promoter caused increase in Zn and Cd, but only under Fe deficiency conditions (Connolly *et al.*, 2002). However, it was surprising that no significant increase in Mn accumulation was observed, particularly as constitutive expression of IRT1 which occurs in the *frd3* mutant background confers a significant accumulation of Mn (Rogers and Guerinot, 2002; see below). An increased tolerance to Mn deficiency has been achieved by overexpression of MnSOD in tobacco plants (Yu *et al.*, 1999). When plants were grown under Mn deficiency conditions, transgenic plants expressing MnSOD in the chloroplast had increased plant growth compared to wild-type controls. This effect was not seen in plants expressing either a chloroplast FeSOD or a mitochondrial MnSOD, indicating that it was overproduction of MnSOD in chloroplasts that gave protection from oxidative stress caused by Mn deficiency. Overexpression of a NA synthase gene in tobacco, leading to a 10-fold increase in NA levels, caused an increase in Fe, Zn and Mn in shoot tissue, but while this strategy could improve growth under Fe limitation, growth on Mn deficiency was not tested (Douchkov *et al.*, 2005).

### 9.2. Mn mutant screening

A mutagenesis approach is a potential means to identify novel components involved in Mn homeostasis, by screening for mutants with altered Mn nutrition. Various mutants with alterations in metal nutrition have been described, some with alterations in Mn levels. The Arabidopsis mutant *manganese accumulator 1* (*man1*) was identified in a screen of an ethyl methanesulfonate (EMS) mutagenised population (Delhaize, 1996). *man1* accumulates high concentrations of Mn compared to wild-type, with a 7.5-fold increase in leaf Mn content and a 10-fold increase in root Mn content. *man1* also has a significant increase in Fe content, particularly in root tissue, and lesser increases in Cu and Zn. The increased Fe accumulation was observed in tandem with high and constitutive ferric reductase activity, which was not dependent on Fe deficiency (Delhaize, 1996; Rogers and Guerinot, 2002). A *ferric reductase defective 3* (*frd3*) mutant, which shows constitutive ferric reductase activity, has subsequently been found to have identical phenotypes and is allelic to *man1* (Rogers and Guerinot, 2002). The *FRD3* gene encodes a member of the MATE transporter family, and does not appear to transport metals directly but is required for regulating root-to-shoot Fe localisation by a citrate transport mechanism (Green and Rogers, 2004; Durrett *et al.*, 2007). *IRT1* is constitutively expressed in the *frd3/man1* mutant and this is the likely cause of the increased Mn accumulation phenotype (Rogers and Guerinot, 2002). The pea mutant *brz* also has constitutive ferric reductase activity, and likewise has increased Fe, Mn and other metal accumulation (Welch and LaRue, 1990). A *Medicago truncatula* mutant *requires additional Zn* (*raz*) has a necrotic leaf phenotype which is similar to that of Zn-deficient plants (Ellis et al., 2003). *raz* has elevated concentrations of Zn, Mn, Cu and Fe, but no alteration in ferric reductase activity. It was suggested that *raz* is a Zn-related mutation and that the accumulation of Mn may be due to enhanced expression of a Zn-regulated metal transporter, such as a *Medicago* ZIP transporter (Ellis et al., 2003; López-Millán et al., 2004). Another Arabidopsis EMS mutant, *IAA-leucine resistant 2* (*ilr2*), was identified in a screen for plants lacking the characteristic root inhibition response to auxin conjugates, in this case IAA-leucine. *ilr2* is resistant to IAA-leucine but also displays slight tolerance to Co and Mn stress (Magidin et al., 2003). Further analysis demonstrated increased ATP-dependent $Mn^{2+}$ transport activity into microsomal membrane vesicles from *ilr2* compared to wild-type. *ILR2* encodes a small soluble protein of unknown function which was proposed to act as a negative regulator of an unknown metal transporter.

Apart from the *cax2*, *mtp11* and *eca1* T-DNA insertion mutants, and the EMS-generated mutants described above, there are currently few examples of Mn homeostasis mutants. However, novel screening techniques, such as a recently described high-throughput, mineral nutrient profiling screen, should identify further interesting mutants and subsequently novel genes involved in metal homeostasis. A collection of fast neutron-mutagenised Arabidopsis plants were assayed for alterations in shoot element levels to determine the effect of mutation on a plant's 'ionome' (Lahner et al., 2003). This was determined by simultaneously measuring 18 elements by using highly sensitive inductively coupled plasma mass spectroscopy (ICP-MS) and ICP atomic emission spectroscopy (ICP-AES) techniques. Approximately 6000 plants were initially analysed of which 51 plants had significant and reproducible elemental alterations. Plants with alterations in various metal concentrations including Mn have been identified, although the mutated genes have yet to be cloned. One of these mutants, which was dominant and had a significant increase in Mn, has been found to be a new allele of *frd3* (Lahner et al., 2003). Very few (11%) of the mutants were altered in a single metal, suggesting that interactions of many metal homeostatic pathways are frequent. Subsequently, over 60000 plant samples have been analysed, including fast-neutron mutants and T-DNA insertion mutants (Salt, 2004; Baxter et al., 2007), and some interesting novel genes involved in mineral nutrition should be identified in the future.

## 10. CONCLUSIONS AND FUTURE PROSPECTS

Since the increased availability of genomic sequence information, there has been a significant increase in the identification of genes that encode putative metal transporters. Analysis of putative plant transporters has begun to determine that some of these proteins have the ability to transport Mn (Table 1). It is apparent that many plant Mn transporters, particularly those involved in cytosolic Mn influx such as some of the ZIP, Nramp and YSL transporters, have broad substrate selectivity, often coupled with Fe transport and are regulated by Fe deficiency. Similarly, mutant plants with altered Mn nutrition such as the *man1* or *raz* mutants also show alterations for other metals, indicating that Mn homeostasis is tightly linked with that of other metal nutrients. It can be argued that as many of these transporters are up-regulated by Fe deficiency, they are transporters for Fe rather than Mn *in planta*. This may be true for some of these transporters; however, a knockout study, such as that of *irt1* has demonstrated that IRT1 is important for Mn accumulation in addition to Fe, during Fe deficiency. Furthermore, the effect of Mn deficiency has not been assessed for most of these genes. However, due to a plant's relatively low requirement for Mn, a specific Mn influx pathway may not be necessary and a plant can utilise a variety of Fe transporters for Mn acquisition. In contrast, some transporters that mediate Mn sequestration into the vacuole, such as the $Mn^{2+}/H^+$ antiporters CAX2 and ShMTP1, appear to be more specific for Mn and are not influenced by Fe nutrition. Broad selectivity is a widespread feature of metal transporters from all organisms and may be partly explained by the similar characteristics of some metals. Since the substrate specificity of many of these transporters was determined by *in vitro* studies, such as by heterologous expression in yeast, further work is needed to determine if the transport of particular metals including Mn occur under physiological conditions in the plant itself.

This review has described many processes in a plant where Mn transport is required, such as during acquisition from the soil, transfer into organelles, loading/unloading into and from the vascular pathway, and mobilisation into various plant organs. In addition, Mn transporters play a key role in providing tolerance to toxic concentrations of Mn, by mediating removal from the cytosol, and have been implicated in adaptive strategies to overcome Mn deficiency, such as by improving the efficiency of Mn acquisition. A future challenge is to identify transporters responsible for these processes. Many other components in mineral nutrient homeostasis remain unknown, such as metal sensing mechanisms, metal chaperone proteins, and metal transport regulators, either transcription factors or post-translational regulators. Now we have begun to identify components in Mn homeostasis, there is potential for genetic manipulation as a strategy to overcome plant Mn toxicity or Mn deficiency stress.

## 11. ACKNOWLEDGEMENT

I am grateful to Dr. Toshiro Shigaki and Miss Clare Edmond for critical reading of the manuscript.

## 12. LITERATURE CITED

Alam, S., Kamei, S. and Kawai, S. (2001) Amelioration of manganese toxicity in barley with iron. *J. Plant Nutr.,* **24**: 1421-1433.

Allen, M.D., Kropat, J., Tottey, S., Del Campo, J.A. and Merchant, S.S. (2007) Manganese deficiency in Chlamydomonas results in loss of Photosystem II and MnSOD function, sensitivity to peroxides, and secondary phosphorus and iron deficiency. *Plant Physiol.,* **143**: 263-277.

Axelsen, K.B. and Palmgren, M.G. (2001) Inventory of P-type ion pumps in Arabidopsis. *Plant Physiol.,* **126**: 696-706.

Bartsevich, V.V. and Pakrasi, H.B. (1996) Manganese transport in the cyanobacterium *Synechocystis* sp PCC 6803. *J. Biol. Chem.,* **271**: 26057-26061.

Baxter, I., Ouzzani, M., Orcun, S., Kennedy, B., Jandhyala, S.S. and Salt, D.E. (2007) Purdue Ionomics Information Management System. An integrated functional genomics platform. *Plant Physiol.,* **143**: 600-611.

Becher, M., Talke, I.N., Krall, L. and Krämer, U. (2004) Cross-species microarray transcript profiling reveals high constitutive expression of metal homeostasis genes in shoots of the zinc hyperaccumulator *Arabidopsis halleri. Plant J.,* **37**: 251-268.

Bereczky, Z., Wang, H.-Y., Schubert, V., Ganal, M. and Bauer, P. (2003) Differential regulation of *nramp* and *irt* metal transporter genes in wild type and iron uptake mutants of tomato. *J. Biol. Chem.,* **278**: 24697-24704.

Bidwell, S.D., Woodrow, I.E., Batianoff, G.N. and Sommer-Knudsen, J. (2002) Hyperaccumulation of manganese in the rainforest tree *Austromyrtus bidwillii* (Myrtaceae) from Queensland, Australia. *Funct. Plant Biol.,* **29**: 899-905.

Bityutskii, N.P., Magnitskiy, S.V., Korobeynikova, L.P., Lukina, E.I., Soloviova, A.N., Patsevitch, V.G., Lapshina, I.N. and Matveeva, G.V. (2002) Distribution of iron, manganese, and zinc in mature grain and their mobilization during germination and early seedling development in maize. *J. Plant Nutr.,* **25**: 635-653.

Cagnac, O., Bourbouloux, A., Chakrabarty, D., Zhang, M.-Y. and Delrot, S. (2004) AtOPT6 transports glutathione derivatives and is induced by primisulfuron. *Plant Physiol.,* **135**: 1378-1387.

Clarkson, D.T. (1988) The uptake and translocation of manganese by plant roots. In: *Manganese in Soils and Plants* (Eds. Graham R.D., Hannam R.J. and Uren N.C.), Kluwer Academic Publishers, Dordrecht, pp 101-111.

Cohen, C.K., Fox, T.C., Garvin, D.F. and Kochian, L.V. (1998) The role of iron-deficiency stress responses in stimulating heavy-metal transport in plants. *Plant Physiol.,* **116**: 1063-1072.

Cohen, C.K., Garvin, D.F. and Kochian, L.V. (2004) Kinetic properties of a micronutrient transporter from *Pisum sativum* indicate a primary function in Fe uptake from the soil. *Planta,* **218**: 784-792.

Colangelo, E.P. and Guerinot, M.L. (2004) The essential basic helix-loop-helix protein FIT1 is required for the iron deficiency response. *Plant Cell,* **16**: 3400-3412.

Connolly, E.L., Fett, J.P. and Guerinot, M.L. (2002) Expression of the IRT1 metal transporter is controlled by metals at the levels of transcript and protein accumulation. *Plant Cell,* **14**: 1347-1457.

Culotta, V.C., Yang, M. and Hall, M.D. (2005) Manganese transport and trafficking: lessons learned from *Saccharmoyces cerevisiae*. Eukaryot. *Cell,* **4**: 1159-1165

Curie, C., Panaviene, Z., Loulergue, C., Dellaporta, S.L., Briat, J.-F. and Walker, E.L. (2001) Maize *yellow stripe1* encodes a membrane protein directly involved in Fe(III) uptake. *Nature,* **409**: 346-349.

Daly, M.J., Gaidamakova, E.K., Matrosova, V.Y., Vasilenko, A., Zhai, M., Venkateswaran, A., Hess, M., Omelchenko, M.V., Kostandarithes, H.M., Makarova, K.S., Wackett, L.P., Fredrickson, J.K. and Ghosal, D. (2004) Accumulation of Mn(II) in *Deinococcus radiodurans* facilitates gamma-radiation resistance. *Science,* **306**: 1025-1028.

Delhaize, E. (1996) A metal-accumulator mutant of *Arabidopsis thaliana. Plant Physiol.,* **111**: 849-855.

Delhaize, E., Gruber, B.D., Pittman, J.K., White, R.G., Leung, H., Miao, Y., Jiang, L., Ryan, P.R. and Richardson, A.E. (2007) A role for the *AtMTP11* gene of Arabidopsis in manganese transport and tolerance. *Plant J.,* **51**: 198-210.

Delhaize, E., Kataoka, T., Hebb, D.M., White, R.G. and Ryan, P.R. (2003) Genes encoding proteins of the cation diffusion facilitator family that confer manganese tolerance. *Plant Cell,* **15**: 1131-1142.

Douchkov, D., Gryczka, C., Stephan, U.W., Hell, R. and Bäumlein, H. (2005) Ectopic expression of nicotianamine synthase genes results in improved iron accumulation and increased nickel tolerance in transgenic tobacco. *Plant Cell Environ.*, **28**: 365-374.

Ducic, T., Leinemann, L., Finkeldey, R. and Polle, A. (2006) Uptake and translocation of manganese in seedlings of two varieties of Douglas fir (*Pseudotsuga menziesii* var. *viridis* and *glauca*). *New Phytol.*, **170**: 11-20.

Dürr, G., Strayle, J., Plemper, R., Elbs, S., Klee, S.K., Catty, P., Wolf, D.H. and Rudolph, H.K. (1998) The medial-Golgi ion pump Pmr1 supplies the yeast secretory pathway with $Ca^{2+}$ and $Mn^{2+}$ required for glycosylation, sorting, and endoplasmic reticulum associated protein degradation. *Mol. Biol. Cell*, **9**: 1149-1162.

Durrett, T.P., Gassmann, W. and Rogers, E.E. (2007) The FRD3-mediated efflux of citrate into the root vasculature is necessary for efficient iron translocation. *Plant Physiol.*, **144**: 197-205.

Eckhardt, U., Marques, A.M. and Buckhout, T.J. (2001) Two iron-regulated cation transporters from tomato complement metal uptake-deficient yeast mutants. *Plant Mol. Biol.*, **45**: 437-448.

Eide, D.J., Clark, S., Nair, T.M., Gehl, M., Gribskov, M., Guerinot, M.L. and Harper, J.F. (2005) Characterisation of the yeast ionome: a genome-wide analysis of nutrient mineral and trace element homeostasis in *Saccharomyces cerevisiae*. *Genome Biol.*, **6**: R77.

Ellis, D.R., López-Millán, A.F. and Grusak, M.A. (2003) Metal physiology and accumulation in a *Medicago truncatula* mutant exhibiting an elevated requirement for zinc. *New Phytol.*, **158**: 207-218.

Fecht-Christoffers, M.M., Maier, P. and Horst, W.J. (2003) Apoplastic peroxidases and ascorbate are involved in manganese toxicity and tolerance of *Vigna unguiculata*. *Physiol. Plant.*, **117**: 237-244.

Fernando, D.R., Bakkaus, E.J., Perrier, N., Baker, A.J.M., Woodrow, I.E., Batianoff, G.N. and Collins, R.N. (2006) Manganese accumulation in the leaf of mesophyll of four tree species: a PIXE/EDAX localisation study. *New Phytol.*, **171**: 751-758.

Foy, C.D., Scott, B.J. and Fisher, J.A. (1988) Genetic differences in plant tolerance to manganese toxicity. In: Manganese in Soils and Plants (Eds. Graham, R.D., Hannam, R.J. and Uren, N.C.), Kluwer Academic Publishers, Dordrecht, pp 293-307.

Ghasemi-Fasaei, R., Ronaghi, A., Maftoun, M., Karimian, N.A. and Soltanpour, P.N. (2003) Influence of FeEDDHA on iron-manganese interaction in soybean genotypes in a calcareous soil. *J. Plant Nutr.*, **26**: 1815-1823.

Ghasemi-Fasaei, R., Ronaghi, A., Maftoun, M., Karimian, N.A. and Soltanpour, P.N. (2005) Iron-manganese interaction in chickpea as affected by foliar and soil application of iron in a calcareous soil. Commun. *Soil Sci. Plant Anal.*, **36**: 1717-1725.

González, A., Koren'kov, V. and Wagner, G.J. (1999) A comparison of Zn, Mn, Cd, and Ca transport mechanisms in oat root tonoplast vesicles. *Physiol. Plant.*, **106**: 203-209.

González, A. and Lynch, J.P. (1999) Subcellular and tissue Mn compartmentation in bean leaves under Mn toxicity stress. *Aust. J. Plant Physiol.*, **26**: 811-822.

González, A., Steffen, K.L. and Lynch, J.P. (1998) Light and excess manganese. Implications for oxidative stress in common bean. *Plant Physiol.*, **118**: 493-504.

Graham, R.D. (1988) Genotypic differences in tolerance to manganese deficiency. In: Manganese in Soils and Plants (Eds. Graham, R.D., Hannam, R.J. and Uren, N.C.), Kluwer Academic Publishers, Dordrecht, pp 261-276.

Green, L.S. and Rogers, E.E. (2004) *FRD3* controls iron localization in Arabidopsis. *Plant Physiol.*, **136**: 2523-2531.

Grotz, N., Fox, T., Connolly, E., Park, W., Guerinot, M.L. and Eide, D. (1998) Identification of a family of zinc transporter genes from Arabidopsis that respond to zinc deficiency. *Proc. Natl. Acad. Sci. USA,* **95**: 7220-7224.

Hall, J.L. and Williams, L.E. (2003) Transition metal transporters in plants. *J. Exp. Bot.,* **54**: 2601-2613.

Hao, Z., Chen, S. and Wilson, D.B. (1999) Cloning, expression, and characterization of cadmium and manganese uptake genes from *Lactobacillus plantarum*. *Appl. Environ. Microbiol.,* **65**: 4746-4752.

Hernandez, L.E., Lozano-Rodriguez, E., Garate, A. and Carpena-Ruiz, R. (1998) Influence of cadmium on the uptake, tissue accumulation and subcellular distribution of manganese in pea seedlings. *Plant Sci.,* **132**: 139-151.

Hirschi, K.D., Korenkov, V.D., Wilganowski, N.L. and Wagner, G.J. (2000) Expression of Arabidopsis CAX2 in tobacco. Altered metal accumulation and increased manganese tolerance. *Plant Physiol.,* **124**: 125-133.

Hocking, P.J., Pate, J.S., Wee, S.C. and McComb, A.J. (1977) Manganese nutrition of *Lupinus* spp. especially in relation to developing seeds. *Ann. Bot.,* **41**: 677-688.

Horsburgh, M.J., Wharton, S.J., Cox, A.G., Ingham, E., Peacock, S. and Foster, S.J. (2002a) MntR modulates expression of the PerR regulon and superoxide resistance in *Staphylococcus aureus* through control of manganese uptake. *Mol. Microbiol.,* **44**: 1269-1286.

Horsburgh, M.J., Wharton, S.J., Karavolos, M. and Foster, S.J. (2002b) Manganese: elemental defence for a life with oxygen? *Trends Microbiol.,* **10**: 496-501.

Horst, W.J. and Maier, P. (1999) Compartmentation of manganese in the vacuoles and in the apoplast of leaves in relation to genotypic manganese leaf-tissue tolerance in *Vignia unguiculata* (L.) Walp. In: *Plant Nutrition – Molecular Biology and Genetics* (Eds. Gissel-Nielsen, G. and Jensen, A.), Kluwer Academic Publishers, Dordrecht, pp 223-234.

Izaguirre-Mayoral, M.L. and Sinclair, T.R. (2005) Soybean genotypic differences in growth, nutrient accumulation and ultrastructure in response to manganese and iron supply in solution culture. *Ann. Bot.,* **96**: 149-158.

Jensen, L.T., Ajua-Alemanji, M. and Culotta, V.C. (2003) The *Saccharomyces cerevisiae* high affinity phosphate transporter encoded by *PHO84* also functions in manganese homeostasis. *J. Biol. Chem.,* **278**: 42036-42040.

Kaiser, B.N., Moreau, S., Castelli, J., Thomson, R., Lambert, A., Bogliolo, S., Puppo, A. and Day, D.A. (2003) The soybean NRAMP homologue, GmDMT1, is a symbiotic divalent metal transporter capable of ferrous iron transport. *Plant J.,* **35**: 295-304.

Kamiya, T., Akahori, T. and Maeshima, M. (2005) Expression profile of the genes for rice cation/$H^+$ exchanger family and functional analysis in yeast. *Plant Cell Physiol.,* **46**: 1735-1740.

Kamiya, T., Akahori, T., Ashikari, M. and Maeshima, M. (2006) Expression of the vacuolar $Ca^{2+}/H^+$ exchanger, OsCAX1a, in rice: cell and age specificity of expression, and enhancement by $Ca^{2+}$. *Plant Cell Physiol.,* **47**: 96-106.

Kamiya, T. and Maeshima, M. (2004) Residues in internal repeats of the rice cation/$H^+$ exchanger are involved in the transport and selection of cations. *J. Biol. Chem.,* **279**: 812-819.

Kehres, D.G., Janakiraman, A., Slauch, J.M. and Maguire, M.E. (2002) Regulation of *Salmonella enterica* Serovar Typhimurium *mntH* transcription by $H_2O_2$, $Fe^{2+}$, and $Mn^{2+}$. *J. Bacteriol.,* **184**: 3151-3158.

Kehres, D.G. and Maguire, M.E. (2003) Emerging themes in manganese transport, biochemistry and pathogenesis in bacteria. *FEMS Microbiol. Rev.,* **27**: 263-290.

Keren, N., Kidd, M.J., Penner-Hahn, J.E. and Pakrasi, H.B. (2002) A light-dependent mechanism for massive accumulation of manganese in the photosynthetic bacterium *Synechocystis* sp. PCC 6803. *Biochemistry,* **41**: 15085-15092.

Kim, D., Gustin, J.L., Lahner, B., Persans, M.W., Baek, D., Yun, D.-J. and Salt, D.E. (2004) The plant CDF family member TgMTP1 from the Ni/Zn hyperaccumulator *Thalspi goesingense* acts to enhance efflux of Zn at the plasma membrane when expressed in *Saccharomyces cerevisiae*. *Plant J.,* **39**: 237-251.

Kim, S.A., Punshon, T., Lanzirotti, A., Li, L., Alonso, J.M., Ecker, J.R., Kaplan, J. and Guerinot, M.L. (2006) Localization of iron in Arabidopsis seed requires the vacuolar membrane transporter VIT1. *Science,* **314**: 1295-1298.

Kobae, Y., Uemura, T., Sato, M.H., Ohnishi, M., Mimura, T., Nakagawa, T. and Maeshima, M. (2004) Zinc transporter of *Arabidopsis thaliana* AtMTP1 is localized to vacuolar membranes and implicated in zinc homeostasis. *Plant Cell Physiol.,* **45**: 1749-1758.

Koike, S., Inoue, H., Mizuno, D., Takahashi, M., Nakanishi, H., Mori, S. and Nishizawa, N.K. (2004) OsYSL2 is a rice metal-nicotianamine transporter that is regulated by iron and expressed in the phloem. *Plant J.,* **39**: 415-424.

Korshunova, Y.O., Eide, D., Clark, W.G., Guerinot, M.L. and Pakrasi, H.B. (1999) The IRT1 protein from *Arabidopsis thaliana* is a metal transporter with a broad substrate range. *Plant Mol. Biol.,* **40**: 37-44.

Lahner, B., Gong, J.M., Mahmoudian, M., Smith, E.L., Abid, K.B., Rogers, E.E., Guerinot, M.L., Harper, J.F., Ward, J.M., McIntyre, L., Schroeder, J.I. and Salt, D.E. (2003) Genomic scale profiling of nutrient and trace elements in *Arabidopsis thaliana*. *Nature Biotechnol.,* **21**: 1215-1221.

Landi, S. and Fagioli, F. (1983) Efficiency of manganese and copper uptake by excised roots of maize genotypes. *J. Plant Nutr.,* **6**: 957-970.

Lanquar, V., Lelièvre, F., Bolte, S., Hamès, C., Alcon, C., Neumann, D., Vansuyt, G., Curie, C., Schröder, A., Krämer, U., Barbier-Brygoo, H. and Thomine, S. (2005) Mobilization of vacuolar iron by AtNramp3 and AtNramp4 is essential for seed germination on low iron. *EMBO J.,* **24**: 4041-4051.

Lapinskas, P.J., Cunningham, K.W., Liu, X.F., Fink, G.R. and Culotta, V.C. (1995) mutations in PMR1 suppress oxidative damage in yeast cells lacking superoxide dismutase. *Mol. Cell. Biol.,* **15**: 1382-1388.

Lapinskas, P.J., Lin, S.J. and Culotta, V.C. (1996) The role of the *Saccharomyces cerevisiae* CCC1 gene in the homeostasis of manganese ions. *Mol. Microbiol.,* **21**: 519-528.

Laurie, S.H., Tancock, N.P., McGrath, S.P. and Sanders, J.R. (1995) Influence of EDTA complexation on plant uptake of manganese (II). *Plant Sci.,* **109**: 231-235.

Le Jean, M., Schikora, A., Mari, S., Briat, J.-F. and Curie, C. (2005) A loss-of-function mutation in *AtYSL1* reveals its role in iron and nicotianamine seed loading. *Plant J.,* **44**: 769-782.

Li, L.T., Chen, O.S., Ward, D.M. and Kaplan, J. (2001) CCC1 is a transporter that mediates vacuolar iron storage in yeast. *J. Biol. Chem.,* **276**: 29515-29519.

Lidon, F.C. (2001) Tolerance of rice to excess manganese in the early stages of vegetative growth. Characterisation of manganese accumulation. *J. Plant Physiol.,* **158**: 1341-1348.

Ling, H.Q., Bauer, P., Bereczky, Z., Keller, B. and Ganal, M. (2002) The tomato *fer* gene encoding a bHLH protein controls iron-uptake responses in roots. *Proc. Natl. Acad. Sci. USA,* **99**: 13938-13943.

Liu, X.F. and Culotta, V.C. (1999) Post-translational control of Nramp metal transport in yeast: role of metal ions and the *BSD2* gene. *J. Biol. Chem.,* **274**: 4863-4868.

Liu, X.F., Supek, F., Nelson, N. and Culotta, V.C. (1997) Negative control of heavy metal uptake by the *Saccharomyces cerevisiae BSD2* gene. *J. Biol. Chem.*, **272**: 11763-11769.

Loneragan, J.F. (1988) Distribution and movement of manganese in plants. In: *Manganese in Soils and Plants* (Eds. Graham, R.D., Hannam, R.J. and Uren, N.C.), Kluwer Academic Publishers, Dordrecht, pp 113-124.

Longnecker, N.E., Marcar, N.E. and Graham, R.D. (1991) Increased manganese content of barley seeds can increase grain yield in manganese-deficient conditions. *Aust. J. Agric. Res.*, **42**: 1065-1074.

López-Millán, A.F., Ellis, D.R. and Grusak, M.A. (2004) Identification and characterization of several new members of the ZIP family of metal ion transporters in *Medicago truncatula*. *Plant Mol. Biol.*, **54**: 583-596.

Luk, E. and Culotta, V.C. (2001) Manganese superoxide dismutase in *S. cerevisiae* acquires its metal co-factor through a pathway involving the Nramp metal transporter, Smf2p. *J. Biol. Chem.*, **276**: 47556-47562.

Luk, E., Carroll, M., Baker, M. and Culotta, V.C. (2003) Manganese activation of superoxide dismutase 2 in *Saccharomyces cerevisiae* requires *MTM1*, a member of the mitochondrial carrier family. *Proc. Natl. Acad. Sci. USA*, **100**: 10353-10357.

Magidin, M., Pittman, J.K., Hirschi, K.D. and Bartel, B. (2003) *ILR2*, a novel gene regulating IAA conjugate sensitivity and metal transport in *Arabidopsis thaliana*. *Plant J.*, **35**: 523-534.

Marschner, H. (1995) Mineral Nutrition of Higher Plants, Academic Press, London.

Mäser, P., Thomine, S., Schroeder, J.I., Ward, J.M., Hirschi, K., Sze, H., Talke, I.N., Amtmann, A., Maathuis, F.J.M., Sanders, D., Harper, J.F., Tchieu, J., Gribskov, M., Persans, M.W., Salt, D.E., Kim, S.A., Guerinot, M.L. (2001) Phylogenetic relationships within cation transporter families of Arabidopsis. *Plant Physiol.*, **126**: 1646-1667.

McCain, D.C. and Markley, J.L. (1989) More manganese accumulates in maple sun leaves than shade leaves. *Plant Physiol.*, **90**: 1417-1421.

Migocka, M. and Klobus, G. (2007) The properties of the Mn, Ni and Pb transport operating at plasma membranes of cucumber roots. *Physiol. Plant.*, **129**: 578-587.

Mizuno, T., Usui, K., Horie, K., Nosaka, S., Mizuno, N. and Obata, H. (2005) Cloning of three ZIP/Nramp transporter genes from a Ni hyperaccumulating plant *Thlaspi japonicum* and their $Ni^{2+}$-transport abilities. *Plant Physiol. Biochem.*, **43**: 793-801.

Montanini, B., Blaudez, D., Jeandroz, S., Sanders, D. and Chalot, M. (2007) Phylogenetic and functional analysis of the Cation Diffusion Facilitator (CDF) family: improved signature and prediction of substrate specificity. *BMC Genomics*, **8**: 107.

Moore, C.M. and Helmann, J.D. (2005) Metal ion homeostasis in *Bacillus subtilis*. *Curr. Opin. Microbiol.*, **8**: 188-195.

Moraghan, J.T. (1992) Iron-manganese relationships in white lupin grown on a calciaquoll. *Soil Sci. Soc. Am. J.*, **56**: 471-475.

Moussavi-Nik, M., Rengel, Z., Pearson, J.N. and Hollamby, G. (1997) Dynamics of nutrient remobilisation from seed of wheat genotypes during imbibition, germination and early seedling growth. *Plant Soil*, **197**: 271-280.

Norvell, W.A., Welch, R.M., Adams, M.L. and Kochian, L.V. (1993) Reduction of Fe(III), Mn(III), and Cu(II) chelates by roots of pea (*Pisum sativum* L.) or soybean (*Glycine max*). *Plant Soil*, **156**: 123-126.

Ogawa, T., Bao, D.H., Katoh, H., Shibata, M., Pakrasi, H.B. and Bhattacharyya-Pakrasi, M. (2002) A two-component signal transduction pathway regulates manganese homeostasis in *Synechocystis* 6803, a photosynthetic organism. *J. Biol. Chem.*, **277**: 28981-28986.

Otegui, M.S., Capp, R. and Staehelin, L.A. (2002) Developing seeds of Arabidopsis store different minerals in two types of vacuoles and in the endoplasmic reticulum. *Plant Cell*, **14**: 1311-1327.
Paidhungat, M. and Garrett, S. (1998) Cdc1 and the vacuole coordinately regulate $Mn^{2+}$ homeostasis in the yeast *Saccharomyces cerevisae*. *Genetics*, **148**: 1787-1798.
Page, V. and Feller, U. (2005) Selective transport of zinc, manganese, nickel, cobalt and cadmium in the root system and transfer to the leaves in young wheat plants. *Ann. Bot.*, **96**: 425-434.
Patzer, S.I. and Hantke, K. (2001) Dual repression of $Fe^{2+}$-Fur and $Mn^{2+}$-MntR of the *mntH* gene, encoding an Nramp-like $Mn^{2+}$ transporter in *Escherichia coli*. *J. Bacteriol.*, **183**: 4806-4813.
Pearson, J.N., Jenner, C.F., Rengel, Z. and Graham, R.D. (1996) Differential transport of Zn, Mn and sucrose along the longitudinal axis of developing wheat grains. *Physiol. Plant.*, **97**: 332-338.
Pearson, J.N. and Rengel, Z. (1994) Distribution and remobilization of Zn and Mn during grain development in wheat. *J. Exp. Bot.*, **45**: 1829-1835.
Pearson, J.N. and Rengel, Z. (1995) Uptake and distribution of $^{65}Zn$ and $^{54}Mn$ in wheat grown at sufficient and deficient levels of Zn and Mn. II. During grain development. *J. Exp. Bot.*, **46**: 841-845.
Pearson, J.N., Rengel, Z., Jenner, C.F. and Graham, R.D. (1995) Transport of zinc and manganese to developing wheat grains. *Physiol. Plant.*, **95**: 449-455.
Pearson, J.N., Rengel, Z., Jenner, C.F. and Graham, R.D. (1998) Dynamics of zinc and manganese movement in developing wheat grains. *Aust. J. Plant Physiol.*, **25**: 139-144.
Pedas, P., Hebbern, C.A., Schjoerring, J.K., Holm, P.E. and Husted, S. (2005) Differential capacity for high-affinity manganese uptake contributes to differences between barley genotypes in tolerance to low manganese availability. *Plant Physiol.*, **139**: 1411-1420.
Peiter, E., Montanini, B., Gobert, A., Pedas, P., Husted, S., Maathuis, F.J.M., Blaudez, D., Chalot, M. and Sanders, D. (2007) A secretory pathway-localised cation diffusion facilitator confers plant manganese tolerance. *Proc. Natl. Acad. Sci USA*, **104**: 8532-8537.
Pich, A. and Scholz, G. (1996) Translocation of copper and other micronutrients in tomato plants (*Lycopersicon esculentum* Mill.): nicotianamine-stimulated copper transport in the xylem. *J. Exp. Bot.*, **47**: 41-47.
Pittman, J.K. (2005) Managing the manganese: molecular mechanisms of manganese transport and homeostasis. *New Phytol.*, **167**: 733-742
Pittman, J.K., Cheng, N.H., Shigaki, T., Kunta, M. and Hirschi, K.D. (2004a) Functional dependence on calcineurin by variants of the *Saccharomyces cerevisiae* $Ca^{2+}/H^+$ exchanger Vcx1p. *Mol. Microbiol.*, **54**: 1104-1116.
Pittman, J.K. and Hirschi, K.D. (2003) Don't shoot the (second) messenger: endomembrane transporters and binding proteins modulate cytosolic $Ca^{2+}$ levels. *Curr. Opin. Plant Biol.*, **6**: 257-262.
Pittman, J.K., Mills, R.F., O'Connor, C.D. and Williams, L.E. (1999) Two additional type IIA $Ca^{2+}$-ATPases are expressed in *Arabidopsis thaliana*: evidence that type IIA subgroups exist. *Gene*, **236**: 137-147.
Pittman, J.K., Shigaki, T., Marshall, J.L., Morris, J.L., Cheng, N.H. and Hirschi K. D. (2004b) Functional and regulatory analysis of the *Arabidopsis thaliana* CAX2 cation transporter. *Plant Mol. Biol.*, **56**: 959-971.
Posey, J.E. and Gherardini, F.C. (2000) Lack of a role for iron in the Lyme disease pathogen. *Science*, **288**: 1651-1653.

Pozos, T.C., Sekler, I. and Cyert, M.S. (1996) The product of *HUM1*, a novel yeast gene, is required for vacuolar $Ca^{2+}/H^+$ exchange and is related to mammalian $Na^+/Ca^{2+}$ exchangers. *Mol. Cell. Biol.,* **16**: 3730-3741.

Qi, B.-S., Li, C.-G., Chen, Y.-M., Lu, P.-L., Hao, F.-S., Shen, G.-M., Chen, J. and Wang, X.-C. (2005) Functional analysis of rice $Ca^{2+}/H^+$ antiporter OsCAX3 in yeast and its subcellular localization in plant. *Prog. Biochem. Biophys.,* **32**: 876-882.

Que, Q. and Helmann, J. D. (2000) Manganese homeostasis in *Bacillus subtilus* is regulated by MntR, a bifunctional regulator related to the diphtheria toxin repressor family of proteins. *Mol. Microbiol.,* **35**: 1454-1468.

Quiquampoix, H., Loughman, B.C. and Ratcliffe, R.G. (1993) A $^{31}P$ NMR study of the uptake and compartmentation of manganese by maize roots. *J. Exp. Bot.,* **44**: 1819-1827.

Reeves, R.D. and Baker, A.J.M. (2000) Metal-accumulating plants. In: *Phytoremediation of Toxic Metals: Using Plants to Clean Up the Environment.* (Eds. Raskin, I. and Ensley, B.D.), John Wiley and Sons, New York, pp 193-229.

Rengel, Z. (2000) Manganese uptake and transport in plants. In: *Metal Ions in Biological Systems, Vol 37* (Eds. Sigel A. and Sigel H.), Marcel Dekker, New York, pp 57-87.

Rengel, Z. and Marschner, P. (2005) Nutrient availability and management in the rhizosphere: exploiting genotypic differences. *New Phytol.,* **168**: 305-312.

Roberts, L.A., Pierson, A.J., Panaviene, Z. and Walker, E.L. (2004) Yellow Stripe1. Expanded roles for the maize iron-phytosiderophore transporter. *Plant Physiol.,* **135**: 112-120.

Rogers, E.E., Eide, D.J. and Guerinot, M.L. (2000) Altered selectivity in an Arabidopsis metal transporter. *Proc. Natl. Acad. Sci. USA,* **97**: 12356-12360.

Rogers, E.E. and Guerinot, M.L. (2002) FRD3, a member of the multidrug and toxin efflux family, controls iron deficiency responses in Arabidopsis. *Plant Cell,* **14**: 1787-1799.

Roomizadeh, S. and Karimian, N. (1996) Manganese-iron relationship in soybean grown in calcareous soils. *J. Plant Nutr.,* **19**: 397-406.

Rosakis, A. and Köster, W. (2005) Divalent metal transport in the green microalga *Chlamydomonas reinhardtii* is mediated by a protein similar to prokaryotic Nramp homologues. *Biometals,* **18**: 107-120.

Sadana, U.S., Sharma, P., Ortiz, N.C., Samal, D. and Claassen, N. (2005) Manganese uptake and Mn efficiency of wheat cultivars are related to Mn-uptake kinetics and root growth. *J. Plant Nutr. Soil Sci.,* **168**: 581-589.

Salt, D.E. (2004) Update on plant ionomics. *Plant Physiol.,* **136**: 2451-2456.

Schaaf, G., Catoni, E., Fitz, M., Schwacke, R., Schneider, A., von Wirén, N. and Frommer, W.B. (2002) A putative role for the vacuolar calcium/manganese proton antiporter AtCAX2 in heavy metal detoxification. *Plant Biol.,* **4**: 612-618.

Schaaf, G., Ludewig, U., Erenoglu, B.E., Mori, S., Kitahara, T. and von Wirén, N. (2004) ZmYS1 functions as a proton-coupled symporter for phytosiderophore- and nicotianamine-chelated metals. *J. Biol. Chem.,* **279**: 9091-9096.

Schmidke, I. and Stephan, U.W. (1995) Transport of metal micronutrients in the phloem of castor bean (*Ricinus communis*) seedlings. *Physiol. Plant.,* **95**: 147-153.

Shigaki, T. and Hirschi, K.D. (2006) Diverse functions and molecular properties emerging for CAX cation/$H^+$ exchangers in plants. *Plant Biol.,* **8**: 419-429.

Shigaki, T., Pittman, J.K. and Hirschi, K.D. (2003) Manganese specificity determinants in the Arabidopsis metal/$H^+$ antiporter CAX2. *J. Biol. Chem.,* **278**: 6610-6617.

Shingles, R., North, M. and McCarty, R.E. (2002) Ferrous iron transport across chloroplast inner envelope membranes. *Plant Physiol.,* **128**: 1022-1030.

Stacey, M.G., Koh, S., Becker, J. and Stacey, G. (2002) AtOPT3, a member of the oligopeptide transporter family, is essential for embryo development in Arabidopsis. *Plant Cell,* **14**: 2799-2811.

Stacey, M.G., Osawa, H., Patel, A., Gassmann, W. and Stacey, G. (2006) Expression analyses of Arabidopsis oligopeptide transporters during seed germination, vegetative growth and reproduction. *Planta,* **223**: 291-305.

Stephan, U.W., Schmidke, I., Stephan, V.W. and Schloz, G. (1996) The nicotianamine molecule is made-to-measure for complexation of metal micronutrients in plants. *Biometals,* **9**: 84-90.

Supek, F., Supekova, L., Nelson, H. and Nelson, N. (1996) A yeast manganese transporter related to the macrophage protein involved in conferring resistance to mycobacteria. *Proc. Natl. Acad. Sci. USA,* **93**: 5105-5110.

Thomine, S., Lelièvre, F., Debarbieux, E., Schroeder, J.I. and Barbier-Brygoo, H. (2003) AtNRAMP3, a multispecific vacuolar metal transporter involved in plant responses to iron deficiency. *Plant J.,* **34**: 685-695.

Thomine, S. and Schroeder, J.I. (2004) Plant metal transporters with homology to proteins of the NRAMP family. In: *The Nramp Family* (Eds. Cellier, M. and Gros, P.), Eurekah.com and Kluwer Academic/Plenum Publishers, Dordrecht, pp 113-123.

Thomine, S., Wang, R.C., Ward, J.M., Crawford, N.M. and Schroeder, J.I. (2000) Cadmium and iron transport by members of a plant metal transporter family in Arabidopsis with homology to Nramp genes. *Proc. Natl. Acad. Sci. USA,* **97**: 4991-4996.

Van Baelen, K., Dode, L, Vanoevelen, J., Callewaert, G., De Smedt, H., Missiaen, L., Parys, J.B., Raeymaekers, L. and Wuytack, F. (2004) The $Ca^{2+}/Mn^{2+}$ pumps in the Golgi apparatus. *Biochim. Biophys. Acta,* **1742**: 103-112.

Vert, G., Grotz, N., Dédaldéchamp, F., Gaymard, F., Guerinot, M.L., Briat, J.F. and Curie, C. (2002) IRT1, an Arabidopsis transporter essential for iron uptake from the soil and for plant growth. *Plant Cell,* **14**: 1223-1233.

Welch, R.M. and LaRue, T.A. (1990) Physiological characteristics of Fe accumulation in the 'bronze' mutant of *Pisum sativum* L., cv 'Sparkle' E107 (*brz brz*). *Plant Physiol.,* **93**: 723-729.

Welch, R.M. and Norvell, W.A. (1993) Growth and nutrient uptake by barley (*Hordeum vulgare* L. cv Herta). Studies using an *N*-(2-hydroxyethyl)ethylenedinitrilotriacetic acid-buffered nutrient solution technique. II. Role of zinc in the uptake and root-leakage of mineral nutrients. *Plant Physiol.,* **101**: 627-631.

White, M.C., Baker, F.D., Chaney, R.L. and Decker, A.M. (1981) Metal complexation in xylem fluid. II. Theoretical equilibrium model and computational computer program. *Plant Physiol.,* **67**: 301-310.

White, P.J. (1998) Calcium channels in the plasma membrane of root cells. *Ann. Bot.,* **81**: 173-183.

Williams, L.E., Pittman, J.K. and Hall, J.L. (2000) Emerging mechanisms for heavy metal transport in plants. *Biochim. Biophys. Acta,* **1465**: 104-126.

Wintz, H., Fox, T., Wu, Y.-Y., Feng, V., Chen, W., Chang, H.-S., Zhu, T. and Vulpe, C. (2003) Expression profiles of *Arabidopsis thaliana* in mineral deficiencies reveal novel transporters involved in metal homeostasis. *J. Biol. Chem.,* **278**: 47644-47653.

Wu, Z.Y., Liang, F., Hong, B.M., Young, J.C., Sussman, M.R., Harper, J.F. and Sze, H. (2002) An endoplasmic reticulum-bound $Ca^{2+}/Mn^{2+}$ pump, ECA1, supports plant growth and confers tolerance to $Mn^{2+}$ stress. *Plant Physiol.,* **130**: 128-137.

Xu, X., Shi, J., Chen, Y., Chen, X., Wang, H. and Perera, A. (2006) Distribution and mobility of manganese in the hyperaccumulator plant *Phytolacca acinosa* Roxb. (Phytolaccaceae). *Plant Soil,* **285**: 323-331.

Xue, S.G., Chen, Y.X., Reeves, R.D., Baker, A.J.M., Lin, Q. and Fernando, D.R. (2004) Manganese uptake and accumulation by the hyperaccumulator plant *Phytolacca acinosa* Roxb. (Phytolaccaceae). *Environ. Pollut.,* **131**: 393-399.

Yamaguchi, K., Suzuki, L., Yamamoto, H., Lyukevich, A., Bodrova, I., Los, D.A., Piven, I., Zinchenko, V., Kanehisa, M. and Murata, N. (2002) A two-component $Mn^{2+}$-sensing system negatively regulates expression of the *mntCAB* operon in *Synechocystis*. *Plant Cell,* **14**: 2901-2913.

Yen, M.-R., Tseng, Y.-H. and Saier, M.H. (2001) Maize *Yellow Stripe 1*, an iron-phytosiderophore uptake transporter, is a member of the oligopeptide transporter (OPT) family. *Microbiology,* **147**: 2881-2883.

Yu, Q., Osborne, L.D. and Rengel, Z. (1999) Increased tolerance to Mn deficiency in transgenic tobacco overproducing superoxide dismutase. *Ann. Bot.,* **84**: 543-547.

Yuan, Y.X., Zhang, J., Wang, D.W. and Ling, H.Q. (2005) *AtbHLH29* of *Arabidopsis thaliana* is a functional ortholog of tomato *FER* involved in controlling iron acquisition in strategy I plants. *Cell Res.,* **15**: 613-621.

# Chapter 8

## SILICON UPTAKE AND TRANSPORT IN HIGHER PLANTS

YONGCHAO LIANG[1,2]
[1]*Institute of Soil and Fertilizer, and Ministry of Agriculture Key Laboratory of Plant Nutrition and Nutrient Cycling, Chinese Academy of Agricultural Sciences, Beijing - 100 081, P.R. China*
[2]*Key Laboratory of Eco-agriculture Shihezi University, Shihezi - 832 003, P.R. China*
E-mail: yliang@caas.ac.cn

## Abstract

Silicon (Si) is the second most abundant element in soils, but it is still not considered as an essential element for higher plants. Si is beneficial for the healthy growth and development of many plant species. Plant species differ greatly in Si contents in the plant tops, ranging from 0.1-10% of dry weight. Such variation in Si content depends on the ability of roots to take up Si. However, mechanisms of Si uptake and transport remain poorly understood. More recently, a breakthrough has been made in cloning, identification and functional analysis of Si transporters in rice. Meanwhile, rapid progress has also been made in characterizing Si uptake and transport in the other Si-accumulators and intermediate type plant species including monocots and dicots using conventional depletion techniques and radioactive isotope methods. This chapter reviews the current knowledge of Si uptake and transport in higher plants and highlights the future research directions.

**Keywords:** silicon uptake, transport, silicon transporters

## 1. INTRODUCTION

Silicon (Si) is the second most abundant element after oxygen both in the Earth's crust and in soil. Although Si is plentiful, most sources of silicon are insoluble and not in a plant-available form (Richmond and Sussman, 2003). The plant-available form of Si is monosilicic acid ($H_4SiO_4$), which is present in the soil solution at concentrations normally ranging from 0.1 to 0.6 mM (Gunnarsson and Arnórsson, 2000), roughly two orders of magnitude higher than the concentrations of phosphorus in soil solutions (Epstein, 1994, 1999). However, Si is contained in varying amounts in all plants ranging from 0.1% to 10% of top dry weight.

---

© CAB International 2008. *Plant Membrane and Vacuolar Transporters* (eds P.K. Jaiwal, R.P. Singh and O.P. Dhankher)

Yongchao Liang

Although the essentiality of Si for all higher plants has not been generally recognized, Si has been proved to be beneficial for the healthy growth of many plant species, particularly gramineaceous plants such as rice and some cyperaceous plants (Epstein, 1994, 1999; Liang, 1999; Ma et al., 2001a). The beneficial effects of Si are particularly distinct in plants subjected to various forms of stress (Epstein, 1994, 1999; Rogalla and Römheld, 2002; Ma, 2004). More recently, Epstein and Bloom (2005) have proposed a near-universally accepted definition of essentiality of elements. Based on this newly-established definition, an element is essential that fulfils either one or both of the two following criteria: (1) the element is part of a molecule which is an intrinsic component of the structure or metabolism of the plant, and (2) the plant can be so severely deficient in the element that it exhibits abnormalities in growth, development, or reproduction, i.e. 'performance', compared to plants with lower deficiency. Based on this new definition, Si will be an essential element for higher plants.

Over last two decades, numerous studies have been focused on better understanding the possible mechanism(s) for Si-mediated alleviation of various forms of stress and the roles Si plays in plant biology, and rapid progress has been achieved (Horst and Marschner, 1978; Horiguchi and Morita, 1987; Hodson and Evans 1995; Liang, 1999; Epstein, 1999; Rogalla and Römheld, 2002; Liang et al., 2005a, 2005b; Gong et al., 2005; Guo et al., 2005). More recently, an Arabidopsis transcriptome analysis has provided direct evidence for the roles Si plays in plant biology (Fauteus et al., 2006). Silicon alone has apparently no effect on the metabolism of Arabidopsis growing in controlled environment (e.g. unstressed), thus confirming its non essentiality in plant growth; supplying Si alleviates the powdery mildew-induced stress. The way Si modulates this stress response in plants is active or at least not solely limited to a mechanical barrier as previously proposed (Fauteus et al., 2006). However, whether this result can extrapolate to other systems such as blast-stressed rice needs elucidation because rice is a typical Si accumulator and Arabidopsis not. However, on the other hand, the quickest and greatest progress has been made in the low silicon rice 1 (*Lsi1*) gene controlling Si accumulation in rice (Ma et al., 2006). Ma and his co-workers have recently reported their pioneer work on isolation, identification and expression of the *Lsi1* gene (Ma et al., 2006). In this paper, current knowledge on Si uptake mechanisms is reviewed, and future research priority is also discussed.

## 2. SILICON UPTAKE AND TRANSPORT PATTERNS IN PLANTS

Plant species differ greatly in Si contents in the plant tops, ranging from 0.1 to 10.0% of dry weight. Based on the analytical results of Si contents in plant tops, plants are classified into Si accumulator, intermediate type, and Si excluder species (Jones and Handreck, 1967; Takahashi et al., 1990). The difference in Si content in plants has been attributed to the difference in mechanisms for Si uptake by the roots (Jones and Handreck, 1967; Jarvis, 1987; Epstein, 1994). Three modes of Si uptake, i.e. active, passive, and rejective uptake, have been suggested for Si accumulator, intermediate type, and Si excluder plants, respectively (Takahashi et *al.*, 1990). Rice, a typical Si accumulator, takes up Si actively. In addition, some graminaceous plants such as wheat (van der Vorm, 1980; Jarvis, 1987; Rafi and Epstein, 1999; Casey *et al.*, 2003), rye (Jarvis, 1987), and barley (Barber and Shone, 1966) also take up Si actively, while some others such as oat take it up passively (Jones and Handreck, 1965). By contrast, some dicots such as strawberry and soybean (Takahashi *et al.*, 1990; Ma *et al.*, 2001b; Mitani and Ma, 2005) absorb Si passively, and tomato (Takahashi *et al.*, 1990), hordeum (Liang *et al.*, 2003), bean (Jones and Handreck, 1967) and faba bean (Liang *et al.*, 2006) exclude Si from uptake. However, such categorizations (accumulator, intermediate and excluder) proposed previously to distinguish Si uptake patterns (Jones and Handreck, 1967; Takahashi *et al.*, 1990) should be viewed with caution (Epstein, 1999) because these modes do not refer or correlate to a molecular uptake mechanism, such as a channel, pump or carrier, but are based upon measurements of silicon concentration and transpiration rates (Richmond and Sussman, 2003). Furthermore, attention should be drawn to the fact that even relatively "low" values for tissue Si contents, such as 0.1%, are still comparable with the low values found for such macronutrient elements as S, P, and Mg (Epstein, 1994).

## 3. ACTIVE SILICON UPTAKE AND TRANSPORT IN GRAMINACEOUS AND DICOTYLEDONOUS PLANTS

It has been well documented that rice (*Oryza sativa*), a typical Si accumulator takes up Si actively and accumulates Si to levels up to 10.0% of shoot dry weight (Okuda and Takahashi, 1962a; Epstein, 1994, 1999). It has also generally recognized that such high accumulation of Si in rice tops is attributed to the high capacity of rice roots to take up Si actively (Okuda and Takahashi, 1962a, Ma *et al.*, 2002, Liang *et al.*, 2006). Si uptake by rice is an energy-dependent process mediated by Si transporters (Ma *et al.*, 2004, 2006), and is inhibited by metabolic inhibitors such as NaCN, NaF and 2,4-dinitrophenol (Okuda and Takahashi, 1962b, Liang *et al.*, 2006) and by low temperature (Ma *et al.*, 2004; Liang *et al.*, 2006), but not affected by transpiration (Okuda and Takahashi, 1962a). Short uptake experiments show that Si concentration is much higher in xylem exudates of the rice than in the external solution (Okuda and Takahashi, 1962a; Ma *et al.*, 2002; Liang *et al.*, 2006), suggesting that xylem loading of Si is a transporter-mediated transmembrane process against a concentration gradient.

There is an increasing body of evidence showing that an active Si uptake process exists in some other graminaceous plants such as wheat (van der Vorm, 1980; Jarvis, 1987; Rafi and Epstein, 1999; Casey *et al.*, 2003; Rains *et al.*, 2006), rye (Jarvis, 1987), barley (Barber and Shone, 1966) and maize (Liang *et al.*, 2006). More recent studies show that Si uptake by wheat follows the Michaelis–Menten equation with Km being 0.086 mM and $V_{max}$ being 4.49 mmol $g^{-1}$FW roots $h^{-1}$. This seems to suggest that Si uptake by wheat is also a Si-transporter-mediated process as has been proved in rice by Ma and his co-workers (Ma *et al.*, 2006). Interestingly, the apparent affinity of the Si transporters is much higher in wheat (Km=0.086 mM) (Rains *et al.*, 2006) than in rice (Km=0.32) (Tamai and Ma, 2003). It has also been reported that Si concentrations in the xylem exudate of wheat exceed the external concentration by more than 400:1 (Casey *et al.*, 2003) and there is a metabolically mediated transport and translocation system for Si in wheat because Si uptake by wheat is severely inhibited by the metabolic inhibitors such as DNP and KCN, competitively depressed by germanium (Ge) but unaffected by phosphate (Rains *et al.*, 2006). More recent investigation shows that like in rice, both Si uptake measured via the depletion method and Si uptake calculated via transpiration streams increased linearly with time in maize but Si uptake measured via the depletion method was significantly higher than Si uptake calculated via transpiration streams at 0.085 or 0.85 mM Si (Liang *et al.*, 2006). Si uptake by maize was also inhibited by the metabolic inhibitors (NaF, NaCN and 2,4-DNP) and by low temperature at all the three external Si concentrations (0.085, 0.85 and 1.7 mM). Furthermore, Si concentration was at least 5-fold higher in the xylem sap than in the external solution when maize grew at 0.85 mM Si, compared to 10-fold higher when maize grew at 0.085 mM Si (Liang *et al.*, 2006). All these findings along with the research on wheat by Epstein's group (Rafi and Epstein, 1999; Casey *et al.*, 2003; Rains *et al.*, 2006) suggest that there is also a Si-transporter-mediated mechanism for Si uptake and translocation in wheat, maize and perhaps in other gramineaceous plants such as barley and sorghum, and there is nothing unique in the silicon nutrition of rice (and sugarcane) and the differences in this regard between these species and other species, such as wheat and maize (Liang *et al.*, 2006), are a matter of degree only.

It has long been established that most dicots generally take up Si passively. For example, strawberry and soybean (Takahashi *et al.*, 1990; Ma *et al.*, 2001b; Mitani and Ma, 2005) absorb Si passively, and tomato (Takahashi *et al.* 1990), bean (Jones and Handreck, 1967) and faba bean (Liang *et al.*, 2005c) exclude Si from uptake. However, recent investigations with cucumber show that (1) Si uptake measured via the depletion method was more than twice as high as calculated from the rate of transpiration, (2) Si uptake was strongly inhibited by low temperature and 2, 4-dinitrophenol, and (3) concentration of Si was several times higher in xylem exudates than in external solution, suggesting that Si uptake and transport is also an energy-dependent active mechanism (Liang *et al.*, 2005c). As early as in 1980, it was also reported that in sunflower treated with higher external Si levels (0.5 or 2.7 mM Si), the ratio of measured Si uptake to Si uptake calculated from transpiration stream was less than 1.0 compared to 5.6 at lower external Si levels (0.013 mM Si) (van der Vorm, 1980). This implies that Si uptake mode depends on external Si concentration and an active Si uptake mechanism may exist in sunflower at lower external Si levels. More recently, it has been reported that active Si uptake prevails both in rice and maize, and that the passive component also exists at both lower (e.g. 0.085 mM Si) and higher (e.g. 0.85 mM Si) external Si concentrations, with its relative contribution being greater at higher external Si concentration (Liang *et al.*, 2006). However, for sunflower and wax gourd, an active component contributes to the Si uptake and translocation through xylem, especially at lower external Si concentration (i.e. 0.085 mM Si) despite much greater contribution of passive uptake, especially at higher external Si concentration (i.e. 0.85 mM Si).

Silicon uptake and transport in higher plants

More importantly, Si uptake by rice and maize (Si-accumulators) and sunflower and wax gourd (intermediate type species) was significantly inhibited by the treatment with either metabolic inhibitors (NaF, NaCN and 2,4-DNP) or low temperature, particularly at lower external Si concentration (0.085 mM). However, it is worth noting that Si uptake was not fully inhibited by the treatment with either metabolic inhibitors or low temperature even in rice, the typical Si-accumulator (Liang et al., 2006). These results unequivocally demonstrate that both active and passive mechanisms are operating in Si uptake and transport in the same Si-accumulator and intermediate type species with their contribution being dependent upon plant species and external Si concentrations. By using $^{68}$Ge radioactive isotope technique, Nikolic et al. (2007) have recently provided direct and unique evidence to prove that metabolically active mechanism(s) are involved in xylem loading of Si not only in rice and barley but also in cucumber, and suggested for the first time that in the Si-excluding plants such as tomato, passive diffusion of $Si(OH)_4$ into the root cortex obviously coexists with a transpoter-mediated Si exclusion from the root cortical cells into the apoplast, which might further explain the rejective type of Si uptake.

## 4. SILICON TRANSPORTER AND GENES RESPONSIBLE FOR ACTIVE SILICON UPTAKE IN HIGHER PLANTS

In the marine diatom, *Cylindrotheca fusiformis*, Hildebrand et al. (1997) identified a cDNA that encodes a Si transporter. This is the first identification of a Si transporter in any organism. However, it is not until recently that a breakthrough has been made in Si uptake by higher plants at molecular level using low-silicon rice 1 (Lsi1) defective in active Si uptake system (Ma et al., 2001b, 2002; Tamai and Ma, 2003; Ma et al., 2004, 2006). These authors agree that Si uptake and transport in rice is an active process mediated by a specific transporter having a low affinity ($K_m$ = 0.32 mM) compared with K (Tamai and Ma, 2003). The first cloning, identification and functional analysis of Si transporter in higher plants have been accomplished using the rice mutant Lsi1 (Ma et al., 2006). A Si gene responsible for xylem loading of Si has been mapped to chromosome 2 of rice using the Lsi1 mutant (formerly GR1 mutant) (Ma et al., 2004). The *Lsi1* gene is localized on the plasma membrane of the distal side of both exodermis and endodermis cells and constitutively expressed in the roots (Ma et al., 2006). The gene consisting of five exons and four introns is predicted to encode a membrane protein similar to water channel proteins (aquaporins). The complementary DNA of this gene is 1,409-base-pairs (bp) long and the deduced protein comprises 298 amino acids. The predicted amino acid sequence has six transmembrane domains and two Asn-Pro-Ala (NPA) motifs, which is well conserved in typical aquaporins (Ma et al., 2006). Furthermore, expression of *Lsi1* in *Xenopus* oocytes showed transport activity for Si only and suppressed expression of *Lsi1* in RNAi transgenic rice lines showed significantly reduced Si uptake compared with vector control plants (Ma et al., 2006). However, no enhanced Si uptake was observed in the transgenic tobacco carrying one of the diatom Si transporter genes, indicating that the Si uptake system in higher plants is different from that in diatoms (Ma, 2003).

The first identification of the Si transporter in higher plants is believed to not only gain better insight into Si uptake mechanisms in plants, but also provide a new genetic approach to enhancing the capacity of Si uptake of the non-Si-accumulating crops to acquire high resistance to various forms of abiotic and biotic stresses.

## 5. CONCLUSIONS AND FURTHER RESEARCH DIRECTIONS

These recent exciting findings from diatoms and higher plants, especially the first identification and functional analysis of *Lsi1* gene in rice should spur plant physiologists, biochemists, and molecular biologists to gain deeper insight into the role silicon plays in higher plants at both biochemical and molecular levels. On the one hand, further investigations should involve characterizing Si uptake and xylem loading in both the Si-accumulating monocots other than rice and the dicots such as cucumber and sunflower at the molecular level because mechanisms for Si uptake and translocation remain poorly understood and characterizing Si uptake and transport is the very basis for better understanding the functions of Si in plant kingdom. On the other hand, further investigations should be focused on elucidating the role Si plays in conferring plants resistance to various forms of abiotic (Al and heavy metal stress, salinity toxicity, drought stress, cold stress etc.) and biotic (plant diseases and pest damage) stress using rice as model plant species at the molecular level because rice is both the typical Si accumulator and the model plant species with its genome having been sequenced. Furthermore, a large collection of 'reverse genetic' gene knockouts has been available for Arabidopsis and will be available for rice in the near future (Krysan *et al.*, 1996; Richmond and Sussman, 2003). Thus it will be interesting to screen the silicon content of tens of thousands of gene-knockout mutants to identify those genes involved in silicon biology (Richmond and Sussman, 2003). Owing to the recent exciting findings of Si uptake system in rice, the day will come soon when silicon biology will become more fully understood.

## 6. ACKNOWLEDGEMENTS

This research is supported by the Distinguished Talent Program of the Chinese Academy of Agricultural Sciences and by Changjiang Scholars Programme granted to Y. Liang. It was also partly supported by an Alexander von Humboldt Foundation Research Fellowship granted to Y. Liang.

## 7. LITERATURE CITED

Barber, D.A. and Shone, M.G.T. (1966). The absorption of silica from aqueous solutions by plants. *J. Exp. Bot.*, **17**: 569-578.

Casey, W.H., Kinrade, S.D., Knight, T.G., Rains, D.W. and Epstein, E. (2003). Aqueous silicate complexes in wheat, *Triticum aestivum* L. *Plant, Cell & Environ.*, **27**: 51-54.

Silicon uptake and transport in higher plants

Epstein, E. (1994). The anomaly of silicon in plant biology. *Proc. Natl. Acad. Sci. USA,* **91:** 11-17.
Epstein, E. (1999). Silicon. *Ann. Rev. Plant Physiol. and Plant Mol. Biol.,* **50:** 641-644.
Epstein, E. and Bloom, A.J. (2005). *Mineral Nutrition of Plants: Principles and Perspectives,* 2nd Ed., Sinauer, Sunderland, Massachusetts, USA.
Fauteux, F., Chain, F., Belzile, F., Menzies, J.G. and Bélanger, R.R. (2006). The protective role of silicon in Arabidopsis-powdery mildew pathosystem. *Proc. Natl. Acad. Sci. USA,* **103:** 17554-17559.
Gong, H.J., Zhu, X.Y., Chen, K.M., Wang, S.M. and Zhang, C.L. (2005). Silicon alleviates oxidative damage of wheat plants in pots under drought. *Plant Sci.,* **169:** 313-321.
Gunnarsson, I. and Arnórsson, S. (2000). Amorphous silica solubility and the thermodynamic properties of $H_4SiO_4$ in the range of 0° to 350° at Psat. *Geochimica Cosmochimica Acta,* **64:** 2295-2307.
Guo, W., Hou, Y.L., Wang, S.G. and Zhu, Y.G. (2005). Effect of silicate on the growth and arsenate uptake by rice (*Oryza sativa* L.) seedlings in solution culture. *Plant Soil,* **272:** 173-181.
Hildebrand, M., Volcani, B.E., Gassmann, W. and Schroeder, J.I. (1997). A gene family of silicon transporters. *Nature,* **385:** 688-689.
Hodson, M.J. and Evans, D.E. (1995). Aluminium/silicon interactions in higher plants. *J. Exp. Bot.,* **46:** 161-171.
Horiguchi, T. and Morita, S. (1987). Mechanism of manganese toxicity and tolerance of plants. VI. Effect of silicon on alleviation of manganese toxicity of barley. *J. Plant Nutr.,* **10:** 2299-2310.
Horst, W.J. and Marschner, H. (1978). Effect of silicon on manganese tolerance of bean plants (*Phaseolus vulgaris* L.). *Plant Soil,* **50:** 287-303.
Jarvis, S.C. (1987). The uptake and transport of silicon by perennial ryegrass and wheat. *Plant and Soil,* **97:** 429-438.
Jones, L.H.P. and Handreck, K.A. (1965). Studies of silica in the oat plant. III. Uptake of silica from soils by the plant. *Plant Soil,* **23:** 79-99.
Jones, L.H.P. and Handreck, K.A. (1967). Silica in soils plants and animals. *Advances in Agronomy,* **19:** 107-149.
Krysan, P.J., Young, J.C., Tax, F. and Sussman, M.R. (1996). Identification of transferred DNA insertions within Arabidopsis genes involved in signal transduction. *Proc. Natl. Acad. Sci. USA,* **93:** 8145-8150.
Liang, Y.C. (1999). Effects of silicon on enzyme activity, and sodium, potassium and calcium concentration in barley under salt stress. *Plant and Soil,* **209:** 217-224.
Liang, Y.C., Chen, Q., Liu, Q., Zhang, W.H. and Ding, R.X. (2003). Exogenous silicon (Si) increases antioxidant enzyme activity and reduces lipid peroxidation in roots of salt-stressed barley (*Hordeum vulgare* L.). *J. Plant Physiol.,* **160:** 1157-1164.
Liang, Y.C., Wong, J.W.C. and Wei, L. (2005a). Silicon-mediated enhancement of cadmium tolerance in maize (*Zea mays* L.) grown in cadmium contaminated soil. *Chemosphere,* **58:** 475-483.
Liang, Y.C., Zhang, W.H., Chen, Q. and Ding, R.X. (2005b). Effects of silicon on tonoplast $H^+$-ATPase and $H^+$-PPase activity, fatty acid composition and fluidity in roots of salt-stressed barley (*Hordeum vulgare* L.). *Environ. Exp. Bot.,* **53:** 29-37.
Liang, Y.C., Si, J. and Römheld, V. (2005c). Silicon uptake and transport is an active process in *Cucumis sativus*. *New Phytol.,* **167:** 797-804.
Liang, Y.C., Hua, H.X., Zhu, Y.G., Zhang, J., Cheng, C.M. and Römheld, V. (2006). Importance of plant species and external silicon concentration to active silicon uptake and transport. *New Phytol.,* **172:** 63-72.

Ma, J.F., Miyake, Y. and Takahashi, E. (2001a). Silicon as a beneficial element for crop plants. In: *Silicon in Agriculture*. (Eds. Datonoff, L., Snyder, G. and Korndorfer, G.) Elsevier Science Publishing, New York, pp. 17-39.

Ma, J.F., Goto, S., Tamai, K. and Ichii, M. (2001b). Role of root hair and lateral roots in silicon uptake by rice. *Plant Physiol.*, **127:** 1773-1780.

Ma, J.F., Tamai, K., Ichii, M. and Wu, G.F. (2002). A rice mutant defective in Si uptake. *Plant Physiol.*, **130:** 2111-2117.

Ma, J.F. (2003). Mechanism of Si uptake in plants. *Fertilizer*, **94:** 26-32.

Ma, J.F. (2004). Role of silicon in enhancing the resistance of plants to biotic and abiotic stresses. *Soil Sci. Plant Nutr.*, **50:** 11-18.

Ma, J.F., Mitani, N., Nagao, S., Konishi, S., Tamai, K., Iwashita, T. and Yano, M. (2004). Characterization of the silicon uptake and molecular mapping of the silicon transporter gene in rice. *Plant Physiol.*, **136:** 3284-3289.

Ma, J.F., Tamai, K., Yamaji, N., Mitani, N., Konishi, S., Katsuhara, M., Ishiguro, M., Murata, Y. and Yano, M. (2006). A Si transporter in rice. *Nature*, **440:** 688-691.

Mitani, N. and Ma, J.F. (2005). Uptake system of silicon in different plant species. *J. Exp. Bot.*, **56:** 1255-1261.

Nikolic, M., Nikolic, N., Liang, Y., Kirkby, E.A. and Römheld, V. (2007) Germanium-68 as an Adequate Tracer for Silicon Transport in Plants. Characterization of Silicon uptake in Different Crop Species. *Plant Physiol.*, **143:** 495-503.

Okuda, A. and Takahashi, E. (1962a). Studies on the physiological role of silicon in crop plants: Part 8. Some specific behaviors of rice plants in silicic acid uptake. *Japanese Journal of Soil Science and Plant Nutrition*, **33:** 217-221 (in Japanese).

Okuda, A. and Takahashi, E. (1962b). Studies on the physiological role of silicon in crop plants. Part 9. Effect of various metabolic inhibitors on the silicon uptake by rice plant. *Japanese Journal of Soil Science and Plant Nutrition*, **33:** 453–455 (in Japanese).

Rafi, M.M. and Epstein, E. (1999). Silicon absorption by wheat (*Triticum aestivum* L.). *Plant and Soil*, **211:** 223-230.

Rains, D.W., Epstein, E., Zasoski, R.J. and Aslam, M. (2006). Active silicon uptake by wheat. *Plant Soil*, **280:** 223-228.

Richmond, K.E. and Sussman, M. (2003). Got silicon? The non-essential beneficial plant nutrient. *Curr. Opin. Plant Biol.*, **6:** 268-272.

Rogalla, H. and Römheld, V. (2002). Role of leaf apoplast in silicon-mediated manganese tolerance of *Cucumis sativus* L. *Plant Cell Environ.*, **25:** 549-555.

Takahashi, E., Ma, J.F. and Miyake, Y. (1990). The possibility of silicon as an essential element for higher plants. *Comments on Agricultural and Food Chemistry*, **2:** 357-360.

Tamai, K. and Ma, J.F. (2003). Characterization of silicon uptake by rice roots. *New Phytologist*, **158:** 431-436.

van der Vorm, P.D.J. (1980). Uptake of Si by five plant species, as influenced by variation in Si-supply. *Plant and Soil*, **56:** 153-156.

# Chapter 9

## HEAVY METAL TRANSPORTERS IN PLANTS

BIBIN PAULOSE[1], PAWAN K. JAIWAL[2] AND OM PARKASH DHANKHER[1]

[1]Department of Plant, Soil and Insect Sciences, University of Massachusetts, Amherst, MA 01002, USA
[2]Advanced Centre for Biotechnology, Maharshi Dayanand University, Rohtak - 124 001, India
E-mail: parkash@psis.umass.edu

**Abstract**

Heavy metals are important environmental pollutants that affect the health of millions of people worldwide. This chapter summarizes the current research and information available on the membrane transporters involved in heavy metal uptake and transport in plants. The essential nutrient metals such as Zn, Cu, Fe, B, Ni etc are required for normal growth and functioning of plants, however, in excess these become toxic to plants. Non-essential toxic heavy metal cations such as Cd, Pb, Hg, Cr, and metalloid oxyanions such as As and Se are also transported via these membrane transporters. Most of the membrane transporter families described here are suggested and/or shown to facilitate the nutrient metal uptake, transport, and homeostasis in plants. These metal transporters have broad substrate range and thus shown to transport more than one metal. Although, a plethora of literature is available on the uptake and efflux of metals in plants, still there is more to unravel on this aspect. Understanding the mechanism of transport of metals in plants can help to develop strategies for engineering food crops for enhanced essential metal nutrients uptake and non-food crops for phytoremediation of the heavy metals contaminating soil, sediment and water.

**Keywords:** heavy metal uptake transporters, metal efflux transporters, arsenic transporters

## 1. INTRODUCTION

Transport of metals and alkali cations across plasma membranes is essential for plant development, growth, nutrition, signal transduction, and for tolerance and detoxification of toxic metals. The response of plants to mineral deficiencies likely involves a complex regulatory cascade ultimately resulting in changes in the expression of key transporters and metal homeostasis proteins as well as by

---

© CAB International 2008. *Plant Membrane and Vacuolar Transporters* (eds P.K. Jaiwal, R.P. Singh and O.P. Dhankher)

inducing changes in their growth patterns. Metals uptake, distribution to different plant organs and cell type, storage of excess metals and re-mobilization is achieved through the metals transporters (Krämer et al., 2007). In the case of all the transition metals, the range of optimal intracellular concentrations is very narrow that requires the metal transport and it's regulation to function with high precision and specificity (Krämer et al., 2007). In recent years, a large number of cation transporter families and homologous gene family members have been characterized, largely due to the use of heterologous complementation screens in yeast (Sacchromyces cerevisiae), functional characterization in Xenopus oocytes, and availability of plant genome sequences (Mäser et al., 2001, Clemens, 2006). Further, the availability and use of a large collection of Arabidopsis thaliana T-DNA mutant lines have proved highly useful for identification and characterization of numerous metal transporters in plants. Lethal phenotypes exhibited by Arabidopsis mutants with defective metal transport emphasize the importance of metal transporter proteins in plants (Abdel-Ghany et al., 2005).

Many of these cation transporters potentially involved in metal ion uptake and homeostasis are grouped into two main categories: metal uptake transporters and metal efflux transporters. The metal uptake transporters families are: ZIP (Zinc-regulated transporter, Iron-regulated transporter Protein) family, NRAMP (Natural Resistance Associated Macrophage Protein) family, and YSL (Yellow-Stripe 1-Like) family. The metal efflux transporter families are: CDF (Cation Diffusion Facilitator) family, P1B-type subfamily of P-type ATPases superfamily, three subfamilies of ABC (ATP-binding cassette) transporter superfamily - the multidrug resistance-associated proteins (MRP), the ABC transporters of the mitochondria (ATM) and the pleiotropic drug resistance (PDR) transporters. Also there are secondary multidrug transporter families such as major facilitator superfamily (MFS), the small multidrug resistance (SMR), the resistance nodulation-cell division (RND), and multidrug and toxic compound extrusion (MATE) family. Additionally, in plants the metalloid oxyanion arsenate ($As^V$) is transported via phosphate transporters, whereas, no transporter for oxyanion arsenite ($As^{III}$) is reported from plants yet. Here we discuss all these metals transporters families in detail with the current knowledge available at this time, though readers are advised to read more recent reviews of metal transporters in other chapters of this book.

## 2. HEAVY METAL TRANSPORTERS FAMILIES

### 2.1. Metal Uptake Transporters

*2.1.1. ZIP family*

The ZIP (Zinc-regulated transporter, Iron-regulated transporter Protein) family is universally present in living organisms including bacteria, fungi, plants and humans (Guerinot, 2000). ZIP proteins have been shown to contribute to metal ions homeostasis via the transport of cations into the cytoplasm. In plants, ZIP transporters have been identified from both dicots including *Arabidopsis thaliana*, pea (*Pisum sativum*), *Thlaspi caerulescens*, *Arabidopsis halleri*, *Medicago truncatula* and monocots including rice (*Oryza sativa*). Members of the *ZIP* family transporter genes are highly regulated in response to metal deficiency (Eide *et al.*, 1996; Korshunova *et al.*, 1999; Grotz *et al.*, 1998; Wintz *et al.*, 2003) and this suggests their important role in regulating metal uptake. Some of the ZIP transporters are not metal-specific and thus transport several divalent metal ions, including Cd, Co, Mn, and Zn with varying affinity. The predicted topology of most of the ZIP proteins contains 8 transmembrane (TM) domains with a histidine-rich loop between TM domains III and IV as shown in Figure 1. The potential metal binding ability of the histidine-rich loop indicates a role in metal transport or its regulation. Site-directed mutagenesis of two histidines to alanines in the histidine-rich loop significantly reduced the activity of a human zinc transporter hZIP1 (Milon *et al.*, 2006). Moreover, the histidines in TM domains IV and V were also reported to be essential for metal transport in hZIP1 as well as in AtIRT1 (Rogers *et al.*, 2000; Colangelo and Guerinot, 2006). These residues are thought to play a role in intermolecular exchange of Zn between Zn ligands and the transporter (Milon *et al.*, 2006).

In *A. thaliana* genome, there are 15 genes encoding ZIP family transporters namely AtZIP1 to AtZIP12 and AtIRT1 to AtIRT3. Transcripts of genes encoding ZIP plant transporters are regulated by Zn or Fe deprivation (Eide *et al.*, 1996; Grotz *et al.*, 1998). Whereas, some *ZIP* family members, such as *IRT* genes, are constitutively expressed but their expression levels increase with iron deprivation (Vert *et al.*, 2001). AtZIP1, AtZIP2 and AtZIP3 were detected only in the roots of Zn deficient plants. However, AtZIP4 was induced in both shoot and root under the same conditions (Grotz *et al.*, 1998). Similar responses of AtZIP2 and At ZIP4 were also reported under Cu deficiency and both complemented a mutant yeast strain lacking high affinity Cu transporter (Wintz *et al.*, 2003). ZIP4 and IRT3 have also been found to be highly expressed in shoots of metal hyperaccumulators *T.*

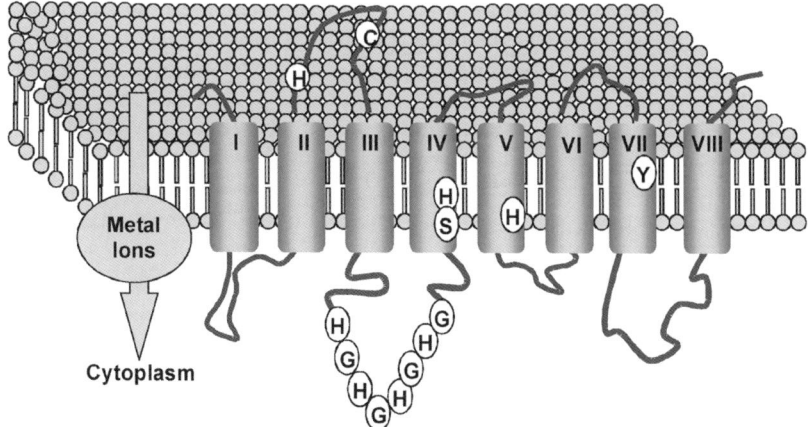

**Figure 1:** General topology of ZIP transporters indicating highly conserved residues and His-rich domain based on description provided by Eide (2006) and Rogers *et al.* (2000)

*caerulescens* and *A. halleri* (Hammond *et al.*, 2006; Talke *et al.*, 2006). Taken together, the expression pattern of these ZIP proteins and the presence of chloroplast targeting motif in ZIP4, it is presumed that the members of the ZIP family transport metals at different locations in plants. Although, the ZIP proteins are considered to have broad substrate specificity, transcript levels of AtZIP1, AtZIP3 and AtZIP4 were found to be unaffected by high and low levels of Fe in the nutrient medium. AtZIP2 and AtZIP4 are also involved in Cu transport. When expressed in mutant yeast strain lacking high affinity copper uptake ($\Delta ctrI$), both AtZIP2 and AtZIP4 restored the yeast growth and caused Cu accumulation (Wintz *et al.*, 2003; Sancenon *et al.*, 2003, 2004). Therefore, from the work of Wintz *et al.* (2003) and Grotz *et al.* (1998), it is concluded that AtZIP2 plays a role in Cu and Zn homeostasis in only roots, whereas, AtZIP4 plays a role in Cu and Zn transport in both roots and shoots.

In *A. thaliana*, IRT1 is the major transporter responsible for high-affinity Fe uptake from the soil (Vert *et al.*, 2002; Colangelo and Guerinot, 2006). IRT1 mediates the uptake of Fe (Eide *et al.*, 1996), Zn, and Mn (Korshunova *et al.*, 1999), when the protein is expressed in yeast. IRT1 also transports Cd as it inhibits the uptake of other metals and imparts increased Cd sensitivity to yeast cells (Rogers *et al.*, 2000). The transcript level of *IRT1* mRNA was induced within 24h after transfer of plants to Fe-deficient conditions, and peaked at 72 h. Whereas, protein level was detectable 48 h after transfer of plants to Fe-deficient conditions and peaking 72 h after transfer. *IRT1* mRNA and protein levels reverted to original conditions when the plants were shifted back to Fe-sufficient conditions (Connolly *et al.*, 2002). Ectopic overexpression of *IRT1* gene under *35S* promoter revealed that the IRT1 protein in the overexpressing lines was present only in the roots when Fe is limiting, even though *IRT1* mRNA was expressed constitutively in whole plants. In Fe-limiting conditions, plants that overexpressed IRT1 accumulated higher levels of Cd and Zn than wild-type plants, suggesting that IRT1 is responsible for the uptake of these metals and higher levels of IRT1 protein were also observed in these plants. These results demonstrated that IRT1 is subject to an additional level of regulation that occurs post-transcriptionally in response to Fe (Connolly *et al.*, 2002).

The mRNA levels of *TcIRT1* and *TcIRT2* genes were induced by Fe deficiency in Zn/Cd hyperaccumulator *T. caerulescens*, nevertheless, only *TcIRT1* responded to Cd exposure (Plaza *et al.*, 2007). ZNT1/TcZIP4 was shown to mediate high-affinity Zn uptake as well as low affinity Cd uptake through functional complementation in yeast (Pence *et al.*, 2000). Two ZIP genes, *TjZNT1* and *TjZNT2* from the Ni hyperaccumulator *Thlaspi japonicum* imparted resistance to yeast cells under higher concentration of Ni (Mizuno *et al.*, 2005), indicating a potential role of these genes in Ni tolerance. A recent microarray study comparing Zn accumulating and non-accumulating plants resulted in the identification of *ZIP6* as a candidate gene for a role in Zn hyperaccumulation in *A. halleri* (Becher *et al.*, 2004).

In the model legume *Medicago truncatula*, six cDNAs encoding ZIP family members (*MtZIPs*) have been cloned and are shown to complement metal uptake in mutant yeast, *zrt1/zrt2* (Lopez-Millan *et al.*, 2004). These genes were able to

Heavy metal transporters in plants

rescue the Zn, Mn and/or Fe uptake mutants, indicating a function in metal transport. A homolog of AtIRT1 in Pea, named as RIT1 was found to be able to complement yeast mutants defective in high affinity and low affinity Fe transport (Cohen et al., 2004). Radiotracer experiments showed that RIT1 is a very high-affinity Fe uptake system ($K_m$ =54–93 nM) and can also mediate a lower affinity Zn and Cd influx ($K_m$ of 4 and 100 µM, for $Zn^{2+}$ and $Cd^{2+}$, respectively) (Cohen et al., 2004). The role of ZIP family members in metal transport has also been elucidated in the model cereal crop, rice (*Oryza sativa*). A rice homolog of AtIRT1, OsIRT1 was reported to be induced in roots under Fe and Cu deficiency. In addition, when expressed in yeast cells, OsIRT1 reversed the growth defect of the Fe uptake mutant (Wintz et al., 2003). Rice OsZIP4 was highly expressed under conditions of Zn deficiency in roots and shoots and complemented a Zn-uptake-deficient yeast (*S. cerevisiae*) mutant, *zrt1zrt2*, indicating that OsZIP4 is a functional transporter of Zn. *In situ* hybridization analysis located OsZIP4 in both shoots and roots, especially in phloem cells of Zn-deficient rice (Ishimaru et al., 2005). OsZIP4-GFP fusion protein transiently expressed in onion epidermal cells was localized to the plasma membrane. These results suggested that OsZIP4 is a Zn transporter responsible for the translocation of Zn within rice plants (Ishimaru et al., 2005). Further studies are needed to elucidate the mechanism of Zn uptake and transport particularly in major food crops like rice, maize and wheat.

*2.1.2. NRAMP family*

The NRAMP family includes a novel family of related proteins whose members have been implicated in the transport of divalent metal ions. NRAMP homologs have been described in all forms of life: bacteria, fungi, mammals and plants. The founding member of this family, mammalian Nramp1 was identified in macrophage populations during a study on natural resistance to infection with intracellular parasites (Cellier et al., 1995), hence the name- Natural Resistance Associated Macrophage Protein (NRAMP). Yeast NRAMP homologs SMFs and mammalian DCT1/Nramp2 were shown to mediate the uptake of a broad range of metals (Supek et al., 1996; Chen et al., 1999). Common features of this family include 12 amphipathic transmembrane domains containing a pair of invariant histidines in TM VI and a highly conserved 20 amino acid stretch (5'-GQSSTITGTYAGQFIMGGFL-3') in the loop connecting TM VIII and TM VIV, especially in eukaryotic members. Complementation studies in yeast and *Xenopus* oocytes indicated that the divalent metal ion transport by the NRAMP family of proteins is pH dependent and has broad substrate specificity (Forbes and Gross, 2001).

Plant NRAMP family members have also been implicated in the transport of several divalent cations (Curie et al., 2000; Kaiser et al., 2003; Thomine et al., 2000). *A. thaliana* genome contains six genes encoding proteins with high homology to mammalian NRAMPs. Phylogenetic analysis divides these six sequences into 2 subfamilies: AtNRAMP1 and 6 forming the first group and AtNRAMP2 through 5 constituting the second group (Mäser et al., 2001). AtNRAMP1, 2, and 6 are located on chromosome I, AtNRAMP3 on chromosome II, AtNRAMP5 on chromosome IV, and AtNRAMP4 on chromosome V (Mäser et al., 2001). These AtNRAMPs encode multi-metal specific transporters that have been shown to transport various metals both in *in planta* and heterologous yeast expression system (Alonso et al., 1999; Curie et al., 2000; Thomine et al., 2000). When expressed in *fet3fet4* yeast mutant deficient for Mn or Fe uptake, *AtNRAMP1*, 3, and 4 complemented these yeast strains (Curie et al., 2000; Thomine et al., 2000), but *AtNRAMP2* failed to complement the mutants. In addition, expression of *AtNRAMP1*, 3, and 4 in yeast increased their Cd sensitivity and Cd accumulation (Thomine et al., 2000). Northern blot analysis revealed that *AtNRAMP1* transcript levels increased in *A. thaliana* roots between 3 to 5 days after Fe starvation. However, both *AtNRAMP1* and *AtNRAMP2* did not respond to Fe status in shoots where their constitutive expression levels remained very low. In contrast to *AtNRAMP1*, *AtNRAMP2* mRNA was highly expressed in Fe sufficient plants and slightly downregulated by Fe starvation. Overexpression of *AtNRAMP1* under *35S* promoter caused an increase in plant resistance to Fe toxicity (Curie et al., 2000).

*AtNRAMP3* and *AtNRAMP4* mRNA expression is also upregulated by Fe deficiency and is shown to mediate the remobilization of Fe from vacuolar stores (Lanquar et al., 2005; Thomine et al., 2003). Although, in normal plants, *AtNRAMP3* and *AtNRAMP4* were expressed in roots and shoots at comparable levels, under Fe starvation *AtNRAMP3* mRNA is upregulated only in roots, whereas *AtNRAMP4* transcripts were increased in both roots and shoots. Both *AtNRAMP3* and *AtNRAMP4* were able to restore the yeast mutant *smf1* that lacks the yeast NRAMP gene *SMF1* (Thomine et al., 2000). Additionally, expression of these genes complemented the yeast mutant *fet3fet4*, but the complementation was negligible at pH values above 6. This pH dependence is in concordance with the behavior of mammalian Nramp1/divalent cation transporter 1 (DCT1), which functions as a proton/metal symporter (Cellier and Gros, 2004). Apart from its possible involvement in Fe transport, *AtNRAMP3* and *AtNRAMP4* increased the Cd sensitivity of wild type yeast strains. Their role in metal transport was further confirmed *in planta* by ectopic overexpression and phenotypic characterization of respective T-DNA insertion lines (Thomine et al., 2003).

Expression of GUS under *AtNRAMP3* promoters was maximal in vascular tissues and the fluorescence from an AtNRAMP3::GFP fusion protein was localized exclusively to the vacuolar membrane in onion epidermal cells indicating its intracellular location (Thomine et al., 2003). However, the overexpression of this fusion protein in yeast showed that the protein is mainly targeted to the vacuolar membrane, but to a small extent, also to the plasma membrane. Interestingly, the overexpression of *AtNRAMP3* down-regulated the primary Fe uptake transporter, IRT1, most probably by mobilizing the metals stored in vacuole into the cytosol, thereby increasing the $Fe^{2+}$ concentration in cytosol. This presumption was further supported by the Fe dependent germination defect of *nramp3/nramp4* double mutant Arabidopsis seeds (Lanquar et al., 2005). The mutant seeds failed to germinate properly under low Fe supply but were rescued by either individual expression of AtNRAMP3 or AtNRAMP4, or by supplying high levels of Fe in nutrient medium (Lanquar et al., 2005). It appears that the *nramp3/nramp4* double mutants were not able to mobilize the Fe from the iron storage pool of vacuolar globoids and expression of any of the functionally redundant AtNRAMP3 or AtNRAMP4 could retrieve the globoid associated Fe from vacuole during germination.

Heterologous expression of NRAMP members of tomato, *LeNRAMP1* and *LeNRAMP3*, demonstrated that these genes encode functional metal transporters (Bereczky *et al.*, 2003). As the localization and expression of their GFP fusion products in yeast were dependent on Fe concentration in the growth media, both these gene products are presumed to be controlled post-transcriptionally or post-translationally in response to the Fe status of the cytoplasm. *LeNRAMP1* was primarily expressed in vascular root parenchyma and was upregulated by Fe deficiency. NRAMP proteins have also been isolated and characterized from metal hyperaccumulators, *A. halleri* and *T. japonicum*. An *AtNRAMP4* homolog, *TjNRAMP4*, has been cloned from *T. japonicum* and predicted to contribute to Ni tolerance (Mizuno *et al.*, 2005). *TjNRAMP4* expression in wild type yeast increased the Ni sensitivity and resulted in Ni accumulation but not other divalent cations of Zn, Cd, or Mn (Mizuno *et al.*, 2005).

Rice homologs of mammalian Nramp1 namely *OsNRAMP1*, *OsNRAMP2*, and *OsNRAMP3* have been characterized. In normal plant tissues, *OsNRAMP1* is a root-asociated transporter and *OsNRAMP2* is confined to leaves, whereas, *OsNRAMP3* is expressed in both shoot and root tissues (Belouchi *et al.*, 1997). Their distinct tissue-specific expression pattern suggests these transporters may have differential regulation and function in certain cellular environments or they may transport different metal cations for particular physiological functions in different parts of the plant. Mimicking their Arabidopsis homologs, *OsNRAMP1*, but not *OsNRAMP2* complemented the *fet3fet4* yeast mutant (Curie *et al.*, 2000). Soybean (*Glycine max*) divalent metal transporter1 (GmDmt1), an NRAMP homolog share significant sequence homology to the Arabidopsis NRAMP proteins (Kaiser *et al.*, 2003). *GmDmt1* mRNA was abundant in root nodules but could barely be detected in roots, leaves or stems. Interestingly, the mRNA level was the highest during the growth period associated with maximum rates of nitrogen fixation. The anti-GmDmt1 antibodies located the proteins in the peribacteroid membrane (Membrane derived from the plasma membrane of plant cell and that surrounds the nitrogen-fixing bacteroids) of the nodule. Notably, the protein was larger than its predicted size, apparently from extensive post-translational modifications, compliant with the human NRAMP proteins that appeared about 40% larger on SDS-PAGE than the predicted size from the amino acid sequence due to glycosylation. As in the case of AtNRAMP1, GmDmt1 complemented the Fe uptake mutant of yeast and in addition, it enhanced the Fe accumulation approximately four-fold over the control cells (Kaiser *et al.*, 2003). Despite these studies, more investigations are needed to elucidate their role on metal transport in plants, particularly their direct role in metal hyperaccumulator as these transporters are highly expressed in plants.

In addition to the IRT/ZIP family of Fe transporters, these above-mentioned studies on AtNRAMPs suggest that AtNRAMPs contribute to Fe homeostasis in plants (Eide *et al.*, 1996; Curie *et al.*, 2000; Thomine *et al.*, 2000; 2003). Further detailed studies regarding their cellular/subcellular expression pattern and substrate specificity are required to understand the roles of the plant NRAMP transporter gene family in metal uptake and transport in plants.

### 2.1.3. YSL Family

YSL (yellow stripe like) proteins mediate the uptake of metals that are complexed with plant-derived phytosiderophores (PS) or nicotianamine (NA) – a secondary amino acid found in plants. Nicotianamine, which occurs in all higher plants, is structurally similar to phytosiderophores and serves as a precursor in their biosynthetic pathway in grasses (Curie et al., 2001). YSL is considered to be a subfamily of the large oligopeptide transporter (OPT) family, even though it does not transport oligopeptides (Curie et al., 2001). However, the PS are amino acid derivatives and both the YSL and the remaining members of the OPT family transport their substrates towards the cytoplasm (Yen et al., 2001). This subfamily got its name from the phenotypic characteristic of the maize mutant, *yellow stripe1*, which is defective in $Fe^{3+}$-phytosiderophore uptake. Transposon-tagging of the *YS1* gene in maize allowed identification of ZmYS1 as a highly hydrophobic protein with 14 putative transmembrane-spanning domains that confers growth of the Fe uptake-defective yeast mutant *fet3fet4* on phytosiderophore-bound Fe, even in the presence of the Fe(II) chelator BPDS1 (Curie et al., 2001). The maize YS1 (ZmYS1) protein is shown to accumulate in roots and leaves of the Fe-starved plants predicted to function as a proton-coupled symporter to transport Fe-PS complexes (Roberts et al., 2004; Schaaf et al., 2004). The transcript levels were upregulated in both roots and shoots under Fe deficient conditions, but not strongly affected by Cu or Zn limitation. ZmYS1 has been shown to complement the yeast Fe uptake mutant specifically on $Fe^{3+}$-phytosiderophore media. However, it did not significantly complement the yeast Zn uptake mutant *zrt1zrt2* (Roberts et al., 2004). In contrast, it is reported to complement another yeast Zn uptake mutant *zap1* (Schaaf et al., 2004).

Electrophysiological analysis in *Xenopus* oocytes showed that ZmYS1 is a proton-coupled symporter and the rate of metal transport depends on the membrane potential (Schaaf et al., 2004). Chelated $Fe^{3+}$-induced currents through voltage clamped *ZmYS1* expressing oocytes were strongly dependent on external pH. Highest currents were obtained at pH 5 and gradually decreased by increasing the pH to 9.5. Replacing the $H^+$ source with $Na^+$ did not stimulate transport indicating that $Na^+$ could not substitute $H^+$ to drive metal-chelate uptake.

In *A. thaliana* genome, there are eight genes (*AtYSL1* through *8*) that are homologous to maize *ZmYS1*. Although, dicotyledonous plants do not depend on phytosiderophores for Fe uptake, they synthesize NA, a strong chelator. Therefore, AtYSLs are most likely transport metal-NA complexes. Recently, two Arabidopsis YSL1 members, AtYSL1 and AtYSL2, are characterized in details. Arabidopsis *ysl1* mutants exhibited increased concentration of nicotianamine in shoots (Le Jean et al., 2005). Furthermore, seeds of the *ysl1* mutant contained less Fe and NA when they were grown under a Fe-sufficient nutrient medium. The slow germination of

## Heavy metal transporters in plants

*ysl1* mutant seeds was corrected either by supplementing with higher concentration of Fe or by ectopically expressing functional *AtYSL1* in the parent mutants. *AtYSL1* is a shoot-specific gene that expressed in xylem parenchyma of leaves and the mRNA levels were increased in response to excess Fe (Le Jean *et al.*, 2005). These observations suggest the role of AtYSL1 in long distance transport of Fe and NA and their delivery to the seed.

On the contrary, AtYSL2 preferred chelated $Fe^{2+}$ over $Fe^{3+}$, as proven by expression studies in *fet3fet4* yeast mutant (DiDonato *et al.*, 2004). When grown under Fe-deficient conditions, *AtYSL2* mRNA transcripts were downregulated in both shoots and roots, whereas, mRNA transcripts were upregulated under conditions of Fe sufficiency and Fe resupply (DiDonato *et al.*, 2004; Schaaf *et al.*, 2004). *AtYSL2* mRNA transcript levels also responded to Cu and Zn. GUS expression driven by the *AtYSL2* promoter was strongest in vascular tissues of both roots and shoots, yet almost every cell type was stained positively but faintly except root cap and meristematic tissues. Taken together, the expression pattern and the localization of AtYSL2::GFP fusion protein to the lateral plasma membrane, the function of AtYSL2 appeared to be in the lateral movement of metals. Further, the T-DNA insertional mutant *ysl2* did not show any phenotypic difference with wild type indicating functional redundancy in the Arabidopsis YSL family. However, the double mutant *ysl1ysl3* showed visible symptoms of Fe deficiency like interveinal chlorosis (Waters *et al.*, 2006). Double mutants displayed lower Fe concentrations in leaves with elevated levels of Mn, Zn and Cu. Seeds contained lower concentrations of Fe, Zn and Cu than wild type and the mutants had reduced fertility due to defective anther and embryo development. In addition, the disruption of both AtYSL1 and AtYSL3 affected the mobilization of metals from leaves during their senescence (Waters *et al.*, 2006).

Rice genome has 19 gene sequences that exhibit significant similarity to the maize Fe transporter *ZmYS1* (Koike *et al.*, 2004). *OsYSL2*, one of the 19 putative *YS1* like genes was induced in leaves by Fe deficiency. Strangely, this gene was not transcribed to detectable level in roots either in Fe-deficient or in Fe-sufficient plants. In onion epidermal cells, the OsYSL2::GFP fusion protein was localized to plasma membrane. Under the *OsYSL2* promoter, GUS reporter gene was expressed in phloem cells in shoots and companion cells in roots. The phloem-specific expression of the *OsYSL2* promoter suggests that OsYSL2 is involved in the phloem transport of Fe. Electrophysiological measurements using *Xenopus* oocytes demonstrated that OsYSL2 transported NA-complexed $Fe^{2+}$ and $Mn^{2+}$ but not the PS-complexed $Fe^{3+}$ (Koike *et al.*, 2004). Recently, YSL transporters have also been identified and cloned from Ni hyperaccumulator *T. caerulescens*. *TcYSL3*, *TcYSL5*, and *TcYSL7* genes were found to be highly expressed in both roots and shoots and further, *TcYSL3* and *TcYSL7* expression was localized around the vascular tissues in roots (Gendre *et al.*, 2006). Interestingly, a recent study showed that TcYSL3 functions as a NA-chelated Ni influx transporter (Gendre *et al.*, 2007). So far, YSL transporters has been shown to play a role in Fe homeostasis, however, their role in homeostasis and transport of Zn, Cd and other divalent metal cations in metal hyperaccumulators is unclear.

## 2.2. Metal Efflux Transporters

### 2.2.1. *CDF family*

CDF (Cation diffusion facilitator) transporter family members are ubiquitous, spanning all three kingdoms of life: archaea, eubacteria and eukaryotes. It has been shown that the substrate for all known CDF transporters is divalent metal cations with ionic radii of 72 ($Zn^{2+}$) to 97 ($Cd^{2+}$) pm (Nies, 2003). Most CDFs are powered by proton or $K^+$ antiport and catalyze the efflux of transition metal cations, including $Zn^{2+}$, $Co^{2+}$, $Fe^{2+}$, $Cd^{2+}$, $Ni^{2+}$ and $Mn^{2+}$ from the cytoplasm to the outside of the cell or into subcellular compartments (Delhaize *et al.*, 2003; Hall and Williams, 2003; Persans *et al.*, 2001; Montanini *et al* 2007). With a few exceptions, all the CDF family members have 3 key features; a signature N-terminal domain, a cation efflux domain and six predicted membrane spanning domains (Kim *et al.*, 2004; Eide, 2006) as shown in figure 2. The signature sequence spans TM II and III with fully conserved glycine (TM III), aspartate (TM II, V and VI) and histidine (TM II and V) residues. Most conserved regions are the amphipathic TM I, II, V, and VI, which are suggested to be involved in metal transfer (Haney *et al.*, 2005). Substitutions of the conserved residues, when substituted individually, resulted in loss of function in PtdMTP1, a CDF member from hybrid Poplar (Blaudez *et al.*, 2003). Most CDF family transporters also contain a histidine-rich region between TM IV and V and could function as potential metal ($Zn^{2+}$, $Co^{2+}$, and/or $Cd^{2+}$) binding domains. Plant CDF members are usually called Metal Tolerance Proteins (MTPs), while vertebrate

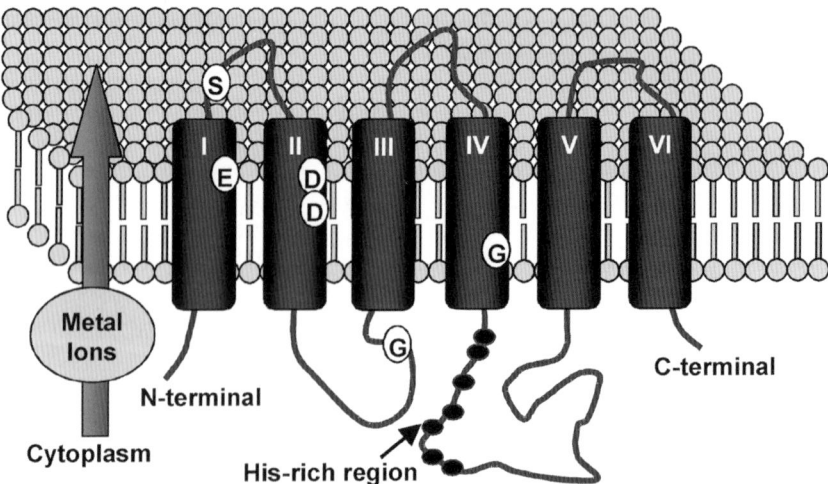

**Figure 2:** General topology of CDF transporters indicating highly conserved residues (signature sequence spans from serine in loop between TMI and TMII and aspartate in loop between TMII and TMIII), and His-rich region as represented by filled circles between TMIV and TMV as described by Eide (2006) and Haney *et al.* (2005)

members are named Zinc Transporter (ZnT) or Solute carrier family 30 (SLC30). Interestingly, a vacuolar Mn transporter ShMTP1 (ShCDF8) isolated from the Mn tolerant plant *Stylosanthes hamata* is a member of CDF family of heavy metal transporters, but has a predicted IV or V TM topology, and does not contain the histidine-rich domain. This transporter increases $Mn^{2+}$ homeostasis and tolerance when expressed in yeast or Arabidopsis plants (Delhaize *et al.*, 2003).

Considering both phylogenetic comparisons and the substrate specificities, the CDF family members can be divided into 3 major groups viz. Zn-, Fe/Zn- and Mn-CDF. The majority of the known plant CDFs belong to the first and third groups. *A. thaliana* genome contains 12 CDF members (Blaudez *et al.*, 2003). The first CDF gene characterized in *A.thaliana* was *AtMTP1*. AtMTP1::GFP fusion protein, when overexpressed in plants was shown to be localized to the vacuolar membranes of leaf and root cells. Its role in Zn transport has been proved by overexpression and knockdown *in planta*. The *mtp1* knockdown plants accumulated less Zn in various tissues and the overexpression of AtMTP1 conferred Zn tolerance to plants. MTP1 has been identified using both transcriptomic and proteomic approaches in metal hyperaccumulator *A. halleri* and the protein was able to complement a Zn-sensitive yeast mutant that lacks the vacuolar Zn resistance conferring (*ZRC1*) and cobalt transporter1 (*COT1*) genes (Drager *et al.*, 2004). *A. halleri* contains three independently segregating MTP1 genes, two of which co-segregate with Zn tolerance in a backcross between hyperaccumulator *A. halleri* and non-accumulator *A. lyrata*. While the AhMTP1::GFP fusion protein localized to vacuolar membrane as observed in the case of AtMTP1, its homolog TgMTP1, from a Ni/Zn hyperaccumulator *Thlapsi goesingense* is found to be localized in plasma membrane when expressed in *A. thaliana*. In *T. goesingense* there are several allelic variants of TgMTP1 and all of them confer resistance to *zrc1* and *cot1* mutant strains of yeast and the expression of TgMTP1 leads to increased efflux of Zn from cells, thereby lowering the Zn concentration in cells (Kim *et al.*, 2004).

The ShMTP8 from *Stylosanthes hamata*, and its paralogs ShMTP2, ShMTP3, and ShMTP4 along with AtMTP11 are members of the Mn-CDF group. In accord with the substrate specificity of this group, MTP8 from *A. halleri* is not Zn-responsive but upregulated in response to Cd and Cu in roots and to Cu in shoots. Both the Arabidopsis and Poplar MTP11 cation diffusion facilitators were able to reinstate the Mn tolerance of a manganese hypersensitive yeast mutant (Peiter *et al.*, 2007). Furthermore, yeast microsomes expressing AtMTP11 were found to have enhanced Mn uptake. These findings were corroborated by the Mn-hypersensitive *mtp11* mutants and hypertolerant MTP11 overexpressing *A. thaiana* plants. As opposed to the general predictions of localization of CDF members in vacuolar or plasma membrane, the MTP11–EYFP fusions localized to a punctate endomembrane compartment that largely coincides with the distribution of the trans-golgi marker sialyl transferase. In reference to this unusual localization, a novel concept of golgi-based Mn accumulation through vesicular trafficking and exocytosis was proposed. The secretary pathway-mediated Mn detoxification is endorsed by the elevated Mn levels in the *atmtp11* mutants (Peiter *et al.*, 2007). Functional characterization of these CDF homologs from agricultural food crops remains to be studied.

### 2.2.2. P1-B type ATPases

Members of the P-type ATPases superfamily are ubiquitous and translocate a diverse set of ions including $H^+$, $Na^+/K^+$, $H^+/K^+$, $Ca^{2+}$ and heavy metals cations. These are active transporters energized from ATP hydrolysis and hence the name "P-type" comes from the phosphorylated intermediate formed by the members of this superfamily during its catalytic cycle. Within this mechanism, the catalytic phosphorylation of the aspartyl group in the consensus sequence 'DKTGT' is the unifying characteristic of all P-type ATPases (Argüello, 2003). The superfamily is divided into five subfamilies and the P1-B type subfamily comprises the heavy metal transporting P-type ATPases from different forms of life. As shown in figure 3, most P1B-ATPases appear to have eight transmembrane domains (TMs), the signature sequence (CPC, CPH or CPS) present in their sixth transmembrane helix (H6) and one or more metal binding domains in the cytoplasmic N terminal region (N-MBD) (Argüello, 2003).

In spite of their structural differences, all N- and C- terminal MBDs appear to control the enzyme turnover rate rather than affecting metal binding to transmembrane transport sites. Mutagenesis of the conserved residues Cys in H6, Asn and Tyr in H7 and Met and Ser in H8 of an archaeal Cu(+)-ATPase has revealed that their side-chains likely coordinate the metals during transport and constitute a central unique component of these enzymes (Argüello *et al.*, 2007). P1B-ATPases were initially named as CPx-ATPases based on their characteristic sequence CP[C/H/S] in TM H6. Most of the P1B-ATPases have N-terminal metal binding domains (N-MBDs) which frequently contain a CXXC metal-binding motif.

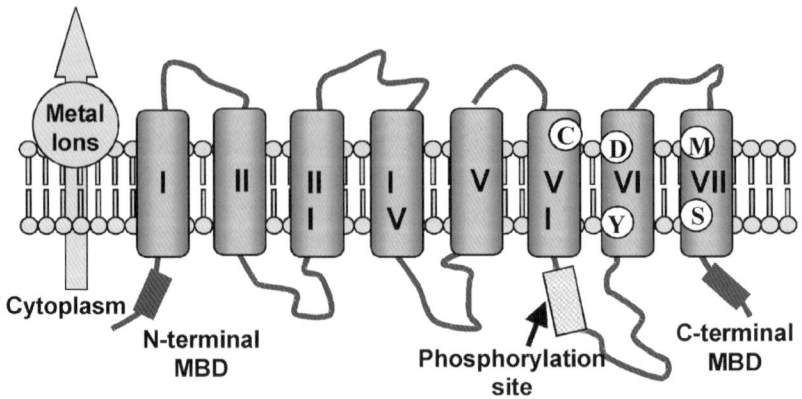

**Figure 3:** Predicted topology of P-1B type ATPases indicating general features as described by Argüello (2003) and Argüello *et al.* (2007)

Heavy metal transporters in plants

P1B-ATPases are divided into six subgroups based on their metal specificity and the conserved sequences in putative substrate binding TMs H6, H7 and H8. Members of the first subgroup (1B-1) possess an invariant signature sequence $CPC(X_6)P$ in H6 and N-terminal MBDs are either two, one or even absent. As suggested by enzymatic and transport assays, they mediate the efflux of $Cu^+$ from cytoplasm. Apart from $Cu^+$, these proteins are also activated by $Ag^+$ ions but not by other divalent ions. The second subgroup (1B-2) has a conserved signature sequence $CPC(X_4)(S/T)XP$ in transmembrane helix (H6) and contain one or two N-MBDs. The conserved residues in TM helices 7 and 8 markedly differ from those of the first subgroup and the members, referred to as Zn/Cd/Pb ATPases, generally transport divalent metals. *A. thaliana* proteins HMA2, HMA3 and HMA4 are members of this subgroup. Interestingly, some proteins in 1B-2 have (HX)n repeats similar to that of the ZIP and CDF families of transporters. The third subgroup (1B-3) has the conserved motif 'CPH' in H6 and their N-MBDs are rich in histidine. These proteins are activated by $Cu^{2+}$, $Cu^+$ and $Ag^+$ ions. However, no eukaryotic protein has been found with these characteristics. The smallest P-type ATPases comes in 1B-4 with only six or seven transmembrane domains and a conserved motif 'SPC' in H6. Most proteins in this subgroup do not have the N-MBDs except a few such as Arabidopsis HMA1. No plant proteins have been identified in subgroups 5 and 6.

Many members of the first subgroup are identified in the *A. thaliana* genome, while the leaves of *A. thaliana* mutant defective in *PAA1* (P-type ATPase of Arabidopsis)/*HMA6* (Heavy metal ATPase) gene retained normal levels of Cu and its chloroplast contained very low concentration of Cu (Shikanai *et al.*, 2003). Moreover, the electron transport defect of this high chlorophyll fluorescence phenotype was evident on the medium containing 1µM Cu, but it was suppressed by the addition of 10µM Cu. These observations suggested a possible role of PAA1 in Cu transport across the chloroplast envelope. PAA2/HMA8, another P-type ATPase with sequence similarity to PAA1 was localized to thylakoids as revealed by its fusion product with GFP (Abdel-Ghany *et al.*, 2005). In addition, Cu delivery to the stroma of plastid was inhibited only in *paa1* mutant. These observations lead to a conclusion that PAA1 and PAA2 function sequentially in Cu transport over the chloroplast peripheral envelope and thylakoid membrane, respectively. Furthermore, the *paa1paa2* double mutant was seedling lethal indicating its role in Cu transport is indispensable for photosynthesis.

*AtHMA5* is primarily expressed in roots and the transcript levels are upregulated by Cu in whole plants. T-DNA insertion lines silenced for *HMA5-1* and *HMA5-2* expression were hypersensitive to Cu but remained unaffected by Fe, Zn and Cd. Copper treated *hma5* mutants exhibited visible symptoms of toxicity such as complete arrest of root growth and emergence of lateral roots from the crown. The mutant lines accumulate higher amounts of Cu in roots as compared to that of wild type plants. Yeast two hybrid assay revealed that the Cu is delivered to N-MBDs of HMA5 by ATX1, an Arabidopsis metallochaperonin (Andres-Colas *et al.*, 2006). Responsive to antagonist1 (RAN1)/AtHMA7 encodes a Cu transporting P-ATPase. Expression of RAN1 complemented *ccc2p* yeast mutant that lacks the Cu transporting

P-ATPase Ccc2p. The phenotype of the *Arabidopsis ran1* mutants was partially suppressed by additions of excess Cu ions to the plant growth medium, but not by Fe (Hirayama *et al.*, 1999).

AtHMA2, AtHMA3 and AtHMA4 are members of subgroup 1B-2. In fact, plants are the only eukaryotes where these subgroup members are identified. AtHMA2 was functionally characterized by heterologous expression in yeast. It interacts with Zn and Cd with higher affinity and is also activated by other divalent metal ions such as $Pb^{2+}$, $Ni^{2+}$ and $Co^{2+}$ (Eren and Argüello, 2004). Similar to AtHMA5, AtHMA2 showed preference to complexed metal ions as its activity was dependent on concentration of Cys in the assay medium. Transport assays using the yeast membrane preparations confirmed that AtHMA2 transports metals outside the cytoplasm. In accord, the *hma2* mutant plants accumulated higher amounts of Zn and Cd as compared to the wild type (Eren and Argüello, 2004). The wild type strain of yeast when transformed with TcHMA4 from *Thlapsi carulescens* showed significant increase in growth at higher concentrations of Cd. However, the TcHMA4-transformed yeast cells accumulated 70% less Cd to that of wild type strains suggesting that it operates at the yeast plasma membrane to pump metals out of the cell. TcHMA4 is primarily a root associated metal transporter and it is induced by both low and high Zn concentrations in the medium. However, unlike the downregulated AtHMA4, TcHMA4 is upregulated by Cd exposure. Similar to AtHMA2 and AtHMA4, TcHMA4 is believed to mediate metal efflux from xylem parenchyma to Xylem vessels (Papoyan and Kochian, 2004).

### 2.2.3. *Multidrug transporters*

Multidrug transporters are found in almost all organisms and mediate the extrusion of a variety of potentially cytotoxic compounds. In general, they are divided into 5 superfamilies: the ATP binding cassette (ABC) superfamily, the major facilitator superfamily (MFS), the small multidrug resistance (SMR) family, the resistance nodulation-cell division (RND) family, and the multidrug and toxin extrusion (MATE) family (Eckardt, 2001). These proteins are able to transport different kinds of structurally unrelated compounds and the kinetic studies indicated the presence of multiple substrate binding sites (Putman *et al.*, 2000).

### 2.2.3.1. ABC transporters

The ABC-transporter superfamily is one of the largest protein family whose members are found in all organisms including bacteria, fungi, plants, and animals (Henikoff *et al.*, 1997). *A. thaliana* genome contains 130 ORFs that encode ABC transporters but the function of most of them is still unknown (Sánchez-Fernandez *et al.*, 2001). Beside their role in normal developmental processes in plants, members of the ABC transporters family have been reported to transport heavy metals (Martinoia *et al.*, 2002; Sánchez-Fernandez *et al.*, 2001). These proteins are characterized by ABC signature sequences in the nucleotide binding folds (NBF) and integral membrane spanning domains (MSD). In plants, animals and yeast, these ABC-transporters can be subdivided into three groups based on their structural properties: A) Full size ABC transporters containing at least two NBFs and MSDs. B) Half size ABC transporters containing one each NBF and MSD. C) Soluble ABC transporters without any MSDs. Their energy source is either Mg-ATP or Mg-GTP and this reaction is typically inhibited by vanadate at micromolar concentrations. There are 3 subfamilies of ABC-transporter superfamily reported to be involved in metal transport. These are the multidrug resistance associated proteins (MRPs), the ABC transporters of mitochondria (ATMs) and the pleiotropic drug resistance transporters (PDRs).

The MRPs are the first class of ABC transporters from plants that have been characterized at both molecular and biochemical levels (Rea, 1999). MRPs are considered to be involved in active transport of glutathionated compounds and other bulky amphipathic anions, thereby participating in detoxification of herbicides and other xenobiotics, the alleviation of oxidative damage, storage of endotoxins, sequestration of heavy metals and the vacuolar storage of secondary metabolites. Yeast cadmium factor 1 (ScYCF1) is an ABC-transporter that has been shown to provide $Cd^{2+}$ resistance by pumping glutathione-$Cd^{2+}$ complexes in the vacuole (Szczypka et al., 1994; Li et al., 1997). Transport studies with yeast membrane preparations containing plant orthologs of ScYCF1, AtMRP1 demonstrated its role in the transport of glutathione conjugates. However, unlike the structurally similar ScYCF1, AtMRP1 was not competent in GSH dependent transport of $Cd^{2+}$ (Lu et al., 1997). Further screening of 15 putative sequences coding for AtMRPs showed that out of 14 mRNA sequences detected in plants, four AtMRPs (AtMRP3, 6, 7, and 14) were upregulated upon Cd treatment (Bovet et al., 2003). AtMRP3 mRNA level was higher in shoots of 7-day old seedlings as compared to that of four week old plants. Interestingly, $Cd^{2+}$ contents in plant tissue were correlated well with the age-dependent expression pattern of AtMRP3 implicating a possible role of AtMRP3 in $Cd^{2+}$ transport (Bovet et al., 2003). Further, AtMRP3 has been shown to partially complement the Cd sensitivity in *ycf1* mutant (Tommasini et al., 1998). For further details on Plant MRPs and their roles, readers are suggested to read a comprehensive review by Rea (1999) and other more recent reviews.

The plant members of the ATM subfamily of ABC transporters in Arabidopsis are termed as half-size transporters, which comprise one transmembrane and one ATP-binding domain. All these proteins contain N-terminal mitochondrial target sequences and are thus most likely to be mitochondrial protein. AtATM1, in *A. thaliana* transports Fe/S cluster precursor from mitochondria to the cytosol (Kushnir et al., 2001). Further, *AtATM3* is upregulated in roots of plants when treated with Cd or Pb (Kim et al., 2006a). *AtATM3* overexpressing plants were comparatively more resistant to Cd whereas the *atatm3* mutant plants were Cd sensitive. The ectopic expression of *AtATM3* in the *atatm3* mutant rescued the Cd sensitivity (Kim et al., 2006a). Moreover, there was a negative correlation between the mRNA levels of *AtATM3* and GSH, suggesting that AtATM3 may mediate transport of glutathione conjugated $Cd^{2+}$ across the mitochondrial membrane (Kim et al., 2006a,b). Recently, it has been shown that proteins with close similarity with AtATM3 transport Cd-conjugates (Hanikenne et al., 2005) and its close homologs in *C. elegans*, CeHMT1, are required for Cd tolerance (Vatamaniuk et al., 2005).

An Arabidopsis homolog of pleiotropic drug resistance protein (PDR), AtPDR12, responds to Pb treatment (Lee *et al.*, 2005). It is upregulated in both shoots and roots of Cd treated Arabidopsis plants indicating its possible function in Pb detoxification. It was further confirmed by the phenotypic screening of its T-DNA insertional mutants and AtPDR12 overexpressing plants. The inverse relation of *AtPDR12* mRNA level with the Pb contents in plants as well as its plasma membrane localization suggests that AtPDR12 functions as a pump to extrude the Pb from cytoplasm (Lee *et al.*, 2005). Furthermore, the transport activity of AtPDR12 was found to be independent of glutathione as the phenotypic difference between the mutant and AtPDR12 overexpressing plants was not affected by the glutathione inhibitor, buthionine sulfoximine (BSO) (Lee *et al.*, 2005).

2.2.3.2. Secondary multidrug transporters

The remaining multidrug transporter families (MFS, SMR, and RND) are called secondary multidrug transporters since they are not energized from direct hydrolysis of ATP but by the existing transmembrane electrochemical gradient of proton or $Na^+$ ions. The SMR and RND transporter families are found only in bacteria (Putnam *et al.*, 2000), whereas, the MFS family consists of membrane proteins found in both prokaryotes and eukaryotes. These transporters contain 12-14 transmembrane domains and are involved in transport of carbohydrates, metabolic intermediates, phosphate esters, antibiotics as well as ligand-complexed heavy metals (Putman *et al.*, 2000). The Arabidopsis genome contains a MFS member designated as Zn-induced facilitator1 (*AtZIF1*) that encodes a membrane transporter. *AtZIF1* mutants showed increased accumulation of Zn in shoots and were more sensitive to Cd but less sensitive to Ni. *AtZIF1* mRNA was highly induced by Zn and to a lesser extent by Mn. GUS expression under *AtZIF1* promoter was strong throughout young seedlings and in the vasculature of mature plants. Its fusion product with GFP was localized to tonoplast suggesting its involvement in vacuolar sequestration of a Zn ligand or Zn-ligand complexes (Haydon and Cobbett, 2007).

2.2.3.3. MATE Protein family

Multidrug and toxic compound extrusion (MATE) proteins comprise the most recently designated family of multidrug transporter proteins. Well characterized prokaryotic MATE transporters mediate the efflux of cationic drugs through proton or $Na^+$ exchange (Omote *et al.*, 2006). *A. thaliana* genome contains 56 ORFs for MATE-type proteins (Li *et al.*, 2002). MATE proteins in Arabidopsis genome can be divided into 5 subfamilies based on phylogenetic analysis. Hydropathy analysis showed all the members encode proteins with 12 TM domains except the third subfamily which contain 8-13 TM domains. Many members of this family are present in tandem arrangements in chromosomes. Genes within a tandem unit show higher homology indicating gene duplication as the most probable cause of their diversion (Li *et al.*, 2002). Functional screening of the *A. thaliana* cDNA library revealed a MATE transporter *AtDTX1* that transports alkaloids, antibiotics and other toxic compounds. Moreover, AtDTX1 was also capable of detoxifying Cd but not other metals such as Cu, Mn, Zn and Al. When the *KAM3* mutant bacterial cells that are defective in *AcrAB* genes failed to grow in the medium containing 10 μM or higher concentration of Cd, these mutant cells transformed with AtDTX1 grown in the presence of 100 μM Cd attained the density of control (without Cd) cells at the end of 24h cultures (Li *et al*, 2002). Further, AtDTX1 is localized on plasma membrane in plant cells thereby mediating the efflux of toxic cations from the cytoplasm. More detailed studies are needed to explore the function of MATE transporters in plants.

## 3. OTHER FAMILIES

Apart from the above described metal transporter families, there are some other metal transporters in plants that do not fall within these transporter families. There are five members of the COPT (copper transporting proteins) family that are putative high affinity Cu transporters in *A. thaliana* genome namely COPT1 to COPT5. COPT1 has been characterized in detail and contains an N-terminal methionine and histidine-rich Cu-binding domain of bacterial copper-binding proteins. AtCOPT1 has successfully complemented the yeast mutant, *ctr1-3*, defective in high affinity Cu uptake (Kampfenkel *et al*., 1995). In Arabidopsis, the *AtCOPT1* is expressed in embryo, trichomes, stomata, pollen and root tips. Knockdown of its expression resulted in anomalies like elongated roots and defective pollen development, which were reversed by Cu addition (Kampfenkel *et al*., 1995).

The members of the IREG family are suggested to be involved in long-distance transport of nutrient metals in plants. In *A. thaliana* genome, there are three IREG proteins (AtIREG1 to 3). These transporters are homologs of mammalian IREG1 protein (Ferroportin1), which is shown to mediate the transport of Fe from the basolateral surface of the enterocytes to the blood stream (McKie *et al*., 2000). Recently, *AtIREG2* has been shown to be co-regulated with *AtIRT1*, the principal Fe transporter in *A. thaliana*. However, complementation studies revealed that AtIREG2 is not involved in Fe transport, instead it complemented the Ni sensitive *cot1* mutant of yeast conferring resistance in high Ni concentration. It localizes to tonoplast in suspension cells and to root cells in the intact plant. Its substrate specificity for Ni is confirmed by both forward and reverse genetic approaches *in planta* especially under Fe deficient conditions (Schaaf *et al*., 2006). An IREG1 homolog has also been identified in the genome of a unicellular red alga, *Cyanidioschyzon merolae* but lacking in *Chlamydomonas reinhardtii* and *S. cerevisiae* genomes (Hanikenne *et al*., 2005). Further, it is hypothesized that the presence of an IREG1-like protein may help in the environmental adaptation of red algae *C. merolae* to cope with high Fe availability of the acidic sulphur-rich hot springs (Hanikenne *et al*., 2005). Song *et al*. (2004) reported that a small Cys-rich membrane protein named as AtPCR1 (plant cadmium resistance1) conferred resistance to yeast cells against high concentrations of cadmium. Its overexpression increased the Cd tolerance by decreasing the Cd content in plants.

## 4. ARSENIC TRANSPORTERS

Like heavy metals, metalloids arsenic (As) is also extremely toxic to plants and other living organisms. The mechanisms of As detoxification and transport have been well characterized in bacteria and yeast but these mechanisms remain largely unknown in plants. A common mechanism by which these microorganisms achieve tolerance to As is by the reduction of arsenate (As$^V$) to arsenite (As$^{III}$), and then the exclusion of toxic oxyanions As$^{III}$ from the cell by inducible and selective transporters (Rosen, 2002; Mukhopadhyay and Rosen, 2002). In bacteria, As$^V$ is transported into the cells across the plasma membrane via phosphate transporters. The As$^V$ is further reduced to As$^{III}$ by arsenate reductase, ArsC (Chen et al., 1986) and the latter is subsequently transported out of the cell by an As$^{III}$ efflux pump, ArsAB (Rosen, 2002), and thus conferring As resistance. Similarly, in yeast As$^V$ is transported into the cells across the plasma membrane via phosphate transporters, where As$^V$ is further reduced to As$^{III}$ by arsenate reductase, ACR2. The As$^{III}$ is either effluxed outside by As$^{III}$ efflux transporter ACR3 or conjugation of As$^{III}$ to glutathione (GSH) and subsequent transport of As$^{III}$-glutathione complexes, As(GS)$_3$, into the vacuole by the ATP-binding cassette (ABC) transporter, YCF1p (Rosen, 2002; Mukhopadhyay and Rosen, 2002). Recently, a plant arsenate reductase (AtACR2/ Ath;CDC25) was isolated and characterized from *A. thaliana* (Dhankher et al., 2006) and from As hyperaccumulator fern, *Pteris vitatta* (PvACR2) (Ellis et al., 2006). RNAi knockdown of AtACR2 in Arabidopsis translocated/transported 15-fold more arsenic from roots to shoots tissues (Dhankher et al., 2006), though the transport mechanism is not known yet. Further, no functional orthologs of As$^{III}$ tranporters have yet been identified in plants.

In bacteria, yeast, and mammals, As$^{III}$ is also transported into cells by aquaglyceroporins (Sanders et al., 1997; Liu et al., 2002; Wysocki et al., 2001). Aquaglyceroporins are the members of the aquaporin or major intrinsic proteins (MIP) superfamily and transport a variety of small neutral solute polyols such as glycerol and urea (Agre, 2007; Liu et al., 2007). Members of the aquaglyceroporin family from bacteria (GlpF), yeast (Fps1p), *Leishmania major* (LmAQP1), and mammals (AQP9) are As$^{III}$ channel proteins (Liu et al., 2002; Sanders et al., 1997; Wysocki et al., 2001). At neutral pH, the predominant species of As in solution is As$^{III}$ (As(OH)$_3$). Thus, physiologically, As$^{III}$ appears to be a polyhydroxylated molecule; therefore, polyol transporters would be reasonable candidates for As$^{III}$ uptake systems. Given the ubiquity of aquaglyceroporins in plants and their ability to transport As$^{III}$ into microorganisms, it is reasonable to speculate that As$^{III}$ transport in plants may occur via aquaporin homologs, however, there is no direct evidence available yet.

In plants, the mechanism of As$^{III}$ uptake and further translocation of As$^{III}$ from roots to shoots remain to be elucidated. There is published evidence that As$^V$ (AsO$_4^{-3}$) and phosphate (PO$_4^{-3}$) are taken up by the same membrane transporters in plant roots (Meharg and Macnair, 1992; Wang et al., 2002). However, the mechanism of As$^{III}$ transport in plants is not known yet and to date no As$^{III}$ transporter has been isolated from plants. There are a few reports indicating that plants actively uptake $As^{III}$. For example, in paddy fields under aquatic conditions, it has been shown that $As^V$ is reduced to $As^{III}$ and rice plants actively take up $As^{III}$ (Abedin et al., 2002). In another study, we grew Arabidopsis plants on $As^{III}$ media and noticed that plants accumulated several-fold more total As in shoot tissues than the total amount of $As^{III}$ in the growth media (Dhankher, 2005; Dhankher et al., unpublished). The preliminary studies in *Brassica juncea* (Pickering et al., 2000) and sunflower (Raab et al., 2005) showed the presence of both $As^V$ and $As^{III}$ in xylem sap, which suggests that plants may have $As^{III}$ tranporters in roots that transpocate $As^{III}$ from roots to shoots. These reports indicate that $As^{III}$ transporters exist in plants, and facilitate $As^{III}$ uptake.

## 5. CONCLUSIONS AND FUTURE PROSPECTS

In the last few years, there has been a significant progress in understanding the uptake, transport and homeostasis of metals in model systems including Arabidopsis. However, with few exceptions, knowledge about the metal transporters in important agricultural food crops including rice, soybean, corn and wheat is lacking. A complete understanding of the mechanisms of heavy metal uptake, transport and detoxification in plants will help to design future strategies for increased efficiency of nutrient metal uptake, efficient phytoremediation of toxic metals from the environment and conversely, to develop strategies for decreased toxic metal uptake in crop plants. This knowledge about the mechanism of heavy metals will not only ensure the development of strategies for decreased toxic metal uptake in crop plants, but enhanced resistance to toxic metal will also result in increased crop yield and biomass. Additionally, enhancing heavy metal tolerance in non-food bioenergy crops such as switchgrass, yellow poplar, willows, industrial rapeseed (*Crambe abyssinica*), and jatropa, will be highly desirable to increase lignocellulosic biomass yield, and further increase acreage by cultivating these crops on heavy metal-contaminated soils. Also, these strategies for developing crop plants with reduced uptake potential of toxic metals will enable the use of inner city land marginally contaminated with low and moderate levels of arsenic and other toxic metals as community gardens. Therefore, further progress on the heavy metals uptake and transport in plants will have a significant impact on human health enhancement by preventing the entry of toxic pollutants into the food chain and by remediation of metal contamination from contaminated soils and water.

## 6. LITERATURE CITED

Abdel-Ghany, S.E., Muller-Moule, P., Niyogi, K.K., Pilon, M. and Shikanai, T. (2005) Two P-type ATPases are required for copper delivery in *Arabidopsis thaliana* chloroplasts. *Plant Cell,* **17**: 1233–1251.

Abedin, M.J., Cresser, M.S., Meharg, A.A., Feldman, J. and Cotter-Howells, J. (2002) Arsenic accumulation and metabolism in rice (*Oryza sativa* L.). *Environ. Sci. Technol.,* **36**: 962-968.

Agre, P. (2007) The aquaporin water channels. *Roc. Am. Thorac. Soc.,* **3**: 5-13.

Alonso, J.M., Hirayama, T., Roman, G., Nourizadeh, S. and Ecker, J.R. (1999) EIN2, a bifunctional transducer of ethylene and stress responses in Arabidopsis. *Science,* **284**: 2148-2152.

Andres-Colas, N., Sancenon, V., Rodriguez-Navarro, S., Mayo, S., Thiele, D.J., Ecker, J.R., Puig, S. and Penarrubia, L. (2006) The Arabidopsis heavy metal P-type ATPase HMA5 interacts with metallochaperones and functions in copper detoxification of roots. *Plant J.,* **45**: 225-236.

Argüello, J.M. (2003) Identification of ion selectivity determinants in heavy metal transport P1B-type ATPases. *J. Membr. Biol.,* **195**: 93-108.

Argüello, J.M., Eren, E. and González-Guerrero, M. (2007) The structure and function of heavy metal transport P(1B)-ATPases. *Biometals.,* **20**: 233-248.

Becher, M., Talke, I.N., Krall, L. and Krämer, U. (2004) Cross-species microarray transcript profiling reveals high constitutive expression of metal homeostasis genes in shoots of the zinc hyperaccumulator *Arabidopsis halleri*. *Plant J.,* **37**: 251-268.

Belouchi, A., Kwan, T. and Gros, P. (1997) Cloning and characterization of the OsNramp family from *Oryza sativa*, a new family of membrane proteins possibly implicated in the transport of metal ions. *Plant Mol. Biol.,* **33**: 1085-1092.

Bereczky, Z., Wang, H.Y., Schubert, V., Ganal, M. and Bauer P. (2003) Differential regulation of *nramp* and *irt* metal transporter genes in wild type and iron uptake mutants of tomato, *J. Biol. Chem.,* **278**: 24697-24704.

Blaudez, D., Kohler, A., Martin, F., Sanders, D. and Chalot, M. (2003) Poplar metal tolerance protein 1 (MTP1) confers zinc tolerance and is an oligomeric vacuolar zinc transporter with an essential leucine zipper motif. *Plant Cell,* **15**: 2911-2928.

Bovet, L., Eggmann, T., Meylan-Bettex, M., Polier, J., Kammer, P., Marin, E., Feller, U. and Martinoia, E. (2003) Transcript levels of AtMRPs after cadmium treatment: induction of AtMRP3. *Plant Cell Environ.,* **26**: 371-381.

Cellier, M. and Gros, P. (2004) *The Nramp Family*, Kluwer Academic, New York.

Cellier, M., Prive, G., Belouchi, A., Kwan, T., Rodrigues, V., Chia, W. and Gros, P. (1995) Nramp defines a family of membrane proteins. *Proc. Natl. Acad. Sci. USA,* **92**: 10089-10093.

Chen, C.-M., Misra, S., Silver, S. and Rosen, B.P. (1986) Nucleotide sequence of the structural genes for an anion pump: the plasmid-encoded arsenical resistance operon. *J. Boil. Chem.,* **261**: 15030-15038.

Chen, X.Z., Peng, J.B., Cohen, A., Nelson, H., Nelson, N. and Hediger, M.A. (1999) Yeast SMF1 mediates $H^+$-coupled iron uptake with concomitant uncoupled cation currents. *J. Biol. Chem.,* **274**: 35089-35094.

Clemens, S. (2006) Toxic metal accumulation, response to exposure and mechanisms of tolerance in plants. *Biochimie,* **88**: 1707-1719.

Cohen, C.K., Garvin, D.F. and Kochian, L.V. (2004) Kinetic properties of a micronutrient transporter from *Pisum sativum* indicate a primary function in Fe uptake from the soil. *Planta,* **218**: 784-792.

Colangelo, E.P. and Guerinot, M.L. (2006) Put the metal to the petal: metal uptake and transport throughout plants. *Curr. Opin. Plant Biol.,* **9**: 322-330.

Connolly, E.L., Fett, J.P. and Guerinot, M.L. (2002) Expression of the IRT1 metal transporter is controlled by metals at the levels of transcript and protein accumulation. *Plant Cell,* **14**: 1347-1357.

Curie, C., Alonso, J.M., Le Jean, M., Ecker, J.R. and Briat, J.F. (2000) Involvement of NRAMP1 from *Arabidopsis thaliana* in iron transport. *Biochem. J.,* **347**: 749-755.

## Heavy metal transporters in plants

Curie, C., Panaviene, Z., Loulergue, C., Dellaporta, S.L., Briat, J.F. and Walker, E.L. (2001) Maize yellow stripe1 encodes a membrane protein directly involved in Fe (III) uptake. *Nature,* **409**: 346-349.

Delhaize, E., Kataoka, T., Hebb, D.M. and Ryan, P.R. (2003) Genes encoding proteins of the cation diffusion facilitator family that confer manganese tolerance. *Plant Cell,* **15**: 1131-1142.

Dhankher, O.P. (2005) Arsenic metabolism in plants: an inside story. *New Phytol.,* **168**: 503-505.

Dhankher, O.P., McKinney, E.C., Rosen, B.P. and Meagher, R.B. (2006) Hyperaccumulation of arsenic in the shoots of *Arabidopsis* silenced for arsenate reductase, ACR2. *Proc. Natl. Acad. Sci. USA,* **103**: 5413-5418.

DiDonato, Jr R.J., Roberts, L.A., Sanderson, T., Eisley, R.B. and Walker, E.L. (2004) Arabidopsis *YELLOW STRIPE-LIKE2 (YSL2)*: a metal-regulated gene encoding a plasma membrane transporter of nicotianamine-metal complexes, *Plant J.,* **39**: 403–414.

Drager, D.B., Desbrosses-Fonrouge, A.G., Krach, C., Chardonnens, A.N., Meyer, R.C., Saumitou-Laprade, P. and Krämer, U. (2004) Two genes encoding *Arabidopsis halleri* MTP1 metal transport proteins co-segregate with zinc tolerance and account for high MTP1 transcript levels. *Plant J.,* **39**: 425–439.

Eckardt, N.A. (2001) Move It on Out with MATEs. *Plant Cell,* **13**: 1477-1479.

Eide, D., Broderius, M., Fett, J. and Guerinot, M.L. (1996) A novel ironregulated metal transporter from plants identified by functional expression in yeast. *Proc. Natl. Acad. Sci. USA,* **93**: 5624–5628.

Eide, D. (2006) Zinc transporters and the cellular trafficking of zinc. *Biochimica et Biophysica Acta,* **1763**: 711–722.

Ellis, D.R., Gumaelius, L., Indriolo, E., Pickering, I.J., Banks, J.A. and Salt, D.E. (2006) A novel arsenate reductase from the arsenic hyperaccumulating fern, *Pteris vittata. Plant Physiol.,* **141**: 1544-1554.

Eren, E. and Argüello, J.M. (2004) Arabidopsis HMA2, a divalent heavy metal-transporting PIB-type ATPase, is involved in cytoplasmic $Zn^{2+}$ homeostasis. *Plant Physiol.,* **136**: 3712–3723.

Forbes, J.R. and Gros, P. (2001). Divalent-metal transport by NRAMP proteins at the interface of host–pathogen interactions. *Trends Microbiol.,* **9**: 397–403.

Gendre, D., Czernic, P., Conejero, G., Pianelli, K., Briat, J., Lebrun, M. and Mari, S. (2006) TcYSL3, a member of the YSL gene family from the hyperaccumulator *Thlaspi caerulescens*, encodes a nicotianamine Ni/Fe transporter. *Plant J.,* **49**: 1–15.

Gendre, D., Czernic, P., Conéjéro, G., Pianelli, K., Briat, J.F., Lebrun, M. and Mari, S. (2007) TcYSL3, a member of the YSL gene family from the hyper-accumulator *Thlaspi caerulescens*, encodes a nicotianamine-Ni/Fe transporter. *Plant J.,* **49**: 1-15.

Grotz, N., Fox, T., Connolly, E., Park, W., Guerinot, M.L. and Eide, D. (1998) Identification of a family of zinc transporter genes from Arabidopsis that respond to zinc deficiency. *Proc. Natl. Acad. Sci. USA,* **95**: 7220–7224.

Guerinot, M.L. (2000). The ZIP family of metal transporters. *Biochim. Biophys. Acta,* **1465**: 190–198.

Hall, J.L. and Williams, L.E. (2003) Transition metal transporters in plants. *J. Exp. Bot.,* **54**: 2601–2613.

Hammond, J.P., Bowen, H.C., White, P.J., Mills, V., Pyke, K.A., Baker, A.J.M., Whiting, S.N., May, S.T. and Broadley, M.R. (2006) A comparison of the *Thlaspi caerulescens* and *Thlaspi arvense* shoot transcriptomes. *New Phytol.,* **170**: 2, 239-260.

Haney, C.J., Grass, G., Franke, S. and Rensing, C. (2005). New developments in the understanding of the cation diffusion facilitator family. *J. Ind. Microbiol. Biotechnol.,* **32**: 215-226.

Hanikenne, M., Motte, P., Wu, M.C.S., Wang, T., Loppes, R., Matagne, R.F. (2005) A mitochondrial half-size ABC transporter is involved in cadmium tolerance in *Chlamydomonas reinhardtii. Plant Cell Environ.,* **28**: 863-873.

Haydon, M.J. and Cobbett, C.S. (2007) A novel major facilitator superfamily protein at the tonoplast influences Zn tolerance and accumulation in Arabidopsis. *Plant Physiol.,* **143**: 1705-1719.

Henikoff, S., Greene, E.A., Pietrokovski, S., Bork, P., Attwood, T.E. and Hood, L. (1997) Gene families: the taxonomy of protein paralogs and chimeras. *Science,* **278**: 609-614.

Hirayama, T., Kieber, J.J., Hirayama, N., Kogan, M., Guzman, P., Nourizadeh, S., Alonso, J.M., Dailey, W.P., Dancis, A. and Ecker, J.R. (1999) Responsive-to-antagonist1, a Menkes/Wilson disease-related copper transporter, is required for ethylene signaling in Arabidopsis. *Cell,* **97**: 383-393.

Ishimaru, Y., Suzuki, M., Kobayashi, T., Takahashi, M., Nakanishi, H., Mori, S. and Nishizawa, N.K. (2005) OsZIP4, a novel zinc-regulated zinc transporter in rice, *J. Exp. Bot.,* **56**: 3207-3214.

Kaiser, B.N., Moreau, S., Castelli, J., Thomson, R., Lambert, A., Bogliolo, S., Puppo, A. and Day, D.A. (2003) The soybean NRAMP homologue, GmDMT1, is a symbiotic divalent metal transporter capable of ferrous iron transport, *Plant J.,* **35**: 295-304.

Kampfenkel, K., Kushnir, S., Babiychuk, E., Inze, D. and van Montagu, M. (1995) Molecular characterization of a putative *Arabidopsis thaliana* copper transporter and its yeast homolog. *J. Biol. Chem.,* **270**: 28479-28486

Kim, D., Gustin, J.L., Lahner, B., Persans, M.W., Baek, D., Yun, D.J. and Salt, D.E. (2004) The plant CDF family member TgMTP1 from the Ni/Zn hyperaccumulator *Thlaspi goesingense* acts to enhance efflux of Zn at the plasma membrane when expressed in *Saccharomyces cerevisiae. Plant J.,* **39**: 237-251.

Kim, D-Y., Bovet, L., Noh, E.W., Kushnir, S., Martinoia, E. and Lee, Y. (2006a) AtATM3 is involved in heavy metal resistance in Arabidopsis, *Plant Physiol.,* **140**: 933-945.

Kim, S.A., Punshon, T., Lanzirotti, A., Li, L., Alonso, J.M., Ecker, J.R., Kaplan, J. and Guerinot, M.L. (2006b) Localization of iron in Arabidopsis seed requires the vacuolar membrane transporter VIT1. *Science,* **314**: 1295-1298.

Koike, S., Inoue, H., Mizuno, D., Takahashi, M., Nakanishi, H., Mori, S. and Nishizawa, N.K. (2004) OsYSL2 is a rice metal-nicotianamine transporter that is regulated by iron and expressed in the phloem. *The Plant J.,* **39**: 415-424.

Korshunova, Y.O., Eide, D., Clark, W.G., Guerinot, M.L. and Pakrasi, H.B. (1999) The IRT1 protein from *Arabidopsis thaliana* is a metal transporter with a broad substrate range. *Plant Mol. Biol.,* **40**: 37-44.

Krämer, U., Talke, I.N. and Hanikenne, M. (2007) Transition metal transport. *FEBS Letters,* **581**: 2263-2272

Kushnir, S., Babiychuk, E., Storozhenko, S., Davey, M.W., Papenbrock, J., Rycke R.D., Engler, G., Stephan, U.W., Lange, H., Kispal, G., Lill, R. and van Montagu, M. (2001) A mutation of the mitochondrial ABC transporter Sta1 leads to dwarfism and chlorosis in the Arabidopsis mutant *starik. Plant Cell,* **13**: 89-100.

Lanquar, V., Lelievre, F., Bolte, S., Hames, C., Alcon, C., Neumann, D., Vansuyt, G., Curie, C., Schroder, A. and Kramer, U. (2005) Mobilization of vacuolar iron by AtNramp3 and AtNramp4 is essential for seed germination on low iron. *EMBO J.,* **24**: 4041-4051.

Lee, M., Lee, K., Lee, J., Noh, E.W. and Lee, Y. (2005) AtPDR12 contributes to lead resistance in Arabidopsis. *Plant Physiol.,* **138**: 827–836.

Le Jean, M., Schikora, A., Mari, S., Briat, J.F. and Curie, C. (2005) A loss-of-function mutation in AtYSL1 reveals its role in iron and nicotianamine seed loading, *Plant J.,* **44**: 769-782.

Li, L., He, Z., Pandey, G.K., Tsuchiya, T. and Luan, S. (2002) Functional cloning and characterization of a plant efflux carrier for multidrug and heavy metal detoxification, *J. Biol. Chem.,* **277**: 5360-5368.

Li, Z.S., Lu, Y.P., Zhen, R.G., Szczypka, M., Thiele, D.J. and Rea, P.A. (1997) A new pathway for vacuolar cadmium sequestration in *Saccharomyces cerevisiae*: YCF1-catalyzed transport of *bis* (glutathionate) cadmium. *Proc. Natl. Acad. Sci. USA,* **94**: 42–47.

Liu, Y., Promeneur, D., Rojek, A., Kumar, N., Frokiaer, J., Nielsen, S., King, L.S., Agre, P. and Carbrey, J.M. (2007) Aquaporin 9 is the major pathway for glycerol uptake by mouse erythrocytes, with implications for malarial virulence. *Proc Natl Acad Sci USA.,* **104**: 12560-4.

Liu, Z., Carbrey, J.M., Agre, P. and Rosen, B.P. (2002) Arsenic trioxide uptake by human and rat aquaglyceroporins. *Biochem. Biophys. Res. Commun.,* **316**: 1178-1185.

Lopez-Millan, A.F., Ellis, D.R. and Grusak, M.A. (2004) Identification and characterization of several new members of the ZIP family of metal ion transporters in *Medicago truncatula*. *Plant Mol. Biol.,* **54**: 583–596.

Lu, Y.P., Li, Z.S. and Rea, P.A. (1997) *AtMRP1* gene of Arabidopsis encodes a glutathione S-conjugate pump: isolation and functional definition of a plant ATP-binding cassette transporter gene. *Proc. Natl. Acad. Sci. USA,* **94**: 8243–8248.

Martinoia, E., Klein, M., Geisler, M., Bovet, L., Forestier, C., Kolukisaoglu, U., Muller-Rober, B. and Schulz, B. (2002) Multifunctionality of plant ABC transporters - more than just detoxifiers *Planta*, 214: 345-355

Mäser, P., Thomine, S., Schroeder, J.I., Ward, J.M., Hirschi, K., Sze, H., Talke, I.N., Amtmann, A., Maathuis, F.J.M., Sanders, D. *et al.* (2001) Phylogenetic relationships within cation transporter families of Arabidopsis. *Plant Physiol.,* **126**: 1646–1667.

McKie, A.T. *et al.* (2000) A novel duodenal iron-regulated transporter, IREG1, implicated in the basolateral transfer of iron to the circulation. *Mol. Cell,* **5**: 299-309.

Meharg, A.A. and Macnair, M.R. (1992). Genetic correlation between arsenate tolerance and the rate of influx of arsenate and phosphate in *Holcus lanatus*. *Heredity,* **69**: 336-341.

Milon, B., Wu, Q., Zou, J., Costello, L.C. and Franklin, R.B. (2006) Histidine residues in the region between transmembrane domains III and IV of hZip1 are required for zinc transport across the plasma membrane in PC-3 cells. *Biomembranes,* **1758**: 1696–1701.

Mizuno, T., Usui, K., Horie, K., Nosaka, S., Mizuno, N. and Obata, H. (2005) Cloning of three ZIP/Nramp transporter genes from a Ni hyperaccumulator plant *Thlaspi japonicum* and their $Ni^{2+}$-transport abilities. *Plant Physiol. Biochem.,* **43**: 793–801.

Montanini, B., Blaudez, D., Jeandroz, S., Sanders, D. and Chalot, M. (2007) Phylogenetic and functional analysis of the Cation Diffusion Facilitator (CDF) family: improved signature and prediction of substrate specificity. *BMC Genomics,* **8**: 107.

Mukhopadhyay, R. and Rosen, B.P. (2002) The phosphatase C(X)5R motif is required for catalytic activity of the *Saccharomyces cerevisiae* Acr2p arsenate reductase. *J. Biol. Chem.,* **28**: 34738-42.

Nies, D.H. (2003) Efflux-mediated heavy metal resistance in prokaryotes. *FEMS Microbiol. Rev.,* **27**: 313–339.

Omote, H., Hiasa, M., Matsumoto, T., Otsuka, M. and Moriyama, Y. (2006) The MATE Proteins as fundamental transporters of metabolic and xenobiotic organic cations. *Trends Pharmacol. Sci.*, **27**: 587-593.

Papoyan, A. and Kochian, L.V. (2004) Identification of *Thlaspi caerulescens* genes that may be involved in heavy metal hyperaccumulation and tolerance. Characterization of a novel heavy metal transporting ATPase. *Plant Physiol.*, **136**: 3814–3823.

Peiter, E., Montanini, B., Gobert, A., Pedas, P., Husted, S., Maathuis, F.J.M., Blaudez, D., Chalot, M. and Sanders, D. (2007) A secretory pathway-localized cation diffusion facilitator confers plant manganese tolerance. *Proc. Natl. Acad. Sci. USA*, **104**: 20, 8532–8537

Pence, N.S., Larsen, P.B., Ebbs, S.D., Letham, D.L.D., Lasat, M.M., Garvin, D.F., Eide, D. and Kochian, V. (2000) The molecular physiology of heavy metal transport in the Zn/Cd hyperaccumulator *Thlaspi caerulescens*. *Proc. Natl. Acad. Sci. USA*, **97**: 4956–4960.

Persans, M.W., Nieman, K. and Salt, D.E. (2001) Functional activity and role of cation-efflux family members in Ni hyperaccumulation in *Thlaspi goesingense*. *Proc. Natl. Acad. Sci. USA*, **98**: 9995–10000.

Pickering, I.J., Prince, R.C., George, M.J., Smith, R.D., George, G.N. and Salt, D.E. (2000) Reduction and coordination of arsenic in Indian mustard. *Plant Physiol.*, **122**: 1171-1177.

Plaza, S., Tearall, K.L., Zhao, F.J., Buchner, P., McGrath, S.P. and Hawkesford, M.J. (2007) Expression and functional analysis of metal transporter genes in two contrasting ecotypes of the hyperaccumulator *Thlaspi caerulescens*. *J. Exp. Bot.*, **58**: 1717-1728.

Putman, M., van Veen, H.W. and Konings, W.N. (2000) Molecular properties of bacterial multidrug transporters. *Microbiol. Mol. Biol. Rev.*, **64**: 672-693.

Raab, A., Schat, H., Meharg, A. and Feldmann, J. (2005) Uptake, translocation and transformation of arsenate and arsenite in sunflower (*Helianthus annuus*): formation of arsenic-phytochelatin complexes during exposure to high arsenic concentrations. *New Phytol.*, **168**: 551-558.

Rea, P.A. (1999) MRP subfamily ABC transporters from plants and yeast. *J. Exp. Bot.*, **50**: 895-913

Roberts, L.A., Pierson, A.J., Panaviene, Z. and Walker, E.L. (2004) Yellow stripe1, Expanded roles for the maize iron-phytosiderophore transporter. *Plant Physiol.*, **135**: 112-120.

Rogers, E.E., Eide, D.J. and Guerinot, M.L. (2000) Altered selectivity in an Arabidopsis metal transporter. *Proc. Natl. Acad. Sci. USA*, **97**: 12356–12360.

Rosen, B.P. (2002) Biochemistry of arsenic detoxification. FEBS Lett. 529, 86-92.

Sancenon, V., Puig, S., Mira, H., Thiele, D.J. and Penarrubia, L. (2003) Identification of a copper transporter family in *Arabidopsis thaliana*. *Plant Mol. Biol.*, **51**: 577–587

Sancenon, V., Puig, S., Mateu-Andres, I., Dorcey, E., Thiele, D.J. and Penarrubia, L. (2004) The Arabidopsis copper transporter COPT1 functions in root elongation and pollen development. *J. Biol. Chem.*, **279**: 15348–15355.

Sánchez-Fernández, R., Davies, T.G.E., Coleman, J.O.D. and Rea, P.A. (2001). The *Arabidopsis thaliana* ABC protein superfamily, a complete inventory. *J. Biol. Chem.*, **276**: 30231–30244.

Sanders, O.I., Rensing, C., Kuroda, M., Mitra, B. and Rosen, B.P. (1997) Antimonite is accumulated by the glycerol facilitator GlpF in *Escherichia coli*. *J. Bacteriol.*, **179**: 3365-3367.

Heavy metal transporters in plants

Schaaf, G., Honsbein, A., Meda, A.R., Kirchner, S., Wipf, D., von Wiren, N. (2006) *AtIREG2* encodes a tonoplast transport protein involved in iron-dependent nickel detoxification in *Arabidopsis thaliana* roots. *J. Biol. Chem.,* **281**: 35, 25532–25540.
Schaaf, G., Ludewig, U., Erenoglu, B.E., Mori, S., Kitahara, T. and von Wiren, N. (2004) ZmYS1 functions as a proton-coupled symporter for phytosiderophore- and nicotianamine-chelated metals. *J. Biol. Chem.,* **279**: 10, 9091–9096.
Shikanai, T., Muller-Moule, P., Munekage, Y., Niyogi, K.K. and Pilon, M. (2003) PAA1, a P-type ATPase of Arabidopsis, functions in copper transport in chloroplasts. *Plant Cell,* **15**: 1333–1346.
Song, W.Y., Martinoia, E., Lee, J., Kim, D., Kim, D.Y., Vogt, E., Shim, D., Choi, K.S., Hwang, I. and Lee, Y. (2004). A novel family of cys-rich membrane proteins mediates cadmium resistance in Arabidopsis. *Plant Physiol.,* **135**: 1027-1039.
Supek, F., Supekova, L., Nelson, H. and Nelson, N. (1996) A yeast manganese transporter related to the macrophage protein involved in conferring resistance to mycobacteria. *Proc. Natl. Acad. Sci. USA,* **93**: 5105-5110.
Szczypka, M.S., Wemmis, J.A., Moye-Rowley, W.S. and Thiele, D.J. (1994) A yeast metal resistance protein similar to human cystic fibrosis transmembrane conductance regulator (CFTR) and multidrug resistance-associated protein. *J. Biol. Chem.,* **269**: 22853–22857.
Talke, I.N., Hanikenne, M. and Krämer, U. (2006). Zinc-dependent global transcriptional control, transcriptional deregulation, and higher gene copy number for genes in metal homeostasis of the hyperaccumulator *Arabidopsis halleri*. *Plant Physiol.,* **142**: 148–167.
Thomine, S., Wang, R.C., Ward, J.M., Crawford, N.M. and Schroeder, J.I. (2000) Cadmium and iron transport by members of a plant metal transporter family in Arabidopsis with homology to *Nramp* genes. *Proc. Natl. Acad. Sci. USA,* **97**: 4991–4996.
Thomine, S., Lelievre, F., Debarbieux, E., Schroeder, J.I. and Barbier-Brygoo, H. (2003) *At*Nramp3, a multispecific vacuolar metal transporter involved in plant responses to iron deficiency. *Plant J.,* **34**: 685–695.
Tommasini, R., Vogt, E., Fromentau, M., Hortensteiner, S., Matile, P., Amrhein, N. and Martinoia, E. (1998) An ABC-transporter of *Arabidopsis thaliana* has both glutathione-conjugate and chlorophyll catabolite transport activity. *Plant J.,* **13**: 773-780.
Vatamaniuk, O.K., Bucher, E.A., Sundaram, M.V. and Rea, P.A. (2005) CeHMT-1, a putative phytochelatin transporter, is required for cadmium tolerance in *Caenorhabditis elegans*. *J. Biol. Chem.,* **280**: 23684-23690.
Vert, G., Briat, J.-F. and Curie, C. (2001) Arabidopsis *IRT2* gene encodes a root-periphery transporter. *Plant J.,* **26**: 181–189.
Vert, G., Grotz, N., Dédaldéchamp, F., Gaymard, F., Guerinot, M.L., Briat, J.-F. and Curie, C. (2002) IRT1, an Arabidopsis transporter essential for iron uptake from the soil and plant growth. *Plant Cell,* **14**: 1223–1233.
Wang, J., Zhao, F.-J., Meharg, A.A., Raab, A., Felfman, J. and McGarth, S.P. (2002) Mechanics of arsenic hyperaccumulation in *Pteris vittata*. Uptake kinetics, interactions with phosphate, and arsenic speciation. *Plant Physiol.,* **130**: 1552-1561.
Waters, B.M., Chu, H.H., Didonato, R.J., Roberts, L.A., Eisley, R.B., Lahner, B., Salt, D.E. and Walker, E.L. (2006) Mutations in Arabidopsis yellow stripe-like1 and yellow stripe-like3 reveal their roles in metal ion homeostasis and loading of metal ions in seeds. *Plant Physiol.,* **141**: 1446–1458.
Wintz, H., Fox, T., Wu, Y.Y., Feng, V., Chen, W., Chang, H.S., Zhu, T. and Vulpe, C. (2003) Expression profiles of *Arabidopsis thaliana* in mineral deficiencies reveal novel transporters involved in metal homeostasis. *J. Biol. Chem.,* **28**: 47644–47653.

Wysocki, R., Chery, C.C., Wawrzycka, D., Van Hulle, M., Cornelis, R., Thevelein, J.M. and Tamas, M.J. (2001) The glycerol channel Fps1p mediates the uptake of arsenite and antimonite in *Saccharomyces cerevisiae*. *Mol. Microbiol.,* **40**: 1391-401.

Yen, M.R., Tseng, Y.H. and Saier, Jr. M.H. (2001) Maize Yellow Stripe1, and iron-phytosiderophore uptake transporter, is a member of the oligopeptide transporter (OPT) family. *Microbiology,* **147**: 2881-2883.

# Chapter 10

## SUGAR AND POLYOL TRANSPORTERS IN PLANTS

KATSUHIRO SHIRATAKE
*Graduate School of Bioagricultural Sciences, Nagoya University, Chikusa, Nagoya 464-8601, Japan*
E-mail: shira@agr.nagoya-u.ac.jp

**Abstract**

At least 69 sugar transporter homologs have been found from Arabidopsis. They fall into eight large families: hexose transporter (STP/HXT), sucrose transporter (SUC/SUT), polyol transporter (PLT), *myo*-inositol transporter (ITR/MIT), plastidic glucose transporter (pGlcT), putative monosaccharide sensing protein (AZT/MSSP), sugar-porter family protein (SFP) and vacuolar glucose transporter (VGT). This chapter reviews the characters, regulations and functions of sugar and polyol transporters in plants. Many more sucrose and hexose transporters have been studied and reviewed than polyol transporters. Most sugar and polyol transporters have been shown or suggested to be in the plasma membrane. Therefore, here, we focus more on polyol transporters and vacuolar sugar transporters.

**Keywords:** plasma membrane, polyol transporter, sugar transporter, vacuolar membrane

## 1. INTRODUCTION

Sugar and polyols are primary, essential molecules of photosynthesis and long distance carbon translocation in plants. In addition to sugar and polyol metabolism, their transport and compartmentalization are important in controlling plant growth and development. Although transport of sugar and polyols by plasmodesmata is dominant in some plants or organs, most long distance translocation and intra- and inter-cellular transport of sugar and polyols depend on transporters.

After the genes encoding glucose transporter of Chlorella (Sauer and Tanner, 1989) and the sucrose transporter of spinach (Reismeier *et al.*, 1992) were identified, many sugar and polyol transporters were isolated from various plant species and were characterized. At least 69 sugar transporter homologs have been found from the Arabidopsis genome. The history of sugar transporter study in plants and information on plant sugar and polyol transporters have been summarized in many reviews (Büttner and Sauer, 2000; Kühn, 2003; Kühn *et al.*, 1999; Lalonde *et al.*, 1999, 2004; Lemoine, 2000; Noiraud *et al.*, 2001a; Ward *et al.*, 1998; Williams *et al.*,

2000; Shiratake, 2007). Lalonde *et al.* (2004) classified sugar transporter homologs into families (which they called "clades") and summarized them well. This chapter also reviews characters, regulations and functions of plant sugar and polyol transporters adding recent information. Here, we focus more attention on polyol transporters.

Although the vacuole is the largest compartment for sugar storage in plant cells, far less information on vacuolar sugar and polyol transporters has been reported compared with plasma membrane transporters. Recent genomic and proteomic studies are helping to advance the study of vacuolar sugar and polyol transporters. This chapter reviews also studies on vacuolar sugar and polyol transporters.

## 2. GENE FAMILY OF SUGAR AND POLYOL TRANSPORTERS

Sugar and polyol transporters are members of a major facilitator superfamily (MFS) and most of them have 12 transmembrane domains (TMDs). Here, 69 sugar transporter homologs were chosen from the Arabidopsis genome and a phylogenetic tree was constructed (Fig. 1). The 69 homologs were classified into eight large families and we call them here by names used commonly: sucrose transporter (SUC/SUT), hexose transporter (STP/HXT), polyol transporter (PLT), *myo*-inositol transporter (ITR/MIT), plastidic glucose transporter (pGlcT), a putative monosaccharide sensing protein (AZT/MSSP), Sugar-porter family protein (SFP) and vacuolar glucose transporter (VGT). The SUC/SUT family was further classified into three subfamilies: SUC2/SUT1, SUC3/SUT2 and SUC4 (Fig. 2).

Lalonde *et al.* (2004) classified sugar transporter homologs of plants, yeast and humans into three sucrose transporter clades and eleven monosaccharide transporter clades. All sugar and polyol transporter families in Fig. 1 correspond to the clades of Lalonde *et al.* (2004): STP/HXT, PLT, ITR/MIT, pGlcT, MSSP/AZT, SFP, and VGT families correspond to the monosaccharide transporter Clade IV, Clade II, Clade V, Clade VI, Clade III, Clade I and Clade VII, respectively. The SUC/SUT family in Fig. 1 corresponds to the sucrose transporter superfamily. SUC2/SUT1, SUC3/SUT2 and SUC4 subfamilies in Fig. 2 correspond to the sucrose transporter Clade I, Clade III and Clade II in the review by Lalonde *et al.* (2004). Here, clades containing no plant transporters (monosaccharide transporter Clade X, Clade XI, Clade XII and Clade XIII) and clades of uncertain transporters (monosaccharide transporter Clade VIII and Clade IX) are omitted.

Some putative sugar transporters do not belong to MFS, such as sucrose binding protein (SBP) and the putative maltose transporter (MEX), but are mentioned in this chapter.

### 2.1. Sucrose transporter (SUC/SUT) family

Most plants synthesize sucrose as a major photosynthetic product and use it for long distance carbon transport. Therefore, sucrose transport, including long distance transport from source to sink, must be highly regulated and sucrose transporters must have indispensable roles in the regulation. Riesmeier *et al.* (1992) identified sucrose transporter cDNA (*SoSUT1*) in the spinach cDNA library by using an elegant screening system using yeast that lacked endogenous apoplastic invertase but had cytosolic sucrose synthase added. The yeast grew if the sucrose transporter was added. After *SoSUT1* was identified, many sucrose transporters were isolated and characterized. Sucrose transporter studies have been well-summarized (Kühn *et al.*, 1999; Lalonde *et al.*, 1999, 2004; Lemoine, 2000; Ward *et al.*, 1998; Williams *et al.*, 2000) and the reviews by Kühn (2003) and Shiratake (2007) are more recent and detailed. Here, I summarize sucrose transporters from a different viewpoint.

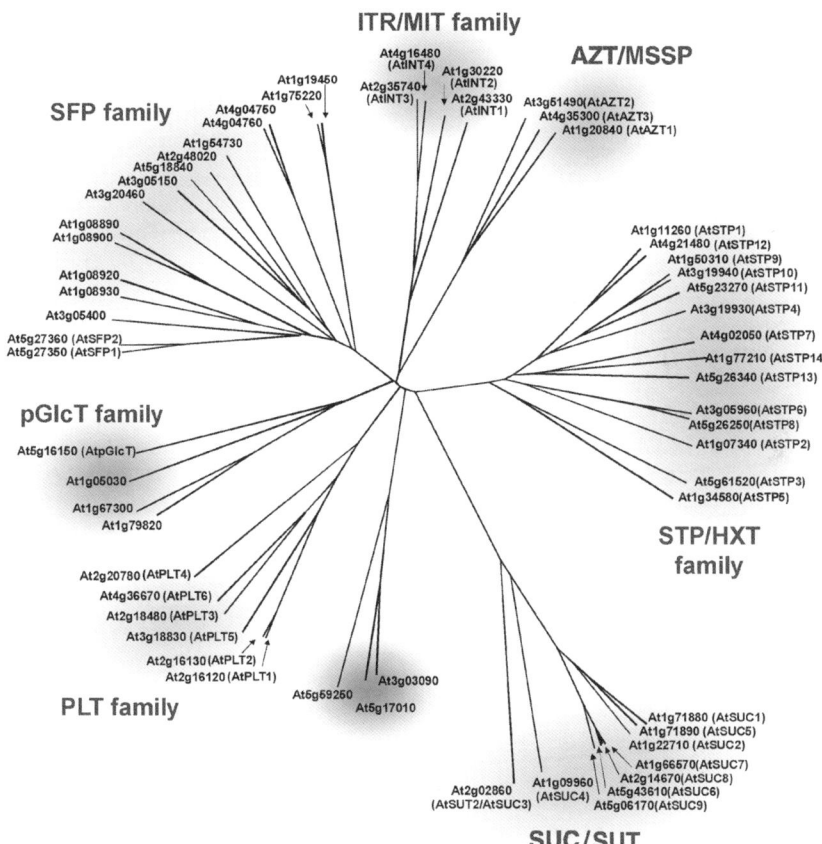

**Figure 1:** Phylogenetic tree of sugar and polyol transporter homologs of Arabidopsis. Sixty-nine sugar and polyol transporter homologs in Arabidopsis classify into eight large families: sucrose transporter (SUC/SUT), hexose transporter (STP/HXT), polyol transporter (PLT), myo-inositol transporter (ITR/MIT), plastidic glucose transporter (pGlcT), putative monosaccharide sensing protein (AZT/MSSP), sugar-porter family protein (SFP) and vacuolar glucose transporter (VGT). An unrooted N-J tree was constructed by using CLUSTAL W (http://align.genome.jp/).

### 2.1.1. Subfamilies in SUC/SUT family

The SUC/SUT family is further divided into three subfamilies based on homology: SUC2/SUT1, SUC3/SUT2 and SUC4 subfamilies (Fig. 2). Interestingly, sucrose transporters of dicotyledons are distributed in all three subfamilies, but those of monocotyledons are only in SUC3/SUT2 and SUC4 subfamilies.

The SUC2/SUT1 subfamily was identified first and is the most well-characterized of the three subfamilies. It includes high-affinity sucrose transporters, and their $Km$ values for sucrose are 100 μM to 2 mM. Figure 2 shows that each plant family

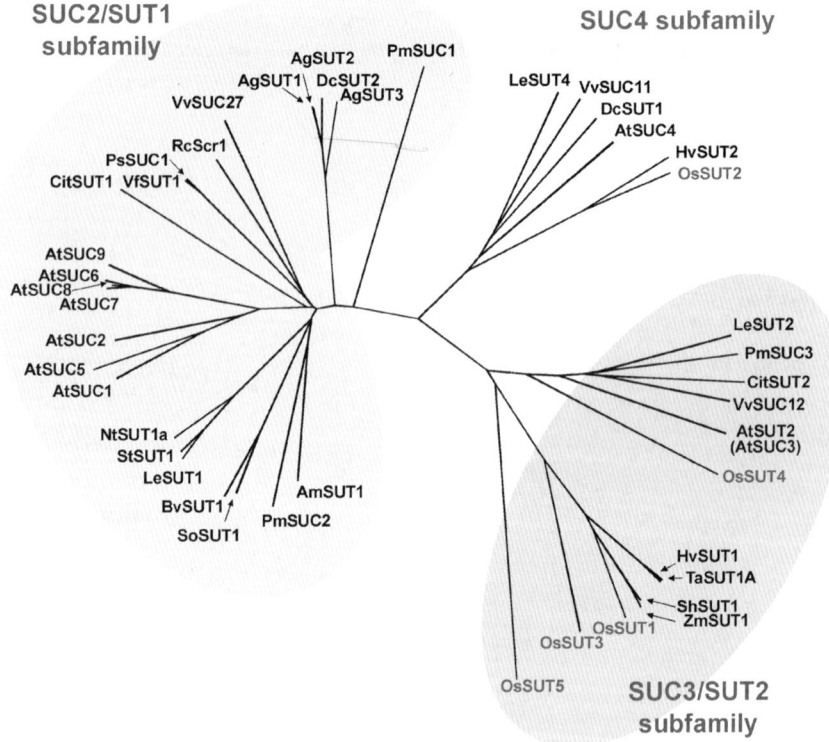

**Figure 2:** Phylogenetic tree of the SUC/SUT family. An unrooted N-J tree was constructed for sucrose transporters of various plants referred in the text by using CLUSTAL W. The SUC/SUT family was further classified into three subfamilies: SUC2/SUT1, SUC3/SUT2 and SUC4 subfamilies. Ag, *Apium graveolens*; Am, *Alonsoa meridionalis*; At, *Arabidopsis thaliana*; Bv, *Beta vulgaris*; Cit, *Citrus sinensis*; Dc, *Daucus carota*; Hv, *Hordeum vulgaris*; Le, *Lycopersicon esculentum*; Nt, *Nicotiana tabacum*; Os, *Oryza sativa*; Pm, *Plantago major*; Ps, *Pisum sativum*; Rc, *Ricinus communis*; Sh, *Saccharum hybrid*; So, *Spinacia oleracea*; St, *Solanum tuberosum*; Ta, *Triticum aestivum*; Vf, *Vicia faba*; Vv, *Vitis vinifera*; Zm, *Zea mays*.

or species in the SUC2/SUT1 subfamily has close paralogs. For instance, Arabidopsis has seven close paralogs.

The SUC4 subfamily, with few exceptions, includes low-affinity sucrose transporters of $Km$ about 5 mM. Some members of the SUC4 subfamily, such as Arabidopsis *AtSUT4* and tomato *LeSUT4*, are expressed in minor veins of source leaves and have roles in phloem loading (Weise *et al.*, 2000). A member of SUC4 subfamily in barley, *HvSUT2*, was expressed in endosperm transfer cells (Weschke *et al.*, 2000). AtSUT4 and HvSUT2 are in the vacuolar membrane (Endler *et al.*, 2006, see the later section "Sugar transporter homologs found by proteomic studies of vacuolar membrane proteins"). The SUC4 subfamily may function in the vacuolar membrane.

The SUC3/SUT2 subfamily, with some exceptions in rice, has extended domains in the central loop and N terminus (Barker *et al.*, 2000). The extended central loop is not required for transport because its deletion does not affect transport activity

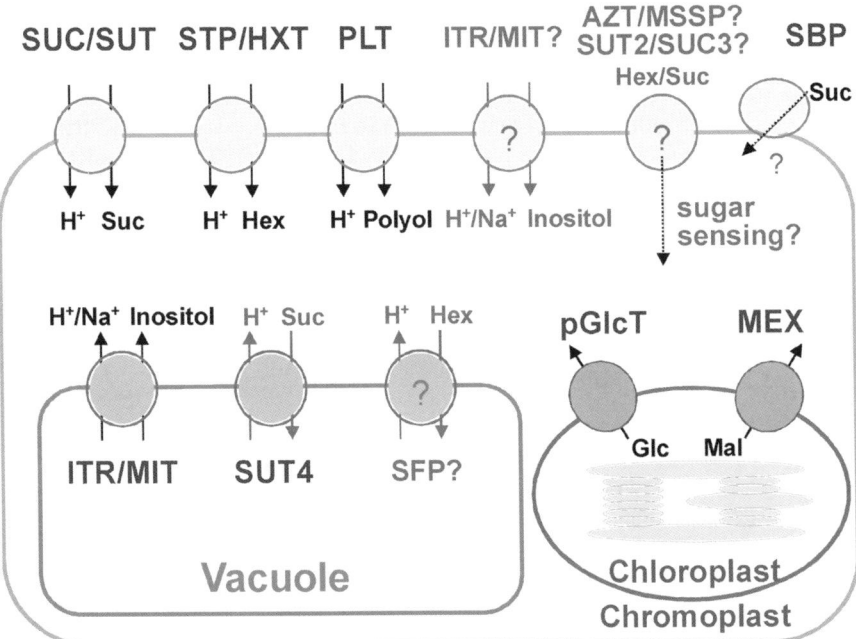

**Figure 3:** Subcellular localization of plant sugar and polyol transporters. The SUC/SUT, STP/HXT and PLT families are in the plasma membrane. Members of the ITR/MIT family are found both in the plasma membrane and in the vacuolar membrane. AZT/MSSP family and SUC3/SUT2 subfamily are suggested to be sugar sensors in the plasma membrane. The SUC4 subfamily and VGT, AZT/MSSP and SFP families reside in the vacuolar membrane. The pGlcT family is in the envelope of chloroplasts and its derivatives. SBP and MEX, putative sugar transporters other than MFS, are in the plasma membrane and chloroplast envelope, respectively.

(Meyer et al., 2000; Schulze et al., 2000). Yeast sugar sensor proteins, such as SNF3 and RGT2, have a hexose transporter-like structure with an extended domain (Özcan et al., 1996, 1998). At first, sucrose transport activity was not detected in the yeast-expressed tomato LeSUT2 (Barker et al., 2000). Therefore, LeSUT2 was hypothesized to act as a sucrose sensor. However, the SUC3/SUT2 subfamily, such as common plantain PmSUC3, rice OsSUT1, barley HvSUT1, show low-affinity sucrose transport activities (Barth et al., 2003; Hirose et al., 1997; Sivitz et al., 2005; Toyofuku et al., 2000). In the preliminary experiment, the knockout mutant of Arabidopsis AtSUT2 has no conspicuous phenotype (Barth et al., 2003) because AtSUT2 does not act as a sensor because a more marked phenotype would be expected for such a sensor. Whether the SUC3/SUT2 subfamily members act as sucrose sensors is debatable (Eckardt, 2003).

*2.1.2. Role of sucrose transporters in phloem loading*

Generally, sucrose concentration in sieve tubes is higher than in photosynthetic cells, and sieve elements (SE) and companion cells (CC) are isolated apoplastically from photosynthetic cells in many sucrose-loading plants (Gamalei, 1989). Thus sucrose transporters take up sucrose from apoplast to SE or CC or both.

In situ hybridization showed that potato *StSUT1* is expressed specifically in the phloem of minor vein in mature leaves (Riesmeier et al., 1993). Both mRNA and protein of StSUT1 are mainly in SE, but SE lacks a nucleus and ribosome (Kühn et al., 1997). Localization of *StSUT1* mRNA is preferentially associated with plasmodesmata between SE and CC, suggesting that *StSUT1* mRNA or StSUT1 protein, or both, may be synthesized in CC and then move to SE by plasmodesmata. Antisense-transgenic potato plants of *StSUT1* under the control of *CaMV35S* promoter or phloem-specific promoter accumulate high amounts of sugars and starch in source leaves and their photosynthetic activity decreases (Kühn et al., 1996, 2003; Lemoine et al., 1996; Riesmeier et al., 1994; Schulz et al., 1998). Growth of transformants decreases and plants have a reduced number of smaller tubers. Thus sucrose transporters in phloem have indispensable roles in sucrose phloem loading and long distance transport of photoassimilate. Leggewie et al. (2003) produced potato plants that overexpressed spinach *SoSUT1* under the control of *CaMV35S* promoter. Sucrose transport activity in plasma membrane vesicles from the transformants was higher than for wild-type plants, but the impact of *SoSUT1* overexpression on photosynthesis and on potato tuber yield was little.

Suppression of phloem-specific sucrose transporters has been reported in other plants. In tobacco, suppression of *NtSUT1* by antisense transformation with *CaMV35S* promoter causes curling, chlorosis and necrosis in leaves, reduced root growth and delayed flowering (Bürkle et al., 1998). Newly fixed $^{14}CO_2$ in transformants cannot export from leaves. Both high- and low-affinity sucrose transporters LeSUT1 and LeSUT4, respectively, localize in the phloem of tomato source leaves (Reinders et al., 2002). Phloem-specific *LeSUT1* antisense plants show a phenotype consistent with an essential role in sucrose phloem loading (Hackel et al., 2006), suggesting that LeSUT1 has a dominant role in phloem loading. Rice *OsSUT1* is expressed in CC of leaf blades (Hirose *et al.*, 1997; Matsukura *et al.*, 2000) and antisense suppression of *OsSUT1* driven by maize ubiquitin-1 promoter leads to impaired grain filling and germination, but does not affect photosynthesis (Scofield *et al.*, 2002).

The *AtSUC2* gene in Arabidopsis is expressed in phloem of green tissues, and AtSUC2 protein localizes specifically in CC (Stadler and Sauer, 1996; Truernit and Sauer, 1995). T-DNA-inserted mutants of *AtSUC2* show similar phenotypes of the transgenic plants that suppress their sucrose transporters in phloem (Gottwald *et al.*, 2000). *AtSUC2* mutants are sterile and decrease growth and development. Leaves of mutants accumulate a great excess of sugars and starch because of a decrease in efficient sucrose transport from source leaves to sink organs. Interestingly, the *AtSUC2* gene was induced in a syncytia cell-complex formed by nematode infection, which suggests the *AtSUC2* gene is induced by pathogen infection (Juergensen *et al.*, 2003). The low-affinity transporter AtSUT4 in Arabidopsis colocalizes with the high-affinity transporter AtSUC2 in minor veins of leaves (Weise *et al.*, 2000). A similar colocalization is in tomato: LeSUT1, a high-affinity transporter, colocalizes with LeSUT4, a low-affinity transporter (Barker *et al.*, 2000; Weise *et al.*, 2000). In phloem loading, high-affinity sucrose transporters may cooperate with low-affinity and high-capacity transporters.

Sucrose transporters are in the phloem of celery (*AgSUT1*, Noiraud *et al.*, 2000) and sugar beet (*BvSUT1*, Vaughn *et al.*, 2002), and rice *OsSUT1* mRNA was detected specifically in CC (Matsukura *et al.*, 2000). Sucrose transporters *PmSUC2* and *PmSUC3* are expressed in the phloem of common plantain. PmSUC2 protein localizes in CC, and PmSUC3 protein localizes in SE (Barth *et al.*, 2003; Stadler *et al.*, 1995). These proteins may have different roles in phloem but is unclear. Sucrose transporters are in the phloem of symplastic phloem loaders, such as *AmSUT1* in *Alonsoa meridionalis* (Knop *et al.*, 2001, 2004), suggesting that sucrose transporters partly contribute to phloem loading in addition to symplastic phloem loading in symplastic phloem loaders.

Sucrose transporters localized in the phloem of minor veins of source leaves must contribute to phloem loading. However, some sucrose transporters, including AtSUC3 (AtSUT2) (Meyer *et al.*, 2000, 2004; Schulze *et al.*, 2000), are most abundant in the phloem of major veins or petioles and may participate more in sucrose retrieval than in phloem loading.

### 2.1.3. *Other physiological roles of sucrose transporters*

Sugar transporters expressed in various sink organs and tissues include *RcScr1* in seedlings of *Ricinus communis* (Bick *et al.*, 1998; Weig and Komor, 1996), *NtSUT3* in pollen of tobacco (Lemoine *et al.*, 1999), *DcSUT1* in green parts and sink tissues of carrot (Shakya and Sturm, 1998), ShSUT1 protein in veins of sugarcane stem (Rae *et al.*, 2005), *OsSUT2*, *OsSUT3*, *OsSUT4*, and *OsSUT5* in various sink organs of rice (Aoki *et al.*, 2003), common plantain *PmSUC1*, pea *PsSUT1*, fava bean *VfSUT1*, barley *HvSUT1* and *HvSUT2*, and wheat *TaSUT1A* in developing seeds (Aoki *et al.*, 2002; Gahrtz *et al.*, 1996; Tegeder *et al.*, 1999; Weber *et al.*, 1997;

Weschke et al., 2000). These transporters have roles in phloem or post-phloem unloading. Sucrose transporter genes identified in fruit trees include *CitSUT1* and *CitSUT2* in citrus (Li et al., 2003) and *VvSUC11*, *VvSUC12* and *VvSUC27* in grapevine (Ageorges et al., 2000; Davies et al., 1999; Manning et al., 2001). These genes are expressed in fruits and play a role in sugar accumulation in fruits.

Sucrose transporter paralogs of Arabidopsis have been detected in sink organs and tissues: *AtSUC3* in guard cells, trichome, germinating pollen, root tip, carpel cell layer, etc. (Meyer et al., 2000, 2004), *AtSUC5* in endosperm (Baud et al., 2005), *AtSUC8* and *AtSUC9* in floral tissues (Sauer et al., 2004). AtSUC5 mediates vitamin H (Ludwig at al., 2000). AtSUC1 protein localizes in connective tissues and funiculi and, interestingly, despite no AtSUC1 protein, large amounts of *AtSUC1* mRNA are in mature pollen and AtSUC1 protein appears inside the pollen after pollination (Stadler et al., 1999). *AtSUC1* mRNA in mature pollen might be available for protein synthesis after pollination.

Tomato low-affinity sucrose transporter *LeSUT2* is expressed predominantly in sink organs, such as immature leaves and fruit (Reinders et al., 2002). Constitutive *LeSUT2* antisense-inhibition markedly decreases fruit and seed development and pollen germination (Hackel et al., 2006). On the other hand, seed-specific overexpression of *StSUT1* in pea increases sucrose uptake and growth rate in developing cotyledons (Rosche et al., 2002). Thus sucrose transporters are important in phloem and post-phloem unloading in sink organs.

*2.1.4. Regulation of sucrose transporter activity*

Sucrose transport activity is regulated at the gene expression level of sucrose transporters and gene expressions are regulated by organ or tissue specificity, developmental cues, circadian rhythm, stresses and environmental factors. Expressions of potato *StSUT1* and Arabidopsis *AtSUC2* are under developmental control and both are induced during sink-to-source transition in leaves (Riesmeier et al., 1993; Truernit and Sauer, 1995). Sugar beet *BvSUT1* is induced by cutting and ageing (Sakr et al., 1997) and Arabidopsis *AtSUC3* is induced by wounding (Meyer et al., 2004). Celery *AgSUT1* expression decreases by salt stress (Noiraud et al., 2000). Maize *ZmSUT1* is expressed in mature leaves and the expression changes diurnally (Aoki et al., 1999), suggesting that its expression changes to regulate the sucrose export rate from leaves. Sucrose level controls expression of sucrose transporters: tomato *LeSUT2* is induced (Barker et al., 2000), but other *SUT* genes are downregulated by sucrose (Chiou and Bush, 1998; Vaughn et al., 2002).

Sucrose transport activity is regulated by post-translational regulations. Many enzymatic activities and transport activities are regulated by phosphorylation of enzymes and transporters. Protein phosphatase inhibitors, such as okadaic acid, reduce sucrose transport activity in plasma membrane vesicles of sugar beet (Ransom-Hodgkins et al., 2003; Roblin et al., 1998), suggesting that protein phosphorylation is included in regulation of sucrose transporters. Homo- and hetero-oligomerization of transporters often affect transporter activities, such as

Sugar and polyol transporters

$K_m$ values and $V_{max}$. Oligomerization was reported for tomato sucrose transporters: LeSUT4 colocalizes with LeSUT1 and LeSUT2 in SE (Barker *et al.*, 2000), and LeSUT4 interacts with LeSUT1 and LeSUT2 (Reinders *et al.*, 2002). The SUC3/ SUT2 subfamily has extended domains in the central loop and N terminus. These extended domains may relate to regulation of sucrose transport activity, including oligomerization.

### 2.2. Hexose transporter (STP/HXT) family

Sucrose is most abundant and important for long distance carbon transport in plants. Another abundant sugar in plants is hexose, and glucose is a fundamental molecule in many metabolic reactions. Hexose transporters have roles, acting together with apoplastic invertase, in phloem and post-phloem unloading: sucrose unloaded from phloem is converted to fructose and glucose by apoplastic invertase and then the hexose transporter take them up into cells. Hexose is a signal molecule for many cellular events. Therefore, in addition to sucrose transport, hexose transport is also important for plant growth and development. Hexose transporters was isolated first from Chlorella (Sauer and Tanner, 1989). Many hexose transporters isolated from plants show low substrate specificity and can transport not only hexose, but also pentose. Hexose transporter studies have been summarized by Büttner and Sauer (2000), Lalonde *et al.* (1999, 2004) and Williams *et al.* (2000).

*2.2.1. STP/HXT in Arabidopsis*

Fourteen members of the STP/HXT family are in Arabidopsis, and AtSTP1 is the best characterized member. Transport activity of AtSTP1 is measured by heterologous expression systems in yeast (Sauer *et al.*, 1990; Stolz *et al.*, 1994) and in *Xenopus* oocytes (Boorer *et al.*, 1992, 1994). Transport experiments showed that AtSTP1 is a low substrate specific $H^+$/hexose transporter and that it can transport 3-O-methylglucose (OMG), glucose, D-galactose and D-mannose, but not D-fructose and L-arabinose. The rate of uptake of OMG and glucose in seedlings of *AtSTP1* knockout mutant is 60% less than the wild type (Sherson *et al.*, 2000), suggesting that AtSTP1 is the major hexose transporter in Arabidopsis seedlings. Hexose transport activity in Arabidopsis seedlings is markedly decreased by exogenous glucose (Sherson *et al.*, 2003), suggesting that AtSTP1 activity is regulated by sugar. *AtSTP1* expression is high in leaves and low in root and flowers (Sauer *et al.*, 1990; Sherson *et al.*, 2003). Relatively high expression of *AtSTP1* is detected in guard cells, and *AtSTP1* mRNA and protein levels are regulated diurnally and by light (Stadler *et al.*, 2003). The function of AtSTP1 may relate to stomatal opening and closing.

Other STP/HXT families in Arabidopsis have been studied. AtSTP2, AtSTP6 and AtSTP11 are high-affinity and low specific transporters and AtSTP9 is a high-affinity and specific transporter of glucose (Schneidereit *et al.*, 2003, 2005; Scholz-Starke *et al.*, 2003; Truernit *et al.*, 1999). Their genes or proteins, or both, are specifically in pollen or the pollen tube. AtSTPs in pollen have a role in transport of glucose derived from callose degradation during pollen maturation, and AtSTPs in the pollen tube have a role in hexose uptake from pistils for pollen tube growth. AtSTP3 is a low specific H$^+$/hexose transporter; *AtSTP3* expression is high in green leaves and floral tissues and is induced by wounding (Büttner *et al.*, 2000). *AtSTP4* expression is detected mainly in floral tissues and roots and is induced by wounding, elicitors and pathogen infection (Truernit *et al.*, 1996). Increase in *AtSTP4* expression is accompanied by an increase in glucose uptake and cell wall invertase activity (Fotopoulos *et al.*, 2003), and meets the carbohydrate demand for defense responses against pathogens. AtSTP13 is a high-affinity and hexose-specific H$^+$-symporter and may be negatively regulated by phosporylation (Nørholm *et al.*, 2006). *AtSTP13* expression is induced by programmed cell death (PCD), and AtSTP13 function may relate to senescence and PCD.

### 2.2.2. STP/HXT in other plants

Hexose transporter genes have been isolated from various plant species. Most STP/HXT genes are expressed in sink organs: tobacco *NtMST1* in root, flowers and young leaves (Sauer and Stadler, 1993), soybean *GmMST1* in root tip and hypocotyl and *GmMST2* in various sink tissues (Dimou *et al.*, 2005), petunia *Pmt1* in mature and germinating pollen (Ylstra *et al.*, 1998), rice *OsMST3* in leaves, callus and root and *OsMST5* in the panicle before pollination (Ngampanya *et al.*, 2003; Toyofuku *et al.*, 2000). Some hexose transporters, such as fava bean *VfSTP1* and barley *HvSTP1*, are expressed in developing seeds isolated from maternal tissue (Borisjuk *et al.*, 2002; Weber *et al.*, 1997; Weschke *et al.*, 2003). Hexose transporters together with apoplastic invertase in developing seeds have a role in the uptake of hexose from maternal tissue to seeds. Gene expression of *Medicago Mtst1* and poplar *MttMST3;1* is increased by mycorrhiza interaction, and, conversely, poplar *MttMST1;2* and *MttMST2;2* is reduced (Grunze *et al.*, 2004; Harrison, 1996). In *Chenopodium* suspension cells, hexose transporters are induced together with cell wall invertase by cytokinins (Ehne and Roitsch, 1997) and may relate to stimulation of cell division by cytokinins. Some fruit, such as tomato and grape berry, accumulate hexose at high concentrations. Hexose transporter genes are isolated from tomato and grape berry (Dibley *et al.*, 2005; Fillion *et al.*, 1999; Gear *et al.*, 2000; Vignault *et al.*, 2005). Hexose transporters expressed in fruit, such as tomato *LeHT1* and *LeHT3* and grape *VvHT1* and *VvHT2*, function in hexose accumulation in fruit. The gene expression pattern and promoter analysis of *VvHT1* shows that *VvHT1* expression is regulated by sugars and abscisic acid (Atanassova *et al.*, 2003; Çakir *et al.*, 2003; Vignault *et al.*, 2005). *VvHT1*cDNA is transformed to tobacco plants under the control of *CaMV35S* promoter in a sense or antisense orientation (Leterrier *et al.*, 2003). Both sense- and antisense-transgenic tobacco plants have lower expression of internal hexose transporter genes (*NtMST1*). Transformation decreases plant growth, and glucose uptake into the leaf discs of sense-transformants decreases to 25% of the value of the wild type. Thus hexose transporters may have an important role in assimilating transport like sucrose transporters.

Sugar and polyol transporters

## 2.3. Polyol transporter (PLT) family

Polyols, also called sugar alcohols, are reduced forms of aldose and ketose. Some polyols, namely sorbitol, mannitol, *myo*-inositol and dulcitol, are in phloem exudate of some plants (Zimmermann and Ziegler, 1975). Polyols participate in not only long distance carbon transport, but also as compatible solutes against salt, drought and low temperature stresses (Loescher and Everard, 1996; Noiraud *et al.*, 2001a; Willamson *et al.*, 2002). Polyols have a potential role in pathogen interaction. The most frequently found polyols with high concentrations in plants are sorbitol and mannitol. Rosaceae trees synthesize sorbitol in source leaves and use it as a major translocating sugar. Celery synthesizes and translocates mannitol. Although the ITR/MIT family participates in polyol transport (see the section below, "ITR/MIT family"), the PLT family participates in polyol transport dominantly in plants and some members have a key role in long distance photoassimilate translocation (Noiraud *et al.*, 2001a).

### 2.3.1. PLTs in Rosaceae

Sorbitol is the major photosynthetic product in Rosaceae, such as apple, pear and peach. Although some plants other than Rosaceae contain polyols, Rosaceae synthesize polyols most actively at high amounts. Sorbitol accounts for 60 to 90% of carbon exported from peach leaves, and sorbitol concentration in phloem sap is 560 mM (Moing *et al.*, 1997; Wallaart, 1980; Willamson *et al.*, 2002). Sorbitol transport activities had been reported for apple fruit tissue (Berüter, 1993; Berüter and Feusi, 1995), for plasma membrane vesicles of peach leaves and buds (Marquat *et al.*, 1996, 1997; Maurel *et al.*, 2004) and for protoplasts and intact vacuoles isolated from apple fruit (Yamaki and Asakura, 1988, 1991).

Sorbitol transporter genes were isolated from Rosaceae based on homology with celery mannitol transporter AgMaT1. Sorbitol transporters PcSOT1 and PcSOT2 were identified in sour cherry that accumulates large quantities of sorbitol (Gao *et al.*, 2003). Gao *et al.*, (2005) cloned cDNA encoding apple sorbitol transporters *MdSOT1* and *MdSOT2*. In the yeast expression system, PcSOT1, PcSOT2, MdSOT1 and MdSOT2 transport sorbitol. Expressions of *PcSOT1* and *PcSOT2* are dominant in fruit and of *MdSOT1* and *MdSOT2* are high in sink organs, including fruit. Although a relatively high expression of *MdSOT1* is detected in leaves, the expression is high in young leaves and decreases with leaf development. These expression patterns suggest that they have more important roles in sorbitol unloading or accumulation or both in sink organs.

Sorbitol transporter cDNAs *MdSOT3*, *MdSOT4* and *MdSOT5* have been cloned from the cDNA library of apple mature leaf (Watari *et al.*, 2004). The yeasts that express MdSOTs take up sorbitol markedly, and transport activities depending on MdSOT3 and MdSOT5 are dominant. Expression of *MdSOT3* is limited in mature leaves and *MdSOT4* and *MdSOT5* are also expressed in mature leaves. Expression of *MdSOTs* increases with leaf development, and *in situ* hybridization shows that *MdSOT* mRNAs localize specifically in the phloem of mature leaves. In Rosaceae, the sorbitol phloem loading mechanism has been debated, that is, both apoplastic and symplastic loading pathways suggested by the structure of phloem (Gamalei, 1989) and measurement of sorbitol concentrations in parenchyma or phloem (Moing et al., 1997). Thus sorbitol is strongly suggested to be loaded in the phloem apoplastically by sorbitol transporters.

### 2.3.2. PLTs in other plants

In celery, mannitol accounts for 10 to 60% of carbon exported from leaves and the mannitol concentrations in phloem sap are 150 to 300 mM (Daie, 1986). Mannitol transport activities have been reported for plasma membrane vesicles of celery (Salmon et al., 1995), intact vacuoles of celery petioles (Greutert et al., 1998) and leaf discs of olive (Flora and Madore, 1993). The first polyol transporter gene *AgMaT1* was identified in the cDNA library of celery petiole phloem (Noiraud et al., 2001b). The yeast that expresses transporter AgMaT1 takes up mannitol markedly. *AgMaT1* is expressed highly in source leaves and phloem, suggesting that AgMaT1 participates in phloem loading in celery.

Common plantain uses sorbitol for long distance transport, but sorbitol accounts for less than 30% of carbon in phloem. Ramsperger-Gleixner et al. (2004) cloned cDNAs *PmPLT1* and *PmPLT2* that encode polyol transporters from the vein-specific cDNA library of common plantain. Sorbitol transport in yeast that expresses PmPLT1 and PmPLT2, show that these transporters are low-affinity and low-specificity polyol symporters; this was confirmed by using *Xenopus* oocytes that express PmPLTs. Peptide antibodies against PmPLT1 and PmPLT2 recognize their proteins specifically in CC, suggesting they have a role in phloem loading.

Although Arabidopsis contains little quantity of polyols, six genes of the PLT family are in the genome (Fig. 1). Only *AtPLT5* was characterized by two studies that obtained similar results (Klepek et al., 2005; Reinders et al., 2005). *AtPLT5* expression is high in sink tissues, such as roots and flowers, and is induced by wounding. AtPLT5 is surprisingly a broad-spectrum $H^+$-symporter and transports sorbitol, xylitol, erythritol, glycerol, *myo*-inositol, hexose and pentose, but not disaccharides. Whether AtPLT5 participates in polyol transport in Arabidopsis plant is not clear.

### 2.3.3. Myo-*inositol transporter (ITR/MIT) family*

Inositol is a ubiquitous cellular component in plants, animals and microorganisms and acts in osmolytes and second messengers or regulators of endo- and exo-cytosis. Although *myo*-inositol is not a major translocating molecule, relatively high concentrations of *myo*-inositol are in the phloem of some plants, such as *Syringa, Eucalyptus, Citrus* and *Celtis* (Zimmermann and Ziegler, 1975). The major carbon in young fruit of *Actinidia arguta* is *myo*-inositol at 60 mg/g dry weight (Klages et al., 1998). Although the PLT family may have roles in inositol transport in plants, the ITR/MIT family is thought to participate more specifically in inositol transport.

MITR1 and MITR2 from *Mesembryanthemum crystallinum* are members of the ITR/MIT family and are homologous to $Na^+$[or $H^+$]/*myo*-inositol transporters of yeast and humans (Chauhan et al., 2000). Peptide antibodies against MITR1 and MITR2 react strongly to vacuolar membrane fractions compared with plasma membrane fractions. The MITR1-complemented yeast mutant of $H^+$/myo-inositol symporter (Itr1) and MITR1 transcripts increase by salt stress in roots. Thus MITR1 may act in removing sodium from vacuoles in roots. Miyazaki et al. (2004) transformed MITR1 (McITR) and its homolog of Arabidopsis (AtITR1) into Itr1-deficient yeasts and detected myo-inositol transport activity coupled with a proton gradient. Myo-inositol transport activity is inhibited more strongly by the methylated inositol derivative D-onitol than by myo-inositol itself. The MITR1 homolog PcITR1 was isolated from pear fruit. The PcITR1-green fluorescent protein (GFP) fusion protein localizes in the plasma membrane when it is expressed in Arabidopsis culture cells (Keta et al., unpublished data). Recently, Schneider et al. (2006) reported that AtINT4 is located in plasma membrane and responsible for $H^+$/myo-inositol symport. Thus the ITR/MIT family resides not only in the vacuolar membrane, but also in the plasma membrane.

## 2.4. Plastidic glucose transporter (pGlcT) family

Chloroplasts accumulate some photoassimilate as starch in the day and the starch is broken down to glucose, which is exported from chloroplasts at night. For glucose export, the glucose transporter should be localized in the chloroplast envelope. Weber et al. (2000) identified 43 kDa proteins in the chloroplast inner envelope of spinach by differential labeling with glucose. By peptide sequencing of the protein, cDNA (pGlcT) was also identified. pGlcT has a structure similar to other sugar transporters and also has a chloroplast signal sequence.

Entire cDNA of the olive pGlcT homolog and partial cDNA of the tomato pGlcT homolog were cloned and were expressed in heterotropic fruit tissues (Butowt et al., 2003). Chloroplasts of olive fruit change to plastids and chromoplasts during fruit maturation, suggesting that pGlcT acts not only in chloroplasts, but also in plastids and chromoplasts.

## 2.5. Other families

AZT/MSSP, SFP and VGT families had hardly been characterized. Very recent reports show their localization in the vacuolar membrane (see "Sugar transporter homologs found by proteomic studies of vacuolar membrane proteins" section).

The AZT/MSSP family has an extended central loop similar to the topology of the SUC3/SUT2 subfamily. Such an extended domain is also in yeast sugar sensors SNF3 and RGT2 (Özcan et al., 1996, 1998), suggesting that the AZT/MSSP family participates in sugar sensing.

Arabidopsis AtSFP1 and AtSFP2 expressions were determined by using Northern blotting and their promoter activities were also checked by using GUS reporter gene assay. Although AtSFP2 is expressed constitutively in almost all organs, AtSFP1 expression is especially high in seedlings and senescent leaves (Quirino et al., 2001).

## 3. SUGAR TRANSPORTERS OTHER THAN THE MFS FAMILY

All sugar and polyol transporters described above are members of the MFS family, but some putative ones, such as SBP and MEX, have no similarity with other sugar and polyol transporters.

SBP was purified as a sucrose-binding protein from microsomes of soybean cotyledons by photoaffinity labeling (Ripp et al., 1998). It localizes specifically in the plasma membrane of mesophyll cells of young leaves, CC of mature phloem and cells of developing cotyledons (Grimes et al., 1992). Yeast- or tobacco cell-expressed SBP takes up sucrose markedly (Delu-Filho et al., 2000; Grimes et al., 1996; Overvoorde et al., 1996). However, surprisingly, SPB is not an integral membrane protein and does not have typical TMDs (Grimes et al., 1992). Antisense-suppression of SBP in tobacco reduces plant growth, but overexpression of SBP accelerates plant growth and flowering (Pedra et al., 2000). Whether SPB can mediate sucrose transport is debatable.

MEX was identified by map-based cloning in Arabidopsis mutants that have high starch and maltose levels (Niittylä et al., 2004). MEX has no similarity with other sugar and polyol transporters, but has multiple TMDs. MEX-yellow fluorescent protein fusion protein localizes specifically in the chloroplast envelope, and MEX allows growth of Escherichia coli that lacks an endogenous maltose transporter (malFmutant) of maltose as a sole carbon source. Thus MEX may be a maltose transporter in chloroplast envelopes and have a role in maltose export from chloroplasts.

## 4. SUGAR AND POLYOL TRANSPORTERS IN THE VACUOLAR MEMBRANE

The vacuole occupies a large space in plant cells, and sugar and polyol are stored in the vacuole temporarily or for long time. Vacuoles in fruit, sugarcane stalk and sugar beet tap root store high concentrations of sugar at concentrations often reaching 1 M. The movement of sugar into vacuoles is believed to depend on transporters, and sugar and polyol transport activities have been detected in isolated vacuoles and vacuolar membrane vesicles. However, little information on genes existed for vacuolar sugar and polyol transporters. Recently, sugar and polyol transporters have been found in the vacuolar membrane by proteomic studies.

### 4.1. Sugar and polyol transport activities in vacuolar membranes

Sugar and polyol transport activities have been reported for intact vacuoles or vacuolar membrane vesicles of many plants. Concentrations of sugar in vacuoles are normally higher than in cytosol, and so active sugar transporters depending on a proton gradient, i.e. $H^+$/sugar antiporters, are thought to exist in the vacuolar membrane. The proton gradient across a vacuolar membrane is generated by two distinct proton pumps: vacuolar $H^+$-ATPase (V-ATPase) and $H^+$-pyrophosphatase (V-PPase) (Maeshima, 2000; Ratajczak, 2000).

Sugar and polyol transporters

Sucrose uptaken into vacuoles or vacuolar membrane vesicles from taproots of red or sugar beets is stimulated by adding ATP (Briskin *et al.*, 1982; Doll *et al.*, 1979; Getz, 1991; Getz and Klein, 1995), and sucrose efflux from vacuolar membrane vesicles of red beet taproot is increased by ATP (Echeverría and Gonzalez, 2000). Active OMG transport into vacuoles of sugarcane cells and ATP-dependent sucrose transport in vacuolar membrane vesicles of sugarcane stalk has been observed (Getz *et al.*, 1991; Thom and Komor, 1984). Conversely, sucrose uptake into vacuoles of sugarcane cells or into vacuolar membrane vesicles of sugarcane stalk is not stimulated by adding ATP (Preisser and Komor, 1991; Williams *et al.*, 1990).

ATP-dependent and -independent sugar, polyol and oligosaccharide uptake has been reported for other plants. Sucrose and stachyose uptake into vacuoles or vacuolar membrane vesicles of Japanese artichoke tuber depends on ATP or a proton gradient (Greutert and Keller, 1993; Keller, 1992; Niland and Schmitz, 1995), and conversely sucrose uptake into vacuolar membrane vesicles of Jerusalem artichoke tuber does not (Pontis *et al.*, 2002). OMG uptake into vacuoles isolated from pea mesophyll and into vacuolar membrane vesicles of corn coleoptiles relates to V-ATPase activity (Guy *et al.*, 1979; Rausch *et al.*, 1987). However, glucose uptake by vacuolar membrane vesicles in barley mesophyll cells does not depend on a proton gradient (Martinoia *et al.*, 1987), and sucrose transport across vacuolar membrane vesicles of pineapple leaves is not stimulated by ATP (McRae *et al.*, 2002). Mannitol transport by vacuoles of storage parenchyma of celery petioles does not depend on a proton gradient (Greutert *et al.*, 1998).

Although the vacuole occupies a large part of the fruit cell volume (Shiratake *et al.*, 1998) and the fruit vacuole accumulates high amounts of sugar, only one ATP-dependent polyol uptaken into fruit vacuoles has been reported. Yamaki (1987) detected ATP-promoted sorbitol transport into vacuoles of apple fruit. However, sucrose uptake into vacuolar membrane vesicles of tomato fruit, hexose uptake into those vesicles of pear fruit and sucrose and hexose uptake into those vesicles of sweet lime juice-cells is not promoted by adding ATP. (Echeverria *et al.*, 1997; Milner *et al.*, 1995; Shiratake *et al.*, 1997). Recently, a sugar transport system into vacuoles independent of a transporter, i.e. "endocytosis", was suggested for sycamore cells and citrus juice-cells (Etxeberria *et al.*, 2005a, 2005b, 2005c). Further studies, including identification of genes for vacuolar sugar transporters are needed to clarify how fruit vacuoles accumulate sugars at such high concentrations.

## 4.2. Putative vacuolar sugar and polyol transporter proteins and genes

Identification of sugar transporter proteins in the vacuolar membrane has been unsuccessful. Thom *et al.* (1992) solubilized proteins from vacuolar membranes of sugarcane parenchyma cells and purified the proteins by using column chromatography. They measured sucrose uptake activity in proteoliposomes and obtained fractions containing higher sucrose transport activity. They prepared monoclonal antibodies against vacuolar membrane proteins and screened for antibodies that inhibit sucrose uptake. By using the antibodies, they detected a protein band of 40 kDa in the purified fractions containing higher sucrose uptake activity. The same group prepared monoclonal antibodies against vacuolar membrane proteins of sugarcane stalk or red beet taproot and screened for antibodies similarly (Getz et al., 1993, 1994). They solubilized vacuolar membrane proteins of sugarcane stalk or red beet taproot, purified the proteins by using column chromatography and found positive fractions by using enzyme-linked immunosolvent assay with the antibodies. After gel filtration, they found an active 55 kDa fraction in sugarcane and 55 to 65 kDa and 110 to 130 kDa fractions in red beet, and they reported only these fractions.

Chiou and Bush (1996) screened for cDNA (*cDNA1*) that encodes a member of the SFP family in sugar beet. The peptide antibody against cDNA1 reacted to the vacuolar membrane fraction of sugar beet. They concluded that cDNA1 localizes in the vacuolar membrane and that it may mediate sugar partitioning between vacuole and cytosol. However, they detected no sugar transport activity in the yeast-expressed cDNA1.

As described above (see "ITR/MIT family" section), a member of the ITR/MIT family, MITR1 from *M. crystallinum* is in the vacuolar membrane and transports *myo*-inositol (Chauhan et al., 2000; Miyazaki et al., 2004).

## 4.3. Sugar transporter homologs found by proteomic studies of vacuolar membrane proteins

Transport activities of many different molecules, including sugars, polyols, organic acids, inorganic ions and secondary metabolites, have been detected from vacuolar membranes, (Barkla and Pantoja, 1996; Martinoia, 1992; Martinoia and Ratajczak, 1997; Martinoia et al., 2000). However, not many genes for those transporters had been identified more than five years before (Maeshima, 2001). The advanced molecular and biochemical analyses including genomics and proteomics have identified more vacuolar membrane transporters in the last five years (Martinoia et al., 2007; Shiratake and Martinoia, 2007).

Proteomic analyses have been done to identify transporters in the vacuolar membrane. The first proteome analysis of vacuolar membrane proteins was by Sazuka et al. (2004) using vacuolar membrane vesicles isolated from Arabidopsis green tissue and sugar transporters AtSTP1 (At1g11260) and AtSTP13 (At5g26340) were discovered. However, the vacuolar membrane in the preparation was not pure enough and doubt exists whether these transporters really localize in the vacuolar membrane. Since the proteomic study by Sazuka et al. (2004), proteome analyses using purer vacuolar membrane preparations were done by Carter et al. (2004), Shimaoka et al. (2004) and Szponarski et al. (2004). Although no sugar transporters were detected by the proteomic study of Szponarski et al. (2004), AtSUC1 (At1g71880) was discovered by the proteomic study of Shimaoka et al. (2004) and AtSTP1 (At1g11260) and members of ITR/MIT family (At2g43330), SFP family (At2g48020) and AZT/MSSP family (At4g35300, AtINT1) were discovered by the proteomic study of Carter et al. (2004). Proteome analysis of vacuolar membrane proteins of barley mesophyll cells identified a sucrose transporter homolog (HvSUT2)

Sugar and polyol transporters

of the SUC4 subfamily and members of SFP and AZT/MSSP families (Endler *et al.*, 2006). GFP fusion proteins of HvSUT2 and its homolog of Arabidopsis (AtSUT4) were expressed in Arabidopsis leaves and in onion epidermis, and GFP-fluorescence was observed in the vacuolar membrane. Wormit *et al.* (2006) identified a member of Arabidopsis AZT/MSSP family (AtTMT1, At1g20840) as a vacuolar glucose transporter. In our preliminary experiment, GFP-fusion protein of a member of tomato SFP family (S1SFP1), which is close to sugar beet cDNA1, localized in the vacuolar membrane (Hioki *et al.*, unpublished data). In addition, Aluri and Büttner (2007) reported that a member of VGT family (AtVGT1, At3g03090) localizes in the vacuolar membrane and yeast-expressed AtVGT1 transport glucose and fructose. Vacuolar sugar transporters had not been identified for a long time. In these few years, it becomes clear that some sugar and polyol transporters are present in the vacuolar membranes. Further investigations of these transporters and their homologs will clarify the mechanism of sugar accumulation in the vacuoles.

## 5. CONCLUSIONS AND FUTURE PROSPECTS

Sugar is one of the most important molecules in plants and large numbers of sugar and polyol transporters have roles in plant growth and development. In the last fifteen years, the study of plant sugar and polyol transporters has advanced dramatically and our understanding has deepened, as described in this chapter. However, our knowledge is still fragmentary. For instance, we know that at least 69 sugar and polyol transporter genes are in the Arabidopsis genome, but we know little of their biochemical characters, localizations, regulations and physiological functions. We need more detailed studies to understand them.

In addition to detailed studies, comprehensive studies are also needed. Many different transporters act together intra- and inter-cellularly in the long-distance transport of sugar and polyol in plants. Now we are just seeing single gene mutation of sugar transporters in Arabidopsis, however other homologs often make up and mask the loss. Double or multiple mutants will tell us the importance of sugar and polyol transporters in plants. Recent post-genome studies, such as transcriptomics, proteomics, metabolomics and phenomics, also help to clarify the roles of sugar and polyol transporters. Transcriptomics and metabolomics are also helpful in comprehensive studies of sugar and polyol transporters.

Carbon fixation, partitioning and storage relate closely to yield and quality of crops, and sugar and polyol transporters have important roles in carbon partitioning in plants. A greater understanding of them will produce crops with greater yield and quality in the future.

## 6. ACKNOWLEDGEMENTS

This study was supported by Grant-in-Aids for Young Scientists (A) (no. 17688002), for Exploratory Research (no. 18658010) and the 21$^{st}$ Century COE Program (no. 14COEA02) from the Ministry of Education, Culture, Sports, Science and Technology of Japan. The study was also supported in part by JSPS and BRAIN.

## 7. LITERATURE CITED

Ageorges, A., Issaly, N., Picaud, S., Delrot, S. and Romieu, C. (2000) Identification and functional expression in yeast of a grape berry sucrose carrier. *Plant Physiol. Biochem.*, **38**: 177-185.

Aluri, S. and Büttner, M. (2007) Identification and functional expression of the *Arabidopsis thaliana* vacuolar glucose transporter 1 and its role in seed germination and flowering. *Proc. Natl. Acad. Sci. USA*, **104**: 2537-2542.

Aoki, N., Hirose, T., Scofield, G.N., Whitfeld, P.R. and Furbank, R.T. (2003) The sucrose transporter gene family in rice. *Plant Cell Physiol.*, **44**: 223-232.

Aoki, N., Hirose, T., Takahashi, S., Ono, K., Ishimaru, K. and Ohsugi, R. (1999) Molecular cloning and expression analysis of a gene for a sucrose transporter in maize (*Zea mays* L.). *Plant Cell Physiol.*, **40**: 1072-1078.

Aoki, N., Whitfeld, P., Hoeren, F., Scofield, G., Newell, K., Patrick, J., Offler, C., Clarke, B., Rahman, S. and Furbank, R.T. (2002) Three sucrose transporter genes are expressed in the developing grain of hexaploid wheat. *Plant Mol. Biol.*, **50**: 453-462.

Atanassova, R., Leterrier, M., Gaillard, C., Agasse, A., Sagot, E., Coutos-Thévenot, P. and Delrot, S. (2003) Sugar-regulated expression of a putative hexose transport gene in grape. *Plant Physiol.*, **131**: 326-334.

Barker, L., Kühn, C., Weise, A., Schulz, A., Gebhardt, C., Hirner, B., Hellmann, H., Schulze, W., Ward, J.M. and Frommer, W.B. (2000) SUT2, a putative sucrose sensor in sieve elements. *Plant Cell*, **12**: 1153-1164.

Barkla, B.J. and Pantoja, O. (1996) Physiology of ion transport across the tonoplast of higher plants. *Annu. Rev. Plant Physiol. Plant Mol. Biol.*, **47**: 159-184.

Barth, I., Meyer, S. and Sauer, N. (2003) PmSUC3: Characterization of a SUT2/SUC3-type sucrose transporter from *Plantago major*. *Plant Cell*, **15**: 1375-1385.

Baud, S., Wuilleme, S., Lenoine, R., Kronenberger, J., Caboche, M., Lepiniec, L. and Rochat, C. (2005) The AtSUC5 sucrose transporter specifically expressed in the endosperm is involved in early seed development in Arabidopsis. *Plant J.*, **43**: 824-836.

Berüter, J. (1993) Characterization of the permeability of excised apple tissue for sorbitol. *J. Exp. Bot.*, **44**: 519-528.

Berüter, J. and Feusi, M.E.S. (1995) Comparison of sorbitol transport in excised tissue discs and cortex tissue of intact apple fruit. *J. Plant Physiol.*, **146**: 95-102.

Bick, J.A., Neelam, A., Smith, E., Nelson, S.J., Hall, J.L. and Williams, L.E. (1998) Expression analysis of a sucrose carrier in the germinating seedling of *Ricinus communis*. *Plant Mol. Biol.*, **38**: 425-435.

Boorer, K.J., Forde, B.G., Leigh, R.A. and Miller, A.J. (1992) Functional expression of a plasma membrane transporter in *Xenopus oocytes*. *FEBS Lett.*, **302**: 166-168.

Boorer, K.J., Loo, D.D.F. and Wright, E.M. (1994) Steady-state and presteady-state kinetics of the $H^+$/hexose cotransporter (STP1) from *Arabidopsis thaliana* expressed in *Xenopus oocytes*. *J. Biol. Chem.*, **269**: 20417-20424.

Borisjuk, L., Walenta, S., Rolletschek, H., Mueller-Klieser, W., Wobus, U. and Weber, H. (2002) Spatial analysis of plant metabolism: Sucrose imaging within *Vicia faba* cotyledons reveals specific developmental patterns. *Plant J.*, **29**: 521-530.

Briskin, D.P., Thornley, W.R. and Wyse, R.E. (1982) Membrane transport in isolated vesicles from sugarbeet taproot: II. evidence for a sucrose/$H^+$-antiport. *Plant Physiol.*, **78**: 871-875.

## Sugar and polyol transporters

Bürkle, L., Hibberd, J.M., Quick, W.P., Kühn, C., Hirner, B. and Frommer, W.B. (1998) The H⁺-sucrose cotransporter NtSUT1 is essential for sugar export from tobacco leaves. *Plant Physiol.,* **118**: 59-68.

Butowt, R., Granot, D. and Rodríguez-García, M.I. (2003) A putative plastidic glucose translocator is expressed in heterotrophic tissues that do not contain starch, during olive (*Olea europea* L.) fruit ripening. *Plant Cell Physiol.,* **44**: 1152-1161.

Büttner, M. and Sauer, N. (2000) Monosaccharide transporters in plants: structure, function and physiology. *Biochim. Biophys. Acta,* **1465**: 263-274.

Büttner, M., Truernit, E., Baier, K., Scholz-Starke, J., Sontheim, M., Lauterbach, C., Huss, V.A.R. and Sauer, N. (2000) AtSTP3, a green leaf-specific, low affinity monosaccharide-H⁺ symporter of *Arabidopsis thaliana*. *Plant Cell Environ.,* **23**: 175-184.

Çakir, B., Agasse, A., Gaillard, C., Saumonneau, A., Delrot, S. and Atanassova, R. (2003) A grape ASR protein involved in sugar and abscisic acid signaling. *Plant Cell,* **15**: 2165-2180.

Carter, C., Pan, S., Zouhar, J., Avila, E.L., Girke, T. and Raikhel, N.V. (2004) The vegetative vacuole proteome of *Arabidopsis thaliana* reveals predicted and unexpected proteins. *Plant Cell,* **16**: 3285-3303.

Chauhan, S., Forsthoefel, N., Ran, Y., Quigley, F., Nelson, D.E. and Bohnert, H.J. (2000) Na⁺/myo-inositol symporters and Na⁺/H⁺-antiport in *Mesembryanthemum crystallinum*. *Plant J.,* **24**: 511-522.

Chiou, T.J. and Bush, D.R. (1996) Molecular cloning, immunochemical localization to the vacuole, and expression in transgenic yeast and tobacco of a putative sugar transporter from sugar beet. *Plant Physiol.,* **110**: 511-520.

Chiou, T.J. and Bush, D.R. (1998) Sucrose is a signal molecule in assimilate partitioning. *Proc. Natl. Acad. Sci. USA,* **95**: 4784-4788.

Daie, J. (1986) Kinetics of sugar transport in isolated vascular bundles and phloem tissue of celery. *J. Amer. Soc. Hort. Sci.,* **111**: 216-220.

Davies, C., Wolf, T. and Robinson, S.P. (1999) Three putative sucrose transporters are differentially expressed in grapevine tissues. *Plant Sci.,* **147**: 93-100.

Delu-Filho, N., Pirovani, C.P., Pedra, J.H.F., Matrangolo, F.S.V., Macedo, J.N.A., Otoni, W.C. and Fontes, E.P.B. (2000) A sucrose binding protein homologue from soybean affects sucrose uptake in suspension-cultured transgenic tobacco cells. *Plant Physiol. Biochem.,* **38**: 353-361.

Dibley, S.J., Gear, M.L., Yang, X., Rosche, E.G., Offler, C.E., McCurdy, D.W. and Patrick, J.W. (2005) Temporal and spatial expression of hexose transporters in developing tomato (*Lycopersicon esculentum*) fruit. *Funct. Plant Biol.,* **32**: 777-785.

Dimou, Maria, Flemetakis, E., Delis, C., Aivalakis, G., Spyropoulos, K.G. and Katinakis, P. (2005) Genes coding for a putative cell-wall invertase and two putative monosaccharide/H⁺ transporters are expressed in roots of etiolated *Glycine max* seedlings. *Plant Sci.,* **169**: 798-804.

Doll, S., Rodier, F. and Willenbrink, J. (1979) Accumulation of sucrose in vacuoles isolated from red beet tissue. *Planta,* **144**: 407-411.

Echeverría, E. and Gonzalez, P.C. (2000) ATP-induced sucrose efflux from red-beet tonoplast vesicles. *Planta,* **211**: 77-84.

Echeverría, E., Gonzalez, P.C. and Brune, A. (1997) Characterization of proton and sugar transport at the tonoplast of sweet lime (*Citrus limmetioides*) juice cells. *Physiol. Plant.,* **101**: 291-300.

Eckardt, N.A. (2003) The function of SUT2/SUC3 sucrose transporters: The debate continues. *Plant Cell,* **15**: 1259-1262.

Ehne, B.R. and Roitsch, T. (1997) Co-ordinated induction of mRNAs for extracellular invertase and a glucose transporter in *Chenopodium rubrum* by cytokinins. *Plant J.*, **11**: 539-548.
Endler, A., Meyer, S., Schelbert, S., Schneider, T., Weschke, W., Peters, S.W., Keller, F., Baginsky, S., Martinoia, E. and Schmidt, U.G. (2006) Identification of a vacuolar sucrose transporter in barley and Arabidopsis mesophyll cells by a tonoplast proteomic approach. *Plant Physiol.*, **141**: 196-207.
Etxeberria, E., Baroja-Fernandez, E., Muñoz, F.J. and Pozueta-Romero, J. (2005a) Sucrose-inducible endocytosis as a mechanism for nutrient uptake in heterotrophic plant cells. *Plant Cell Physiol.*, **46**: 474-481.
Etxeberria, E., Gonzalez, P. and Pozueta-Romero, J. (2005b) Sucrose transport into citrus juice cells: Evidence for an endocytic transport system. *J. Amer. Soc. Hort. Sci.*, **130**: 269-274.
Etxeberria, E., Gonzalez, P., Tomlinson, P. and Pozueta-Romero, J. (2005c) Existence of two parallel mechanisms for glucose uptake in heterotrophic plant cells. *J. Exp. Bot.*, **56**: 1905-1912.
Fillion, L., Ageorges, A., Picaud, S., Coutos-Thevenot, P., Lemoine, R., Romieu, C. and Delrot, S. (1999) Cloning and expression of a hexose transporter gene expressed during the ripening of grape berry. *Plant Physiol.*, **120**: 1083-1093.
Flora, L.L. and Madore, M.A. (1993) Stachyose and mannitol transport in olive (*Olea europaea* L.). *Planta*, **189**: 484-490.
Fotopoulos, V., Gilbert, M.J., Pittman, J.K., Marvier, A.C., Buchanan, A.J., Sauer, N., Hall, J.L. and Williams, L.E. (2003) The monosaccharide transporter gene, *AtSTP4*, and the cell-wall invertase, Atßfruct1, are induced in Arabidopsis during infection with the fungal biotroph *Erysiphe cichoracearum*. *Plant Physiol.*, **132**: 821-829.
Gahrtz, M., Schmelzer, E., Stolz, J. and Sauer, N. (1996) Expression of the *PmSUC1* sucrose carrier gene from *Plantago major* L. is induced during seed development. *Plant J.*, **9**: 93-100.
Gamalei, Y. (1989) Structure and function of leaf minor veins in trees and herbs: a taxonomic review. *Trees*, **3**: 96-110.
Gao, Z., Jayanty, S., Beaudry, R. and Loescher, W. (2005) Sorbitol transporter expression in apple sink tissues: Implications for fruit sugar accumulation and watercore development. *J. Amer. Soc. Hort. Sci.*, **130**: 261-268.
Gao, Z., Maurousset, L., Lemoine, R., Yoo, S.D., van Nocker, S. and Loescher, W. (2003) Cloning, expression, and characterization of sorbitol transporters from developing sour cherry fruit and leaf sink tissues. *Plant Physiol.*, **131**: 1566-1575.
Gear, M.L., McPhillips, M.L., Patrick, J.W. and McCurdy, D.W. (2000) Hexose transporters of tomato: Molecular cloning, expression analysis and functional characterization. *Plant Mol. Biol.*, **44**: 687-697.
Getz, H.P. (1991) Sucrose transport in tonoplast vesicles of red beet roots is linked to ATP hydrolysis. *Planta*, **185**: 261-268.
Getz, H.P. and Klein, M. (1995) Characteristics of sucrose transport and sucrose-induced $H^+$ transport on the tonoplast of red beet (*Beta vulgaris* L.) storage tissue. *Plant Physiol.*, **107**: 459-467.
Getz, H.P., Grosclaude, J., Kurkdjian, A., Lelievre, F., Maretzki, A. and Guern, J. (1993) Immunological evidence for the existence of a carrier protein for sucrose transport in tonoplast vesicles from red beet (*Beta vulgaris* L.) root storage tissue. *Plant Physiol.*, **102**: 751-760.
Getz, H.P., Thom, M. and Maretzki, A. (1991) Proton and sucrose transport in isolated tonoplast vesicles from sugarcane stalk tissue. *Physiol. Plant.*, **83**: 404-410.

## Sugar and polyol transporters

Getz, H.P., Thom, M. and Maretzki, A. (1994) Monoclonal antibodies as tools for the identification of the tonoplast sucrose carrier from sugarcane stalk tissue. *J. Plant Physiol.,* **144**: 525-532.
Gottwald, J.R., Krysan, P.J., Young, J.C., Evert, R.F. and Sussman, M.R. (2000) Genetic evidence for the *in planta* role of phloem specific plasma membrane sucrose transporters. *Proc. Natl. Acad. Sci. USA,* **97**: 13979-13984.
Greutert, H. and Keller, F. (1993) Further evidence for stachyose and sucrose/$H^+$ antiporters on the tonoplast of Japanese artichoke (*Stachys sieboldii*) tubers. *Plant Physiol.,* **101**: 1317-1322.
Greutert, H., Martinoia, E. and Keller, F. (1998) Mannitol transport by vacuoles of storage parenchyma of celery petioles operates by facilitated diffusion. *J. Plant Physiol.,* **153**: 91-96.
Grimes, H.D. and Overvoorde, P.J. (1996) Functional characterization of sucrose binding protein-mediated sucrose uptake in yeast. *J. Exp. Bot.,* **47**: 1217-1222.
Grimes, H.D., Overvoorde, P.J., Ripp, K., Franceschi, V.R. and Hitz, W.D. (1992) A 62-kDa sucrose binding protein is expressed and localized in tissues actively engaged in sucrose transport. *Plant Cell,* **4**: 1561-1574.
Grunze, N., Willmann, M. and Nehls, U. (2004) The impact of ectomycorrhiza formation on monosaccharide transporter gene expression in poplar roots. *New Phytologist,* **164**: 147-155.
Guy, M., Reinhold, L. and Michaeli, D. (1979) Direct evidence for a sugar transport mechanism in isolated vacuoles. *Plant Physiol.,* **64**: 61-64.
Hackel, A., Schauer, N., Carrari, F., Fernie, A.R., Grimm, B. and Kühn, C. (2006) Sucrose transporter LeSUT1 and LeSUT2 inhibition affects tomato fruit development in different ways. *Plant J.,* **45**: 180-192.
Harrison, M.J. (1996) A sugar transporter from *Medicago truncatula*: Altered expression pattern in roots during vesicular-arbuscular (VA) mycorrhizal associations. *Plant J.,* **9**: 491-503.
Hirose, T., Imaizumi, N., Scofield, G.N., Furbank, R.T. and Ohsugi, R. (1997) cDNA cloning and tissue specific expression of a gene for sucrose transporter from rice (*Oryza sativa* L.). *Plant Cell Physiol.,* **38**: 1389-1396.
Juergensen, K., Scholz-Starke, J., Sauer, N., Hess, P., van Bel, A.J.E. and Grundler, F.M.W. (2003) The companion cell-specific Arabidopsis disaccharide carrier AtSUC2 is expressed in nematode-induced syncytia. *Plant Physiol.,* **131**: 61-69.
Keller, F. (1992) Transport of stachyose and sucrose by vacuoles of Japanese artichoke (*Stachys sieboldii*) tubers. *Plant Physiol.,* **98**: 442-445.
Klages, K., Donnison, H., Boldingh, H. and Macrae, E. (1998) *Myo*-Inositol is the major sugar in *Actinidia arguta* during early fruit development. *Aust. J. Plant Physiol.,* **25**: 61-67.
Klepek, Y.S., Geiger, D., Stadler, R., Klebl, F., Landouar-Arsivaud, L., Lemoine, R., Hedrich, R. and Sauer, N. (2005) Arabidopsis POLYOL TRANSPORTER5, a new member of the monosaccharide transporter-like superfamily, mediates $H^+$-symport of numerous substrates, including *myo*-inositol, glycerol, and ribose. *Plant Cell,* **17**: 204-218.
Knop, C., Stadler, R., Sauer, N. and Lohaus, G. (2004) AmSUT1, a sucrose transporter in collection and transport phloem of the putative symplastic phloem loader *Alonsoa meridionalis*. *Plant Physiol.,* **134**: 204-214.
Knop, C., Voitsekhovskaja, O. and Lohaus, G. (2001) Sucrose transporters in two members of the Scrophulariaceae with different types of transport sugar. *Planta,* **213**: 80-91.

Kühn, C. (2003) A comparison of the sucrose transporter systems of different plant species. *Plant Biol.,* **5**: 215-232.
Kühn, C., Barker, L., Bürkle, L. and Frommer, W.B. (1999). Update on sucrose transport in higher plants. *J. Exp. Bot.,* **50**: 935-953.
Kühn, C., Franceschi, V.R., Schulz, A., Lemoine, R. and Frommer, W.B. (1997) Macromolecular trafficking indicated by localization and turnover of sucrose transporters in enucleate sieve elements. *Science,* **275**: 1298-1300.
Kühn, C., Hajirezaei, M.R., Fernie, A.R., Roessner-Tunali, U., Czechowski, T., Hirner, B. and Frommer, W.B. (2003) The sucrose transporter StSUT1 localizes to sieve elements in potato tuber phloem and influences tuber physiology and development. *Plant Physiol.,* **131**: 102-113.
Kühn, C., Quick, W.P., Schulz, A., Riesmeier, J.W., Sonnewald, U. and Frommer, W.B. (1996) Companion cell-specific inhibition of the potato sucrose transporter SUT1. *Plant Cell Environ.,* **19**: 1115-1123.
Lalonde, S., Boles, E., Hellmann, H., Barker, L., Patrick, J.W., Frommer, W.B. and Ward, J.M. (1999) The dual function of sugar carriers: Transport and sugar sensing. *Plant Cell,* **11**: 707-726.
Lalonde, S., Wipf, D. and Frommer, W.B. (2004) Transport mechanisms for organic forms of carbon and nitrogen between source and sink. *Annu. Rev. Plant Biol.,* **55**: 341-372.
Leggewie, G., Kolbe, A., Lemoine, R., Roessner, U., Lytovchenko, A., Zuther, E., Kehr, J., Frommer, W.B., Riesmeier, J.W., Willmitzer, L. and Fernie, A.R. (2003) Overexpression of the sucrose transporter SoSUT1 in potato results in alterations in leaf carbon partitioning and in tuber metabolism but has little impact on tuber morphology. *Planta,* **217**: 158-167.
Lemoine, R. (2000) Sucrose transporters in plants: update on function and structure. *Biochim. Biophys. Acta,* **1465**: 1246-1262.
Lemoine, R., Bürkle, L., Barker, L., Sakr, S., Kühn, C., Regnacq, M., Gaillard, C., Delrot, S. and Frommer, B.W. (1999) Identification of a pollen-specific sucrose transporter-like protein NtSUT3 from tobacco. *FEBS Lett.,* **454**: 325-330.
Lemoine, R., Kühn, C., Thiele, N., Delrot, S. and Frommer, W.B. (1996) Antisense inhibition of the sucrose transporter in potato: Effects on amount and activity. *Plant Cell Environ.,* **19**: 1124-1131.
Leterrier, M., Atanassova, R., Laquitaine, L., Gaillard, C., Coutos-Thévenot, P. and Delrot, S. (2003) Expression of a putative grapevine hexose transporter in tobacco alters morphogenesis and assimilate partitioning. *J. Exp. Bot.,* **54**: 1193-1204.
Li, C.Y., Shi, J.X., Weiss, D. and Goldschmidt, E.E. (2003) Sugars regulate sucrose transporter gene expression in citrus. *Biochem. Biophys. Res. Commu.,* **306**: 402-407.
Loescher, W.H. and Everard, J.D. (1996) Sugar alcohol metabolism in sinks and sources. In: *Photoassimilate Distribution in Plants and Crops* (Eds. Zamski, E. and Schaffer, A.A.) Marcel Dekker Inc., New York, pp 185-207.
Ludwig, A., Stolz, J. and Sauer, N. (2000) Plant sucrose-$H^+$ symporters mediate the transport of vitamin H. *Plant J.,* **24**: 503-509.
Maeshima, M. (2000) Vacuolar $H^+$-pyrophosphatase. *Biochim. Biophys. Acta,* **1465**: 37-51.
Maeshima, M. (2001) Tonoplast transporters: Organization and function. *Ann. Rev. Plant Physiol. Plant Mol. Biol.,* **52**: 469-497.
Manning, K., Davies, C., Bowen, H.C. and White, P.J. (2001) Functional characterization of two ripening-related sucrose transporters from grape berries. *Ann. Bot.,* **87**: 125-129.

## Sugar and polyol transporters

Marquat, C., Pétel, G. and Gendraud, M. (1996) Study of $H^+$-nutrient co-transport in peach-tree and the approach to their involvement in the expression of vegetative bud growth capability. *J. Plant Physiol.,* **149**: 102-108.

Marquat, C., Pétel, G. and Gendraud, M. (1997) Saccharose and sorbitol transporters from plasmalemma membrane vesicles of peach tree leaves. *Biol. Plant,* **39**: 369-378.

Martinoia, E. (1992) Transport processes in vacuoles of higher plants. *Bot. Acta,* **105**: 232-245.

Martinoia, E. and Ratajczak, R. (1997) Transport of organic molecules across the tonoplast. *Adv. Bot. Res.,* **25**: 365-400.

Martinoia, E., Kaiser, G., Schramm, M.J. and Heber, U. (1987) Sugar transport across the plasmalemma and the tonoplast of barley mesophyll protoplasts: evidence for different transport systems. *J. Plant Physiol.,* **131**: 467-478.

Martinoia, E., Massonneau, A. and Frangne, N. (2000) Transport processes of solutes across the vacuolar membrane of higher plants. *Plant Cell Physiol.,* **41**: 1175-1186.

Martinoia, E., Maeshima, M. and Neuhaus, H.E. (2007) Vacuolar transporters and their essential role in plant metabolism. *J. Exp. Bot.,* **58**: 83-102.

Matsukura, C.A., Saitoh, T., Hirose, T., Ohsugi, R., Perata, P. and Yamaguchi, J. (2000) Sugar uptake and transport in rice embryo. Expression of companion cell-specific sucrose transporter (*OsSUT1*) induced by sugar and light. *Plant Physiol.,* **124**: 85-93.

Maurel, K., Sakr, S., Gerbe, F., Guilliot, A., Bonhomme, M., Rageau, R. and Pétel, G. (2004) Sorbitol uptake is regulated by glucose through the hexokinase pathway in vegetative peach-tree buds. *J. Exp. Bot.,* **55**: 879-888.

McRae, S.R., Christopher, J.T., Smith, J.A.C. and Holtum, J.A.M. (2002) Sucrose transport across the vacuolar membrane of *Ananas comosus. Funct. Plant Biol.,* **29**: 717-724.

Meyer, S., Lauterbach, C., Niedermeier, M., Barth, I., Sjolund, R.D. and Sauer, N. (2004) Wounding enhances expression of AtSUC3, a sucrose transporter from Arabidopsis sieve elements and sink tissues. *Plant Physiol.,* **134**: 684-693.

Meyer, S., Melzer, M., Truernit, E., Hummer, C., Besenbeck, R., Stadler, R. and Sauer, N. (2000) AtSUC3, a gene encoding a new Arabidopsis sucrose transporter, is expressed in cells adjacent to the vascular tissue and in a carpel cell layer. *Plant J.,* **24**: 869-882.

Milner, I.D., Ho, L.C. and Hall, J.L. (1995) Properties of proton and sugar transport at the tonoplast of tomato (*Lycopersicon esculentum*) fruit. *Physiol. Plant.,* **94**: 399-410.

Miyazaki, S., Rice, M., Quigley, F. and Bohnert, H.J. (2004) Expression of plant inositol transporters in yeast. *Plant Sci.,* **166**: 245-252.

Moing, A., Carbonne, F., Zipperlin, B., Svanella, L. and Gaudillère, J.P. (1997) Phloem loading in peach: symplastic or apoplastic? *Physiol. Plant.,* **101**: 489-496.

Ngampanya, B., Sobolewska, A., Takeda, T., Toyofuku, K., Narangajavana, J., Ikeda, A. and Yamaguchi, J. (2003) Characterization of rice functional monosaccharide transporter, *OsMST5. Biosci. Biotechnol. Biochem.,* **67**: 556-562.

Niittylä, T., Messerli, G., Trevisan, M., Chen, J., Smith, A.M. and Zeeman, S.C. (2004) A previously unknown maltose transporter essential for starch degradation in leaves. *Science,* **303**: 87-89.

Niland, S. and Schmitz, K. (1995) Sugar transport into storage tubers of *Stachys sieboldii* Miq.: Evidence for symplastic unloading and stachyose uptake into storage vacuoles by an $H^+$-antiport mechanism. *Bot. Acta,* **108**: 24-33.

Noiraud, N., Delrot, S. and Lemoine, R. (2000) The sucrose transporter of celery. Identification and expression during salt stress. *Plant Physiol.,* **122**: 1447-1455.

Noiraud, N., Maurousset, L. and Lemoine, R. (2001a) Transport of polyols in higher plants. *Plant Physiol. Biochem.,* **39**: 717-728.

Noiraud, N., Maurousset, L. and Lemoine, R. (2001b) Identification of a mannitol transporter, AgMaT1, in celery phloem. *Plant Cell*, **13**: 695-705.

Nørholm, M.H.H., Nour-Eldin, H.H., Brodersen, P., Mundy, J. and Halkier, B.A. (2006) Expression of the *Arabidopsis* high-affinity hexose transporter STP13 correlates with programmed cell death. *FEBS Lett.*, **580**: 2381-2387.

Overvoorde, P.J., Frommer, W.B. and Grimes, H.D. (1996) A soybean sucrose binding protein independently mediates nonsaturable sucrose uptake in yeast. *Plant Cell*, **8**: 271-280.

Özcan, S., Dover, J. and Johnston, M. (1998) Glucose sensing and signaling by two glucose receptors in yeast *Saccharomyces cerevisiae*. *EMBO J.*, **17**: 2566-2573.

Özcan, S., Dover, J., Rosenwald, A.G., Wölfel, S. and Johnston, M. (1996) Two glucose transporters in *Saccharomyces cerevisiae* are glucose sensors that generate a signal for induction of gene expression. *Proc. Natl. Acad. Sci. USA*, **93**: 12428-12432.

Pedra, J.H.F., Delu-Filho, N., Pirovani, C.P., Contim, L.A.S., Dewey, R.E., Otoni, W.C. and Fontes, E.P.B. (2000) Antisense and sense expression of a sucrose binding protein homologue gene from soybean in transgenic tobacco affects plant growth and carbohydrate partitioning in leaves. *Plant Sci.*, **152**: 87-98.

Pontis, H.G., Gonzalez, P. and Etxeberria, E. (2002) Transport of 1-kestose across the tonoplast of Jerusalem artichoke tubers. *Phytochemistry*, **59**: 241-247.

Preisser, J. and Komor, E. (1991) Sucrose uptake into vacuoles of sugarcane suspension cells. *Planta*, **186**: 109-114.

Quirino, B.F., Reiter, W.D. and Amasino, R.D. (2001) One of two tandem *Arabidopsis* genes homologous to monosaccharide transporters is senescence-associated. *Plant Mol. Biol.*, **46**: 447-457.

Rae, A.L., Perroux, J.M. and Grof, C.P.L. (2005) Sucrose partitioning between vascular bundles and storage parenchyma in the sugarcane stem: a potential role for the ShSUT1 sucrose transporter. *Planta*, **220**: 817-825.

Ramsperger-Gleixner, M., Geiger, D., Hedrich, R. and Sauer, N. (2004) Differential expression of sucrose transporter and polyol transporter genes during maturation of common plantain companion cells. *Plant Physiol.*, **134**: 147-160.

Ransom-Hodgkins, W.D., Vaughn, M.W. and Bush, D.R. (2003) Protein phosphorylation plays a key role in sucrose-mediated transcriptional regulation of a phloem-specific proton-sucrose symporter. *Planta*, **217**: 483-489.

Ratajczak, R. (2000) Structure, function and regulation of the plant vacuolar $H^+$-translocating ATPase. *Biochim. Biophys. Acta*, **1465**: 17-36.

Rausch, T., Bütcher, D.N. and Taiz, L. (1987) Active glucose transport and proton pumping in tonoplast membrane of *Zea mays* L. coleoptiles are inhibited by anti-$H^+$-ATPase antibodies. *Plant Physiol.*, **85**: 996-999.

Reinders, A., Panshyshyn, J.A. and Ward, J.M. (2005) Analysis of transport activity of Arabidopsis sugar alcohol permease homolog AtPLT5. *J. Biol. Chem.*, **280**: 1594-1602.

Reinders, A., Schulze, W., Kühn, C., Barker, L., Schulz, A., Ward, M.J. and Frommer, W.B. (2002) Protein-protein interactions between sucrose transporters of different affinities colocalized in the same enucleate sieve element. *Plant Cell*, **14**: 1567-1577.

Riesmeier, J.W., Hirner, B. and Frommer, W.B. (1993) Potato sucrose transporter expression in minor veins indicates a role in phloem loading. *Plant Cell*, **5**: 1591-1598.

Riesmeier, J.W., Willmitzer, L. and Frommer, W.B. (1992) Isolation and characterization of a sucrose carrier cDNA from spinach by functional expression in yeast. *EMBO J.*, **11**: 4705-4713.

Sugar and polyol transporters

Riesmeier, J.W., Willmitzer, L. and Frommer, W.B. (1994) Evidence for an essential role of the sucrose transporter in phloem loading and assimilate partitioning. *EMBO J.*, **13**: 1-7.
Ripp, K.G., Viitanen, P.V., Hitz, W.D. and Franceschi, V.R. (1988) Identification of a membrane protein associated with sucrose transport into cells of developing soybean cotyledons. *Plant Physiol.*, **88**: 1435-1445.
Roblin, G., Sakr, S., Bonmort, J. and Delrot, S. (1998) Regulation of a plant plasma membrane sucrose transporter by phosphorylation. *FEBS Lett.*, **424**: 165-168.
Rosche, E., Blackmore, D., Tegeder, M., Richardson, T., Schroeder, H., Higgins, T.J.V., Frommer, W.B., Offler, C.E. and Patrick, J.W. (2002) Seed-specific overexpression of a potato sucrose transporter increases sucrose uptake and growth rates of developing pea cotyledons. *Plant J.*, **30**: 165-175.
Sakr, S., Noubahni, M., Bourbouloux, A., Riesmeier, J., Frommer, W.B., Sauer, N. and Delrot, S. (1997) Cutting, ageing and expression of plant membrane transporters. *Biochim. Biophys. Acta*, **1330**: 207-216.
Salmon, S., Lemoine, R., Jamai, A., Bouché-Pillon, S. and Fromont, J.C. (1995) Study of sucrose and mannitol transport in plasma-membrane vesicles from phloem and non-phloem tissues of celery (*Apium graveolens* L.) petioles. *Planta*, **197**: 76-83.
Sauer, N. and Stadler, R. (1993) A sink-specific $H^+$/monosaccharide co-transporter from *Nicotiana tabacum*: cloning and heterologous expression in baker's yeast. *Plant J.*, **4**: 601-610.
Sauer, N. and Tanner, W. (1989) The hexose carrier from *Chlorella* cDNA cloning of a eucaryotic $H^+$-cotransporter. *FEBS Lett.*, **259**: 43-46.
Sauer, N., Friedländer, K. and Gräml-Wicke, U. (1990) Primary structure, genomic organization and heterologous expression of a glucose transporter from *Arabidopsis thaliana*. *EMBO J.*, **9**: 3045-3050.
Sauer, N., Ludwig, A., Knoblauch, A., Rothe, P., Gahrtz, M. and Klebl, F. (2004) AtSUC8 and AtSUC9 encode functional sucrose transporters, but the closely related AtSUC6 and AtSUC7 genes encode aberrant proteins in different Arabidopsis ecotypes. *Plant J.*, **40**: 120-130.
Sazuka, T., Keta, S., Shiratake, K., Yamaki, S. and Shibata, D. (2004) Identification of membrane-bound proteins from a vacuolar membrane-enriched fraction of *Arabidopsis thaliana*. *DNA Res.*, **11**: 101-113.
Schneider, S., Schneidereit, A., Konrad, K.R., Hajirezaei, M.R., Gramann, M., Hedrich, R. and Sauer, N. (2006) Arabidopsis INOSITOL TRANSPORTER4 mediates high-affinity $H^+$ symport of myoinositol across the plasma membrane. *Plant Physiol.*, **141**: 565-577.
Schneidereit, A., Scholz-Starke, J. and Büttner, M. (2003) Functional characterization and expression analyses of the glucose-specific AtSTP9 monosaccharide transporter in pollen of Arabidopsis. *Plant Physiol.*, **133**: 182-190.
Schneidereit, A., Scholz-Starke, J., Sauer, N. and Büttner, M. (2005) AtSTP11, a pollen tube-specific monosaccharide transporter in Arabidopsis. *Planta*, **221**: 48-55.
Scholz-Starke, J., Büttner, M. and Sauer, N. (2003) AtSTP6, a new pollen-specific $H^+$-monosaccharide symporter from Arabidopsis. *Plant Physiol.*, **131**: 70-77.
Schulz, A., Kühn, C., Riesmeier, J.W. and Frommer, W.B. (1998) Ultrastructural effects in potato leaves due to antisense-inhibition of the sucrose transporter indicate an apoplasmic mode of phloem loading. *Planta*, **206**: 533-543.
Schulze, W., Weise, A., Frommer, W.B. and Ward, J.M. (2000) Function of the cytosolic N-terminus of sucrose transporter AtSUT2 in substrate affinity. *FEBS Lett.*, **485**: 189-194.

Scofield, G.N., Hirose, T., Gaudron, J.A., Upadhyaya, N.M., Ohsugi, R. and Furbank, R.T. (2002) Antisense suppression of the rice sucrose transporter gene, *OsSUT1*, leads to impaired grain filling and germination but does not affect photosynthesis. *Funct. Plant Biol.,* **29**: 815-826.

Shakya, R. and Sturm, A. (1998) Characterization of source- and sink-specific sucrose/$H^+$ symporters from carrot. *Plant Physiol.,* **118**: 1473-1480.

Sherson, S.M., Alford, H.L., Forbes, S.M., Wallace, G. and Smith, S.M. (2003) Roles of cell-wall invertases and monosaccharide transporters in the growth and development of Arabidopsis. *J. Exp. Bot.,* **54**: 525-531.

Sherson, S.M., Hemmann, G., Wallace, G., Forbes, S., Germain, V., Stadler, R., Bechtold, N., Sauer, N. and Smith, S.M. (2000) Monosaccharide/proton symporter AtSTP1 plays a major role in uptake and response of Arabidopsis seeds and seedlings to sugars. *Plant J.,* **24**: 849-857.

Shimaoka, T., Ohnishi, M., Sazuka, T., Mitsuhashi, N., Hara-Nishimura, I., Shimazaki, K.I., Maeshima, M., Yokota, A., Tomizawa, K.I. and Mimura, T. (2004) Isolation of intact vacuoles and proteomic analysis of tonoplast from suspension-cultured cells of *Arabidopsis thaliana. Plant Cell Physiol.,* **45**: 672-683.

Shiratake, K. (2007) Genetics of sucrose transport in plants. *Genes, Genomes and Genomics,* **1**: 73-80.

Shiratake, K. and Martinoia, E. (2007) Transporters in fruit vacuole. *Plant Biotechnol.,* **24**: 127-133.

Shiratake, K., Kanayama, Y., Maeshima, M. and Yamaki, S. (1998) Changes in tonoplast protein and density with the development of pear fruit. *Physiol. Plant.,* **103**: 312-319.

Shiratake, K., Kanayama, Y. and Yamaki, S. (1997) Characterization of hexose transporter for facilitated diffusion of the tonoplast vesicles from pear fruit. *Plant Cell Physiol.,* **38**: 910-916.

Sivitz, A.B., Reinders, A. and Ward, J.M. (2005) Analysis of the transport activity of barley sucrose transporter HvSUT1. *Plant Cell Physiol.,* **46**: 1666-1673.

Stadler, R. and Sauer, N. (1996) The *Arabidopsis thaliana* AtSUC2 gene is specifically expressed in companion cells. *Bot. Acta,* **109**: 299-306.

Stadler, R., Brandner, J., Schulz, A., Gahrtzv, M. and Sauer, N. (1995) Phloem loading by the PmSUC2 sucrose carrier from *Plantago major* occurs into companion cells. *Plant Cell,* **7**: 1545-1554.

Stadler, R., Büttner, M., Ache, P., Hedrich, R., Ivashikina, N., Melzer, M., Shearson, S.M., Smith, S.M. and Sauer, N. (2003) Diurnal and light-regulated expression of AtSTP1 in guard cells of Arabidopsis. *Plant Physiol.,* **133**: 528-537.

Stadler, R., Truernit, E., Gahrtz, M. and Sauer, N. (1999) The AtSUC1 sucrose carrier may represent the osmotic driving force for anther dehiscence and pollen tube growth in Arabidopsis. *Plant J.,* **19**: 269-278.

Stolz, J., Stadler, R., Opekarova, M. and Sauer, N. (1994) Functional reconstitution of the solubilized *Arabidopsis thaliana* STP1 monosaccharide-$H^+$ symporter in lipid vesicles and purification of the histidine tagged protein from transgenic *Saccharomyces cerevisiae. Plant J.,* **6**: 225-233.

Szponarski, W., Sommerer, N., Boyer, J.C., Rossignol, M. and Gibrat, R. (2004) Large-scale characterization of integral proteins from Arabidopsis vacuolar membrane by two-dimensional liquid chromatography. *Proteomics,* **4**: 397-406.

Tegeder, M., Wang, X.D., Frommer, W.B., Offler, C.E. and Patrick, J.W. (1999) Sucrose transport into developing seeds of *Pisum sativum* L. *Plant J.,* **18**: 151-161.

Thom, M. and Komor, E. (1984) $H^+$-sugar antiport as the mechanism of sugar uptake by sugarcane vacuoles. *FEBS Lett.,* **173**: 1-4.

Sugar and polyol transporters

Thom, M., Getz, H.P. and Maretzki, A. (1992) Purification of a tonoplast polypeptide with sucrose transport properties. *Physiol. Plant.*, **86**: 104-114.
Toyofuku, K., Kasahara, M. and Yamaguchi, J. (2000) Characterization and expression of monosaccharide transporters (*OsMSTs*) in rice. *Plant Cell Physiol.*, **41**: 940-947.
Truernit, E. and Sauer, N. (1995) The promoter of the *Arabidopsis thaliana* SUC2 sucrose-H$^+$ symporter gene directs expression of beta-glucuronidase to the phloem: Evidence for phloem loading and unloading by SUC2. *Planta*, **196**: 564-570.
Truernit, E., Schmid, J., Epple, P., Illig, J. and Sauer, N. (1996) The sink-specific and stress-regulated Arabidopsis STP4 gene: Enhanced expression of a gene encoding a monosaccharide transporter bywounding, elicitors, and pathogen challenge. *Plant Cell*, **8**: 2169-2182.
Truernit, E., Stadler, R., Baier, K. and Sauer, N. (1999) A male gametophyte-specific monosaccharide transporter in Arabidopsis. *Plant J.*, **17**: 191-201.
Vaughn, M.W., Harrington, G.N. and Bush, D.R. (2002) Sucrose-mediated transcriptional regulation of sucrose symporter activity in the phloem. *Proc. Natl. Acad. Sci. USA*, **99**: 10876-10880.
Vignault, C., Vachaud, M., Cakir, B., Glissant, D., Dedaldechamp, F., Buettner, M., Atanassova, R., Fleurat-Lessard, P., Lemoine, R. and Delrot, S. (2005) VvHT1 encodes a monosaccharide transporter expressed in the conducting complex of the grape berry phloem. *J. Exp. Bot.*, **56**: 1409-1418.
Wallaart, R.A.M. (1980) Distribution of sorbitol in Rosaceae. *Phytochemistry*, **19**: 2603-2610.
Ward, J.M., Kühn, C., Tegeder, M. and Frommer, W.B. (1998) Sucrose transport in higher plants. *Int. Rev. Cytol.*, **178**: 41-71.
Watari, J., Kobae, Y., Yamaki, S., Yamada, K., Toyofuku, K., Tabuchi, T. and Shiratake, K. (2004) Identification of sorbitol transporters expressed in the phloem of apple source leaves. *Plant Cell Physiol.*, **8**: 1032-1041.
Weber, A., Servaites, J.C., Geiger, D.R., Kofler, H., Hille, D., Gröner, F., Hebbeker, U. and Flügge, U.I. (2000) Identification, purification, and molecular cloning of a putative plastidic glucose transocator. *Plant Cell*, **12**: 787-801.
Weber, H., Borisjuk, L., Heim, U., Sauer, N. and Wobus, U. (1997) A role for sugar transporters during seed development: Molecular characterization of a hexose and a sucrose carrier in fava bean seeds. *Plant Cell*, **9**: 895-908.
Weig, A. and Komor, E. (1996) An active sucrose carrier (Scr1) that is predominantly expressed in the seedling of *Ricinus communis* L. *J. Plant Physiol.*, **147**: 685-690.
Weise, A., Barker, L., Kühn, C., Lalonde, S., Buschmann, H., Frommer, W.B. and Ward, J.M. (2000) A new subfamily of sucrose transporters, SUT4, with low affinity/high capacity localized in enucleate sieve elements of plants. *Plant Cell*, **12**: 1345-1355.
Weschke, W., Panitz, R., Gubatz, S., Wang, Q., Radchuk, R., Weber, H. and Wobus, U. (2003) The role of invertases and hexose transporters in controlling sugar ratios in maternal and filial tissues of barley caryopses during early development. *Plant J.*, **33**: 395-411.
Weschke, W., Panitz, R., Sauer, N., Wang, Q., Neubohn, B., Weber, H. and Wobus, U. (2000) Sucrose transport into barley seeds: molecular characterization of two transporters and implications for seed development and starch accumulation. *Plant J.*, **21**: 455-467.
Williams, L., Thom, M. and Maretzki, A. (1990) Characterization of a proton translocating ATPase and sucrose uptake in a tonoplast-enriched vesicle fraction from sugarcane. *Physiol. Planta.*, **80**: 169-176.

Williams, L.E., Lemoine, R. and Sauer, N. (2000) Sugar transporters in higher plants: a diversity of roles and complex regulation. *Trends Plant Sci.,* **5**: 283-290.

Williamson, J.D., Jennings, D.B., Guo, W.W., Pharr, D.M. and Ehrenshaft, M. (2002) Sugar alcohols, salt stress, and fungal resistance: Polyols-multifunctional plant protection? *J. Amer. Soc. Hort. Sci.,* **127**: 467-473.

Wormit, A., Trentmann, O., Feifer, I., Lohr, C., Tjaden, J., Meyer, S., Schmidt, U., Martinoia, E. and Neuhaus, H.E. (2006) Molecular identification and physiological characterization of a novel monosaccharide transporter from Arabidopsis involved in vacuolar sugar transport. *Plant Cell,* **18**: 3476-3490.

Yamaki, S. (1987) ATP-promoted sorbitol transport into vacuoles isolated from apple fruit. *Plant Cell Physiol.,* **28**: 557-564.

Yamaki, S. and Asakura, T. (1988) Energy coupled transport of sorbitol and other sugars into the protoplast isolated from apple fruit flesh. *Plant Cell Physiol.,* **29**: 961-967.

Yamaki, S. and Asakura, T. (1991) Stimulation of the uptake of sorbitol into vacuoles from apple fruit flesh by abscisic acid and into protoplasts by indoleacetic acid. *Plant Cell Physiol.,* **32**: 315-318.

Ylstra, B., Garrido, D., Busscher, J. and van Tunen, A.J. (1998) Hexose transport in growing petunia pollen tubes and characterization of a pollen-specific, putative monosaccharide transporter. *Plant Physiol.,* **118**: 297-304.

Zimmermann, M.H. and Ziegler, H. (1975) List of sugar and sugar alcohols in sieve-tube exdudates. In: *Transport in Plants I: Phloem Transport, Encyclopedia of Plant Physiology, New Series, Vol. 1* (Ed. Zimmermann, M.H.), Springer-Verlag, Berlin, pp 480-503.

# Chapter 11

## AMINO ACID TRANSPORTERS IN PLANTS

### UWE LUDEWIG AND WOLFGANG KOCH
*Center for Plant Molecular Biology, Plant Physiology, University of Tübingen, Auf der Morgenstelle 1, D-72076 Tübingen, Germany*
E-mail: uwe.ludewig@zmbp.uni-tuebingen.de

**Abstract**

Plants require large amounts of nitrogen for optimal growth and development. Nitrogen is mainly acquired from the soil in inorganic form such as $NO_3^-$ and $NH_4^+$. Inorganic nitrogen is partially assimilated in roots or translocated via the vasculature and is then reduced to organic nitrogen and assimilated into amino acids. Amino acids are translocated to the sink tissues via the phloem and xylem. The loading and unloading steps require amino acid transport proteins. During plant development, the nitrogen requirement for most organs changes drastically. In the earlier stages of development, growing leaves and roots are major sinks for carbon and nitrogen. At later stages, reproductive organs, such as seeds or tubers, become the major sinks. In the model plant Arabidopsis, but similarly in other plants, several amino acid transporters with specific expression patterns and transport activities have been molecularly identified and characterized in the past years. Related genes are found in the databases of genome projects of various plant species. The potential role of these transporters in the plant adaptation to meet the nitrogen requirements of different plant tissues and under varying environmental conditions is discussed and summarized here.

**Keywords:** amino acid transport, long distance transport, uptake and distribution of organic nitrogen

## 1. INTRODUCTION

Nitrogen is quantitatively the most important mineral nutrient for plant growth and often limits optimal development. Under most conditions nitrogen nutrition mainly depends on inorganic nitrate and ammonium. Once taken up, inorganic nitrogen is either directly assimilated into amino acids in the roots or translocated to the leaves, which frequently have larger nitrogen assimilation capacity. In certain ecosystems with slow mineralization rates, however, organic nitrogen sources such as peptides and amino acids can play a crucial role in nitrogen nutrition. It has

---

© CAB International 2008. *Plant Membrane and Vacuolar Transporters* (eds P.K. Jaiwal, R.P. Singh and O.P. Dhankher)

been shown that plants in boreal forests, arctic tundra and alpine regions take up and utilize organic nitrogen in the form of amino acids (Meline and Nilsson, 1953; Nasholm et al., 1998). Amino acid transport may also play an important role in the communication of plant roots with fungal or bacterial symbiotic partners; the molecular nature of these amino acid (or even peptide) transporters, however, has not yet been identified.

In contrast to animals and humans, whose diet requires the uptake of several essential amino acids, plants have the ability to synthesize all amino acids. However, this capacity differs between organs and varies for specific amino acids. The amino acids (AA) are precursors not only of protein biosynthesis, but of multiple primary and secondary metabolites. From the site of biosynthesis, amino acids are distributed together with water and other solutes between different organs via apoplastic and symplastic pathways. Transport in the xylem is an unidirectional flow along the transpiration stream. Generally all amino acids are found in the xylem and phloem, but the xylem amino acid concentration is ~20-fold lower than that of the phloem. This is in contrast to the distribution of carbohydrates, such as sucrose, which are not identified in the xylem. Solute transport within the symplast, including the phloem, is by osmotically driven pressure flow.

The most prominent amino acids in the phloem are species specific, but glutamine, glutamate, aspartate and asparagine often dominate (Lam et al., 1995). These amino acids are therefore called transport amino acids and follow mainly the 'sink' pathway of carbohydrates to developing organs such as young developing leaves, roots or fruits. The total free amino acid concentration varies between night and day, reflecting the transcriptional regulation of assimilating enzymes by light and the varying availability of metabolic precursors. Especially the availability of the carbohydrate skeleton derived from photosynthesis has an impact on amino acid biosynthesis and the amount of the transport amino acids glutamine and asparagine (Lam et al., 1996). The concentration of amino acids in the phloem and in the mesophyll cytoplasm is significantly higher than the concentration found in the apoplasm, suggesting an active import of amino acids into the phloem and permanent transport activities to maintain the gradient (Lohaus and Heldt, 1995). Tracer experiments with radiolabelled amino acids have shown that a major amount of amino acids that is added to the xylem appears later in the phloem sap, indicating that amino acids can be exchanged between xylem and phloem and cycle within the whole plant (Pate et al., 1975; Atkins, 2000).

Physiological studies of amino acid transport in several plants and isolated plasma membrane vesicles suggest the existence of multiple amino acid carrier systems which do not have specificity to amino acids in general, but show little preference among specific amino acids. All physiological studies on plants or plant materials have indicated that amino acids are transported across the plasma membranes by co-transport with protons (Kinraide, 1980; Mounoury et al., 1984; Li and Bush, 1990a,b). The first kinds of plant plasma membrane amino acid transporters were molecularly and functionally identified by the complementation of yeast mutants defective in amino acid uptake (Frommer et al., 1993; Hsu et al.,

## Amino acid transport in plants

1993; Kwart *et al.*, 1993; Fischer *et al.*, 1995; Frommer *et al.*, 1995; Rentsch *et al.*, 1996; Chen and Bush, 1997; Chen *et al.*, 2001). Within the entire genome of a single plant, such as the model plant Arabidopsis, genes coding for more than 53 putative amino acid carriers have been identified (Wipf *et al.*, 2002). In every sequenced plant genome, orthologs to the Arabidopsis transporters are found, and an increasing number of these transporters have functionally been characterized as amino acid transporters. The first functionally identified amino acid transporters were members of the AAP-subfamily (amino acid permeases), a subgroup belonging to the ATF family (amino acid transporter family). A phylogenetic tree of characterized ATF transporters from various species is shown in Figure 1. Structurally and functionally related proteins were later identified in yeast and animals (Fischer *et al.*, 1998). The ATF superfamily contains plant-specific sub-branches and branches that are more structurally related to mammalian transporters (Wipf *et al.*, 2002). Within the plant specific sub-branches, the best-characterized members are the AAPs (Figure 1). Several AAPs have been isolated from Arabidopsis by complementation of yeast transport mutants with plant cDNAs (Frommer *et al.*, 1993; Hsu *et al.*, 1993; Kwart *et al.*, 1993; Fischer *et al.*, 1995; Rentsch *et al.*, 1996). AAPs mediate H$^+$-coupled uptake of a wide spectrum of amino acids (AA) when expressed in *Xenopus* oocytes (Fischer *et al.*, 2002). *At*AAP1–5 represent low affinity transporters with low selectivity towards AA side chains; only AAP3 and 5 transport basic AA efficiently. AAP6 has a ~10-fold higher affinity to all substrates when compared with other AAPs. For glutamine, glutamate and asparagine, the major transport forms of organic nitrogen, these transporters have $K_m$-values that are in a physiological range with respect to the concentrations of these AA found in the phloem and xylem. Aspartate is transported by most AAPs with an extremely low affinity and efficiency. Only AAP6 and AAP8 have a $K_m$-value for aspartate that is in a physiological range (Fischer *et al.*, 2002; Okumoto *et al.*, 2002).

Phylogenetic analysis reveals two additional major sub-groups with functionally characterized transporters: the ProT-family (Grallath *et al.*, 2005), that transports proline and compatible solutes, including AtGAT1 which transports rather specifically γ-aminobutytric acid (Meyer *et al.*, 2006). Another sub-family was named after the founding member LHT1, which was initially described as a lysine and histidine specific transporter (Chen *et al.*, 1997). Recent studies characterized both LHT1 and LHT2 as high affinity transporters for acidic and neutral amino acids (Lee and Tegeder, 2004; Hirner *et al.*, 2006).

ANT1, the only characterized member of a branch more distantly related to the AAPs mediates the transport of aromatic and neutral amino acids (Chen *et al.*, 2001). An auxin importer AUX1 is also suggested to be related to the amino acid transporter (Bennett *et al.*, 1996). Given the structural similarities of the plant hormone auxin and tryptophan, it is not too surprising to identify auxin transporters within the ATF family.

A second unrelated protein family with functionally characterized proteins is the APC family, containing the CATs (cationic amino acid transporters) and additional putative amino acid transporters (LATs). Genes encoding proteins of this group are

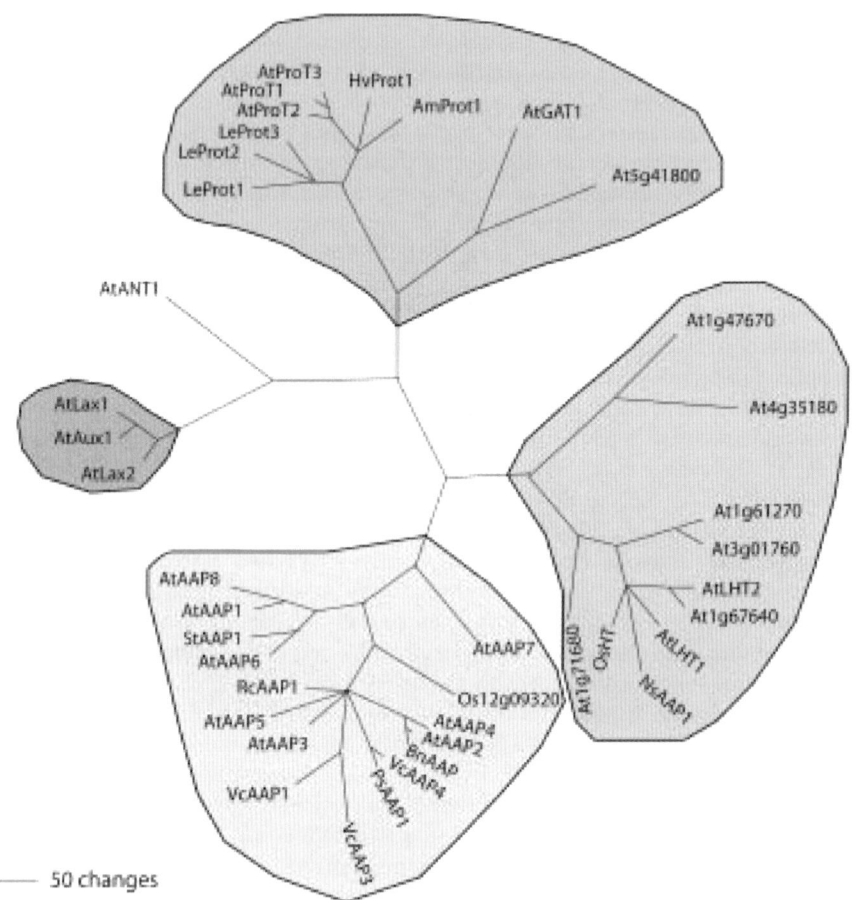

**Figure 1:** Phylogenetic analysis of selected members of the ATF superfamily from various plant species. Maximum parsimony analysis was performed using PAUP 4.0b4a (Swofford, 1998) with all DNA characters unweighted and gaps scored as missing characters. The complete alignment was based on 503 sites. 469 were phylogenetically informative. Sequences from different plant species cluster together in the major clades of the ATF-family: The AAPs, LHTs and the compatible solute transporters (ProTs). These families harbour functional characterized transporters from Arabidopsis (At), tomato (Le), potato (St), canola (Bn), bean (Vc), pisum (Ps), castor plant (Rc), barley (Hv), tobacco (Ns), rice (Os) and mangrove (Am). Numerous additional putative amino acid transporters are present in the Arabidopsis genome and related sequences to AtANT (aromatic and neutral amino acid transporter and AUX1 (auxin transporter 1) can be found in every sequenced genome. These proteins have typically 9-11 transmembrane domains.

found in prokaryotes, fungi and mammals in addition to plants, but in the plant kingdom, only some members have been characterized in Arabidopsis so far (Frommer et al., 1995; Su et al., 2004; Hammes et al., 2006).

The two major techniques used to characterize the transport activities of the amino acid tranporters are shown in Figure 2. These techniques involve yeast complementation and/or expression in *Xenopus* oocytes. For characterization of amino acid transport in yeast, several mutant strains with compromised amino acid uptake have been used. While these strains are unable to grow on specific amino acids as sole nitrogen sources, heterologous expression of plant amino acid transport proteins allows these strains to grow. In the following pages, we try to describe how this huge number of transporters with overlapping substrate affinities is intergrated in a nitrogen allocation network in the plant.

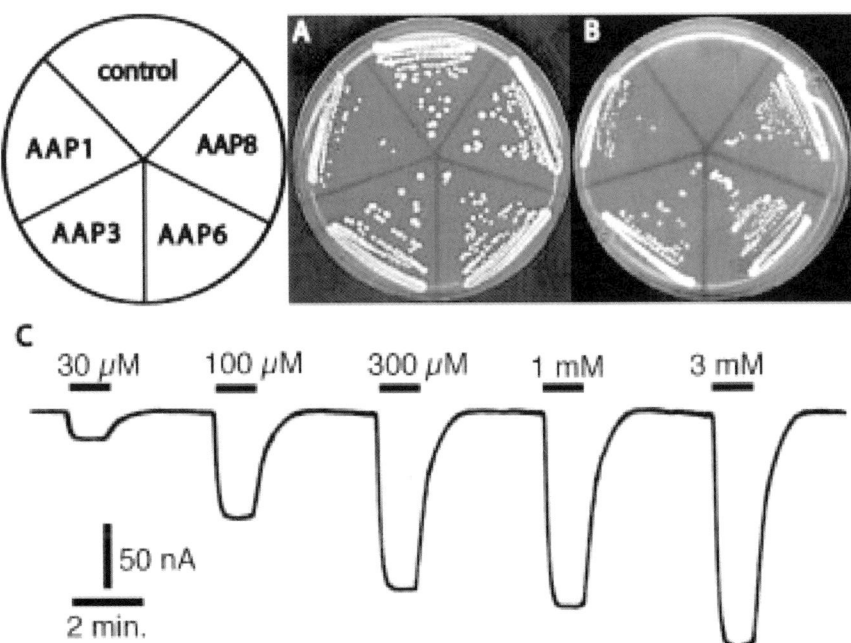

**Figure 2:** Widespread methods to characterize amino acid transporters. Analysis of transport activity of AAPs in a yeast mutant strain. The yeast strain is unable to grow on amino acids (proline, glutamate, aspartate, citrulin and GABA) as N-source. The strain, however, can grow on 10 mM $(NH_4)_2SO_4$ and yeast transformed with vector control and AAPs is shown in (A). On limiting N-conditions (3 mM proline) only yeast transformed with the plant amino acid permeases are able to grow (B). Analysis of transport activity by expression of an AAP in *Xenopus* oocytes (C). Currents elicited with variable amounts of arginine in AtAAP5-expressing oocytes at external pH 5.0.

## 2. POTENTIAL ROLE OF THE SUBGROUPS OF THE ATF-FAMILY

### 2.1. AAPs

*2.1.1. Transporters involved in long distance transport of amino acids and seed development*

Since plants take up nitrate and ammonium and also AA as a nitrogen source and, more importantly, since AA are the principal long distance transport forms of organic nitrogen, it is necessary to localize the transporters to identify their specific function. In accordance with the necessity to distribute amino acids throughout

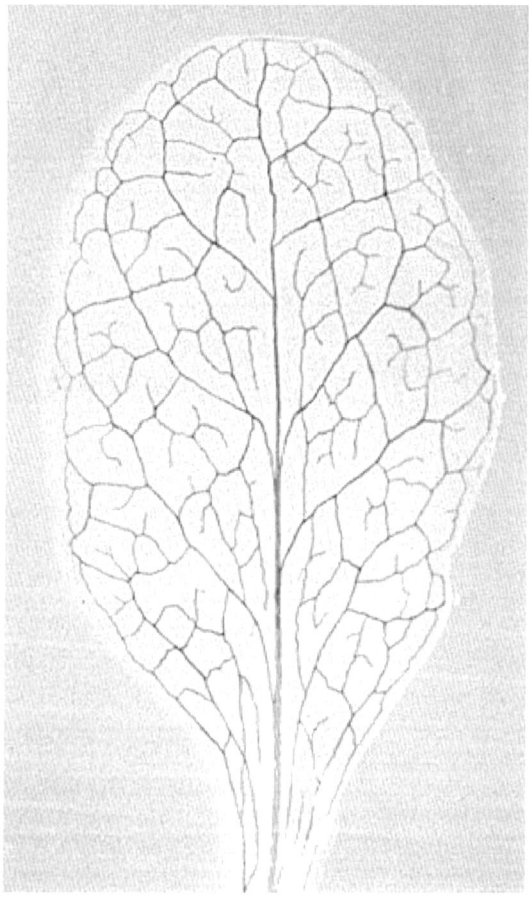

**Figure 3:** Expression of the GUS gene under control of the promoter of an amino acid transporter. The GUS expression shows promoter activity in the vasculature of a mature Arabidopsis leaf.

Amino acid transport in plants

the whole plant, reporter gene analyses have shown that the members of the broad specific amino acid transporters, the AAP-family, are associated with the vascular tissues or reproductive organs (Fig. 3).

Several members of the AAP-family, e.g. AAP1 and AAP2 were found in the seed and the vasculature of the siliques (Hirner et al., 1998). AAP8 has been localized to young developing seeds and later in development the expression switches to the vascular tissue of the silique (Okumoto et al., 2002). Expression of the high affinity transporter AAP6 was found in the xylem parenchyma, suggesting a role in the discussed xylem to phloem transfer of the amino acids. AAP6 is thus thought to transfer amino acids from the apoplastic space of the xylem to the surrounding xylem parenchyma cells. A possible candidate for the loading of amino acids taken up by roots into the phloem and supplying growing roots with amino acids is AAP3. The expression of AAP3 is restricted to the companion cells of root phloem and the endodermis (Okumoto et al., 2004). It has also been shown that AAP3 is localized to the plasmamembrane, the golgi apparatus and small vesicles. This probably reflects the fact that the transporter is most probably delivered to the membrane via golgi derived vesicles (Lee and Tegeder, 2004). The first direct evidence that AAPs are directly involved in long distance transport was shown by the *antisense* repression of the potato amino acid permease StAAP1. *StAAP1* is preferentially expressed in mature source leaves. A reduction of *StAAP1* mRNA levels in leaves leads to reduced amino acid content in potato tubers (Koch et al., 2003). The data suggested a role for StAAP1 in phloem loading in the leaves and demonstrate that repression of an amino acid transporter influences long distance transport. It is important to note that not only transport within the vascular system per se, but exchange between the xylem and phloem is expected to have high relevance for several organs of the plant. This is because nutrients, such as amino acids, have to be transferred to the rapidly growing, but exceedingly slowly transpiring sinks such as seeds, fruit and meristems, whose supply is mainly through the phloem.

*2.1.2. Seed development*

The import of AA is an essential prerequisite for the seed development, since the accumulation of storage proteins must be preceded by an AA import. It could be shown that the expression of the AA transporters in seeds, indeed, precedes the expression of the storage protein *At*S1 in Arabidopsis (Hirner et al., 1998). A similar increase in the expression of a legume AA transporter has been shown in *Vicia faba* (Miranda et al., 2001). In pea, it has been demonstrated that *Ps*AAP1 is expressed in the transfer cell layer of cotyledons and might be responsible for taking up AA released from the seed coat (Tegeder et al., 2000). This apoplasic gap between the maternal and filial tissue is a region where an intense exchange of nutrients requires import transporters at the filial part to deliver nutrients into the growing embryo. The important role of transporters at this site was shown by ectopic overexpression of an AA permease in pea seeds, which increased AA uptake and storage protein content by 40-50%. The results suggest that nitrogen supply limits protein synthesis and storage in the seed (Rolletschek et al., 2005). Although amino acid importers have been molecularly identified in seeds, the symplastic isolation of the developing embryo and reproductive cells from their maternal tissues suggests that exporters exist in the maternal tissues. As the phloem in source tissues is at least partially loaded via an apoplastic pathway, such exporters would also be expected to participate in AA export from mesophyll cells or export amino acids into the root xylem. However, such export transporters have not yet been identified. A putative component of an exporter includes the small membrane protein GDU1 that when overexpressed, leads to amino acid accumulation and glutamine secretion at the hydathodes (Pilot et al., 2004). However, GDU1 has only a single membrane spanning domain, making it unlikely to function as an amino acid exporter protein itself. All functionally characterized plant amino acid transporters are relatively large hydrophobic proteins with a molecular mass around 30-40 kDa, contain 9-14 membrane spans and function as proton symporters.

### 2.2. LHTs

From the second plant ATF-family, the LHTs (lysine histidin transporter, Fig. 1), two members in Arabidopsis and one in rice have been functionally characterized so far. *At*LHT1 has been described as a transporter specific for lysine and histidine, but other AA are also recognized (Chen and Bush, 1997). LHT1 is present in all tissues and its expression is strongest in young leaves, flowers and siliques of Arabidopsis. Expression of LHT1 is found in roots which could make it a candidate for amino acid uptake from the soil. In agreement with expression in roots, mutants of LHT1 are unable to grow on certain amino acids as sole nitrogen source and display reduced amino acid uptake activity (Hirner et al., 2006; Svennerstam et al., 2007). A second potential function for LHT1 is the uptake of amino acids into mesophyll cells from the apoplast probably mediating cell to cell transport in cooperation with an unidentified exporter (Hirner et al., 2006). Recent characterization of a related protein, LHT2, shows that this transporter is more likely to be a general AA permease with a high affinity for proline ($K_m \sim 10$ μM) and aspartate ($K_m \sim 72$ μM). Inhibition studies indicate potentially similar affinities to most AA, except the basic AA histidine, lysine and arginine (Lee and Tegeder, 2004). Expression of the gene in the tapetum tissue of the anther suggests a role in the transfer of organic nitrogen to the tapetum tissue and in supplying the pollen with amino acids. OsHT, the first transporter of this group described from rice, transports histidine like LHT1, but other amino acids have not been tested. The protein is probably localized to the plasma membrane (Liu et al., 2005).

*2.2.1. Redundancy of transporters or specificity?*

The AA concentrations in phloem, xylem, and supposedly in mesophyll cells, change during daytime subject to carbon supply, inorganic N and light. Given the metabolic variations, it is likely that the relatively immobile plants can respond to variations in AA concentrations with a multitude of transporters that have different $K_m$ values and transportation speed.

For many enzymatic reactions, including the principle pathway of nitrogen assimilation by glutamine synthetase (GS), L-glutamate dehydrogenase (GDH) and glutamate synthase (GOGAT), several isoenzymes are identified in plants. In spite of this, screens for auxotrophic mutants have revealed several mutants involved in amino acid biosynthesis, but these screens did not identify transport systems for amino acids, suggesting some redundancy of transport. The partially overlapping expression profiles prove this redundancy, and make it absolutely necessary to localize each of the individual proteins on cellular and subcellular levels in order to assign specific functions to each of the AA transporter genes. Nevertheless, the described properties of members of the ATF and APC families in relation to localization and transport, match the activities expected for transport proteins involved in moving a wide spectrum of AA in the phloem and xylem. The data described, so far, are consistent with the properties described for uptake kinetics measured in plasma membrane vesicles of leaves (Li and Bush, 1990a, b).

## 2.3. ProTs

### 2.3.1. Salt stress and desiccation

A special case in terms of amino acid transport is the transport of proline and glycine betain by members of the ProT-family (Fig. 1). These transporters belong to the ATF family but their transport activity is more specific. In contrast to the plant carriers just described, ProTs preferentially transport the compatible solutes proline ($K_m$ values: 0.4 mM, 0.5 mM and 1 mM AtProT1-3 respectively), betaine ($K_m$ values: 0.1mM, 0.3 mM and 0.3 mM AtProT 1, 2, 3) and γ-aminobutyric acid (4.5 mM, 4 mM and 5 mM) (Grallath et al., 2005; Rentsch et al., 1996). AtProT1 is expressed in phloem parenchyma cells, whereas AtProT2 expression is restricted to the epidermis and cortex cells of roots. Finally, AtProT3 is only expressed in the aerial parts of the plant, in the epidermal cells of the leaves. Induction of the expression of the AtProTs upon salt stress implies a role in stress adaptation or a function in organs of the plants that desiccate like pollens and seeds (Rentsch et al., 1996). In tomato, expression of LeProT1 is restricted to pollen and most likely contributes to the high proline content in mature pollen (up to 70% of free amino acid content) by importing proline into pollen (Schwacke et al., 1999). In mangroves, permanently exposed to high salt concentrations, the transport of compatible solutes is thought to be an important factor for adaptation to the environment. A functional proline transporter (AmProT1) grouping into the ProT family was isolated and characterized from mangroves (Waditee et al., 2002). In barley, a proline transporter was isolated from roots after salt stress treatment (Ueda et al., 2001). The regulation and properties of the ProTs support a function of the ProTs in accumulating osmoprotectans like proline and glycin betaine in tissues that are exposed to changing water supply like roots or desiccate, like seeds and pollen.

## 2.4. APCs

An unrelated group of amino acid transporters, which have relatives in many other organisms, including animals and prokaryotes, have only been fragmentarily analyzed. In Arabidopsis, 14 members of this APC-family (amino acid-polyamine-choline) of amino acid transporters have been identified (Fig. 4). Of these, two major branches, the CAT and LAT-branch can be distinguished. Members of the LAT-branch are entirely uncharacterized, but these proteins contain 12 putative transmembrane domains. In contrast, the members of the CAT branch have a relatively high similarity to cationic amino acid transporters in humans and are proposed to have 14 transmembrane domains. In all transporters from this APC-family, the carboxy- and amino-terminal tails are proposed to be on the same, probably cytosolic side. The first transporter characterized from this family was initially called AtAAT1, but was later renamed AtCAT1, because of a name conflict with the amino acid transferase enzyme one, that was also abbreviated AAT1. Heterologous expression in yeast showed that this transporter was indeed cation-specific, transported histidine and had a high affinity to arginine and lysine (Frommer et al., 1995). Similarly, the closely related surface expressed AtCAT5 transported the positively charged amino acids arginine, lysine and histidine, and was saturated

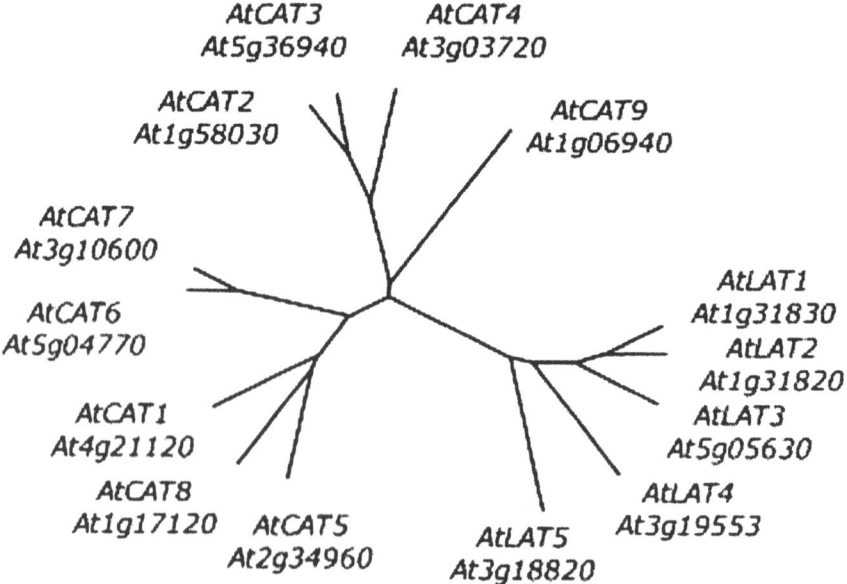

**Figure 4:** Phylogenetic analysis of the APC-family of amino acid transporters in the plant Arabidopsis. Two major clades are observed, the LAT-branch and the CAT-branch. Transporters of the LAT-branch are predicted to contain 12 transmembrane spans, while the transporters of the CAT-branch have 14 putative transmembrane domains.

with a $K_m$=12 µM for arginine (Su *et al.*, 2004). Cationic amino acids are found in relatively low concentration in the apoplast, thus to supply tissues with all amino acids, high affinity transport systems for cationic amino acids appear essential. The broad specific transporters that have no preferences may only weakly transport these rare amino acids. However, another transporter from the same branch (*At*CAT8) appears to have a different amino acid preference, which was evident from the import of toxic amino acid analogs, such as L-methionine sulfoximine, which resembles glutamate. The organ specific expression pattern of *At*CAT8 is evident from promoter-reporter plants in which the *At*CAT8 promoter drives the reporter β-glucuronidase (GUS) gene (Fig. 5). Enzymatic staining was observed in rapidly dividing tissues, such as the root tips and emerging leaves. This suggests a function of *At*CAT8 in amino acid supply to the major nitrogen sinks in young plants. In contrast, *At*CAT6 was expressed in lateral root primordia, flowers and seeds. *At*CAT6 transported large, neutral and cationic amino acids in preference to other amino acids (Hammes *et al.*, 2006).

## 3. INTRACELLULAR TRANSPORT

Interestingly, translational fusions with green fluorescent protein (GFP) have identified that members of the CAT family localize to different subcellular compartments (Fig. 6). While *At*CAT5 has been shown to be surface expressed and

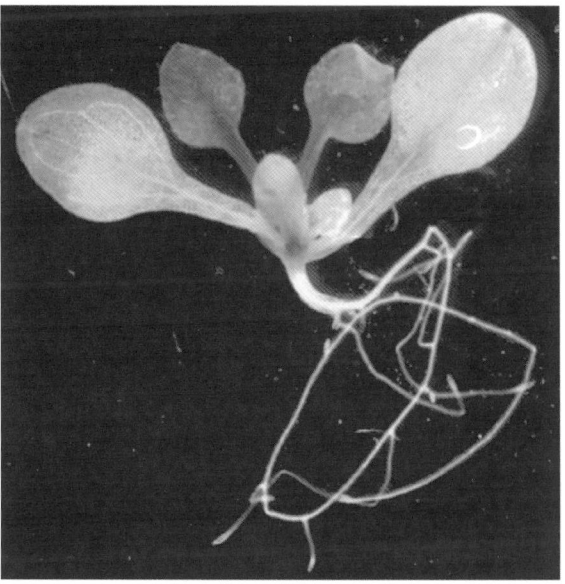

**Figure 5:** Analysis of the expression pattern of AtCAT8 using a promoter-GUS reporter construct expressing plant. Staining indicates the site of promoter activity and AtCAT8 expression.

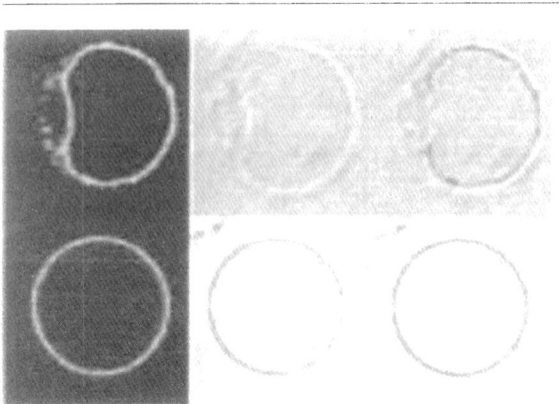

**Figure 6:** Subcellular localization of CAT-amino acid transporter translational fusions with GFP transiently expressed in Arabidopsis plant protoplasts. GFP-Fluorescence (left), bright field image (top right) and overlay (bottom right). Upper panel: *At*CAT2-GFP, lower panel: *At*CAT5-GFP.

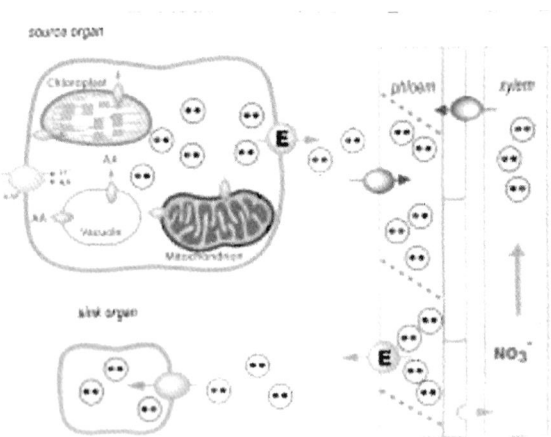

**Figure 7:** Proposed model illustrating transfer steps where transport proteins are necessary to mediate amino acid transfer across membranes. Amino acids are shown as ⊝ symbols. Amino acids are taken up into mesophyll cells and phloem by proton coupled importers in both sink and source organs. Exporters (E) are expected, but have not yet been molecularly identified, when amino acids exit the mesophyll cell of source organs to enter the apoplastic space prior to uptake into the phloem. Amino acid compartmentation and release from the vacuole, from chloroplasts and mitochondria is also mediated by amino acid transporters.

mediates amino acid transport across the plasma membrane, *At*CAT2 was localized to the vacuolar membrane, the tonoplast, when transiently expressed as a GFP-fusion protein in protoplasts (Su *et al.*, 2004). Vacuolar/vesicular transporters are expected to participate in amino acid storage and re-mobilization. Several amino acid transport activities have been biochemically, but not all of them molecularly, identified on the tonoplast of plant cells. These include facilitators that appear relatively broad specific (and transport many amino acids) and those whose transport is coupled to the presence or hydrolysis of ATP.

In barley and spinach, a non-aqueous fractionation analysis identified that despite the large vacuolar volume, the total amount of amino acids in the vacuole is generally rather low. The concentration of amino acids in the vacuole appears roughly 10-100-fold lower than those in the cytosol (Winter *et al.*, 1994). This suggests that the major transport activities in the tonoplast are frequently involved in amino acid export from the vacuole, rather than import from the cytoplasm. Vacuolar protease degradation of storage proteins in seeds is followed by amino acid release into the cytoplasm, suggesting an important role for vacuolar amino acid transporters during early germination, where nutrition relies almost completely on the stored nutrient content, and this is deposited in the vacuole. However, in storage compartments such as potato tubers, similar AA concentrations in the cytoplasm and in the vacuole can be found, suggesting equilibrative transport in such organs (Farre *et al.*, 2001).

In addition to vacuolar transporters, organellar im- and exporters are required. This reflects in part the structural complexity in plants, with compartmented amino acid biosynthesis in chloroplasts or mitochondria. Amino acid biosynthesis in plants is more complex than in unicellular microorganisms. Several transport activities in organelles have been reported, but relatively little is yet known about the molecular identity of the respective proteins. In plastids, the glutamate/malate translocator DiT2, has been molecularly identified and in mitochondrial membranes an arginine/ornithine shuttle.

## 4. CONCLUSIONS AND FURTHER PERSPECTIVES

Plants have abundant genes encoding amino acid transporters in their genome, which probably represents the complex requirements for the distribution of amino acids within plants. A summary of transport steps involved in optimal distribution of AA within the plant is schematically shown in Fig. 7. Two major gene families have been implicated in several of the transfer steps shown, but the list of genes to be characterized is still long. Direct assignment of individual physiological functions to specific genes remains a task for extensive future work.

## 5. LITERATURE CITED

Atkins, C.A. (2000). Biochemical aspects of assimilate transfers along the phloem path: N-solutes in lupins. *Aust. J. Plant Physiol.*, **27**: 531-537.

Bennett, M.J., Marchant, A., Green, H.G., May, S.T., Ward, S.P., Millner, P.A., Walker, A.R., Schulz, B. and Feldmann, K.A. (1996). Arabidopsis *AUX1* gene: a permease-like regulator of root gravitropism. *Science,* 273: 948-950.

Chen, L. and Bush, D.R. (1997). LHT1, a lysine- and histidine-specific amino acid transporter in Arabidopsis. *Plant Physiol.,* 115: 1127-1134.

Chen, L., Ortiz-Lopez, A., Jung, A. and Bush, D.R. (2001). ANT1, an aromatic and neutral amino acid transporter in Arabidopsis. *Plant Physiol.,* 125: 1813-1820.

Farre, E.M., Tiessen, A., Roessner, U., Geigenberger, P., Trethewey, R.N. and Willmitzer, L. (2001). Analysis of the compartmentation of glycolytic intermediates, nucleotides, sugars, organic acids, amino acids, and sugar alcohols in potato tubers using a nonaqueous fractionation method. *Plant Physiol.,* 127: 685-700

Fischer, W.N., Andre, B., Rentsch, D., Krolkiewicz, S., Tegeder, M., Breitkreuz, K. and Frommer, W.B. (1998). Amino acid transport in plants. *Trends in Plant Science,* 3: 188-195.

Fischer, W.N., Kwart, M., Hummel, S. and Frommer, W.B. (1995). Substrate specificity and expression profile of amino acid transporters (AAPs) in Arabidopsis. *J. Biol. Chem.,* 270: 16315-16320.

Fischer, W.N., Loo, D.D., Koch, W., Ludewig, U., Boorer, K.J., Tegeder, M., Rentsch, D., Wright, E.M. and Frommer, W.B. (2002). Low and high affinity amino acid $H^+$-cotransporters for cellular import of neutral and charged amino acids. *Plant J.,* 29: 717-731.

Frommer, W.B., Hummel, S. and Riesmeier, J.W. (1993). Expression cloning in yeast of a cDNA encoding a broad specificity amino acid permease from *Arabidopsis thaliana*. *Proc. Natl. Acad. Sci. USA,* 90: 5944-5948.

Frommer, W.B., Hummel, S., Unseld, M. and Ninnemann, O. (1995). Seed and vascular expression of a high-affinity transporter for cationic amino acids in Arabidopsis. *Proc. Natl. Acad. Sci. USA,* 92: 12036-12040.

Grallath, S., Weimar, T., Meyer, A., Gumy, C., Suter-Grotemeyer, M., Neuhaus, J.M. and Rentsch, D. (2005). The *At*ProT family. Compatible solute transporters with similar substrate specificity but differential expression patterns. *Plant Physiol.,* 137: 117-126.

Hammes, U.Z., Nielsen, E., Honaas, L.A., Taylor, C.G. and Schachtman, D.P. (2006) *At*CAT6, a sink-tissue-localized transporter for essential amino acids in Arabidopsis. *Plant J.,* 48: 416-426.

Hirner, A., Ladwig, F., Stransky, H., Okumoto, S., Keinath, M., Harms, A., Frommer, W.B. and Koch, W. (2006). Arabidopsis LHT1 is a high affinity transporter for cellular amino acid uptake in both root epidermis and leaf mesophyll. *Plant Cell,* 18: 1931-1946.

Hirner, B., Fischer, W.N., Rentsch, D., Kwart, M. and Frommer, W.B. (1998) Developmental control of $H^+$/amino acid permease gene expression during seed development of Arabidopsis. *Plant J.,* 14: 535-544.

Hsu, L.C., Chiou, T.J., Chen, L. and Bush, D.R. (1993). Cloning a plant amino acid transporter by functional complementation of a yeast amino acid transport mutant. *Proc. Natl. Acad. Sci. USA,* 90: 7441-7445.

Kinraide, T. (1980). Electrical evidence for different mechanisms of uptake of basic, neutral and acidic amino acids in oat. *Plant Physiol.,* 65: 1085-1089.

Koch, W., Kwart, M., Laubner, M., Heineke, D., Stransky, H., Frommer, W.B. and Tegeder, M. (2003). Reduced amino acid content in transgenic potato tubers due to antisense inhibition of the leaf $H^+$/amino acid symporter StAAP1. *Plant J.,* 33: 211-220.

Amino acid transport in plants

Kwart, M., Hirner, B., Hummel, S. and Frommer, W.B. (1993). Differential expression of two related amino acid transporters with differing substrate specificity in *Arabidopsis thaliana*. *Plant J.*, **4**: 993-1002.
Lam, H.M., Coschigano, K., Schultz, C., Melo-Oliveira, R., Tjaden, G., Oliveira, I., Ngai, N., Hsieh, M.H. and Coruzzi, G. (1995). Use of Arabidopsis mutants and genes to study amide amino acid biosynthesis 11. *Plant Cell*, **7**: 887-898.
Lam, H.M., Coschigano, K.T., Oliveira, I.C., Melo-Oliveira, R. and Coruzzi, G.M. (1996). The molecular-genetics of nitrogen assimilation into amino acids in higher plants. *Annu. Rev. Plant Physiol Plant Mol. Biol.*, **47**: 569-593.
Lee, Y.H. and Tegeder, M. (2004). Selective expression of a novel high-affinity transport system for acidic and neutral amino acids in the tapetum cells of Arabidopsis flowers. *Plant J.*, **40**: 60-74.
Li, Z.C. and Bush, D.R. (1990a). pH-dependent amino acid transport into plasma membrane vesicles isolated from sugar beet leaves. I. Evidence for carrier-mediated, electrogenic flux through multiple transport systems. *Plant Physiol.*, **94**: 268-277.
Li, Z.C. and Bush, D.R. (1990b). pH-dependent amino acid transport into plasma membrane vesicles isolated from sugar beet leaves. II. Evidence for multiple aliphatic, neutral amino acid symports. *Plant Physiol.*, **96**: 1338-1344.
Liu, D., Gong, W., Bai, Y. Luo, J.C. and Zhu, Y.X. (2005). OsHT, a rice gene encoding for a plasma-membrane localized histidine transporter. *J. Integrative Plant Biol.*, **47**: 92-99
Lohaus, G. and Heldt, H.W. (1995). Further studies of the phloem loading process in leaves of barley and spinach. The comparison of metabolite concentrations in the apoplastic compartment with those in the cytosolic compartment and in the sieve tubes. *Bot. Acta*, **108**: 270-275.
Meline, E. and Nilsson, H. (1953). Transfer of labelled nitrogen from glutamic acid to pine seedlings through the mycelium of *Boletus variegatus* (Sw.) Fr. *Nature*, **171**: 134.
Meyer, A., Eskandari, S., Grallath, S. and Rentsch, D. (2006). AtGAT1, a high affinity transporter for gamma-aminobutyric acid in *Arabidopsis thaliana*. *J. Biol. Chem.*, **281**: 7197-7204.
Miranda, M., Borisjuk, L., Tewes, A., Heim, U., Sauer, N., Wobus, U. and Weber, H. (2001). Amino acid permeases in developing seeds of *Vicia faba* L.: expression precedes storage protein synthesis and is regulated by amino acid supply. *Plant J.*, **28**: 61-71.
Mounoury, G., Delrot, S. and Bonnemain, J. (1984). Energetics of threonine uptake by pod wall tissues of *Vicia faba*. *Planta*, **161**: 178-185.
Nasholm, T., Ekblad, A., Nordin, A., Giesler, R., Hogberg, M. and Hogberg, P. (1998). Boreal forest plants take up organic nitrogen. *Nature*, **392**: 914-916.
Okumoto, S., Schmidt, R., Tegeder, M., Fischer, W.N., Rentsch, D., Frommer, W.B. and Koch, W. (2002). High affinity amino acid transporters specifically expressed in xylem parenchyma and developing seeds of Arabidopsis. *J. Biol. Chem.*, **277**: 45338-45346.
Okumoto, S., Koch, W., Tegeder, M., Fischer, W.N., Biehl, A., Leister, D., Stierhof, Y.D., and Frommer, W.B. (2004). Root phloem-specific expression of the plasma membrane amino acid proton co-transporter AAP3. *J. Exp. Bot.*, **55**: 2155-2168.
Pate, J.S., Sharkey, P.J. and Lewis, O.A.M. (1975). Xylem to phloem transfer of solutes in fruiting shoots of legumes, studied by a phloem bleeding technique. *Planta*, **122**: 11-26.

Pilot, G., Stransky, H., Bushey, D.F., Pratelli, R., Ludewig, U., Wingate, V.P. and Frommer, W.B. (2004). Overexpression of GLUTAMINE DUMPER1 leads to hypersecretion of glutamine from hydathodes of Arabidopsis leaves. *Plant Cell*, **16**: 1827-40.

Rentsch, D., Hirner, B., Schmelzer, E. and Frommer, W.B. (1996). Salt stress-induced proline transporters and salt stress-repressed broad specificity amino acid permeases identified by suppression of a yeast amino acid permease-targeting mutant. *Plant Cell*, **8**: 1437-1446.

Rolletschek, H., Hosein, F., Miranda, M., Heim, U., Gotz, K.P., Schlereth, A., Borisjuk, L., Saalbach, I., Wobus, U. and Weber, H. (2005). Ectopic Expression of an amino acid transporter (VfAAP1) in seeds of *Vicia narbonensis* and pea increases storage proteins. *Plant Physiol.*, **137**: 1236-1249.

Schwacke, R., Grallath, S., Breitkreuz, K.E., Stransky, E., Stransky, H., Frommer, W.B., and Rentsch, D. (1999). LeProT1, a transporter for proline, glycine betaine, and gamma-amino butyric acid in tomato pollen. *Plant Cell*, **11**: 377-392.

Su, Y.H., Frommer, W.B. and Ludewig, U. (2004). Molecular and functional characterization of a family of amino acid transporters from Arabidopsis. *Plant Physiol.*, **136**: 3104-3113.

Svennerstam, H., Ganeteg, U., Bellini, C. and Nasholm, T. (2007). Comparative screening of Arabidopsis mutants suggests the lysine histidine transporter 1 to be involved in plant uptake of amino acids. *Plant Physiol.*, **143**: 1853-1860.

Swofford, D.L. (1998). PAUP*. Phylogenetic Analysis Using Parsimony (*and Other Methods). Version 4. Sinauer Associates, Sunderland, Massachusetts.

Tegeder, M., Offler, C.E., Frommer, W.B. and Patrick, J.W. (2000). Amino acid transporters are localized to transfer cells of developing pea seeds. *Plant Physiol.*, **122**: 319-326.

Ueda, A., Shi, W., Sanmiya, K., Shono, M. and Takabe, T. (2001). Functional analysis of salt-inducible proline transporter of barley roots. *Plant Cell Physiol.*, **42**: 1282-1289.

Waditee, R., Hibino, T., Tanaka, Y., Nakamura, T., Incharoensakdi, A., Hayakawa, S., Suzuki, S., Futsuhara, Y., Kawamitsu, Y., Takabe, T. and Takabe, T. (2002). Functional characterization of betaine/proline transporters in betaine-accumulating mangrove 27. *J. Biol. Chem.*, **277**: 18373-18382.

Winter, H., Robinson, D.G. and Heldt, H.W. (1994). Subcellular volumes and metabolite concentrations in spinach leaves. *Planta*, **193**: 530-535.

Wipf, D., Ludewig, U., Tegeder, M., Rentsch, D., Koch, W. and Frommer, W.B. (2002). Conservation of amino acid transporters in fungi, plants and animals. *Trends Biochem. Sci.*, **27**: 139-147.

# Chapter 12

## MEMBRANE TRANSPORT OF SECONDARY METABOLITES IN PLANTS

NOBUKAZU SHITAN AND KAZUFUMI YAZAKI
*Research Institute for Sustainable Humanosphere, Kyoto University, Gokasho, Uji 611-0011, Japan*
E-mail: yazaki@rish.kyoto-u.ac.jp

**Abstract**

Higher plants produce a large number of secondary metabolites, such as alkaloids, terpenoids, phenolic compounds, and many further compounds have combined structures of those groups. They are often accumulated in particular sink organs, and some are translocated from source cells via long distance transport. The membrane transport of plant secondary metabolites is a newly developing research area. Recent progress in genome sequencing projects and expressed sequence tag (EST) databases has revealed that many genes coding for transporters and channels exist in the plant genome. Studies on phenotype analyses of many mutants by various criteria have identified transporter molecules responsible for the membrane transport of plant secondary metabolites. Characterizations of such transporter genes have clarified that the membrane transport for each secondary metabolite is fairly specific and highly regulated. Not only genes involved in the biosynthesis of secondary metabolites but also genes relevant to their transport will be important for the high accumulation of those metabolites. Hence, knowledge about the membrane transport mechanism is necessary for systematic metabolic engineering aimed at increasing the productivity of valuable secondary metabolites *in planta*.

**Keywords:** secondary metabolite; alkaloid; terpenoid; phenol; plant ABC transporter; primary transport

## 1. INTRODUCTION

Higher plants produce a vast number of secondary metabolites, in addition to primary metabolites, via complex pathways, which are regulated in highly sophisticated manners (Yazaki, 2004). Based on their chemical structure, secondary metabolites are roughly classified into four groups: alkaloids, terpenoids, and phenolic compounds including flavonoids, and quinones. Alkaloids were originally defined as plant-derived nitrogen-containing basic compounds of low molecular

© CAB International 2008. *Plant Membrane and Vacuolar Transporters* (eds P.K. Jaiwal, R.P. Singh and O.P. Dhankher)

weight. As many alkaloids exhibit divergent biological and pharmacological activities, plants containing them are used in the medical field, either directly as crude drugs or as resources to isolate those alkaloids for medicinal purposes. Terpenoids are formed from the isoprene unit of five carbon atoms, and they display great diversity in their chemical structures, from the simplest C5 skeleton (hemiterpene) to large-molecular-weight polymers. Many important flavour components in essential oils and saponins belong to this group. Phenolic compounds, such as flavonoids, the representative secondary metabolites responsible for flower color, as well as phenylpropanoids involved in lignin formation, are also very widely distributed in the plant kingdom.

In addition to some alkaloids, many other secondary metabolites also show strong biological activities, e.g. inhibition of DNA and protein synthesis, inhibition of the nerve system, cardiac activity, modulation of microtuble structures, etc. Bioactive natural metabolites are often clinically used in modern medicines, and such medicinal plants containing these compounds are also used as herbal or crude drugs in many recipes in traditional medicines (Zuin and Vilegas, 2000; Rios and Recio, 2005). In most cases these bioactive natural compounds are found in particular organs, called the "medicinal part" in pharmacognosy, and their contents in such organs are often seasonally regulated (Rocha *et al.*, 2005).

The physiological roles of these secondary metabolites in plants have not yet been clearly elucidated, but reasonable explanations have been made for some cases, i.e. they may function as biological protectants from herbivores, pathogen attacks and abiotic environmental stresses such as UV irradiation (Harborne, 1990; Bouwmeester *et al.*, 2003). For instance, nicotine, a pyrrolidine alkaloid of *Nicotiana* species, and caffeine, a purine derivative from the coffee tree, are reported to act as strong insecticides (Kircher and Lieberman, 1967; Ogita *et al.*, 2003). Some secondary metabolites are known to function as mediators necessary for interaction with other organisms, as being allelopathic substances or insect attractants to facilitate pollination (Hoballah *et al.*, 2005). To achieve these functions, accumulation or secretion of these compounds has to be highly regulated, for instance, flavonoids acting as UV protectants are specifically accumulated in epidermal cells (Schmitz-Hoerner and Weissenbock, 2003), and insect attractants are emitted from flower petals (Kolosova *et al.*, 2001). Biosynthetic genes responsible for the formation of those secondary metabolites may be highly expressed in such tissues where the metabolites are mainly accumulated, while the translocation of natural compounds among plant organs often occurs as well, e.g. biosynthetic genes for nicotine are mostly expressed in root tissues (source organ), whereas it is transported to the aerial part and accumulated in leaves (sink organ) (Shoji *et al.*, 2000). Recent progress in molecular biology has enabled us to study transporter proteins for these natural products in plants (Yazaki, 2005). In this chapter, we introduce the development of this research field over the past few years, providing an overview of proteins that are involved in the membrane transport of secondary metabolites.

For the transport of secondary metabolites, two major mechanisms are proposed, i.e. directly energized primary transport that is mediated by ATP-binding cassette

Plant transporters of secondary metabolites

(ABC) transporters (Martinoia *et al.*, 2002), and H$^+$-gradient-dependent secondary transport via H$^+$-antiport. ABC transporters constitute a large protein family that is found in a range of organisms from bacteria to humans. From intensive studies on the roles of ABC transporters in multidrug resistance in animal cancer cells, it has been long believed that they exhibit broad substrate specificity. However, recent studies have demonstrated that the function of ABC transporters is not restricted to detoxification processes (Martinoia *et al.*, 2002), but they are also involved in many specific biological events, such as cell to cell signaling, in which strict substrate specificity has been shown (Geisler *et al.*, 2005; Terasaka *et al.*, 2005). Moreover, other divergent physiological functions have been reported, such as guard cell regulation (Klein *et al.*, 2003), and secretion of wax components to the cuticle (Pighin *et al.*, 2004). Therefore, it has been suggested in recent years that the ABC transporter family might have evolved according to the need to transport specific substrates in each organism, and not as drug efflux pumps (Sheps *et al.*, 2004).

## 2. TRANSPORTERS OF SECONDARY METABOLITES

### 2.1. Alkaloids

Alkaloids are nitrogen-containing low-molecular-weight compounds, which are found in about 20% of plant species. This diverse group implies most bioactive metabolites among four secondary metabolite groups, and approximately 12,000 compounds have been elucidated to date (Croteau *et al.*, 2000). Because of their strong biological activities, significant phytochemical and pharmacological research has been performed, e.g. many studies have been published for a representative alkaloid morphine from opium poppy, *Papaver somniferum* (Papaveraceae), which is one of the most effective analgesics ever known. The structures of most alkaloids are biosynthesized from aromatic compounds containing a nitrogen atom that provides the basic feature of this group. The nitrogen atom is usually derived from the amino residue of an amino acid, i.e. the fundamental biosynthetic precursors of plant alkaloids are amino acids.

Bioactive alkaloids, which influence, for example, the stability of chromosome structures or inhibit DNA duplication, and are often used as anti-cancer or antibacterial medicines, can be potentially toxic to plant cells but the producer plants seem to be insensitive to their own metabolites. For instance, when berberine, an isoquinoline alkaloid was added to various plant cell cultures, it showed strong cytotoxicity to berberine-non-producing plant species like tobacco, while *Thalictrum minus* as well as *Coptis japonica*, both berberine producers, exhibited clear tolerance to this endogenous alkaloid (Sato *et al.*, 1990; Terasaka *et al.*, 2003a). Moreover, *C. japonica* cells revealed an ability to take up berberine from the medium against the concentration gradient when exogenously added to the culture medium, and the absorbed berberine was exclusively accumulated in the vacuoles (Sato *et al.*, 1993). In the cellular transport of berberine by *C. japonica* cells, two transport events are involved, i.e. uptake of berberine at the plasma membrane and efflux from the cytosol into the vacuolar lumen at the tonoplast. Inhibitor experiments suggested the possible involvement of an ABC transporter in the cellular transport of berberine by cultured cells (Sakai *et al.*, 2002). A multidrug resistance (MDR, or ABCB)-type ABC transporter was then cloned from *C. japonica* cell cultures via homology-based RT-PCR as a candidate berberine transporter (Yazaki *et al.*, 2001b). Functional analyses of the ABC transporter designated CjMDR1 were performed with *Xenopus* oocytes, showing that this ABC transporter recognized berberine as its substrate and transported it in an inward direction (Shitan *et al.*, 2003). In *C. japonica* cells CjMDR1 was shown to be localized at the plasma membrane of *C. japonica*. This was the first example of an eukaryotic ABC transporter mediating the uptake and not the efflux of a substance.

In the intact plant, berberine is biosynthesized in root tissues, and then translocated to the rhizome and trapped by plasma membrane-localized CjMDR1, resulting in its accumulation in the rhizome. Since the rhizome is the sink organ also for starch, this plant accumulates the alkaloid having strong antimicrobial activity as a chemical defense against soil-borne microorganisms. Contrary to the plasma membrane, the vacuolar transport of berberine in this plant cell was sensitive to the disruption of $H^+$-gradient across the tonoplast, but insensitive to vanadate (Otani *et al.*, 2005), suggesting that the vacuolar transport of berberine is not mediated by an ABC transporter, but probably a $H^+$-antiporter is responsible for the berberine transport in vacuoles of *C. japonica* (Fig. 1).

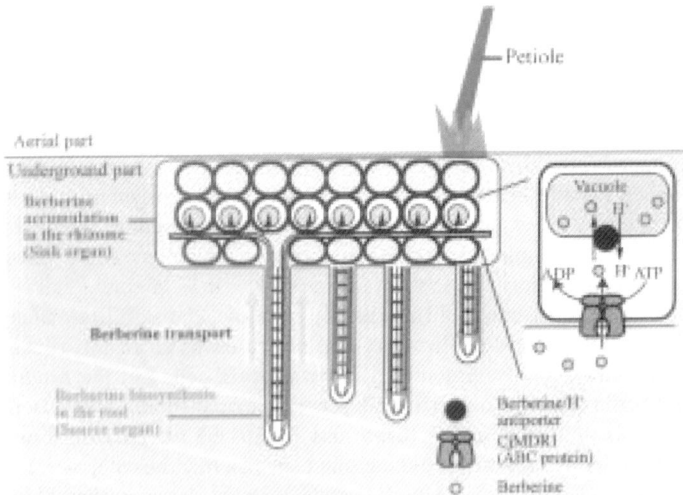

**Figure 1:** Model of berberine translocation and accumulation in intact *Coptis japonica*. Berberine, synthesized in the root, is translocated to the rhizome via the xylem. CjMDR1, an ABC transporter localized to the plasma membrane in the rhizome, is involved in berberine uptake from xylem. In the cytosol, the $H^+$/ berberine antiporter effluxes berberine from the cytosol to the vacuole via proton gradient. By these two transporters, *Coptis* plants can accumulate berberine in the rhizome.

Plant transporters of secondary metabolites

In another berberine producer, *T. minus* cell cultures, which secreted berberine to the culture medium, the possible involvement of an ABC transporter in berberine secretion was demonstrated (Terasaka *et al.*, 2003a; Terasaka *et al.*, 2003b). Interestingly the identified ABC transporter shared high similarity with CjMDR1. The regulatory mechanism which determines the direction of transportation is now under investigation.

For the transport of berberine, a vesicle-mediated mechanism was also proposed in a different plant (Bock *et al.*, 2002). The terminal steps of berberine biosynthesis take place exclusively in specific intracellular vesicles in *Berberis*, which are probably derived from the endoplasmic reticulum (ER) and later fuse with the central vacuole. This scheme fits plants that produce and accumulate berberine in the same cells, but the carrier-mediated mechanism is appropriate for the plants whose sink and source organs are distant, as in *C. japonica* (Shitan *et al.*, 2003).

Berberine can also be transported by a $H^+$-antiport mechanism in heterologous plants that do not produce this alkaloid. A detoxifying efflux carrier, AtDTX1 belonging to the MATE (Multidrug And Toxic compound Extrusion) family, was identified in Arabidopsis as a gene that conferred resistance to several exogenous toxic compounds including berberine (Li *et al.*, 2002). MATE is a large gene family with at least 56 members encoded by the Arabidopsis genome (Hvorup *et al.*, 2003). AtDTX1 is localized to the plasma membrane and is proposed to mediate the efflux of xenobiotics by mostly a proton-motive force.

One of the most well-known examples of long-distance transport is nicotine alkaloids in *Nicotiana* species. Nicotine is biosynthesized in root tissues, where it is specifically increased in the response to attacks by pathogens and herbivores, and the produced nicotine is translocated to the aerial part for accumulation (Shoji *et al.*, 2000). Considering the translocation, this alkaloid should be loaded into xylem tissue and unloaded at mesophyll cells where nicotine is finally accumulated in the vacuoles (Hashimoto and Yamada, 2003). This process implies the transport of nicotine across at least three different membranes, plasma membranes in the root and in the leaf, and vacuolar membrane of mesophyll cells, while no specific nicotine transporter has been identified so far. An antiporter and/or an ABC-type protein may play a role in a membrane transport event of this alkaloid.

Another isoquinoline alkaloid morphine, a major alkaloid in the latex of opium poppy, is accumulated in the large membranous vesicles of such latex. Immunofluorescence analyses using antibodies specific for five enzymes of alkaloid formation in opium poppy were recently reported (Weid *et al.*, 2004). In the capsule and stem, two *O*-methyltransferases and an *O*-acetyltransferase were found predominantly in parenchyma cells within the vascular bundle, while codeinone reductase was localized to laticifers. Another group reported that three of those biosynthetic enzymes of morphine were localized in sieve elements of this plant (Bird *et al.*, 2003). In either case, the transport of the intermediate from a specific cell-type of vascular tissue to the laticifer was proposed, where the involvement of ABC transporter might be possible.

Early work by Zenk indicated that the vacuolar transport of indole alkaloids was mediated by a $H^+$-antiporter (Deus-Neumann and Zenk, 1984), although no endogenous transporter gene for indole alkaloids has been, to our knowledge, isolated so far. Recent studies demonstrated that indole-3-acetic acid transport is mediated by ABC transporters of MDR (ABCB)-type (Noh et al., 2001; Geisler et al., 2005; Santelia et al., 2005; Terasaka et al., 2005). Moreover, inhibitory activity specific to auxin was reported in indole alkaloids, such as brucine and yohimbine, as competitors (Jambois et al., 2004). These papers suggest that an ABC transporter is involved in the transport of indole alkaloids in plants.

## 2.2. Terpenoids

Terpenoids are probably the most divergent secondary metabolites in chemical structure. To date, more than 25,000 compounds have been isolated and their structures were elucidated (Croteau et al., 2000). Terpenoids are biosynthesized by condensation of the monomeric C5 unit, dimethylallyl diphosphate and isopentenyl diphosphate, and are classified according to the degree of condensation as hemiterpenes ($C_5$), monoterpenes ($C_{10}$), sesquiterpenes ($C_{15}$), diterpenes ($C_{20}$), sesterterpenes ($C_{25}$), triterpenes ($C_{30}$) and tetraterpenes ($C_{40}$, carotenoid). Plant steroids are also biosynthetically classified as terpenoids because their basic structures originate from the common precursor of other terpenes, but steroids are biosynthesized from squalene, which is a dimerized intermediate of farnesyl diphosphate, in a head-to-head manner. Polymers such as gutta-percha and isoprene rubber can also be recognized as terpenoids in a broad sense.

Contrary to photochemical and biosynthetic studies, the membrane transport of terpenoids is still largely unknown, except for the diterpene compound, sclareol. This antifungal compound is a dicyclic natural metabolite synthesized by *Nicotiana* species. From a study on plasma membrane proteins that are inducible by sclareolide, an antifungal analog, a pleiotropic drug resistance (PDR)-type ABC transporter was identified (Jasinski et al., 2001). The gene expression of this PDR member (NpABC1) was strongly induced in response to both sclareol and sclareolide in the leaves of *N. plumbaginifolia*, and the possible excretion of these diterpene derivatives to the leaf surface was suggested by inhibitor experiments. An ortholog was isolated from tobacco (Sasabe et al., 2002), in which the PDR-type ABC transporter also showed close relevance to the pathogen response. The *Arabidopsis* genome has 15 members of the PDR subfamily, one of which, AtPDR12 was reported to be strongly induced by elicitor treatment, suggesting its direct involvement in pathogen resistance processes by transporting antimicrobial metabolites (Campbell et al., 2003). Similar inducibility of a PDR ortholog was also reported in *Spirodela polyrrhiza* and resistance against sclareol was revealed in both Arabidopsis (Smart and Fleming, 1996; Campbell et al., 2003) and *S. polyrrhiza* PDRs by a germination assay in which root elongation was evaluated (van den Brule et al., 2002). These data are indicative that PDR members of Arabidopsis and *Spirodela* may recognize sclareol or other natural compounds of similar structure as the substrate and transport it in an outward direction; however, it is as yet unknown whether this diterpene is actually biosynthesized in those plant species.

Plant transporters of secondary metabolites

The accumulation of terpene compounds has been described in many plants, for example, monoterpenes in Labiatae plants are biosynthesized in secretory cells and accumulate in the epicuticular cavity of glandular trichomes (Lange and Croteau, 1999), while terpenoids of woody plants are secreted into the resin duct (Martin *et al.*, 2004). For volatile mono- and sesquiterpenenoids, their emission from flowers of Arabidopsis (Chen *et al.*, 2003), snapdragon (Dudareva *et al.*, 2003), and from leaves of woody plants (Martin *et al.*, 2003) has been reported. Furthermore, a dramatic increase in the emission of volatile terpenoids was demonstrated by insect attacks in maize leaves (Schmelz *et al.*, 2003) and cotton flower buds (Rose and Tumlinson, 2004), where biosynthetic genes were strongly induced under these conditions. Excretion of higher terpenoids is also known, i.e. a hydrophobic triterpene, bryonolic acid, is highly accumulated in the apoplastic space of some plant cell cultures of Cucurbitaceae, and is probably attached to the cell wall (Tabata *et al.*, 1993). To our knowledge, however, the transporter molecules involved in the secretion of those terpene compounds have not been identified yet. In addition, for isoprene, a highly volatile hemiterpene (C5) emitted in a large amount from leaves of some plant species like poplar, no transporter seemed to be required for emission (Niinemets *et al.*, 2004).

## 2.3. Phenolic compounds

Phenolic secondary metabolites involve simple phenylpropanoids including coumarins and lignans, flavonoids, and also polyphenols of high molecular weight such as tannins. Many of these phenolic secondary metabolites are involved in plant pathogen interaction, protectants against abiotic stress, and in the formation of structural components like lignins. Phenylpropanoids and flavonoids are one of the most intensively studied plant secondary metabolites, not only for their chemical structures but also for their biological activities and biosynthesis (Wang *et al.*, 1998; Harborne and Williams, 2000; Winkel-Shirley, 2001), particularly in the context of plant defense.

Many phenolic compounds are detected in glycosylated form in plants, and glycosidation plays a key role in the detoxification of endogenous secondary metabolites and also xenobiotics in plants. The major sugar moiety is glucose, and these glucosides often end up accumulated in the vacuoles. Multidrug resistance-associated protein (MRP or ABCC)-type ABC transporters are reported to be involved in the vacuolar sequestration of such glucosides, in addition to glucuronides and glutathione conjugates (Bartholomew *et al.*, 2002). According to studies by Martinoia's group, there seemed to be an apportionment of transporter types either for endogenous or exogenous substrates. For instance, a flavonoid glucoside, isovitexin, a native C-glucoside in barley, was transported into the isolated vacuoles of barley via electrochemical gradient-dependent secondary transport, whereas a herbicide glucoside of hydroxyprimisulfuron was taken up by a directly energized primary transport mechanism (Klein *et al.*, 1996). They also reported that the uptake of the main barley flavonoid saponarin, an apigenin glucoside, into barley vacuoles occurred via $H^+$-antiport, whereas the transport of saponarin into Arabidopsis vacuoles, a heterologous plant that did not produce this metabolite, displayed typical characteristics of an ABC transporter (Frangne *et al.*, 2002).

In contrast to glycosylated flavonoids transported to vacuole, some flavonoid aglycones are secreted out to the rhizosphere, some of which are indispensable for the establishment of symbiosis with soil microbes in legume plants. Biochemical analyses showed that a soybean ABC transporter was involved in the secretion of genistein, a signal isoflavone, at the plasma membrane of the roots, which was recognized by the specific symbiotic *Rhizobium, Bradyrhizobium japonicum* (Sugiyama *et al.*, 2007). The expression analysis of ABC proteins in the model legume plant *Lotus japonicus* revealed that PDR-type ABC proteins were likely to be involved in the nodulation process (Sugiyama *et al.*, 2006). These data suggested that ABC proteins were involved in the secretion of flavonoid molecules leading to the formation of symbiosis in legume plants (Kosslak *et al.*, 1987).

Studies on the transport mechanism of phenolic compounds have probably been most actively performed for anthocyanins as they play a central role in the coloring of flowers. Most anthocyanins are glycosylated and accumulated in vacuoles, except for some anthocyanins found in the apoplastic space, like riccionidin A in *Rhus javanica* (Taniguchi *et al.*, 2000). The involvement of MRPs (ABCC) in the vacuolar transport of such phenolic glucosides was suggested in the *bronze-2* (*bz2*) mutant of maize (Marrs *et al.*, 1995). This mutant, in which *bz2* encodes a glutathione *S*-transferase, was defective in the accumulation of anthocyanin in the vacuole. As MRPs have a substrate preference for glutathione conjugates, and since their transport activity is often stimulated in the presence of glutathione, the involvement of MRPs in the vacuolar transport of anthocyanin was presumed. Similar results were also reported in dicots, such as petunia (Alfenito *et al.*, 1998), Arabidopsis (Kitamura *et al.*, 2004), and carnation (Larsen *et al.*, 2003). Further strong evidence for the involvement of MRP proteins in anthocyanin accumulation was provided via reverse-genetic studies by Goodman *et al.* (2004). The maize ABC transporter, ZmMRP3, is localized to the tonoplast, and is required for the anthocyanin accumulation process in maize.

Contrary to those reports, the possible involvement of $H^+$-gradient-dependent transport for anthocyanin accumulation was also reported. The Arabidopsis gene *tt12* showed strong reduction in the proanthocyanidin deposition in vacuoles of seed endothelial cells (Debeaujon *et al.*, 2001). The gene product of TT12 was a secondary transporter-like protein belonging to the MATE family, suggesting that this protein might be responsible for the vacuolar transport of proanthocyanidin and maybe of anthocyanin via an antiport mechanism. A similar MATE-protein MTP77 was also reported in tomato (Mathews *et al.*, 2003), whereas further biochemical evidence is needed to prove the direct involvement of these antiporters in anthocyanin transport.

The preference for a certain hydrophilic moiety of conjugated phenolic compounds for vacuolar transport, either glucose or glutathione, was analyzed using vacuolar membrane vesicles purified from red beet (*Beta vulgaris*)

Plant transporters of secondary metabolites

(Bartholomew et al., 2002). Whereas two phenol glucosides of p-hydroxycinnamic acid and p-hydroxybenzoic acid were transported apparently by a $H^+$-gradient-dependent mechanism, the glutathione conjugate of a herbicide chlorsulfuron analog appeared to be transported by an ABC transporter. Another experiment with phenylpropanoid derivatives showed that a glutathione conjugate of cinnamic acid was transported into the tonoplast vesicles via a GS-X pump, i.e. MRP-type ABC transporter (Walczak and Dean, 2000). These data suggested that the sugar moiety was a 'tag' to be recognized by secondary transporters, while a glutathione moiety was a preferred 'tag' for MRP proteins functioning as primary transporters (Ishikawa et al., 1997; Rea et al., 1998; Kolukisaoglu et al., 2002), although some glucosides seemed to be recognized by MRP-type ABC transporters. This indicates that the combination between substrates and preferred transporter types may depend on plant species.

Sucrose transport is known to be mediated by a $H^+$-symporter. A survey of the substrate specificity of the sucrose transporter AtSUC2 of Arabidopsis has provided new insight into secondary metabolite transporters in plants (Chandran et al., 2003). In an assay using AtSUC2 expressed in *Xenopus* oocytes, large inward currents were observed for eight glucosides, including arbutin and salicin, out of the 24 sugars and glucosides analyzed. This suggests that a sucrose transporter of low substrate specificity might participate in transporting glucosides, adding to its main function of sucrose transportation into the phloem.

Secondary metabolites may act as endogenous modulators of plant ABC transporters. Some MDR (PGP or ABCB) members were reported as auxin transporters (Noh et al., 2001; Geisler et al., 2005; Terasaka et al., 2005), while flavonoid aglycones, such as kaempferol and quercetin, appeared to act as negative regulators of auxin transport in Arabidopsis (Brown et al., 2001; Buer and Muday, 2004; Terasaka et al., 2005). A possible function of flavonoids as endogenous modulators of plant MDRs is suggested.

## 2.4. Wax

The plant body is covered by the cuticle, which is composed of cutin, polysaccharide and wax. The wax component is made of very long chain fatty acids and their derivatives. A recent finding showed that a half-size ABC transporter AtWBC12 (ABCG-type) in Arabidopsis was involved in wax secretion on the stem surface (Pighin et al., 2004). This member is in the reverse-oriented subfamily of the ABC transporter, and in the mutant plant (*cer5*) the wax components on the epidermal surface decreased to half compared to the wild type. Its localization at the plasma membrane was also revealed with GFP fusion protein. It is, however, still to be clarified how ABC protein plays a critical role in wax secretion, because wax components are solid at the cuticle. Since the substrates are very lipophilic, a vesicle-mediated transport mechanism is also proposed for wax secretion. The putative interaction between vesicle transport and ABC transporter will be a hot topic in the near future.

## 2.5. Vesicle transport

Little is known about the transport mechanism for many lipophilic secondary metabolites, such as triterpenes and phytosterols. One model of the lipophilic secondary metabolite transportation is found in the shikonin production system in *Lithospermum erythrorhizon* cell and hairy root cultures (Yazaki *et al.*, 2001a). After their biosynthesis in the ER, shikonin derivatives, which are red naphthoquinones, are accumulated in red granules attached to the cell surface (Tsukada and Tabata, 1984; Tabata, 1996). There, the intracellular movement of shikonin is managed via vesicle transport, and two hypotheses are presented for the mechanism of this transport: first, direct transfer of lipids from ER to the plasma membrane and, second, Golgi-mediated exocytosis, as proposed for cuticular wax transport (Kunst and Samuels, 2003).

Another model of vesicle transport has been reported in maize BMS (Black Mexican Sweet) cells. The induction of the *P1* transcription factor caused the accumulation of green auto-fluorescent bodies in the maize cytoplasm, which later appeared to fuse with the plasma membrane and thus be secreted out of the cytoplasm to the cell wall (Lin *et al.*, 2003). This system will provide material that might be applicable for the study of vesicle-mediated transport of secondary metabolites.

## 3. TRANSPORT OF SECONDARY METABOLITES BY NON-PLANT TRANSPORTERS

Fungal pathogens are exposed to a variety of fungitoxic secondary metabolites produced by plants during pathogenesis (Del Sorbo *et al.*, 2000). ABC transporters can play an essential role in protection against those plant defense compounds during invasion. In the fungal pathogen *Magnaporthe grisea*, which caused rice blast disease, an ABC transporter similar to the yeast multidrug resistance pump was identified (Urban *et al.*, 1999). The insertional mutant of this gene arrested growth and the hypha died shortly after penetrating the epidermal cells of rice or barley, indicating that this ABC transporter was a pathogenicity factor. Its expression was indeed inducible by drugs and rice phytoalexin. Another ABC transporter gene acting as virulence factor *MgAtr4* was identified in a wheat pathogen *Mycosphaerella graminicola* out of five similar genes of this fungi (Stergiopoulos *et al.*, 2003). A disruption strain of *MgAtr4* displayed reduced intercellular growth in wheat leaves and less efficient colonization of substomatal cavities. The native substrate has not been identified in both transporters yet, but the gene expression response of the fungal ABC transporter to plant metabolites might offer clues to find transport substrates. When the gene expression was analyzed, two other members *MgAtr1* and *MgAtr2* responded to a phenylpropanoid eugenol and an indole alkaloid reserpine in their gene expression in a similar way as to azole fungicides, suggesting that they were broad substrate drug efflux pumps (Zwiers and De Waard, 2000).

Multispecific ABC transporters are particularly relevant to plant pathogens that have a broad host range since they are exposed to many plant defense compounds.

Plant transporters of secondary metabolites

*Botrytis cinerea* is an example of such a pathogen. A PDR-type ABC transporter BcatrB was isolated as a candidate drug efflux pump of broad substrate specificity (Schoonbeek *et al.*, 2001). BcatrB expression was upregulated by the grapevine phytoalexin resveratrol, a stilbene, as well as fungicides, and the gene replacement mutant became more sensitive to resveratrol. In gene expression analyses, some other fungal ABC transporters were also shown to respond to plant secondary metabolites (Del Sorbo *et al.*, 1997; Nakaune *et al.*, 1998).

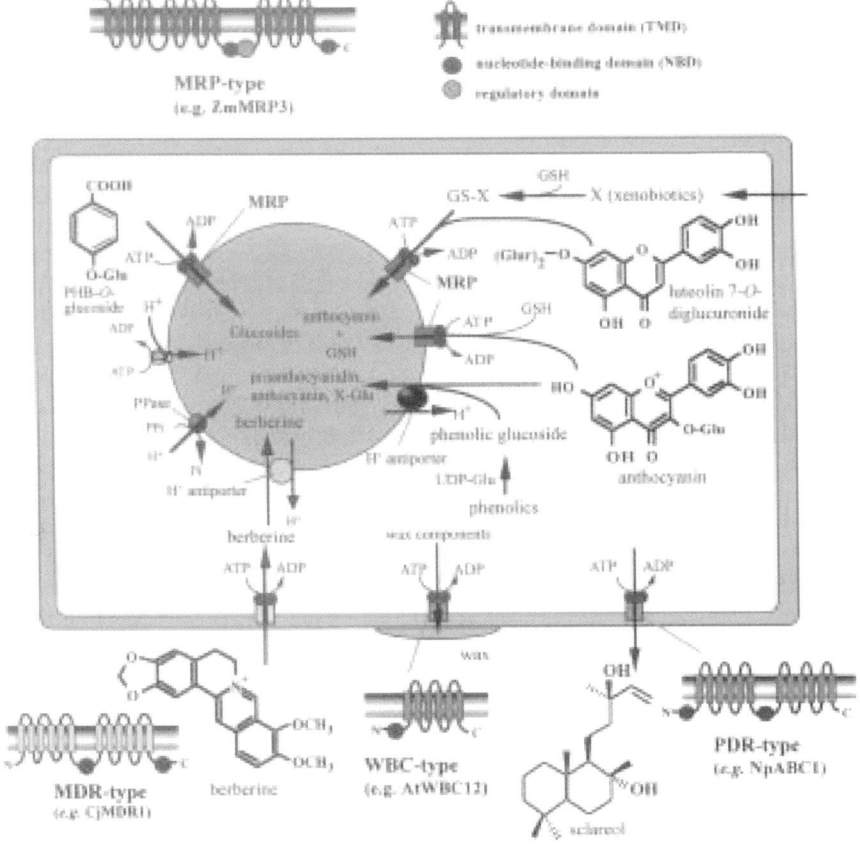

**Figure 2:** Membrane transport of secondary metabolites and various involved transporters in plant cells. Proton antiporters utilize the proton gradient across the tonoplast to accumulate natural products into the vacuolar lumen (secondary transport). ABC transporters are directly energized by ATP hydrolysis and pump secondary metabolites independently from the proton gradient (primary transport). It is believed that the nucleotide-binding domain of ABC proteins responsible for ATP hydrolysis faces the cytosol. Representative secondary metabolites are shown in this figure.

A herbivorous tobacco hornworm possesses a detoxification mechanism for nicotine. Transport activity similar to MDR1 (ABCB1) was reported in the Malpighian tubules of this insect, which excreted the alkaloid from tissues (Gaertner et al., 1998). Nicotine transport was inhibited by atropine, while vinblastine transport was suppressed by nicotine, indicative that the alkaloid transporter at the excretory Malpighian tubules recognized other alkaloids of a different type. By immunostaining, the existence of a similar ABC transporter for nicotine excretion at the blood-brain barrier of insects was also suggested (Murray et al., 1994).

## 4. SUBSTRATE RECOGNITION OF ABC TRANSPORTERS

From intensive studies on the roles of some mammalian ABC transporters in multidrug resistance in cancer cells, the simple assumption that ABC transporters could generally exhibit broad substrate specificity is widely accepted. However, recent studies have demonstrated that their functions are not only restricted to detoxification processes (Martinoia et al., 2002), but are also involved in many specific biological activities, such as the translocation of endogenous metabolites and cell signaling, in which they show narrow substrate specificity (van den Brule et al., 2002; Shitan et al., 2003; Geisler et al., 2005; Terasaka et al., 2005), and other divergent physiological functions (Klein et al., 2003; Pighin et al., 2004). It has recently been suggested that the ABC transporter family has evolved because of the necessity of transporting specific substrates in each organism, and not as drug efflux pumps (Sheps et al., 2004).

The molecular mechanism of substrate recognition is still largely unknown. The amino acid sequence identity between human MDR1 (ABCB1) and MRP1 (ABCC1) is only 17%, although they show overlapping in the substrates to a large extent. On the other hand, human MDR1 and MDR2 share 75% amino acid identity but they show very different functions, i.e. the former is a multiple drug efflux pump whereas the latter is a flippase for phosphatidyl choline while MDR1 cannot transport this phospholipid (Smit et al., 1993). Comparing CjMDR1, a fairly specific alkaloid transporter for endogenous berberine, to human MDR1 recognizing many plant alkaloids, there is 35% amino acid identity with strong similarity in the hydropathy profile, whereas no significant feature is found to explain their difference in substrate specificity and the transport direction. To argue these points, three-dimensional structure analyses of ABC transporters will be necessary.

## 5. CONCLUSIONS AND FUTURE PROSPECTS

A schematic drawing of the transport processes of secondary metabolites is shown in Figure 2. Molecular analysis of the membrane transport of plant secondary metabolites is a fairly new field in plant biology. The dispersed localizations of both the end-products discussed in this chapter and their biosynthetic enzymes indicate that biosynthetic intermediates might move among organelles during the biosynthesis of secondary metabolites (Shitan and Yazaki, 2007; De Luca and St Pierre, 2000; Sanchez-Fernandez et al., 2001; Zhao et al., 2003). Although simple diffusion might be sufficient for hydrophilic intermediates, transporters localized in

Plant transporters of secondary metabolites

organelles might take part in the regulation of transportation, especially that of lipophilic intermediates (Ohara et al., 2004). Metabolic engineering has become a popular means of increasing the production of secondary metabolites by overexpressing biosynthetic enzymes (Yazaki, 2004). The introduction of accumulation mechanisms by engineering transport systems should be an effective way to increase the production of secondary metabolites.

## 6. LITERATURE CITED

Alfenito, M.R., Souer, E., Goodman, C.D., Buell, R., Mol, J., Koes, R., and Walbot, V. (1998). Functional complementation of anthocyanin sequestration in the vacuole by widely divergent glutathione S-transferases. *Plant Cell,* **10:** 1135-1149.

Bartholomew, D.M., Van Dyk, D.E., Lau, S.M., O'Keefe, D.P., Rea, P.A., and Viitanen, P.V. (2002). Alternate energy-dependent pathways for the vacuolar uptake of glucose and glutathione conjugates. *Plant Physiol.,* **130:** 1562-1572.

Bird, D.A., Franceschi, V.R., and Facchini, P.J. (2003). A tale of three cell types: alkaloid biosynthesis is localized to sieve elements in opium poppy. *Plant Cell,* **15:** 2626-2635.

Bock, A., Wanner, G., and Zenk, M.H. (2002). Immunocytological localization of two enzymes involved in berberine biosynthesis. *Planta,* **216:** 57-63.

Bouwmeester, H.J., Matusova, R., Zhongkui, S., and Beale, M.H. (2003). Secondary metabolite signalling in host-parasitic plant interactions. *Curr. Opin. Plant Biol.,* **6:** 358-364.

Brown, D.E., Rashotte, A.M., Murphy, A.S., Normanly, J., Tague, B.W., Peer, W.A., Taiz, L., and Muday, G.K. (2001). Flavonoids act as negative regulators of auxin transport *in vivo* in Arabidopsis. *Plant Physiol.,* **126:** 524-535.

Buer, C.S., and Muday, G.K. (2004). The *transparent testa4* mutation prevents flavonoid synthesis and alters auxin transport and the response of Arabidopsis roots to gravity and light. *Plant Cell,* **16:** 1191-1205.

Campbell, E.J., Schenk, P.M., Kazan, K., Penninckx, I.A., Anderson, J.P., Maclean, D.J., Cammue, B.P., Ebert, P.R., and Manners, J.M. (2003). Pathogen-responsive expression of a putative ATP-binding cassette transporter gene conferring resistance to the diterpenoid sclareol is regulated by multiple defense signaling pathways in Arabidopsis. *Plant Physiol.,* **133:** 1272-1284.

Chandran, D., Reinders, A., and Ward, J.M. (2003). Substrate specificity of the *Arabidopsis thaliana* sucrose transporter AtSUC2. *J. Biol. Chem.,* **278:** 44320-44325.

Chen, F., Tholl, D., D'Auria, J.C., Farooq, A., Pichersky, E., and Gershenzon, J. (2003). Biosynthesis and emission of terpenoid volatiles from Arabidopsis flowers. *Plant Cell,* **15:** 481-494.

Croteau, R., Kutchan, T.M., and Lewis, N.G. (2000). *Natural Products (Secondary Metabolites): Biochemistry and Molecular Biology of Plants* (Eds. Buchanan, B., Gruissem, W. and Jones, R.), American Society of Plant Physiologists, Maryland, pp 1250-1318.

Debeaujon, I., Peeters, A.J., Leon-Kloosterziel, K.M., and Koornneef, M. (2001). The *TRANSPARENT TESTA12* gene of Arabidopsis encodes a multidrug secondary transporter-like protein required for flavonoid sequestration in vacuoles of the seed coat endothelium. *Plant Cell,* **13:** 853-871.

Del Sorbo, G., Schoonbeek, H., and De Waard, M.A. (2000). Fungal transporters involved in efflux of natural toxic compounds and fungicides. *Fungal Genet. Biol.,* **30:** 1-15.

Del Sorbo, G., Andrade, A.C., Van Nistelrooy, J.G., Van Kan, J.A., Balzi, E., and De Waard, M.A. (1997). Multidrug resistance in *Aspergillus nidulans* involves novel ATP-binding cassette transporters. *Mol. Gen. Genet*, **254**: 417-426.
De Luca, V. and St Pierre, B. (2000). The cell and developmental biology of alkaloid biosynthesis. *Trends Plant Sci.*, **5**: 168-173.
Deus-Neumann, B., and Zenk, M.H. (1984). A highly selective alkaloid uptake system in vacuoles of higher plants. *Planta*, **162**: 250-260.
Dudareva, N., Martin, D., Kish, C.M., Kolosova, N., Gorenstein, N., Faldt, J., Miller, B., and Bohlmann, J. (2003). (E)-beta-ocimene and myrcene synthase genes of floral scent biosynthesis in snapdragon: function and expression of three terpene synthase genes of a new terpene synthase subfamily. *Plant Cell*, **15**: 1227-1241.
Frangne, N., Eggmann, T., Koblischke, C., Weissenbock, G., Martinoia, E., and Klein, M. (2002). Flavone glucoside uptake into barley mesophyll and Arabidopsis cell culture vacuoles. Energization occurs by $H^+$-antiport and ATP-binding cassette-type mechanisms. *Plant Physiol.*, **128**: 726-733.
Gaertner, L.S., Murray, C.L., and Morris, C.E. (1998). Transepithelial transport of nicotine and vinblastine in isolated malpighian tubules of the tobacco hornworm (*Manduca sexta*) suggests a P-glycoprotein-like mechanism. *J. Exp. Biol.*, **201**: 2637-2645.
Geisler, M., Blakeslee, J.J., Bouchard, R., Lee, O.R., Vincenzetti, V., Bandyopadhyay, A., Titapiwatanakun, B., Peer, W.A., Bailly, A., Richards, E.L., Ejendal, K.F., Smith, A.P., Baroux, C., Grossniklaus, U., Muller, A., Hrycyna, C.A., Dudler, R., Murphy, A.S., and Martinoia, E. (2005). Cellular efflux of auxin catalyzed by the Arabidopsis MDR/PGP transporter AtPGP1. *Plant J.*, **44**: 179-194.
Goodman, C.D., Casati, P., and Walbot, V. (2004). A multidrug resistance-associated protein involved in anthocyanin transport in *Zea mays*. *Plant Cell*, **16**: 1812-1826.
Harborne, J.B. (1990). Role of secondary metabolites in chemical defence mechanisms in plants. *Ciba Found Symp.*, **154**: 126-134; discussion 135-129.
Harborne, J.B., and Williams, C.A. (2000). Advances in flavonoid research since 1992. *Phytochemistry*, **55**: 481-504.
Hashimoto, T., and Yamada, Y. (2003). New genes in alkaloid metabolism and transport. *Curr. Opin. Biotechnol.*, **14**: 163-168.
Hoballah, M.E., Stuurman, J., Turlings, T.C., Guerin, P.M., Connetable, S., and Kuhlemeier, C. (2005). The composition and timing of flower odour emission by wild *Petunia axillaris* coincide with the antennal perception and nocturnal activity of the pollinator *Manduca sexta*. *Planta*, **222**: 141-150.
Hvorup, R.N., Winnen, B., Chang, A.B., Jiang, Y., Zhou, X.F., and Saier, M.H., Jr. (2003). The multidrug/oligosaccharidyl-lipid/polysaccharide (MOP) exporter superfamily. *Eur. J. Biochem.*, **270**: 799-813.
Ishikawa, T., Li, Z.S., Lu, Y.P., and Rea, P.A. (1997). The GS-X pump in plant, yeast, and animal cells: structure, function, and gene expression. *Biosci. Rep.*, **17**: 189-207.
Jambois, A., Ditengou, F.A., Kawano, T., Delbarre, A., and Lapeyrie, F. (2004). The indole alkaloids brucine, yohimbine, and hypaphorine are indole-3-acetic acid-specific competitors which do not alter auxin transport. *Physiol. Plant*, **120**: 501-508.
Jasinski, M., Stukkens, Y., Degand, H., Purnelle, B., Marchand-Brynaert, J., and Boutry, M. (2001). A plant plasma membrane ATP binding cassette-type transporter is involved in antifungal terpenoid secretion. *Plant Cell*, **13**: 1095-1107.
Kircher, H.W., and Lieberman, F.V. (1967). Toxicity of tobacco smoke to the spotted alfalfa aphid *Therioaphis maculata* (Buckton). *Nature*, **215**: 97-98.

Kitamura, S., Shikazono, N., and Tanaka, A. (2004). *TRANSPARENT TESTA 19* is involved in the accumulation of both anthocyanins and proanthocyanidins in Arabidopsis. *Plant J.,* **37**: 104-114.

Klein, M., Weissenbock, G., Dufaud, A., Gaillard, C., Kreuz, K., and Martinoia, E. (1996). Different energization mechanisms drive the vacuolar uptake of a flavonoid glucoside and a herbicide glucoside. *J. Biol. Chem.,* **271**: 29666-29671.

Klein, M., Perfus-Barbeoch, L., Frelet, A., Gaedeke, N., Reinhardt, D., Mueller-Roeber, B., Martinoia, E., and Forestier, C. (2003). The plant multidrug resistance ABC transporter AtMRP5 is involved in guard cell hormonal signalling and water use. *Plant J.,* **33**: 119-129.

Kolosova, N., Sherman, D., Karlson, D., and Dudareva, N. (2001). Cellular and subcellular localization of *S*-adenosyl-L-methionine:benzoic acid carboxyl methyltransferase, the enzyme responsible for biosynthesis of the volatile ester methylbenzoate in snapdragon flowers. *Plant Physiol.,* **126**: 956-964.

Kolukisaoglu, H.U., Bovet, L., Klein, M., Eggmann, T., Geisler, M., Wanke, D., Martinoia, E., and Schulz, B. (2002). Family business: the multidrug-resistance related protein (MRP) ABC transporter genes in *Arabidopsis thaliana*. *Planta,* **216**: 107-119.

Kosslak, R.M., Bookland, R., Barkei, J., Paaren, H.E. and Appelbaum, E.R. (1987). Induction of *Bradyrhizobium japonicum* common nod genes by isoflavones isolated from *Glycine max*. *Proc. Natl. Acad. Sci. USA,* **84**: 7428-7432.

Kunst, L., and Samuels, A.L. (2003). Biosynthesis and secretion of plant cuticular wax. *Prog. Lipid Res.,* **42**: 51-80.

Lange, B.M., and Croteau, R. (1999). Genetic engineering of essential oil production in mint. *Curr. Opin. Plant Biol.,* **2**: 139-144.

Larsen, E.S., Alfenito, M.R., Briggs, W.R., and Walbot, V. (2003). A carnation anthocyanin mutant is complemented by the glutathione S-transferases encoded by maize *Bz2* and petunia *An9*. *Plant Cell Rep.,* **21**: 900-904.

Li, L., He, Z., Pandey, G.K., Tsuchiya, T., and Luan, S. (2002). Functional cloning and characterization of a plant efflux carrier for multidrug and heavy metal detoxification. *J. Biol. Chem.,* **277**: 5360-5368.

Lin, Y., Irani, N.G., and Grotewold, E. (2003). Sub-cellular trafficking of phytochemicals explored using auto-fluorescent compounds in maize cells. *BMC Plant Biol.,* **3**: 10.

Marrs, K.A., Alfenito, M.R., Lloyd, A.M., and Walbot, V. (1995). A glutathione S-transferase involved in vacuolar transfer encoded by the maize gene *Bronze-2*. *Nature,* **375**: 397-400.

Martin, D.M., Gershenzon, J., and Bohlmann, J. (2003). Induction of volatile terpene biosynthesis and diurnal emission by methyl jasmonate in foliage of Norway spruce. *Plant Physiol.,* **132**: 1586-1599.

Martin, D.M., Faldt, J., and Bohlmann, J. (2004). Functional characterization of nine Norway Spruce *TPS* genes and evolution of gymnosperm terpene synthases of the TPS-d subfamily. *Plant Physiol.,* **135**: 1908-1927.

Martinoia, E., Klein, M., Geisler, M., Bovet, L., Forestier, C., Kolukisaoglu, U., Muller-Rober, B., and Schulz, B. (2002). Multifunctionality of plant ABC transporters— more than just detoxifiers. *Planta,* **214**: 345-355.

Mathews, H., Clendennen, S.K., Caldwell, C.G., Liu, X.L., Connors, K., Matheis, N., Schuster, D.K., Menasco, D.J., Wagoner, W., Lightner, J., and Wagner, D.R. (2003). Activation tagging in tomato identifies a transcriptional regulator of anthocyanin biosynthesis, modification, and transport. *Plant Cell,* **15**: 1689-1703.

Murray, C.L., Quaglia, M., Arnason, J.T., and Morris, C.E. (1994). A putative nicotine pump at the metabolic blood-brain barrier of the tobacco hornworm. *J. Neurobiol.*, **25**: 23-34.

Nakaune, R., Adachi, K., Nawata, O., Tomiyama, M., Akutsu, K., and Hibi, T. (1998). A novel ATP-binding cassette transporter involved in multidrug resistance in the phytopathogenic fungus *Penicillium digitatum*. *Appl. Environ. Microbiol.*, **64**: 3983-3988.

Niinemets, U., Loreto, F., and Reichstein, M. (2004). Physiological and physicochemical controls on foliar volatile organic compound emissions. *Trends Plant Sci.*, **9**: 180-186.

Noh, B., Murphy, A.S., and Spalding, E.P. (2001). Multidrug resistance-like genes of Arabidopsis required for auxin transport and auxin-mediated development. *Plant Cell*, **13**: 2441-2454.

Ogita, S., Uefuji, H., Yamaguchi, Y., Koizumi, N., and Sano, H. (2003). Producing decaffeinated coffee plants. *Nature*, **423**: 823.

Ohara, K., Kokado, Y., Yamamoto, H., Sato, F., and Yazaki, K. (2004). Engineering of ubiquinone biosynthesis using the yeast *coq2* gene confers oxidative stress tolerance in transgenic tobacco. *Plant J.*, **40**: 734-743.

Otani, M., Shitan, N., Sakai, K., Martinoia, E., Sato, F., and Yazaki, K. (2005). Characterization of vacuolar transport of the endogenous alkaloid berberine in *Coptis japonica*. *Plant Physiol.*, **138**: 1939-1946.

Pighin, J.A., Zheng, H., Balakshin, L.J., Goodman, I.P., Western, T.L., Jetter, R., Kunst, L., and Samuels, A.L. (2004). Plant cuticular lipid export requires an ABC transporter. *Science*, **306**: 702-704.

Rea, P.A., Li, Z.S., Lu, Y.P., Drozdowicz, Y.M., and Martinoia, E. (1998). From vacuolar (GS-X) pumps to multispecific ABC transporters. *Annu. Rev. Plant Physiol. Plant Mol. Biol.*, **49**: 727-760.

Rios, J.L., and Recio, M.C. (2005). Medicinal plants and antimicrobial activity. *J. Ethnopharmacol.*, **100**: 80-84.

Rocha, L.G., Almeida, J.R., Macedo, R.O., and Barbosa-Filho, J.M. (2005). A review of natural products with antileishmanial activity. *Phytomedicine*, **12**: 514-535.

Rose, U.S., and Tumlinson, J.H. (2004). Volatiles released from cotton plants in response to *Helicoverpa zea* feeding damage on cotton flower buds. *Planta*, **218**: 824-832.

Sakai, K., Shitan, N., Sato, F., Ueda, K., and Yazaki, K. (2002). Characterization of berberine transport into *Coptis japonica* cells and the involvement of ABC protein. *J. Exp. Bot.*, **53**: 1879-1886.

Sanchez-Fernandez, R., Davies, T.G., Coleman, J.O., and Rea, P.A. (2001). The *Arabidopsis thaliana* ABC protein superfamily, a complete inventory. *J. Biol. Chem.*, **276**: 30231-30244.

Santelia, D., Vincenzetti, V., Azzarello, E., Bovet, L., Fukao, Y., Duchtig, P., Mancuso, S., Martinoia, E., and Geisler, M. (2005). MDR-like ABC transporter AtPGP4 is involved in auxin-mediated lateral root and root hair development. *FEBS Lett.*, **579**: 5399-5406.

Sasabe, M., Toyoda, K., Shiraishi, T., Inagaki, Y., and Ichinose, Y. (2002). cDNA cloning and characterization of tobacco ABC transporter: *NtPDR1* is a novel elicitor-responsive gene. *FEBS Lett.*, **518**: 164-168.

Sato, H., Tanaka, S., and Tabata, M. (1993). Kinetics of alkaloid uptake by cultured cells of *Coptis japonica*. *Phytochemistry*, **34**: 697-701.

Sato, H., Kobayashi, Y., Fukui, H., and Tabata, M. (1990). Specific differences in tolerance to exogenous berberine among plant cell cultures. *Plant Cell Rep.*, **9**: 133-136.

Schmelz, E.A., Alborn, H.T., and Tumlinson, J.H. (2003). Synergistic interactions between volicitin, jasmonic acid and ethylene mediate insect-induced volatile emission in *Zea mays*. *Physiol. Plant.*, **117**: 403-412.

Schmitz-Hoerner, R., and Weissenbock, G. (2003). Contribution of phenolic compounds to the UV-B screening capacity of developing barley primary leaves in relation to DNA damage and repair under elevated UV-B levels. *Phytochemistry*, **64**: 243-255.

Schoonbeek, H., Del Sorbo, G., and De Waard, M.A. (2001). The ABC transporter *BcatrB* affects the sensitivity of *Botrytis cinerea* to the phytoalexin resveratrol and the fungicide fenpiclonil. *Mol. Plant Microbe Interact.*, **14**: 562-571.

Sheps, J.A., Ralph, S., Zhao, Z., Baillie, D.L., and Ling, V. (2004). The ABC transporter gene family of *Caenorhabditis elegans* has implications for the evolutionary dynamics of multidrug resistance in eukaryotes. *Genome Biol.*, **5**: R15.

Shitan, N., Bazin, I., Dan, K., Obata, K., Kigawa, K., Ueda, K., Sato, F., Forestier, C., and Yazaki, K. (2003). Involvement of CjMDR1, a plant multidrug-resistance-type ATP-binding cassette protein, in alkaloid transport in *Coptis japonica*. *Proc. Natl. Acad. Sci. USA*, **100**: 751-756.

Shitan, N. and Yazaki, K. (2007). Accumulation and membrane transport of plant alkaloids. *Curr. Pharm. Biotech.*, **8**: 244-252.

Shoji, T., Yamada, Y., and Hashimoto, T. (2000). Jasmonate induction of putrescine N-methyltransferase genes in the root of *Nicotiana sylvestris*. *Plant Cell Physiol.*, **41**: 831-839.

Smart, C.C., and Fleming, A.J. (1996). Hormonal and environmental regulation of a plant PDR5-like ABC transporter. *J. Biol. Chem.*, **271**: 19351-19357.

Smit, J.J., Schinkel, A.H., Oude Elferink, R.P., Groen, A.K., Wagenaar, E., van Deemter, L., Mol, C.A., Ottenhoff, R., van der Lugt, N.M., van Roon, M.A., and *et al.* (1993). Homozygous disruption of the murine *mdr2* P-glycoprotein gene leads to a complete absence of phospholipid from bile and to liver disease. *Cell*, **75**: 451-462.

Stergiopoulos, I., Zwiers, L.H., and De Waard, M.A. (2003). The ABC transporter MgAtr4 is a virulence factor of *Mycosphaerella graminicola* that affects colonization of substomatal cavities in wheat leaves. *Mol. Plant Microbe Interact*, **16**: 689-698.

Sugiyama, A., Shitan, N., Sato, S., Nakamura, Y., Tabata, S. and Yazaki, K. (2006). Genome-wide analysis of ATP-binding cassette (ABC) proteins in a model legume plant, *Lotus japonicus*: comparison with Arabidopsis ABC protein family. *DNA Research*, **13**: 205-228.

Sugiyama, A., Shitan, N. and Yazaki, K. (2007). Involvement of a soybean ATP-binding cassette-type transporter in the secretion of genistein, a signal flavonoid in legume-*Rhizobium* symbiosis. *Plant Physiol.*, **144**: 2000-2008.

Tabata, M. (1996). The mechanism of shikonin biosynthesis in cell *Lithospermum* cultures. *Plant Tissue Culture Letters*, **13**: 117-125.

Tabata, M., Tanaka, S., Cho, H.J., Uno, C., Shimakura, J., Ito, M., Kamisako, W., and Honda, C. (1993). Production of an anti-allergic triterpene bryonolic acid, by plant cell cultures. *J. Nat. Prod.*, **56**: 165-174.

Taniguchi, S., Yazaki, K., Yabuuchi, R., Kawakami, K., Ito, H., Hatano, T., and Yoshida, T. (2000). Galloylglucoses and riccionidin A in *Rhus javanica* adventitious root cultures. *Phytochemistry*, **53**: 357-363.

Terasaka, K., Sakai, K., Sato, F., Yamamoto, H., and Yazaki, K. (2003a). *Thalictrum minus* cell cultures and ABC-like transporter. *Phytochemistry*, **62**: 483-489.

Terasaka, K., Shitan, N., Sato, F., Maniwa, F., Ueda, K., and Yazaki, K. (2003b). Application of vanadate-induced nucleotide trapping to plant cells for detection of ABC proteins. *Plant Cell Physiol.*, **44**: 198-200.

Terasaka, K., Blakeslee, J.J., Titapiwatanakun, B., Peer, W.A., Bandyopadhyay, A., Makam, S.N., Lee, O.R., Richards, E.L., Murphy, A.S., Sato, F., and Yazaki, K. (2005). PGP4, an ATP binding cassette P-glycoprotein, catalyzes auxin transport in *Arabidopsis thaliana* roots. *Plant Cell,* **17**: 2922-2939.

Tsukada, M., and Tabata, M. (1984). Intracelluar localization and secretion of naphthoquinone pigment in cell cultures of *Lithospermum erythrorhizon. Planta Med.,* 338-341.

Urban, M., Bhargava, T., and Hamer, J.E. (1999). An ATP-driven efflux pump is a novel pathogenicity factor in rice blast disease. *EMBO J.,* **18**: 512-521.

van den Brule, S., Muller, A., Fleming, A.J., and Smart, C.C. (2002). The ABC transporter SpTUR2 confers resistance to the antifungal diterpene sclareol. *Plant J.,* **30**: 649-662.

Walczak, H.A., and Dean, J.V. (2000). Vacuolar transport of the glutathione conjugate of trans-cinnamic acid. *Phytochemistry,* **53**: 441-446.

Wang, H.K., Xia, Y., Yang, Z.Y., Natschke, S.L., and Lee, K.H. (1998). Recent advances in the discovery and development of flavonoids and their analogues as antitumor and anti-HIV agents. *Adv. Exp. Med. Biol.,* **439**: 191-225.

Weid, M., Ziegler, J., and Kutchan, T.M. (2004). The roles of latex and the vascular bundle in morphine biosynthesis in the opium poppy, *Papaver somniferum. Proc. Natl. Acad. Sci. USA,* **101**: 13957-13962.

Winkel-Shirley, B. (2001). Flavonoid biosynthesis. A colorful model for genetics, biochemistry, cell biology, and biotechnology. *Plant Physiol.,* **126**: 485-493.

Yazaki, K. (2004). Natural Products and Metabolites. *Handbook of Plant Biotechnology, Vol. 2* (Eds. Christou, P. and Klee, H.) John Wiley & Sons Ltd., New York, pp 811-857.

Yazaki, K. (2005). Transporters of secondary metabolites. *Curr. Opin. Plant Biol.,* **8**: 301-307.

Yazaki, K., Matsuoka, H., Shimomura, K., Bechthold, A., and Sato, F. (2001a). A novel dark-inducible protein, LeDI-2, and its involvement in root-specific secondary metabolism in *Lithospermum erythrorhizon. Plant Physiol.,* **125**: 1831-1841.

Yazaki, K., Shitan, N., Takamatsu, H., Ueda, K., and Sato, F. (2001b). A novel *Coptis japonica* multidrug-resistant protein preferentially expressed in the alkaloid-accumulating rhizome. *J. Exp. Bot.,* **52**: 877-879.

Zhao, P., Inoue, K., Kouno, I., and Yamamoto, H. (2003). Characterization of leachianone G 2"-dimethylallyltransferase, a novel prenyl side-chain elongation enzyme for the formation of the lavandulyl group of sophoraflavanone G in *Sophora flavescens* Ait. cell suspension cultures. *Plant Physiol.,* **133**: 1306-1313.

Zuin, V.G., and Vilegas, J.H. (2000). Pesticide residues in medicinal plants and phytomedicines. *Phytother. Res.,* **14**: 73-88.

Zwiers, L.H., and De Waard, M.A. (2000). Characterization of the ABC transporter genes *MgAtr1* and *MgAtr2* from the wheat pathogen *Mycosphaerella graminicola. Fungal Genet. Biol.,* **30**: 115-125.

# Chapter 13

## PROTEOMIC ANALYSIS OF THE VACUOLAR MEMBRANE

TETSURO MIMURA[1], MIWA OHNISHI[1], TAISE SHIMAOKA[2] AND KEN-ICHI TOMIZAWA[2]

[1]*Department of Biology, Graduate School of Science, Kobe University, Nada, Kobe 657-8501, Japan*
[2]*Plant Research Group, Research Institute of Innovative Technology for the Earth, Kizugawadai, Kizu-cho, Soraku-gun, Kyoto 619-0292, Japan*
E-mail: mimura@kobe-u.ac.jp

**Abstract**

A large number of proteins in the vacuolar membrane (tonoplast), including pumps, carriers, ion channels and receptors support the various activities of the plant vacuole. Molecular analysis of these proteins is an essential step in understanding how vacuoles function. However, few proteins involved in these activities have been identified at the molecular level. In part this is due to the difficulty in detecting the low levels of protein activity in the vacuolar membrane, and in part due to the relative lack of mutants related to physiological functions of the vacuole. Proteomic analysis of Arabidopsis and barley has been used by several groups to try to identify new vacuolar membrane proteins. A primary requirement of any organelle analysis by proteomics is that the purity of the isolated organelle needs to be high so that its composition can be unambiguously analyzed by mass spectrometry. Proteins identified so far include previously well-characterized proteins such as V-type $H^+$-ATPases and V-type $H^+$-PPases, along with a number of novel proteins. Functions of some of these newly identified proteins seem reasonable for their location in the membrane; for most others though, function has not been established. In this chapter, we examine recent advances in our understanding of vacuolar proteomics, focusing on important aspects of methodology and on newly identified proteins.

**Keywords:** Arabidopsis, Channel, Membrane, Proteome, Pump, Tonoplast, Transporter, Vacuole

## 1. INTRODUCTION

### 1.1. Roles of the vacuole in plant cells and analysis of its functions

The vacuole is the largest organelle in a plant cell. Its main functions include:

---

© CAB International 2008. *Plant Membrane and Vacuolar Transporters* (eds P.K. Jaiwal, R.P. Singh and O.P. Dhankher)

1. void filling of the cell, 2. storing of metabolites and inorganic nutrients, 3. maintenance of cytoplasmic homeostasis under environmental stresses, and 4. degradation of wasting products (Deepesh, 2000). These roles are achieved through the integration of morphology with a broad range of plant physiological processes. Recent research has focused on specific vacuolar functions, such as the accumulation of storage proteins (Vitale and Hinz, 2005), pH or ion homeostasis (Dietz et al., 2001), the separation of adverse ions and heavy metals (Hall, 2002), the coloring of flowers (Yoshida et al., 2005), or the degradation of small organelles and proteins (Thompson and Vierstra, 2005). How these activities are coordinated and controlled in plant cells is one of the central questions of plant organelle biology. A number of experimental techniques have been developed specifically for vacuolar analysis, such as the preparation of avacuolated cells (mini protoplasts) (Sonobe, 1997), replacement of *in situ* vacuolar content (Shimmen et al., 1994), isolation of vacuoles (Shimaoka et al., 2004), visualization of vacuolar substances, etc. Mutation of vacuole-related genes is an important tool for analyzing vacuolar functions at the molecular level. There are a few reports dealing with vacuolar mutants (Gogarten et al., 1992; Schumacher et al., 1999; Gaxiola et al., 2002), but generation of such mutants is difficult because of the paucity of knowledge concerning genes encoding vacuolar proteins. Proteomic analysis provides a direct mechanism for detecting and identifying a large number of such proteins. In this section, the current status of proteomic analysis of plant vacuoles is discussed.

## 2. PROTEINS OF VACUOLES

Plant vacuoles can be categorized into two groups - vegetative vacuoles and protein storage vacuoles. The latter are present mainly in storage tissues such as seeds, fruits or tubers. Albumin and globulin are well-studied proteins that are stored in protein storage vacuoles. The molecular mechanisms of synthesis, vesicle transport and accumulation of these proteins in vacuoles have been extensively studied (Chrispeels and Herman, 2000; Vitale and Hinz, 2005). In vegetative vacuoles, proteases have received much attention.

Most activities within the vacuole are dependent upon transport across the vacuolar membrane. Transport proteins are therefore crucial to the proper functioning of the vacuole, while signaling molecules on the membrane surface contribute to vacuolar homeostasis. It is therefore important to understand how these proteins operate. Energization of the vacuolar membrane by $H^+$-ATPase or $H^+$-PPase has been well studied (Dietz et al., 2001; Maeshima, 2000), as have some ion transporters such as those involved in calcium transport (Hirschi, 2004). Unfortunately, we have little knowledge about most other vacuolar membrane proteins (Martinoia et al., 2000; Maeshima, 2001), for two main reasons – the amount of these proteins is usually very small, and the transport activity is generally very low and therefore hard to detect. Targeting of single proteins is therefore difficult. Proteomic analysis provides a means of simultaneously examining a large number of proteins.

## 3. METHODOLOGY

### 3.1. General proteomic analysis

The first step in proteomic analysis is isolation of the proteins. In general, after isolation of targeted proteins, they are separated with two-dimensional protein electrophoresis. The spots are then analyzed by a range of procedures (mass spectroscopy, amino acid sequence analyzer etc) and the resulting data on molecular weight or amino acid sequences are used to identify the protein using various databases (e.g. MASCOT (http://www.matrixscience.com/) or SEQUEST (http://fields.scripps.edu/sequest/index.html). One of the critical issues in the analysis of organelles is ensuring the purity of the isolated organelle fraction since contamination from cytoplasm or other organelles can lead to spurious assignment of protein function. Proteomic analysis of chloroplasts and mitochondria has been relatively successful, because these organelles are easily purified and contain a high density of proteins (Ferro et al., 2002; Ferro et al., 2003; Heazlewood et al., 2003; Peltier et al., 2002; Schubert et al., 2002). For vacuoles or vacuolar membranes, a number of isolation methods have been employed and in general, the purity is dependent on the method. A problem with membrane proteins from vacuoles is that they are generally too hydrophobic to separate with two-dimensional electrophoresis methods. The high protein contents of chloroplasts or mitochondrial membranes have the advantage that the hydrophobicities of these proteins are much lower compared with membrane proteins of other organelles. For vacuolar membrane analysis, we must resort to other means of separation.

### 3.2. Isolation of vacuole and vacuolar membrane

As far as we are aware, only five papers dealing with the whole vacuolar membrane proteome have been published (Sazuka et al., 2004; Szponarski et al., 2004; Shimaoka et al., 2004; Carter et al., 2004; Endler et al., 2006). Endler et al. (2006) analyzed barley leaves; all other reports deal with Arabidopsis. Arabidopsis is usually chosen as the experimental material because its genome has been fully sequenced and it was therefore easier to relate molecular information to protein function. Sazuka et al. (2004) and Carter et al. (2004) used Arabidopsis leaf mesophyll cells, and Szponarski et al. (2004) and Shimaoka et al. (2004) used suspension-cultured cells. Sazuka et al. (2004) and Szponarski et al. (2004) used a membrane fraction purified by sucrose density gradient and liquid chromatographic separation. This method is suitable for large amounts of sample and can generate larger amounts of protein for subsequent analysis than other methods. The drawback is that there is inevitably significant contamination from other membranes. The other three groups (Shimaoka et al., 2004; Carter et al., 2004; Endler et al., 2006) first isolated intact vacuoles from protoplasts then purified membrane proteins by a combination of rupturing intact vacuoles and ultracentrifugation, resulting in a much purer vacuolar membrane fraction (Fig. 1). The disadvantage of this method is that it is not easy to collect large amounts of protein. However, recent improvements in protein analysis have increased sensitivity so that less protein is now needed than previously.

Purity of membrane fractions can be confirmed with Western blot analysis against proteins that are known to locate in the other organelle membranes. P-type $H^+$-ATPase or PIP can be used as indicators of plasma membrane contamination, Bip for ER, COX or AOX for mitochondria, and HSP93 or CF-type $H^+$-ATPase (alpha

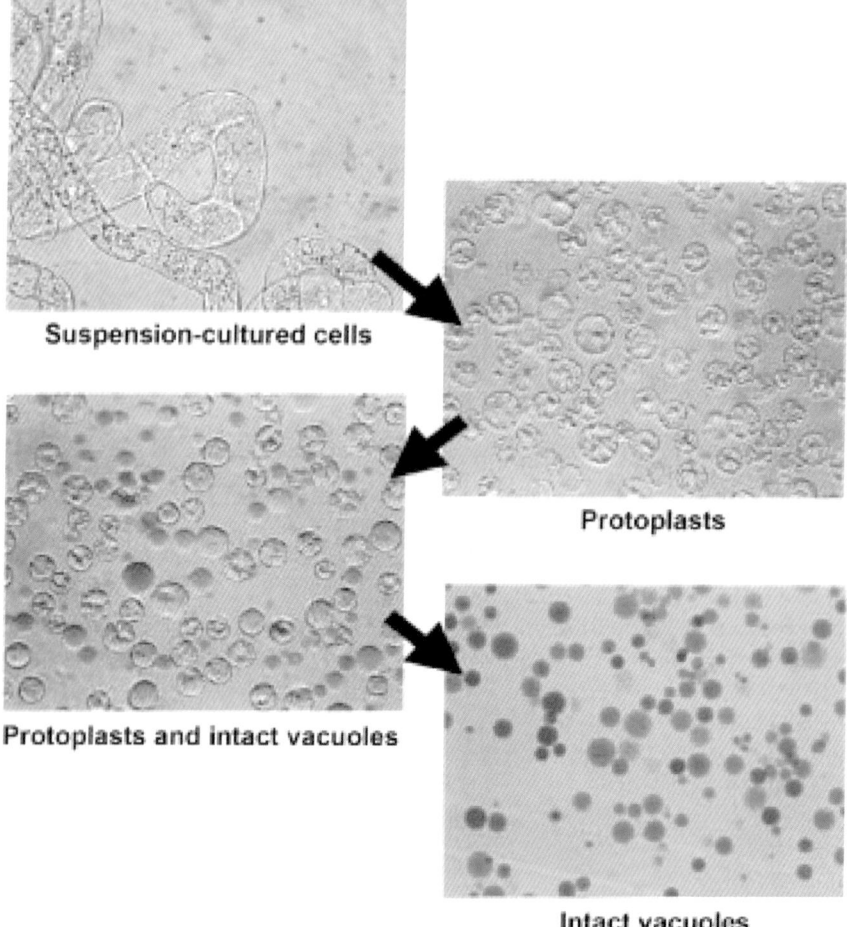

**Figure 1:** Isolation of intact vacuoles from suspension-cultured *Catharanthus* cells. First, protoplasts were prepared by enzyme digestion with a combination of hemicellulase and cellulase. Then, protoplasts were gently ruptured by osmotic shock or sheer force treatments. By rupturing of protoplasts, some of them released intact vacuoles which are stained with neutral red. The mixture of protoplasts and intact vacuoles was put on the density gradient, and then intact vacuoles were purified.

subunit) for chloroplasts. Alternatively, activities of enzymes associated with other organelles can be measured to give an indication of the degree of contamination.

### 3.3. Extraction and analysis of membrane proteins

Various methods have been used for the separation of membrane proteins. The membrane fraction contains both peripheral and membrane-spanning proteins and methods of purification need to take into account the differing chemical nature of the two types. While peripheral proteins may have important functions for vacuolar biological activities, membrane-spanning proteins are likely to have more critical roles in determining exchanges between the cytoplasm and the vacuole. It is therefore desirable to separate peripheral proteins from membrane intrinsic proteins. The peripheral protein fraction is also more likely to contain non-vacuole contaminant proteins. In order to wash peripheral proteins, membrane samples can be treated with high ionic strength and/or alkaline solutions. For example, treatment with 50 mM KCl and 0.05% deoxycholate removes surface proteins while subsequent treatment with 100 mM $Na_2CO_3$ and 500 mM KSCN at 4°C for 30 min removes tightly bound peripheral proteins. To extract membrane-spanning proteins from the remainder, membrane samples are often treated with detergent or organic solvents, for example CHAPS, dodecyl maltoside, octyl glucoside, or chloroform/methanol. Szponarski et al. (2004) used docecyl maltoside solubilization of membrane proteins; Sazuka et al. (2004) used a combination of urea, CHAPS, and $N$-decyl-$N,N$-dimethyl-3-ammonio-1-propanesulfonate. Other groups (Carter et al., 2004; Shimaoka et al., 2004; Endler et al., 2006) have avoided extraction altogether. In our preliminary work, we experimented with various detergents and organic solvents, but some proteins were always precipitated. As mentioned above, vacuolar membrane proteins are very hydrophobic, and it is not easy to resolve all proteins from the purified membrane fraction, especially with conventional two-dimensional electrophoresis. This is well illustrated by our demonstration of the failure of $H^+$-PPase, an abundant intrinsic protein, to enter the gel for the first step of the two dimensional electrophoresis without denaturing.

In order to separate membrane-spanning proteins, two different methods have been used (Fig. 2). In the most popular, proteins solubilized in SDS solution are separated with conventional 1D polyacrylamide gel electrophoresis (SDS-PAGE). After SDS-PAGE, the stained bands and smearing parts are cut off at appropriate intervals, e.g. 1 mm. The excised gel pieces are washed, reduced, alkylated and digested with trypsin. Digestion of proteins to small polypeptides with trypsin is essential for the next mass spectrometry analysis. Some proteins always sit at the start of the gel and therefore cannot be analyzed. The alternative method is to simply treat the protein sample in buffered solution with trypsin directly. Some of the proteins that cannot be electrophoresed, are possibly digested with trypsin and can then be analyzed.

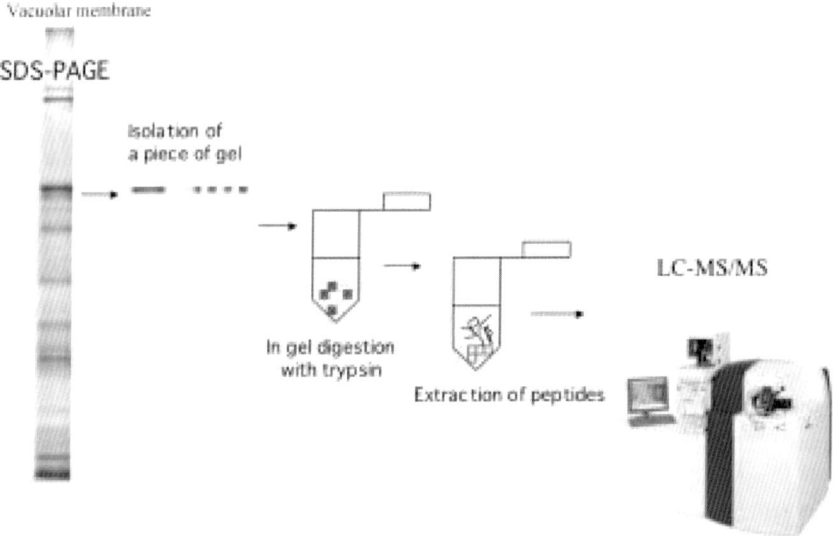

**Figure 2:** An example of protein identification by mass spectrometer. Proteins solubilized in SDS solution are separated with conventional 1D PAGE. After SDS-PAGE, the stained bands and smearing parts were cut off at proper intervals. The excised gel pieces were washed, reduced, alkylated and digested with trypsin. The tryptic peptides were injected into a capillary HPLC system which was coupled to a mass spectrometer.

### 3.4. Analysis by mass spectroscopy

After concentrating, the tryptic peptides are usually injected into a capillary HPLC system where they are separated on a column coupled to a mass spectrometer. The MS and MS/MS data are acquired and processed and the results compared with databases using commercially or non-commercially supplied programs such as MASCOT or SEQUEST, which also give simple explanations about MS analysis. In the case of MS analysis, a finger peptide map with molecular weight distribution is obtained. Although it is possible to identify some proteins with MS analysis data alone using the peak strength and genome data, some ambiguities of identification remain. With MS/MS data, it is possible to identify the amino acid sequence of each peptide more exactly. In MS/MS analysis, it is possible not only to identify peptides, but to predict proteins by homology without recourse to genome data. However, in any cases, for identification of proteins, the peptide sequence is needed to be checked often manually. Sazuka et al. (2004) used Edman sequencing for determination of amino acid sequences in spite of MS analysis, because more accurate determination of amino acid sequences is possible. Thus, although it is possible to determine amino acid sequences without mass spectrometry, this is not a popular choice at present.

A recent review of membrane proteomic analysis (Rolland et al., 2006) provides some useful information relating to methodology.

### 3.5. Confirmation of protein localization

Proteomic analysis is a powerful tool for determining intracellular protein localization, provided that care is taken to avoid contamination of proteins from other sources. Despite the most careful efforts, we find that there are inevitably some proteins in the vacuolar fraction that are known to be related to activities in other organelles. While they may be true contaminants, it must be remembered that plant vacuoles also have a lytic function, and may therefore contain components of other organelles at various stages of degradation. In order to confirm whether candidate proteins are native to the vacuolar membrane, it is necessary to verify their true location with other methods. The easiest way is to use GFP-fusion proteins, as demonstrated by Endler et al. (2006), although there is some suggestion that overexpression of GFP proteins results in expression in other locations. If it is possible to raise an antibody to a target protein, immunoelectronmicroscopy is a more reliable method.

## 4. PROTEOMIC ANALYSIS OF VACUOLAR MEMBRANE

### 4.1. Overview

As shown in Table 1, each of the studies found around 60-200 proteins which have at least one transmembrane domain (TMD). Carter et al. (2004) conducted proteomic analysis of not only membrane proteins but vacuolar soluble proteins and found approximately 400 proteins in total. Table 1 also shows the number of vacuolar $H^+$ pumps and the number of proteins with more than one TMD.

About 20 to 100 proteins that were judged to be obvious contaminants were excluded. There were some contaminants from the plasma membrane or other organelle membranes, as well as ribosomal proteins which were found with the peripheral membrane fraction. This may be due to washing the vacuolar membrane with solutions containing high ionic strength or detergent. Overall, however, all studies found many new candidate vacuolar membrane proteins. In the next section, we discuss vacuolar proteins that have at least one TMD; we will not discuss membrane peripheral proteins or soluble proteins.

Most membrane proteins ranged between 20 to 200 kDa. Figure 3 shows the distribution of molecular weights of proteins shown in Table 1. It can be seen that most of the proteins in the vacuolar membrane fall in the range 20-70 kDa. It is not easy to analyze the smaller molecular weight proteins because they are poorly separated using SDS-PAGE. A more detailed examination in the lower molecular weight range might reveal some interesting proteins.

Figure 4 shows the distribution of the number of predicted TMDs for the proteins detected. The precise number of TMD varies with the prediction program, especially in deciding if a protein contains one domain or not, a distinction that is important in separating membrane proteins from soluble proteins. Although many proteins are grouped in TMD = 1, many others had more than 2 TMDs.

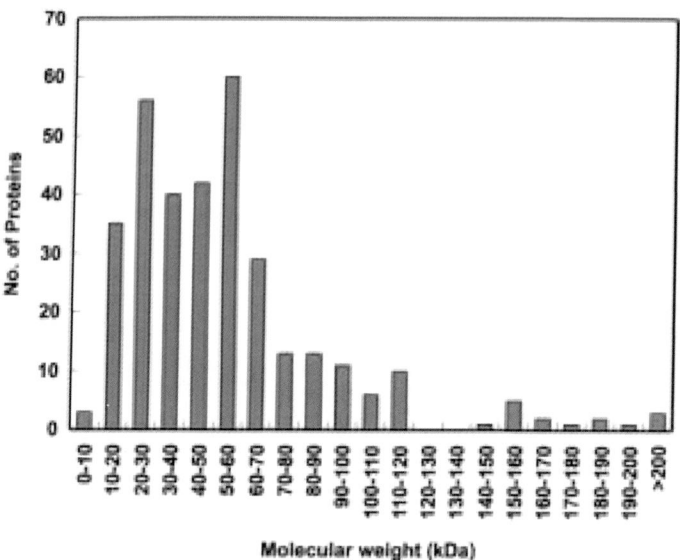

**Figure 3:** Distribution of molecular weight of identified vacuolar membrane proteins that have two or more transmembrane domains.

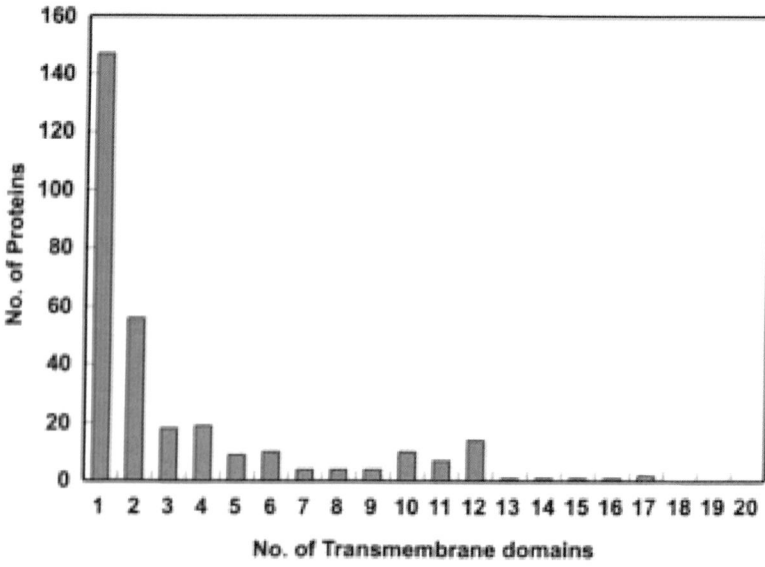

**Figure 4:** No. of transmembrane domains of identified vacuolar membrane proteins that have more than one transmembrane domain.

Table 1. Overview of proteomic analysis of the vacuolar membrane.

| Reference | Material Species | Cell source | Purification Method | Protein Identification | Total | V pump | TMD≥2 | Transporter | Membrane remodelling | Stress response | Others | Unknown | TMD=1 |
|---|---|---|---|---|---|---|---|---|---|---|---|---|---|
| Szponarski et al., 2004 | Arabidopsis | Suspension-cultured cell | Gel filtration, anion exchange chromatography | MALDI-TOF-MS | 70 | 6 | 35 | 6 | 8 | 0 | 11 | 10 | 11 |
| Shimaoka et al., 2004 | Arabidopsis | Suspension-cultured cell | Rupture of intact vacuole, SDS PAGE, LC | Q-TOF-MS/MS | 163 | 19 | 35 | 18 | 0 | 1 | 2 | 14 | 28 |
| Sazuka et al., 2004 | Arabidopsis | Shoot | Sucrose density gradient, LC | Edman sequencing | 64 | 5 | 15 | 10 | 1 | 1 | 1 | 2 | 3 |
| Carter et al., 2004 | Arabidopsis | Mesophyll cell | Rupture of intact vacuole, SDS PAGE, LC | Q-TOF-MS/MS | 402 | 18 | 92 | 25 | 7 | 9 | 28 | 23 | 111 |
| Endler et al., 2006 | Barley | Mesophyll cell | Rupture of intact vacuole, SDS PAGE, LC | ESI-MS/MS | 101 | 14 | 31 | | | | | | 10 |

1) Number of transmembrane domains is as reported from each study. In Carter et al. (2004), a prediction by HMMTOP (http://www.enzim.hu/hmmtop/) was adopted.

**Table 2.** $H^+$ pumps of the vacuolar membrane detected in proteome analyses.

| | Locus | MW | TMD | Szponarski et al. | Shimaoka et al. | Sazuka et al. | Carter et al. |
|---|---|---|---|---|---|---|---|
| **V-type $H^+$-ATPase** | | | | | | | |
| AtVHA-A | At1g78900 | 68795 | 0 | O | O | O | O |
| AtVHA-B1 | At1g76030 | 54090 | 0 | | O | | O |
| AtVHA-B2 | At4g38510 | 54288 | 0 | | O | | O |
| AtVHA-B3 | At1g20260 | 36341 | 0 | | O | | O |
| AtVHA-C | At1g12840 | 42602 | 0 | O | O | | O |
| AtVHA-D | At3g58730 | 29041 | 0 | O | O | | O |
| AtVHA-E1 | At4g11150 | 26042 | 0 | O | O | | O |
| AtVHA-E2 | At3g08560 | 26853 | 0 | | | | |
| AtVHA-E3 | At1g64200 | 27067 | 0 | | O | | O |
| AtVHA-F | At4g02620 | 14242 | 2 | | O | | O |
| AtVHA-G1 | At3g01390 | 12379 | 0 | | O | | O |
| AtVHA-G2 | At4g23710 | 11724 | 0 | | | | O |
| AtVHA-G3 | At4g25950 | 12116 | 0 | | | | |
| AtVHA-H | At3g42050 | 50267 | 0 | | O | | O |
| AtVHA-a1 | At2g21410 | 93089 | 6 | | O | O | O |
| AtVHA-a2 | At4g39080 | 92817 | 6 | O | O | O | O |
| AtVHA-a3 | At2g28520 | 93416 | 6 | | | | |
| AtVHA-c1[a)] | At4g34720 | 16554 | 4 | | O | O | |
| AtVHA-c2 | At1g19910 | 16625 | 3 | | | | |
| AtVHA-c3 | At4g38920 | 16554 | 4 | | | | O |
| AtVHA-c4 | At1g75630 | 16668 | 4 | | | | |

**Table 2.** Continued

|  | Locus | MW | TMD | Szponarski et al. | Shimaoka et al. | Sazuka et al. | Carter et al. |
|---|---|---|---|---|---|---|---|
| AtVHA-c5 | At2g16510 | 16554 | 4 | | | | |
| AtVHA-c"1 | At3g32530 |  | 4 | | | | |
| AtVHA-c"2 | At2g25610 | 18201 | 4 | | | | |
| AtVHA-d1 | At3g28710 | 40774 | 0 | | O | | O |
| AtVHA-d2 | At3g28715 | 40770 | 0 | | O | | O |
| AtVHA-e1 | At5g55290 | 7726 | | | | | |
| AtVHA-e2 | At4g26710 | 7686 | | | | | |
| **V-type H$^+$-PPase** | | | | | | | |
| AVP1 | At1g15690 | 80803 | 16 | O | O | O | O |
| AVPL1[b)] | At1g16780 | 85332 | 17 | | O | | |
| AVP2 | At1g78920 | 85116 | 17 | | O | | |

a) AtVHA-c was detected, but the gene locus was not identified
b) Only consensus fragments among the other isoforms were detected

Significantly, there were many proteins with TMD = 5-7 or 11-13 which are characteristics of membrane transporters.

There are now many plant proteomic databases that compile topology and intracellular targeting predictions, and this information can then be utilized to gain insight into the likely function of membrane proteins (Schwacke et al., 2004). In the following sections, the proteins detected by proteomic analysis of vacuoles are grouped according to likely function and number of TMDs.

### 4.2. Proton pumps

It was expected that the major proteins on the vacuolar membrane would be V-type H$^+$-ATPases, V-type H$^+$-PPases and aquaporins (Maeshima, 2001). The amounts of these two H$^+$-pumps greatly exceeded the amounts of other proteins, and were easily detected. Most of the subunits of V-type H$^+$-ATPase and V-type H$^+$-PPase were found in one or more of the studies (Table 2). Aquaporin proteins were not detected in cultured cells (Table 3), which is consistent with evidence suggesting

Table 3. Putative Arabidopsis vacuolar membrane transporters and channels

| Locus | MW | TMD | Szponarski et al. | Shimaoka et al. | Sazuka et al. | Carter et al. | Annotation |
|---|---|---|---|---|---|---|---|
| **ABC protein** | | | | | | | |
| At1g30400 | 181911 | 15 | O | O | O | O | glutathione S-conjugate transporting ATPase (AtMRP1) |
| At1g59870 | 165084 | 13 | | | O | | ABC transporter |
| At2g34660 | 182114 | 15 | | O | | | ABC transporter (AtMRP2) |
| At2g47800 | 169064 | 16 | O | O | | O | glutathione-conjugate transporter AtMRP4 |
| At3g21250 | 143430 | 10 | O | | | | ABC transporter family protein |
| At3g62700 protein | 172121 | 17 | | O | | O | ABC transporter-like |
| At5g39040 protein | 69086 | 6 | | O | | O | ABC transporter-like |
| **Pumps** | | | | | | | |
| At2g41560 | 112733 | 9 | | | | O | calcium-transporting ATPase 4 |
| At3g57330 | 111928 | 10 | O | O | | O | $Ca^{2+}$-transporting ATPase-like protein |

**Table 3.** Continued

| Locus | MW | TMD | Szpon-arski et al. | Shimaoka et al. | Sazuka et al. | Carter et al. | Annotation |
|---|---|---|---|---|---|---|---|
| **Other transporters and channel-like proteins** | | | | | | | |
| At1g11260 | 57593 | 12 | | | O | O | glucose transporter |
| At1g20840 | 79468 | 11 | | | | O | transporter-related |
| At1g53210 | 63399 | 10 | O | | | | sodium/calcium exchanger family protein |
| At1g71880 | 54841 | 12 | | O | | | sucrose transport protein SUC1 |
| At1g76520 | 42615 | 10 | | O | | | auxin efflux carrier family protein |
| At1g76530 | 45568 | 10 | | O | | | auxin efflux carrier family protein |
| At1g76670 | 37694 | 7 | | | O | | transporter-related |
| At1g80310 | 49534 | 10 | | | | O | sulfate transporter-related |
| At2g02040 | 64404 | 11 | | O | | O | histidine transport protein (PTR2-B) |
| At2g21160 | 28150 | 2 | | O | | | putative signal sequence receptor |
| At2g23150 | 56121 | 12 | | O | | | putative metal ion transporter (NRAMP) |

**Table 3.** Continued

| Locus | MW | TMD | Szponarski et al. | Shimaoka et al. | Sazuka et al. | Carter et al. | Annotation |
|---|---|---|---|---|---|---|---|
| At2g26690 | 63573 | 12 | | | O | | nitrate transporter |
| At2g26975 | 15788 | 2 | | | | O | copper transport protein-related |
| At2g28900 | 15464 | 4 | | | | O | membrane channel protein-related |
| At2g29410 | 42341 | 5 | | | O | | zinc transporter |
| At2g32830 | 59214 | 10 | | | O | | phosphate transporter (AtPht1;5) |
| At2g38170 | 50159 | 10 | | | | O | calcium exchanger (CAX1) |
| At2g43330 | 54795 | 12 | | | | O | sugar transporter family |
| At2g46800 | 43810 | 6 | O | | | | putative zinc transporter |
| At2g48020 | 49682 | 12 | | | | O | sugar transporter, putative |
| At3g01280 | 29408 | ? | | O | | O | porin-related |
| At3g03720 | 63621 | 14 | | | | O | cationic amino acid transporter -related |
| At3g16240 | 25009 | 7 | | | | O | delta tonoplast integral protein (delta-TIP) |

**Table 3.** Continued

| Locus | MW | TMD | Szponarski et al. | Shimaoka et al. | Sazuka et al. | Carter et al. | Annotation |
|---|---|---|---|---|---|---|---|
| At3g26520 | 25831 | 6 | | | | O | tonoplast intrinsic protein, putative |
| At3g26590 | 54304 | 12 | | | | O | MATE efflux protein |
| At3g30390 | 49500 | 11 | | | | O | amino acid transporter |
| At3g58810 | 47428 | 6 | | O | | | zinc transporter |
| At4g03560 | 84856 | 12 | O | | | O | two-pore calcium channel (TPC1) |
| At4g13510 | 53577 | 9 | | | O | | High affinity ammonium transporter (AtAMT1;1) |
| At4g35300 | 79709 | 10 | | | | O | transporter-related low similarity to hexose transporter |
| At5g12080 | 83015 | 5 | | O | | | putative protein similarity to mechano-sensitive ion channel |
| At5g15090 | 29193 | 0 | | | | O | voltage-dependent anion-selective channel protein hsr2 |
| At5g26340 | 57420 | 12 | | | O | | hexose transporter |

**Table 3.** Continued

| Locus | MW | TMD | Szponarski et al. | Shimaoka et al. | Sazuka et al. | Carter et al. | Annotation |
|---|---|---|---|---|---|---|---|
| At5g43340 | 56248 | 10 | | | O | | phosphate transporter (AtPht1;6) |
| At5g62890 | 57563 | 12 | | O | | | permease 1-like protein |
| At5g67330 | 56368 | ? | | O | | O | NRAMP metal ion transporter 4 (NRAMP4) |

that aquaporins are not strongly expressed in liquid suspension-cultured cells (Kobae et al., 2006).

The V-type $H^+$-ATPase consists of a $V_1$ sector and a $V_0$ sector (Sze et al., 2002). The $V_1$ sector is composed of eight subunits, A to H, while the $V_0$ sector is composed of five subunits, a, c, c", d and e. The proposed 28 genes encoding the subunits of the V-type $H^+$-ATPase in the *Arabidopsis thaliana* genome are shown in Table 2 (Sze et al., 2002). Most subunits of the $V_1$ sector and the $V_0$ sector were identified in all analyses. Three genes, *AVP1*, *AVP2* and *AVPL1*, in the *Arabidopsis thaliana* genome encode V-type $H^+$-PPase (Drozdowicz et al., 2000). AVP1 was reported in all studies, but AVP2 and AVPL1 only in the study by Shimaoka et al. (2004).

### 4.3. Transporters and channels

The super family of ABC transporters consists of 115 genes in *Arabidopsis thaliana* (Arabidopsis Genome Initiative 2000). Many ABC transporters are located on the vacuolar membranes or plasma membranes and contribute to the transport of secondary metabolites across these membranes. In the present analyses, seven ABC transporters were identified (Table 3). At1g30400, At2g47800 were detected in all measurements and At3g62700, At5g39040 in two studies.

Two $Ca^{2+}$ pumps were identified along with 36 other transporters and channel-like proteins. Although some of these may be contaminants, most of them are likely to occur in the vacuolar membrane. Unlike the $H^+$ pumps or ABC proteins, these other transporter proteins were mostly detected in only one or other of the studies. Only five proteins (At1g11260, At2g02040, At3g01280, At4g03560, At5g67330) were detected in two studies. This no doubt reflects the low content of these proteins in vacuolar membranes, and the variation in sensitivity of detection according to the experimental conditions used by each group. For most of these putative

**Table 4.** Putative Arabidopsis vacuolar membrane proteins which have two or more transmembrane domains.

| Locus | MW | TM | Szponarski et al. | Shimaoka et al. | Sazuka et al. | Carter et al. | Annotation |
|---|---|---|---|---|---|---|---|
| **Membrane remodeling** | | | | | | | |
| At1g28580 | 43007 | 4 | | | | O | lipase, putative |
| At1g31480 | 106317 | 2 | O | | | | similar to phospholipase |
| At1g31550 | 43079 | 3 | | | | O | lipase, putative |
| At1g32400 | 31538 | 5 | | | | O | senescence-associated protein family |
| At1g43620 | 68327 | 3 | | | | O | sterol glucosyl-transferase-related |
| At1g65320 | 46538 | 2 | | | | O | CBS domain containing protein |
| At1g75010 | 82584 | 4 | O | | | | phosphatidyl-inositol-4-phosphate 5-kinase family protein |
| At2g21160 | 28150 | 2 | | | | O | signal sequence receptor, alpha subunit |
| At3g11130 | 193248 | 2 | O | | | | clathrin heavy chain |
| At3g11430 | 56082 | 2 | | | O | | 1-acylglycerol-3-phosphate O-acyltransferase |
| At3g46060 | 23817 | 2 | | | | O | Ras family GTP-binding protein |

**Table 4.** Continued

| Locus | MW | TM | Szponarski et al. | Shimaoka et al. | Sazuka et al. | Carter et al. | Annotation |
|---|---|---|---|---|---|---|---|
| At4g17170 | 23165 | 2 | O | | | | Rab2-like GTP-binding protein (RAB2) |
| At5g11150 | 25321 | 2 | O | | | | synaptobrevin/vesicle-associated membrane protein 713 (VAMP713) |
| At5g22360 | 25317 | 2 | O | | | | synaptobrevin family protein |
| At5g22460 | 39099 | 4 | O | | | | esterase/lipase/thioesterase family protein |
| At5g23670 | 54290 | 3 | O | | | | serine C-palmitoyltransferase (LCB2) |

**Stress response**

| Locus | MW | TM | Szponarski et al. | Shimaoka et al. | Sazuka et al. | Carter et al. | Annotation |
|---|---|---|---|---|---|---|---|
| At1g24450 | 20724 | 2 | | | | O | ribonuclease III -related |
| At1g30360 | 81918 | 10 | | | O | O | ERD4 protein |
| At1g32400 | 31538 | 5 | | | | O | senescence-associated protein family |
| At1g65820 | 16568 | 3 | | O | | O | microsomal glutathione s-transferase, putative |
| At1g71695 | 39542 | 2 | | | | O | peroxidase, putative |
| At2g15970 | 21587 | 5 | | | | O | cold acclimation protein |

**Table 4.** Continued

| Locus | MW | TM | Szponarski et al. | Shimaoka et al. | Sazuka et al. | Carter et al. | Annotation |
|---|---|---|---|---|---|---|---|
| At4g25090 | 96846 | 5 | | | | O | respiratory burst oxidase (NADPH oxidase), putative |
| At5g49760 | 104691 | 3 | | | | O | receptor protein kinase-related |
| At5g58150 | 86594 | 2 | | | | O | leucine-rich repeat transmembrane protein kinase, putative |
| **Others** | | | | | | | |
| At1g04710 | 46594 | 3 | | | | O | acetyl-CoA C-acyltransferase (3-ketoacyl-CoA thiolase), putative |
| At1g10980 | 58141 | 7 | | | | O | membrane protein PTM1 precursor isolog |
| At1g11910 | 54596 | 2 | | | | O | aspartic proteinase-related |
| At1g15530 | 73128 | 2 | | | | O | receptor lectin kinase-related |
| At1g19450 | 52770 | 12 | | | | O | integral membrane protein, putative |
| At1g19670 | 34837 | 2 | | | | O | coronatine-induced protein 1 (CORI1) |

**Table 4.** Continued

| Locus | MW | TM | Szponarski et al. | Shimaoka et al. | Sazuka et al. | Carter et al. | Annotation |
|---|---|---|---|---|---|---|---|
| At1g32410 | 15007 | 4 | | | | O | expressed protein (vacuolar protein sorting 55 family protein) |
| At1g56075 | 93892 | 2 | O | | | | elongation factor 2 |
| At1g64720 | 43758 | 2 | O | | | | membrane related protein CP5 |
| At1g73950 | 52892 | 5 | | | | O | C3HC4-type zinc finger protein |
| At1g78670 | 39865 | 2 | | | | O | gamma glutamyl hydrolase-related |
| At1g78680 | 38625 | 2 | | | | O | gamma-glutamyl hydrolase (gamma-Glu-X carboxypeptidase/conjugase) |
| At2g29120 | 104243 | 3 | | | | O | glutamate receptor family (GLR2.7) |
| At2g36830 | 25602 | 7 | | | | O | major intrinsic protein (MIP) |
| At2g46370 | 64351 | 2 | O | | | | auxin-responsive GH3 family protein |
| At2g47650 | 49946 | 2 | O | | | | NAD-dependent epimerase/dehydratase family protein |

**Table 4.** Continued

| Locus | MW | TM | Szponarski et al. | Shimaoka et al. | Sazuka et al. | Carter et al. | Annotation |
|---|---|---|---|---|---|---|---|
| At3g01930 | 63500 | 13 | | O | | | nodulin family protein |
| At3g07980 | 151139 | 4 | O | | | | member of MEKK subfamily |
| At3g43190 | 93004 | 2 | O | | | | sucrose synthase |
| At3g55260 | 61212 | 3 | | | | O | glycosyl hydrolase family 20, similar to beta-hexosaminidase A |
| At4g10050 | 38563 | 3 | O | | | | hydrolase, alpha/beta fold family protein |
| At4g17340 | 25062 | 6 | | | | O | major intrinsic protein (MIP) |
| At4g21380 | 96467 | 3 | O | | | | putative receptor-like serine/threonine protein kinases |
| At4g21410 | 75721 | 3 | O | | | | protein kinase family protein |
| At4g25570 | 25847 | 6 | | O | | O | putative protein similarity to cytochrome b561 |
| At4g32920 | 153973 | 6 | | | | O | glycine-rich protein |
| At4g33280 | 52980 | 2 | | | | O | transcriptional factor B3 |
| At4g33360 | 37861 | 2 | O | | | O | terpene cyclase/mutase-related |

**Table 4.** Continued

| Locus | MW | TM | Szponarski et al. | Shimaoka et al. | Sazuka et al. | Carter et al. | Annotation |
|---|---|---|---|---|---|---|---|
| At4g38350 | 116959 | 11 | | | | O | patched protein |
| At5g11700 | 151531 | 8 | | | | O | glycine-rich protein |
| At5g14120 | 63132 | 12 | | | | O | nodulin family protein |
| At5g15350 | 19417 | 2 | | | | O | plastocyanin-like domain containing protein |
| At5g20230 | 20036 | 2 | | | | O | plastocyanin-like domain containing protein |
| At5g28540 | 73612 | 2 | | | | O | luminal binding protein 1 precursor (BiP-1) (AtBP1) |
| At5g42020 | 73544 | 2 | | | | O | luminal binding protein 2 precursor (BiP-2) |
| At5g48410 | 97026 | 3 | | | O | | glutamate receptor family protein (GLR1.3) |
| At5g61790 | 60486 | 2 | O | | | O | calnexin 1 (CNX1) |
| At5g64870 | 52816 | ? | | | | O | nodulin-related |
| At5g66680 | 48727 | 2 | | | | O | dolichyl-diphosphooligosaccharide-protein glycosyltransferase |

**Table 5.** Unannotated proteins which have two or more transmembrane domains detected in Arabidopsis vacuolar membrane proteome analysis.

| Locus | MW | TMD | Szponarski et al. | Shimaoka et al. | Sazuka et al. | Carter et al. | Annotation |
|---|---|---|---|---|---|---|---|
| AAF87033 |  | 2 | O |  |  |  | expressed protein |
| At1g08480 | 15795 | 2 |  |  |  | O | expressed protein |
| At1g16010 | 50403 | 2 |  |  |  | O | expressed protein |
| At1g16290 | 47941 | 4 | O |  |  |  | expressed protein |
| At1g22750 | 27309 | 6 |  |  |  | O | expressed protein |
| At1g29820 | 61051 | 2 |  | O |  |  | unknown protein |
| At1g32400 | 31538 | 3 |  | O |  |  | senescence-associated protein family |
| At1g53210 | 63399 | 11 |  |  |  | O | expressed protein |
| At1g55190 | 21054 | 2 |  | O |  |  | hypothetical protein |
| At1g58030 | 67098 | 15 |  |  |  | O | expressed protein |
| At1g61000 | 111643 | 5 |  | O |  |  | unknown protein Nuf2 family protein |
| At1g61890 | 55116 | 12 |  | O |  |  | hypothetical protein |
| At1g65920 | 111114 | 3 | O |  |  |  | regulator of chromosome condensation (RCC1) family protein |

**Table 5.** Continued

| Locus | MW | TMD | Szpon-arski et al. | Shimaoka et al. | Sazuka et al. | Carter et al. | Annotation |
|---|---|---|---|---|---|---|---|
| At1g73650 | 32942 | 8 | | | | O | expressed protein |
| At2g04690 | 23576 | 2 | | | | O | expressed protein |
| At2g07698 | 85916 | 3 | | | | O | expressed protein |
| At2g20230 | 29704 | 4 | | O | | | unknown protein |
| At2g20230 | 29704 | 4 | | | | O | expressed protein |
| At2g21080 | 46792 | 6 | O | | | | expressed protein |
| At2g25680 | 48374 | 11 | | | | O | hypothetical protein |
| At2g26890 | 279076 | 8 | O | | | | GRV2 has sequence similarity to the *C. elegans* protein RME-8 |
| At2g41470 | 42094 | 2 | | | | O | hypothetical protein |
| At2g42570 | 43166 | 2 | O | | | | expressed protein |
| At3g11530 | 12454 | 3 | | | | O | expressed protein |
| At3g13000 | 62611 | 2 | | | | O | expressed protein |
| At3g21690 | 54934 | 12 | | O | | | integral membrane protein |

**Table 5.** Continued

| Locus | MW | TMD | Szpon-arski et al. | Shimaoka et al. | Sazuka et al. | Carter et al. | Annotation |
|---|---|---|---|---|---|---|---|
| At3g24480 | 54685 | 2 | | O | | | disease resistance protein |
| At3g44330 | 62185 | 2 | | | | O | expressed protein |
| At3g56430 | 48535 | 2 | O | | | | expressed protein |
| At4g11720 | 75505 | 3 | | | O | | expressed protein |
| At4g12650 | 59710 | 9 | O | | | | endomembrane protein 70 |
| At4g18810 | 65468 | 2 | O | | | | expressed protein |
| At4g23630 | 30511 | 4 | | | | O | expressed protein |
| At4g27520 | 35046 | 2 | | | | O | expressed protein |
| At4g28770 | 30959 | 4 | | O | | O | expressed protein |
| At4g33625 | 22596 | 4 | | | | O | expressed protein |
| At4g38350 | 116959 | 11 | O | O | | | putative protein Niemann-Pick C disease protein |
| At5g06390 | 50391 | 2 | | | | O | expressed protein |
| At5g11560 | 108383 | 2 | | | | O | expressed protein |
| At5g11700 | 151531 | 8 | | O | | | glycine-rich predicted protein |

**Table 5.** Continued

| Locus | MW | TMD | Szponarski et al. | Shimaoka et al. | Sazuka et al. | Carter et al. | Annotation |
|---|---|---|---|---|---|---|---|
| At5g20650 | 15766 | 2 | | | | O | expressed protein |
| At5g20660 | 100077 | 9 | | O | | | 24 kDa vacuolar protein |
| At5g23920 | 25786 | 5 | | | | O | expressed protein |
| At5g47020 | 152296 | 4 | O | | | | glycine-rich protein |
| At5g54170 | 50230 | 2 | | | O | | expressed protein |
| At5g61820 | 53054 | 2 | | | | O | expressed protein |
| At5g62200 | 21053 | 2 | | O | | | putative protein embryo-specific protein 3 |

transporters and channels, a function has not been established. Further characterization will be needed to decide their importance in vacuolar physiology.

### 4.4. Other annotated and unannotated proteins with two or more transmembrane domains

There were 92 other proteins with two or more TMDs, 69 of which could be annotated. As shown in Table 4, these proteins were associated with a diversity of activities. Many of the proteins appeared to be involved in either membrane remodeling or stress response. The membrane remodelling proteins included lipid metabolism-related proteins, GTP-binding proteins, and SNARE proteins.

Twenty-three of the proteins detected were not annotated (Table 5). Some of them have more than 10 TMDs, and may therefore have important roles in the vacuolar membrane.

### 4.5. Annotated proteins with one transmembrane domain

In total, 147 proteins were categorized as having one or no TMD. Most of these were found in the peripheral fraction and their degree of interaction with the

**Table 6.** Putative Arabidopsis vacuolar membrane proteins with one transmembrane domain.

| Locus | MW | TM | Szponarski et al. | Shimaoka et al. | Sazuka et al. | Carter et al. | Annotation |
|---|---|---|---|---|---|---|---|
| **Membrane remodeling** | | | | | | | |
| At1g03860 | 31793 | 1 | | | | O | prohibitin 2-related B-cell receptor associated protein |
| At1g16920 | 24020 | 1 | O | | | | Ras-related GTP-binding protein, putative |
| At1g21250 | 81194 | 1 | | | | O | wall-associated kinase 1 |
| At1g25570 | 68955 | 1 | | | | O | leucine rich repeat protein-related |
| At1g67360 | 26407 | 1 | | | | O | stress related protein |
| At1g75040 | 25234 | 1 | | | | O | pathogenesis-related protein 5 (PR-5) |
| At2g30870 | 24212 | 1 | | | | O | glutathione transferase, putative |
| At2g33380 | 26582 | 1 | | | | O | RD20 protein, induced by abscisic acid during dehydration |
| At2g39780 | 29135 | 1 | | | | O | S-like ribonuclease RNS2 |
| At3g05630 | 118811 | 1 | O | | | | putative phospholipase D |

Table 6. Continued

| Locus | MW | TM | Szponarski et al. | Shimaoka et al. | Sazuka et al. | Carter et al. | Annotation |
|---|---|---|---|---|---|---|---|
| At3g22060 | 27846 | 1 | | | | O | receptor protein kinase-related |
| At3g27280 | 30620 | 1 | | | | O | prohibitin - related |
| At3g49110 | 38924 | 1 | | | | O | peroxidase |
| At3g49120 | 38814 | 1 | | | | O | peroxidase, putative |
| At3g57020 | 41440 | 1 | | | | O | strictosidine synthase-related |
| At4g16500 | 12538 | 1 | | | | O | cysteine proteinase inhibitor like protein |
| At5g01820 | 50280 | 1 | | | | O | CBL-interacting protein kinase 14 |
| At5g24770 | 29825 | 1 | | | | O | vegetative storage protein Vsp2 |
| At5g24780 | 30244 | 1 | | | | O | vegetative storage protein Vsp1 |
| At5g40770 | 30382 | 1 | | | | O | prohibitin (gb|AAC 49691.1) |
| At5g44020 | 31039 | 1 | | | | O | vegetative storage protein-related |
| At5g45750 | 23877 | 1 | O | | | | Rab-type small GTP-binding protein-like |

**Table 6.** Continued

| Locus | MW | TM | Szpon-arski et al. | Shimaoka et al. | Sazuka et al. | Carter et al. | Annotation |
|---|---|---|---|---|---|---|---|
| At5g46860 | 29482 | 1 | O | | | O | Syntaxin-related protein (AtVam3) |
| **Stress** | | | | | | | |
| At1g03860 | 31793 | 1 | | | | O | prohibitin 2-related B-cell receptor associated protein |
| At1g21250 | 81194 | 1 | | | | O | wall-associated kinase 1 |
| At1g25570 | 68955 | 1 | | | | O | leucine rich repeat protein-related |
| At1g67360 | 26407 | 1 | | | | O | stress related protein |
| At1g75040 | 25234 | 1 | | | | O | pathogenesis-related protein 5 (PR-5) |
| At2g30870 | 24212 | 1 | | | | O | glutathione transferase, putative |
| At2g33380 | 26582 | 1 | | | | O | RD20 protein, induced by abscisic acid during dehydration |
| At2g39780 | 29135 | 1 | | | | O | S-like ribonuclease RNS2 |
| At3g09440 | 71148 | 1 | O | | | | Heat-shock protein (At-hsc70-3) |

**Table 6.** Continued

| Locus | MW | TM | Szponarski et al. | Shimaoka et al. | Sazuka et al. | Carter et al. | Annotation |
|---|---|---|---|---|---|---|---|
| At3g22060 | 27846 | 1 | | | | O | receptor protein kinase-related |
| At3g27280 | 30620 | 1 | | | | O | prohibitin - related |
| At3g49110 | 38924 | 1 | | | | O | peroxidase |
| At3g49120 | 38814 | 1 | | | | O | peroxidase, putative |
| At3g57020 | 41440 | 1 | | | | O | strictosidine synthase-related |
| At4g16500 | 12538 | 1 | | | | O | cysteine proteinase inhibitor-like protein |
| At5g01820 | 50280 | 1 | | | | O | CBL-interacting protein kinase 14 |
| At5g24770 | 29825 | 1 | | | | O | vegetative storage protein Vsp2 |
| At5g24780 | 30244 | 1 | | | | O | vegetative storage protein Vsp1 |
| At5g40770 | 30382 | 1 | | | | O | prohibitin (gb|AAC 49691.1) |
| At5g44020 | 31039 | 1 | | | | O | vegetative storage protein-related |
| **Others** | | | | | | | |
| At1g04410 | 35553 | 1 | | | | O | malate dehydrogenase, cytosolic, putative |

**Table 6.** Continued

| Locus | MW | TM | Szponarski et al. | Shimaoka et al. | Sazuka et al. | Carter et al. | Annotation |
|---|---|---|---|---|---|---|---|
| At1g04635 | 16999 | 1 | | | | O | Probable lipoate-protein ligase B-related |
| At1g09210 | 48139 | 1 | | | | O | calreticulin 2 (CRT2) |
| At1g14700 | 42034 | 1 | | | | O | calcineurin-like phosphoesterase |
| At1g33590 | 51724 | 1 | | O | | | disease resistance protein-related (LRR) |
| At1g44170 | 53141 | 1 | | O | | O | aldehyde dehydrogenase, putative (ALDH) |
| At1g54000 | 43176 | 1 | | O | | | myrosinase-associated protein |
| At1g54010 | 43126 | 1 | | O | | | myrosinase-associated protein |
| At1g54020 | 41783 | 1 | | O | | | myrosinase-associated protein |
| At1g54220 | 58450 | 1 | | O | | | dihydrolipoamide S-acetyltransferase-related |
| At1g63940 | 52484 | 1 | | O | | | monodehydroascorbate reductase |

**Table 6.** Continued

| Locus | MW | TM | Szponarski et al. | Shimaoka et al. | Sazuka et al. | Carter et al. | Annotation |
|---|---|---|---|---|---|---|---|
| At2g21160 | 28168 | 1 | | | O | | translocon-associated protein alpha (TRAP alpha) family protein |
| At2g27490 | 25747 | 1 | O | | | | dephospho-CoA kinase family |
| At2g32810 | 99181 | 1 | | O | | | glycosyl hydrolase family 35 (beta-galactosidase) |
| At2g32920 | 47737 | 1 | | | | O | protein disulfide isomerase family |
| At3g06480 | 119540 | 1 | | | | O | DEAD box RNA helicase, putative |
| At3g12500 | 34591 | 1 | | O | | | glycosyl hydrolase family 19 (basic endochitinase) |
| At3g14415 | 40289 | 1 | | | | O | glycolate oxidase |
| At3g14990 | 41840 | 1 | | | | O | putative 4-methyl-5 (b-hydroxyethyl)-thiazole monophosphate biosynthesis protein |
| At3g16580 | 44414 | 1 | O | | | | F-box family protein |

**Table 6.** Continued

| Locus | MW | TM | Szponarski et al. | Shimaoka et al. | Sazuka et al. | Carter et al. | Annotation |
|---|---|---|---|---|---|---|---|
| At3g18430 | 20014 | 1 | | | | O | calcineurin B subunit, putative |
| At3g19820 | 65377 | 1 | | O | | O | cell elongation protein (DWARF1) (DIMINUTO) (DIM) |
| At3g20250 | 106978 | 1 | | | | O | pumilio-family RNA-binding protein, putative |
| At3g26720 | 115203 | 1 | | O | | | glycosyl hydrolase family 38 (alpha-mannosidase) |
| At3g45780 | 111672 | 1 | | | | O | nonphototropic hypocotyl 1 |
| At3g56310 | 48345 | 1 | | O | | | glycosyl hydrolase family 27 (alpha-galactosidase/melibiase) |
| At3g63520 | 60891 | 1 | | | | O | 9-cis-epoxy-carotenoid dioxygenase |
| At4g09320 | 18796 | 1 | | | | O | nucleoside-diphosphate kinase |
| At4g11400 | 65667 | 1 | O | | | | ARID/BRIGHT DNA-binding domain-containing protein |

**Table 6.** Continued

| Locus | MW | TM | Szponarski et al. | Shimaoka et al. | Sazuka et al. | Carter et al. | Annotation |
|---|---|---|---|---|---|---|---|
| At4g20830 | 63542 | 1 | | O | | | FAD-binding domain-containing protein |
| At4g23200 | 72607 | 1 | | | | O | serine/threonine kinase-like protein |
| At4g23650 | 59319 | 1 | | | | O | calcium-dependent protein kinase, putative (CDPK) |
| At4g24640 | 19918 | 1 | | | | O | AppB protein |
| At4g29130 | 53690 | 1 | | | | O | hexokinase |
| At4g29260 | 28888 | 1 | | | | O | acid phosphatase-related protein |
| At4g29680 | 54661 | 1 | | | | O | nucleotide pyrophosphatase-like protein |
| At4g29700 | 51570 | 1 | | | | O | nucleotide pyrophosphatase-related |
| At4g31140 | 52698 | 1 | | O | | | 1,3-beta-glucanase-like protein |
| At4g32070 | 91237 | 1 | | | | O | octicosapeptide/Phox/Bem1p (PB1) domain-/tetratricopeptide repeat (TPR)-containing protein |

**Table 6.** Continued

| Locus | MW | TM | Szponarski et al. | Shimaoka et al. | Sazuka et al. | Carter et al. | Annotation |
|---|---|---|---|---|---|---|---|
| At4g35000 | 31572 | 1 | O | | | | L-Ascorbate peroxidase |
| At4g35800 | 204673 | 1 | | | | O | DNA-directed RNA polymerase (EC 2.7.7.6) II largest chain |
| At5g05520 | 58502 | 1 | | O | | | outer membrane OMP85 family protein |
| At5g09660 | 37352 | 1 | | | | O | malate dehydrogenase, glyoxysomal |
| At5g19010 | 64895 | 1 | | | | O | mitogen-activated protein kinase (MAPK), putative |
| At5g28540 | 73630 | 1 | | | O | | luminal binding protein 1 (BiP-1) (BP1) |
| At5g34850 | 54992 | 1 | | | | O | calcineurin-like phosphoesterase family |
| At5g42020 | 73562 | 1 | | | O | | luminal binding protein 1 (BiP-1) (BP1) |
| At5g48580 | 17640 | 1 | | | | O | immunophilin / FKBP-type peptidyl-prolyl cis-trans isomerase, putative |

**Table 6.** Continued

| Locus | MW | TM | Szponarski et al. | Shimaoka et al. | Sazuka et al. | Carter et al. | Annotation |
|---|---|---|---|---|---|---|---|
| At5g53560 | 15066 | 1 | | | | O | cytochrome b5 |
| At5g54500 | 21796 | 1 | O | | | | 1,4-Benzoquinone reductase-like |
| At5g58860 | 58536 | 1 | | | | O | cytochrome P450 86A1 |
| At5g64440 | 66091 | 1 | | | | O | expressed protein, similarity to glutamyl-tRNA amidotransferase subunit A |

**Proteinases**

| Locus | MW | TM | Szponarski et al. | Shimaoka et al. | Sazuka et al. | Carter et al. | Annotation |
|---|---|---|---|---|---|---|---|
| At1g02305 | 40016 | 1 | | O | | O | expressed protein (cathepsin B-like cysteine protease) |
| At1g04040 | 31078 | 1 | | O | | | unknown protein similar to acid phosphatase |
| At1g09850 | 48056 | 1 | | | | O | cysteine protease XBCP3 |
| At1g47128 | 50949 | 1 | | | | O | cysteine proteinase RD21A |
| At1g62290 | 55731 | 1 | | O | | O | aspartic protease-related |
| At2g24200 | 54492 | 1 | | O | | | leucine aminopeptidase-related |

**Table 6.** Continued

| Locus | MW | TM | Szponarski et al. | Shimaoka et al. | Sazuka et al. | Carter et al. | Annotation |
|---|---|---|---|---|---|---|---|
| At3g14067 | 81800 | 1 | | | | O | subtilisin-like serine protease, putative |
| At3g28220 | 42869 | 1 | | | | O | expressed protein (meprin and TRAF homology domain-containing protein) |
| At3g61540 | 57568 | 1 | | O | | | peptidase family similar to prolyl amino-peptidase (proline iminopeptidase) |
| At4g01610 | 39400 | 1 | | | | O | cathepsin B-like cysteine protease, putative |
| At4g12910 | 54211 | 1 | | | | O | serine carboxy-peptidase-related |
| At4g16190 | 41246 | 1 | | | | O | cysteine proteinase |
| At4g30810 | 53038 | 1 | | | | O | serine carboxy-peptidase - related |
| At4g30920 | 61290 | 1 | | O | | | leucyl amino-peptidase-like |
| At4g36195 | 54756 | 1 | | O | | O | expressed protein (serine carboxypeptidase S28 family) |

**Table 6.** Continued

| Locus | MW | TM | Szpon-arski et al. | Shimaoka et al. | Sazuka et al. | Carter et al. | Annotation |
|---|---|---|---|---|---|---|---|
| At5g19740 | 74722 | 1 | | O | | | glutamate hydroxy-peptidase ileal peptidase |
| At5g60160 | 52408 | 1 | | O | | | aspartyl aminopeptidase-like |
| At5g60360 | 38941 | 1 | | | | O | cysteine proteinase AALP |
| At5g65760 | 58555 | 1 | | | | O | hydrolase, alpha/beta fold family |
| **Unknown** | | | | | | | |
| At1g02816 | 18478 | 1 | | | | O | expressed protein |
| At1g07080 | 29680 | 1 | | | | O | expressed protein |
| At1g14250 | 53487 | 1 | | | | O | hypothetical protein |
| At1g20225 | 26034 | 1 | | | | O | expressed protein |
| At1g22060 | 228938 | 1 | | | | O | expressed protein |
| At1g55265 | 19119 | 1 | | | | O | expressed protein |
| At1g73200 | 87034 | 1 | | | | O | hypothetical protein |
| At2g07707 | 18193 | 1 | | | | O | hypothetical protein |
| At2g27730 | 11930 | 1 | | | | O | expressed protein |

**Table 6.** Continued

| Locus | MW | TM | Szpon-arski et al. | Shimaoka et al. | Sazuka et al. | Carter et al. | Annotation |
|---|---|---|---|---|---|---|---|
| At2g34585 | 8553 | 1 | | | | O | expressed protein |
| At2g43950 | 38818 | 1 | | O | | | unknown protein |
| At3g07460 | 19421 | 1 | | | | O | expressed protein |
| At3g07470 | 18761 | 1 | | | | O | expressed protein |
| At3g11700 | 50775 | 1 | | | | O | expressed protein |
| At3g11780 | 16787 | 1 | | | | O | expressed protein |
| At3g14920 | 68979 | 1 | | | | O | expressed protein |
| At3g20370 | 43431 | 1 | | O | | | expressed protein |
| At3g23760 | 22719 | 1 | | | | O | expressed protein |
| At3g29210 | 67907 | 1 | O | | | | hypothetical protein |
| At3g44100 | 16052 | 1 | | | | O | expressed protein |
| At3g52640 | 51955 | 1 | | O | | | hypothetical protein |
| At4g02370 | 18338 | 1 | | | | O | expressed protein |
| At4g14100 | 23912 | 1 | | | | O | expressed protein |
| At4g27020 | 59664 | 1 | | | | O | expressed protein |

**Table 6.** Continued

| Locus | MW | TM | Szponarski et al. | Shimaoka et al. | Sazuka et al. | Carter et al. | Annotation |
|---|---|---|---|---|---|---|---|
| At4g39730 | 20118 | 1 | | | | O | expressed protein |
| At5g12950 | 96219 | 1 | | | | O | expressed protein |
| At5g13640 | 74139 | 1 | | | | O | expressed protein |
| At5g19860 | 20425 | 1 | | | | O | expressed protein |
| At5g23820 | 17890 | 1 | | | | O | expressed protein |
| At5g26280 | 39428 | 1 | | O | | | expressed protein |
| At5g39410 | 49663 | 1 | | | | O | expressed protein |
| At5g54870 | 59718 | 1 | | | | O | expressed protein |
| At5g57655 | 53702 | 1 | | | | O | expressed protein |

membrane is often hard to judge. In many cases, there were regions of hydrophobicity which may or may not indicate a TMD. Different prediction programs assigned varying numbers of TMDs. Furthermore, the preparative methods for separating peripheral and integral proteins are not perfect and some proteins without TMDs were found in the integral fraction as well. Proteins without TMDs will not be considered here.

Most of the annotated proteins, which had one TMD, have not previously been shown to be associated with vacuoles (Table 6). However, some of them have a suggested vacuolar role, for example SNARE-related proteins or the FAD-linked oxidoreductase family protein (Shimaoka et al., 2004). There were also many metabolism-related proteins, such as proteinases which may have a lytic role in the vacuole. There were various other proteins whose functions do not seem to be relevant to vacuoles, plus others whose function is unknown.

Carter et al. (2004) analyzed proteins including soluble ones to clarify vacuole-targeting signals. They found some possible signals in amino acid sequences but exactly how the membrane proteins are sorted to the vacuole remains a mystery.

## 5. CONCLUSIONS AND FUTURE PROSPECTS

By comprehensive proteomic analysis of purified vacuolar membrane, many proteins have now been identified. They include almost all subunits of the major vacuolar $H^+$-pumps, the V-type $H^+$-ATPase and the V-type $H^+$-PPase, as well as a broad range of other transporters. There are also numerous proteins with two or more TMDs whose function could be predicted from existing databases. This data is an important first step in characterizing the roles of these proteins in the overall physiology of the vacuole. During the preparation of this manuscript, a new work on vacuole proteome (Jaquinod et al., 2007) appeared using vacuoles isolated from cultured cells and nano LC-ESI-MS/MS. They confirmed the same proteins as reported in the present article and found some new proteins, too. Also, Dunkley et al. (2006) showed a new trial to identify protein localization in the cell using a gradient quantitative detection of isolated organelles. They found some proteins in the vacuolar fraction, too.

## 6. ACKNOWLEDGMENT

We greatly appreciate Dr. Rob Reid (University of Adelaide, Adelaide, Australia) for his kind discussion and correction of this manuscript.

## 7. LITERATURE CITED

Arabidopsis Genome Initiative (2000) Analysis of the genome sequence of the flowering plant *Arabidopsis thaliana*. *Nature,* **408**: 796-815.

Carter, C., Pan, S., Zouhar, J., Avila, E.L., Girke, T. and Raikhel, N.V. (2004) The vegetative vacuole proteome of *Arabidopsis thaliana* reveals predicted and unexpected proteins. *Plant Cell,* **16**: 3285-303.

Chrispeels, M.J. and Herman, E.M. (2000) Endoplasmic reticulum-derived compartments function in storage and as mediators of vacuolar remodeling via a new type of organelle, precursor protease vesicles. *Plant Physiol.,* **123**: 1227-1234.

Deepesh, N. De (2000) Plant Cell Vacuoles An Introduction. CSIRO Publishing, Clayton South, Victoria, Australia.

Dietz, K.J., Tavakoli, N., Kluge, C., Mimura, T., Sharma, S.S., Harris, G.C., Chardonnens, A.N. and Golldack, D. (2001) Significance of the V-type ATPase for the adaptation to stressful growth conditions and its regulation on the molecular and biochemical level. *J. Exp. Bot.,* **52**: 1969-1980.

Drozdowicz, Y.M., Kissinger, J.C. and Rea, P.A. (2000) AVP2, a sequence-divergent, $K^{(+)}$-insensitive $H^{(+)}$-translocating inorganic pyrophosphatase from Arabidopsis. *Plant Physiol.,* **123**: 353-362.

Dunkley, T.P., Hester, S., Shadforth, I.P., Runions, J., Weimer, T., Hanton, S.L., Griffin, J.L., Bessant, C., Brandizzi, F., Hawes, C., Watson, R.B., Dupree, P. and Lilley, K.S. (2006) Mapping the Arabidopsis organelle proteome. *Proc. Natl. Acad. Sci. USA,* **103**: 6518-6523.

Endler, A., Meyer, S., Schelbert, S., Schneider, T., Weschke, W., Peters, S.W., Keller, F., Baginsky, S., Martinoia, E. and Schmidt, U.G. (2006) Identification of a vacuolar sucrose transporter in *Hordeum vulgare* and *Arabidopsis thaliana* mesophyll cells by a tonoplast proteomic approach. *Plant Physiol.*, **141**: 196-207.

Ferro, M., Salvi, D., Riviere-Rolland, H., Vermat, T., Seigneurin-Berny, D., Grunwald, D., Garin, J., Joyard, J. and Rolland, N. (2002) Integral membrane proteins of the chloroplast envelope: identification and subcellular localization of new transporters. *Proc. Natl. Acad. Sci. USA*, **99**: 11487-11492.

Ferro, M., Salvi, D., Brugiere, S., Miras, S., Kowalski, S., Louwagie, M., Garin, J., Joyard, J. and Rolland, N. (2003) Proteomics of the chloroplast envelope membranes from *Arabidopsis thaliana*. *Mol. Cell Proteomics*, **2**: 325-345.

Gaxiola, R.A., Fink, G.R. and Hirschi, K.D. (2002) Genetic manipulation of vacuolar proton pumps and transporters. *Plant Physiol.*, **129**: 967-973.

Gogarten, J., Fichmann, J., Braun, Y., Morgan, L., Styles, P., Taiz, S., DeLapp, K. and Taiz, L. (1992) The use of antisense mRNA to inhibit the tonoplast $H^+$-ATPase in carrot. *Plant Cell*, **4**: 851-864.

Hall, J.L. (2002) Cellular mechanisms for heavy metal detoxification and tolerance. *J. Exp. Bot.*, **53**: 1-11.

Heazlewood, J., Howell, K., Whelan, J. and Millar, A. (2003) Towards an analysis of the rice mitochondrial proteome. *Plant Physiol.*, **132**: 230-242.

Hirschi, K.D. (2004) The calcium conundrum. Both versatile nutrient and specific signal. *Plant Physiol.*, **136**: 2438-2442.

Jaquinod, M., Villiers, F., Kieffer-Jaquinod, S., Hugouvieux, V., Bruley, C., Garin, J. and Bourguignon, J. (2007) A proteomics dissection of *Arabidopsis thaliana* vacuoles isolated from cell culture. *Mol. Cell Proteomics*, **6**: 394-412.

Kobae, Y., Mizutani, M., Segami, S. and Maeshima, M. (2006) Immunochemical analysis of aquaporin isoforms in Arabidopsis suspension-cultured cells. *Biosci. Biotechnol. Biochem.*, **70**: 980-987.

Maeshima, M. (2000) Vacuolar $H^{(+)}$-pyrophosphatase. *Biochim. Biophys. Acta*, **1465**: 37-51.

Maeshima, M. (2001) Tonoplast transporters: Organization and function. *Annu. Rev. Plant Mol. Biol.*, **52**: 469-497.

Martinoia, E., Massonneau, A. and Frangne, N. (2000) Transport processes of solutes across the vacuolar membrane of higher plants. *Plant Cell Physiol.*, **41**: 1175-1186.

Peltier, J., Emanuelsson, O., Kalume, D., Ytterberg, J., Friso, G., Rudella, A., Liberles, D., Soderberg, L., Roepstorff, P., von Heijne, G. and van Wijk, K. (2002) Central functions of the lumenal and peripheral thylakoid proteome of Arabidopsis determined by experimentation and genome-wide prediction. *Plant Cell*, **14**: 211-236.

Rolland, N., Ferro, M., Ephritikhine, G., Marmagne, A., Ramus, C., Brugiere, S., Salvi, D., Seigneurin-Berny, D., Bourguignon, J., Barbier-Brygoo, H., Joyard, J. and Garin, J. (2006) A versatile method for deciphering plant membrane proteomes. *J. Exp. Bot.*, **57**: 1579-1589.

Sazuka, T., Keta, S., Shiratake, K., Yamaki, S. and Shibata, T. (2004) Identification of membrane-bound proteins from a vacuolar membrane-enriched fraction of *Arabidopsis thaliana*. *DNA Res.*, **11**: 101-113.

Schubert, M., Petersson, U., Haas, B., Funk, C., Schroder, W. and Kieselbach, T. (2002) Proteome map of the chloroplast lumen of *Arabidopsis thaliana*. *J. Biol. Chem.*, **277**: 8354-8365.

Schumacher, K., Vafeados, D., McCarthy, M., Sze, H., Wilkins, T. and Chory, J. (1999) The Arabidopsis det3 mutant reveals a central role for the vacuolar $H^{(+)}$-ATPase in plant growth and development. *Genes Dev.,* **13**: 3259-3270.

Schwacke, R., Flugge, U.I. and Kunze, R. (2004) Plant membrane proteome databases. *Plant Physiol. Biochem.,* **42**: 1023-1034.

Shimaoka, T., Ohnishi, M., Sazuka, T., Mitsuhashi, N., Hara-Nishimura, I., Shimazaki, K., Maeshima, M., Yokota, A., Tomizawa, K. and Mimura, T. (2004) Isolation of intact vacuoles and proteomic analysis of tonoplast from suspension-cultured cells of *Arabidopsis thaliana. Plant Cell Physiol.,* **45**: 672-683.

Shimmen, T., Mimura, T., Kikuyama, M. and Tazawa, M. (1994) Characean cells as a tool for studying electrophysiological characteristics of plant cells. *Cell Struct. Funct.,* **19**: 263-278.

Sonobe, S. (1997) Cell model systems in plant cytoskeleton studies. *Int. Rev. Cytol.,* **175**: 1-27.

Sze, H., Schumacher, K., Muller, M., Padmanaban, S. and Taiz, L. (2002) A simple nomenclature for a complex proton pump: VHA genes encode the vacuolar $H^{(+)}$-ATPase. *Trends Plant Sci.,* **7**: 157-161.

Szponarski, W., Sommerer, N., Boyer, J., Rossignol, M. and Gibrat, R. (2004) Large-scale characterization of integral proteins from Arabidopsis vacuolar membrane by two-dimensional liquid chromatography. *Proteomics,* **4**: 397-406.

Thompson, A.R. and Vierstra, R.D. (2005) Autophagic recycling: lessons from yeast help define the process in plants. *Curr. Opin. Plant Biol.,* **8**: 165-173.

Vitale, A. and Hinz, G. (2005) Sorting of proteins to storage vacuoles: how many mechanisms? *Trends Plant Sci.,* **10**: 316-323.

Yoshida, K., Kawachi, M., Mori, M., Maeshima, M., Kondo, M., Nishimura, M. and Kondo, T. (2005) The involvement of tonoplast proton pumps and $Na^+$ $(K^+)/H^+$ exchangers in the change of petal color during flower opening of Morning Glory, *Ipomoea tricolor* cv. Heavenly Blue. *Plant Cell Physiol.,* **46**: 407-415.

# Chapter 14

## ELEMENTAL BIOFORTIFICATION OF CROP PLANTS

SAVITA DAHIYA[1], DARSHNA CHAUDHARY[1], RANJANA JAIWAL[1], OM PARKASH DHANKHER[2], RANA P. SINGH[3] AND PAWAN K. JAIWAL[1]
[1]*Advanced Centre for Biotechnology, M.D. University, Rohtak - 124 001, India*
[2]*Department of Plant, Soil and Insect Sciences, University of Massachusetts, Amherst, MA 01002, USA*
[3]*Department of Environmental Sciences, BBA University, Lucknow - 226 025, India*
E-mail: savi_28484@rediffmail.com

**Abstract**

Plants serve as major sources for all essential minerals required by humans. Unfortunately, major staple food crops are deficient in some of the micronutrients. The population depending on staple food crops (cereals) or those inhabiting regions where soil mineral imbalances occur, often suffer from mineral malnutrition (Fe, Zn, I or Se deficiencies) which is damaging the health of 3 billion (half of the world's population) people especially in developing countries where a diversified diet is not affordable for the majority. The strategies aimed at reducing mineral deficiencies, i.e. supplementation and fortification of food, are also not accessible to the rural poor. An alternative approach is to increase minerals in the edible crops (biofortification) through mineral fertilization, use of mycorrhiza inoculants, and plant breeding and/or transgenic strategies. The first two approaches are costly and non-sustainable. Breeding for enhanced micronutrient concentration in edible portions of crop plants is particularly very difficult. Recent studies have shown that gene technology can complement breeding for developing crops with produce rich in micronutrients to curb malnutrition in a cost effective and sustainable manner. This chapter will provide an updated account of the different approaches used in obtaining biofortified crops to overcome mineral malnutrition.

**Keywords:** mineral malnutrition, food fortification, biofortification

## 1. INTRODUCTION

The micronutrients which are essential for good health, iron, zinc and iodine have been reported to be most at the risk of malnutrition (Welch and Graham, 2004).

---

© CAB International 2008. *Plant Membrane and Vacuolar Transporters* (eds P.K. Jaiwal, R.P. Singh and O.P. Dhankher)

More than half the world's populations throughout the economic spectrum, in both urban and rural settings, do not consume enough of these nutrients in their diet (Underwood, 2003). The situation of mineral malnutrition is more drastic in developing countries as compared to developed countries. In developed countries people spend only 10% of their income on food and are still free from any kind of nutrient deficiencies (Eichholzer, 2003). The reason being that they consume a range of non-staple foods, such as animal and fish products, fruits, pulses and vegetables which are rich in bio-available vitamins and minerals. On the other hand, in developing countries people spend two third of their income on food, yet they suffer from malnutrition and ill health. People in developing countries mostly consume cereals as staple crops but cereals contain very low amounts of these micronutrients, most of which are lost during processing of food (Cheng and Hardy, 2003).

Estimates suggest that about one third of the world's population is affected by vitamin A, iron and iodine deficiencies (Welch *et al.*, 1997; WHO, 1999; World Bank, 1994). Clinical manifestations of childhood and maternal death, lowered immune response, blindness, mental retardation and anemia etc. cause suffering for over half a billion people (Caballero, 2002; Underwood, 2003). But this human devastation is only the tip of the iceberg. Another two billion people, in both urban and rural settings, are marginally deficient in micronutrients and unable to achieve their mental and physical potential. Correcting deficiencies in micronutrient deficient populations can improve the population-wide IQ by 10-15 IQ points, reduce maternal deaths by one third, decrease infant and childhood mortality by 40%, increase a strength and work capacity by 40%, eliminate nutritional blindness and endemic cretinism, and dramatically reduce birth defects and congenital deafness (Administrative Committee on Co-ordination, Subcommittee on Nutrition and International Food Policy Research Institute, 2000). Investments in education will not be maximized unless school children have grown up with adequate amounts of micronutrients. Their learning capability and educational achievements depend upon children being vital and operating with full intellectual capacity (WHO, 2002).

There can be different approaches to overcome the problem of micronutrient malnutrition (hidden hunger). Traditional strategies to increase bioavailable concentration of minerals include supplementation, food fortification and dietary diversification. Fortification involves the process of addition of vitamins and minerals in food during processing, just after the food has left the farm and before it is distributed for consumer consumption. Unfortunately, most malnourished people living in developing countries have no access to these strategies aimed at reducing micronutrient deficiency and have no possibility of improving their nutrition by diversifying their diets (Graham *et al.*, 2001; Bouis, 2003; Timmer, 2003; Lyons *et. al.*, 2003). Recently, an alternative solution to mineral malnutrition termed 'biofortification' has been proposed (Bouis, 2003). Biofortification has been defined as the process of increasing the bioavailable concentrations of essential elements in edible portions of crop plants through agronomic intervention or genetic selection (White and Broadley, 2005). The biofortification has the potential to become sustainable, cost-effective and reach remote rural populations (Bouis *et al.*, 2003;

Elemental biofortification of crop plants

Nestel et al., 2006). Development of mineral dense lines requires one time investment. Once such lines have developed, there will be little additional cost in incorporating them into ongoing breeding programs (Welch and Graham, 2002; Bouis et al., 2003; Timmer, 2003, White and Broadley, 2005). High levels of minerals in seeds contribute to stronger and hardier plants leading to improved pest and drought resistance and increased productivity even in soils deficient in minerals (Bouis et al., 2003). Benefits of agriculture productivity and the improvement of environment will help in adoption of biofortified crop varieties by poor farmers for food security and nutrition, and in increasing their income. Biofortification can be achieved either by mineral fertilization (Nube and Voortman, 2006), use of mycorrhiza (He and Nara, 2007), plant breeding (Bouis, 2002, Bouis et al., 2003, Welch and Graham, 2004; Misra et al., 2004, Poletti et al., 2004; White and Broadley, 2005; Nestel et al., 2006) or by using transgenic strategies (Lonnerdal, 2003, Poletti et al., 2004; Poletti and Sautter, 2005; Lucca et al., 2006, Ghandilyan et al., 2006, Broadley et al., 2007, Zhu et al., 2007). The improvement in calcium content of agriculturally important crops has been reviewed in chapter 2. In this chapter, the current status and future prospects of biofortification of edible crops with minerals, iron, zinc and selenium has been described.

## 2. HIDDEN HUNGER IN HUMANS

Nearby half of the world population depending largely on staple food cereals is suffering from "hidden hunger", the chronic lack of micronutrients (iron, zinc, selenium) in the diet which is more prevalent among infants, children, adolescent and child bearing women thus adversely affecting their health and capacity to work. Humans require at least 17 mineral nutrients to meet their metabolic needs. Some are required in large amounts, but others, such as Fe, Zn, I and Se are required in low amount as higher concentrations can be harmful (Welch and Graham, 2004; Grusak and Cakmak, 2005). Thus, inadequate consumption of any one of these nutrients will result in adverse metabolic disturbances leading to sickness, poor health and large economic costs to society (Branca and Ferrari, 2002; Ramakrishnan et al., 1999; Golden, 1991). Required daily intake for some of these nutrients for adults has been reported by the Food and Agricultural Organization of the United Nations and the World Health Organization (FAO/WHO, 2003) (Table 1). The primary sources of all these nutrients are agricultural products. But, unfortunately agricultural systems such as cereals fail to provide adequate amounts of essential nutrients leading to mineral malnutrition (Welch and Graham, 2004; Schneeman, 2001; McGuire, 1993). Today, over 3 billion people are suffering from micronutrient malnutrition and the numbers are continuously increasing (FAO/WHO, 2003). Almost two thirds of child deaths are associated with nutritional deficiencies, many from micronutrient deficiencies (Caballero, 2002). It is estimated that of the world's total population, 60-80% are Fe deficient and about 15% are Se deficient (White and Broadley, 2005; Kennedy, 2003). In the next 12 months, micronutrient malnutrition will cause 1 million children to die before the age of five, 50,000 women to die from childbirth, and 100,000 infants to be born with

Table 1. The most essential mineral elements required by humans, their recommended daily requirement, safe upper intake levels (data from FAO/WHO, 2003) and antinutrients and promoters in food that affect the bioavailability of minerals (after Graham et al., 2001).

| Elements | Maximum adult RDA[a] | Safe upper intake[b] | Antinutrients | Promoters |
|---|---|---|---|---|
| Ca | 1000-1200 mg | 2 x | Oxalate, phytate, tannins, fibre | Inulin |
| Fe | 10-15 mg | 5 x | Phytate, tannins, oxalic acid, hemagglutinins, fibre | Phyto-ferritin, riboflavin, ascorbate, β-carotene, cysteine, histidine, lysine, fumarate, malate, citrate |
| Zn | 10-12 mg | 10 x | Phytate, tannins, hemagglutinins, fibre | Palmitic acid, riboflavin, ascorbate, cysteine, methionine, histidine, lysine |
| Se | 70 µg | 13 x | - | - |
| I | 150 µg | 13 x | Goitrogens | Selenium |

[a]Recommended dietary allowances per day. Values present are the highest RDA either for male or female adults,
[b]the safe upper intake limit is associated with a low probability of adverse side effects and tolerance to high levels.

unpreventable physical defects. By any measure, micronutrient malnutrition is currently at alarming proportions in many developing countries (Mason and Garcia, 1993; WHO, 2004).

## 3. MICRONUTRIENTS AT THE RISK OF MALNUTRITION

Most widespread micronutrient deficiencies in humans are Fe, Zn and Se (Welch and Graham, 2004; White and Broadley, 2005). Identification and isolation of genes required for the synthesis and accumulation of a target mineral are needed to modify their levels in staple crops for desired dietary change. Before attempting to manipulate nutritional components in food crops it should be evaluated whether excessive dietary intake could have unintended negative health consequences. By clinical and epidemiological evidence, it is clear that iron, zinc, selenium and iodine are the mineral targets which play a significant role in maintenance of optimal health and are limiting in diets worldwide (Poletti et al., 2004; Grusak and Cakmak,

2005). The upper safe levels of intake for these minerals range 2 to 13 times the readily available dose (RDA), allowing a greater range of their manipulation in the dietary components (Table 1).

### 3.1. Iron deficiency

Iron deficiency is the most widespread nutritional disorder in the world including India (Misra *et al.*, 2004). It is estimated that out of the world's 6 billion people, 60-80% are Fe deficient (White and Broadley, 2005). One major cause of iron deficiency is its poor absorption from cereals rich in phytic acid. Phytic acid is a potent inhibitor of iron absorption (Hurrell *et al.*, 1992). In addition, the intake of foods that enhance iron absorption, such as fruits, vegetables or muscle tissue is often limited. Fe deficiency causes anemia, a condition in which the blood contains low levels of red blood cells (World Health Organization [WHO], 2006). Asia has the highest prevalence of anemia (United Nations Administrative Committee on Coordination/Sub Committee on Nutrition [ACC/SCN], 2005). Most severely affected are women at the reproductive age and children. In children, anemia is associated with impaired physical growth, mental development and learning capacity. In adults, anemia impairs immunity. Significant human and economic losses for economies are thus consequences of iron deficiencies. So, in developing countries such as India, where per capita GNP is US$450 and over 30% of its population lives in poverty (WHO, 2002), reducing iron deficiency is an important means to increase productivity and reduce poverty.

### 3.2. Zinc deficiency

Zinc deficiency is the second largest mineral deficiency after iron. More than 30% of people are Zn deficient (Kennedy, 2003; Combs, 2001). In biological systems, Zn is involved in the activity of more than 300 enzymes. In these enzymes, Zn plays either catalytic, co-catalytic or structural roles. Zn also plays a critical role in the synthesis of proteins and metabolism of DNA and RNA. Several zinc-containing proteins are present in plants that affect gene expression (Takatsuji, 1988). The human body relies on zinc to heal wounds, grow and repair body tissue, properly clot blood and ensure sound fetal development (Hotz and Brown, 2004; Pathak and Kapil, 2004). Zn deficiency occurs due to low intake of meat and high intake of cereals, which contain high amounts of anti-nutritional compounds (Hunt, 2002). Zinc deficiency results in a multitude of health problems such as impairment in linear growth, sexual immaturities, impairment of learning ability and immune functions and malformations of the central nervous system (Welch, 2001). The clearest indicator of zinc deficiency is stunting in children (FAO, 2003).

### 3.3. Selenium deficiency

Selenium (Se) is a trace mineral that has been recognized as an essential nutrient for humans and animals or as environmental toxicant; the boundary between the two is narrow. Selenium protects cells from free-radical damage (antioxidant), enables the thyroid to produce thyroid hormone, improves immune response, prevents heart disease, specific cancers and arthritis and is antiviral (see Lyons, 2003). The selenium sources for humans and animals are plants whose Se content depends on the Se content of soil which is low in Europe, New Zealand, and some parts of China (Lyons, et al., 2003, 2004). Se deficiency probably affects at least a billion (15% of world population) people (Combs, 2001). The US Recommended Daily Allowance (RDA) of Se is 55 µg/day, while in the UK; it is as low as 35 µg/day (Broadley et al., 2006). In Australia, it is 70-85 µg/day but below an estimated optimal intake. Inorganic Se forms (selenate, selenite) undergo reductive metabolism, yielding hydrogen selenide, which is incorporated into selenoprotein. Successive methylation of hydrogen selenide detoxifies excess Se. Selenomethionins (more bioavailable form than selenate and selenite) can be incorporated non-specifically into proteins in place of methionine, and serve as a vehicle for selenium storage in organs and tissues. Selenocysteine is catabolyzed directly to hydrogen selenide. It is thus better utilized for selenoenzymes but not retained as selenomethionine (Combs, 2001). Many people do not consume enough Se to support maximum expression of solenoenzymes (80-90 µg/day). Se intake of 200-300µg/day is required to reduce cancer risk (Combs, 2001). Under normal conditions, a Se intake of less than 1000 µg/day (or 15 µg/Kg bodyweight) does not cause toxicity (Poirier, 1994).

## 4. MICRONUTRIENT ACCUMULATION AND BIOAVAILABILITY

Before discussing possible strategies to overcome mineral malnutrition, understanding about micronutrient accumulation in edible portions of plant and its bioavailability to people is necessary. In order to increase micronutrient accumulation in plant edible portions, a complex integrated system of tissues and membrane processes is required. But there are a few barriers to this type of system (Welch, 1995). The first and most important barrier to micronutrient absorption resides at the root-soil interface. The available level of micronutrients at the root-soil interface must be increased to allow more absorption by root cells. It could be achieved by changing root morphology and by stimulating certain root cell processes that modify micronutrient solubility and movement such as by stimulating metal complexing compounds and by increasing root absorptive surface area such as number and extent of fine roots. Secondly, absorption mechanisms (e.g. transporters and ion channels) must be sufficiently active and specific enough to allow for accumulation of micronutrient metals once they enter the apoplasm of root cells. Thirdly, once taken up by root cells, the micronutrients must be efficiently translocated to and accumulated in edible plant organs (Welch, 1986).

### 4.1. Bioavailability

To be effective, the micronutrient metal species accumulated in edible portions must be bioavailable to people that consume the seeds (Fairweather-Tait and Hurrell, 1996; Welch, 1986, 2001). But, unfortunately only a very small proportion of accumulated mineral is bioavailable to the consumer. The greatest proportion of micronutrient is lost during processing of food or feed (Cheng and Hardy, 2003). The bioavailability of dietary non-heme Fe, Zn and other micronutrients to humans

Elemental biofortification of crop plants

is also affected by the presence of antinutrients and promoters in plant foods. Antinutrients such as phytate and tannins limit the absorption of Fe, Zn and Ca by the gut (Mendoza, 2002; Welch and Graham, 2004). In cereals most of the phosphorus is stored in the form of phytate (*myo*-inositol hexa*kis*phosphate) which readily forms complexes with nutritionally important minerals, Fe, Zn and Se (Loewus, 2002). As monogastric animals including man do not synthesize a phytate degrading enzyme, phytase in their digestive tract, most of the phytate and associated minerals from ingested seeds are not absorbed but excreted (Lott *et al.*, 2000). Reduction of antinutrients such as phytate in grain can improve the bioavailability of iron absorbed from the diet. Phytic acid has been completely degraded in cereals used for baby food by adding commercial exogenous phytases (Davidson *et al.*, 1997) or by activating the native phytases by a combination of soaking, germinating and fermenting (Marero *et al.*, 1991).

Some organic compounds are also known which stimulate the absorption of essential mineral elements by human beings called promoters. These include ascorbate, β-carotene, protein, cysteine and various organic and amino acids (Frossard *et al.*, 2000). Hence, by using current plant molecular biological and genetic modification approaches, it is now possible to reduce or eliminate antinutrients from staple plant foods, or to increase the levels of promoter substances in these foods (Welch, 2001).

## 5. APPROACHES TO ASSURE ACCESS TO MICRONUTRIENTS

Traditional interventions to address mineral malnutrition mainly focus on supplementation, food fortification and dietary diversification. Dietary modification is the final permanent solution to hidden hunger. But in developing countries people are in no position to afford pulses, fruits, fish and animal products rich in micronutrients (Timmer, 2003; Graham *et al.*, 2001). Meanwhile, fortification of food can be another approach to overcome mineral malnutrition (Darnton-Hill and Nalubola, 2002).

### 5.1. Food fortification

Fortification of food has been responsible for eradicating most of the mineral deficiencies in developed countries (Darnton-Hill and Nalubola, 2002; Sichert-Hellert *et al.*, 2000). Fortification includes addition of minerals to a particular food vehicle; for example iodination of salt or fluor fortification of toothpaste and tap water. Salt fortification by iodine or sodium selenite works well in developing countries. Fortification is particularly difficult for Fe owing to its rapid oxidation (Boccio and Iyengar, 2003). Even encapsulation of Fe to reduce oxidation does not help (Shrestha *et al.*, 2003), and it increases loss of iodine (Wegmuller *et al.*, 2003). The problem of Fe fortification can be solved to some extent by cooking food in iron pots (Prinsen *et al.*, 2003) or by adding elementary Fe to the diet (Swain *et al.*, 2003). But in spite of all this commercial food fortification is still appealing because if the right food is selected, coverage takes care of itself. So with changing dietary habits fortification is likely to be feasible in the near future in most countries.

The two most important determinants for success in fortification programmes are selection of right food(s) to fortify and industry compliance with fortification regulations (Mora, 2003).

But unfortunately, none of the strategies aimed at reducing mineral deficiencies, i.e. diet diversification, food supplementation and fortification have been universally successful (Bouis et al., 2003; Lyons et al., 2003). Recently, a complimentary solution to mineral malnutrition has been proposed that is getting plants to do the work of fortification, referred to as "Biofortification" (Graham et al., 2001; Bouis, 2003).

## 5.2. Biofortification

An alternative approach to fortification through agricultural management and food processing is the accumulation of micronutrients directly in cereal seeds using conventional breeding on targeted genetic engineering (Tucker, 2003; Zimmermann, 2002). Biofortification is a process that nutritionally enhances or biofortifies staple crop varieties with higher levels of bioavailable vitamins and minerals. The process holds great potential to improve the health of poor in developing countries, particularly in rural areas. The added benefit of biofortification is its cost-effectiveness and long term sustainability. Once the plants are developed, there are no costs each year in buying fortificants and adding them to food supply during processing, and farmers will be driven by a good profit (Gibson, 2004). Mineral-rich seeds have greater seedling vigor, which can be associated with higher yield (Bouis, 2003, Lucca et al., 2006). Hence for subsistence farming biofortification is the approach of choice (Gibson, 2004).

### 5.2.1. Use of micronutrients as fertilizer to increase mineral content of grains

Micronutrient malnutrition in humans is derived from deficiencies of micronutrients in soils and foods. Notably 50% of cultivated soils in India and Turkey, a third of cultivated soils in China and most soils in Western Australia are deficient in Zn (Yang et al., 2007; see Broadley et al., 2007). The application of micronutrients as fertilizer to crops should result in both higher yields and higher crop micronutrient contents. The possibility of combing higher yields with higher crop micronutrient contents is limited, for those micronutrients that are not essential for plants, e.g. iodine and selenium. Yet, application of iodine or selenium to soils, or to irrigation water, or to foliar may result in higher crop contents of these micronutrients (Gupta and Gupta, 2000). In wheat, the addition of as little as 10 kg/ha of sodium selenate can increase grain Se levels by up to 400 µg/kg. Such a strategy has been adopted in Australia, Finland and the UK (Broadley et al., 2006). Pea (Smrkolj et al., 2006) and barley (Gibson et al., 2007) have been biofortified with Se in a similar manner to wheat but this did not affect the malting quality of the barley grain. However, the Se levels in brewing products are yet to be determined (Gibson et al., 2007). The application of Zn to crops can result both in considerably higher crop yields and also in higher crop Zn contents. This has been shown for major food crops such as wheat, rice and maize (Cakmak, 2004; Ozturk et al., 2006).

## Elemental biofortification of crop plants

In most soils iron is present in large quantities. However, most iron in soils is unavailable for plant absorption (Meng *et al.*, 2005). Iron deficiency is common in soils with a high pH or with high levels of elements such as copper, zinc, manganese and phosphate that are iron antagonists and can reduce iron uptake by plants. Yields of iron deficient crops can, in principle, be increased through application of iron to soils. However, as the uptake of iron from soils is highly complex, improving crop yields through fertilization with iron has been shown to be difficult (Schulte, 2004). For example, application of iron to soils in the form of ferrosulfate ($FeSO_4$) has limited effects on crop yields (Frossard *et al.*, 2000). Other forms in which iron is added to soils (e.g. as chelates) are possibly more effective but also expensive, and generally too costly for use on low value staple crops (Goos *et al.*, 2004). Foliar application of iron compounds has been shown to have some positive effects on yields of grains (Rengel *et al.*, 1999; Sadana and Nayyar, 2000; Goos *et al.*, 2004). As regards the effects of iron fertilizer on iron contents of crops, available information is very scarce. Iron contents of leafy vegetables have been significantly increased by fortifying the soil with iron (Reddy and Bhatt, 2001).

A direct application of micronutrients, either on a one to one basis or in well-balanced combinations of a few micronutrients has been found to be a more effective approach than the enrichment of commonly used fertilizers with micronutrients. Further mineral fertilization of crops increases leaf mineral contents and improves yield but does not always increase mineral concentrations in fruit, seed or grain to the desired levels (Frossard *et al.*, 2000; Graham *et al.*, 2001; Chen *et al.*, 2002, Dai *et al.*, 2004; White and Broadley, 2005). The cost of mineral fertilizers can be low but their application, in both economic and environment terms can be costly as they must be reapplied regularly.

*5.2.2. Use of mycorrhizae for mineral biofortification*

The majority of higher land plants (e.g. all major grain crops, almost all vegetables and fruits) are associated with arbuscular mycorrhizae (AM) that provide an extensive surface area or network for the direct uptake of essential mineral elements (N, P, K and micronutrients) from soil therefore enhancing plant growth, productivity and improve nutrient in plant products (Smith and Read, 1997). However, an extensive use of fertilizers and biocides in conventional breeding for high output is detrimental to fungal colonization which could be the major cause of an unbalanced micronutrients in plant. The alternative agricultural systems that are based on low-input or organic farming (use of organic substrates coupled with no use of synthetic fertilizers, biocides etc.) or on farm use of AM inocula are required for the beneficial effects of arbuscular mycorrhizae on crop nutrition and quality. The selection of AM fungal strains for the target elements as well as selection or breeding of new plant varieties that can better build a symbiosis with AM fungi could further contribute to elemental biofortification in a more practical, effective and sustainable manner to curb global human malnutrition (He and Nara, 2007).

*5.2.3. Breeding strategies to manipulate plant micronutrient content*

Breeding strategies to develop micronutrient-enriched staple food crops will be successful only if they meet certain criteria. These are 1) crop productivity must be increased or maintained to the farmer's acceptance, 2) increased micronutrient levels should have significant impact on human health, 3) micronutrient enrichment traits must be stable across various environments, 4) the minerals must be absorbed by the human body and 5) the varieties must be accepted by the consumers. These issues are discussed in detail in a recent review article by Nestel *et al.* (2006). Adequate natural genetic variation for Fe, Zn and other minerals in shoot or seed exists within genera, species and cultivars, making selection of appropriate breeding materials possible. HarvestPlus initiative http://www.harvestplus.org) is focused on traditional breeding for increasing micronutrients in edible tissues of staple crops such as wheat, rice, maize, common bean, cassava and sweet potato of the developing world to curb micronutrient malnutrition with minimal costs in a sustainable manner.

5.2.3.1. The use of natural variation in mineral concentrations

**Iron, Zinc and Selenium:** Genetic variation for accumulation of minerals in accessions of common bean, wheat, rice, maize, pearl millet, cassava and Arabidopsis is observed (Graham *et al.*, 2001; Welch and Graham, 2004; Lyons *et al.*, 2005; Vreugdenhil *et al.*, 2005; White and Broadley, 2005; Ghandilyan *et al.*, 2006; Chhunja *et al.*, 2006; Velu *et al.*, 2007; Broadley *et al.*, 2006, 2007). Cereals (rice, wheat and maize) have accessions with at least double the Fe and Zn contents in grain than the most widely grown varieties. In wheat, Fe and Zn concentrations showed less genetic variability in the cultivated tetraploid (ssp. *durum*) and hexaploid (ssp. *aestivum*) varieties than among wild diploid (ssp. *boeoticum*, ssp. *monococcum*) and tetraploid (ssp. *dicoccoides*) species and relatives (*Aegilops tauschii*, *A. kotschyi*) of wheat (Grusak and Cakmak, 2005; Chhuneja *et al.*, 2006). Furthermore, Fe and Zn concentrations in grain are positively correlated, micronutrient density traits are independent of environment and can be combined with improved agronomic traits (Gregorio *et al.*, 2000; Banziger and Long, 2000; Maziya-Dixon *et al.*, 2000; Welch *et al.*, 2005). Zn-dense wheat varieties have been developed and are being grown on commercial basis in Australia (Rengel and Graham, 1995). Seeds of legumes, such as common bean and pea, root crops, such as cassava and yam (*Dioscorea* species), leafy vegetables such as spinach (*Spinacia oleracea*) and brassica, have higher Fe and Zn concentrations than cereal and show considerable genetic variation (see White and Broadley, 2005). The knowledge about the genetics of the observed variations and insight into the genotype by environment (G x E) interaction are relevant for efficient breeding (Farnham *et al.*, 2007)

**Antinutrients and promoters:** Intra-specific variation in antinutrients, such as phytate and polyphenols and promoter compounds, such as ascorbate, β-carotene, some amino- and organic acids in edible parts of plants is available (see White and Broadley, 2005), which can be exploited either to lower the levels of antinutrients or to increase the level of promoters in plants in order to enhance bioavailability of minerals in humans (Welch and Graham, 2004). Mutation breeding has also been used to generate low phytic acid (lpa) maize, rice, barley and soybean (Raboy et al., 2001). Phytic acid content in seeds of these mutants is reduced by 50 to 80% compared to non-mutant seeds (Raboy, 1996). Reduction in phytic acid is accompanied by an equivalent molar increase in inorganic phosphorus, leaving the total phosphorus content unaffected by mutation. The low phytic acid (lpa) maize has been found to lead to better iron and zinc absorption in a small scale human feeding experiment (Mendoza et al., 1998).

### 5.2.3.2. Marker-assisted selection and QTL

Marker-assisted selection (MAS), i.e. the identification of molecular markers linked to the loci determining variation for the trait can be used to select the most favorable genotypes without a need for determining mineral levels by relatively complex and expensive assays in all breeding generations. The marker associated with the genes underlying the variation of the trait facilitates in locating it on the chromosome. These genes can be best identified using the knowledge on metal homeostasis that is rapidly accumulating for model species such as *Arabidopsis thaliana* (Ghandilyan et al., 2006). The amount of minerals in seeds is a polygenic trait as it depends on a number of processes such as mobilization from soil, uptake by the roots, translocation and redistribution within the plant, import and deposition in the seeds, etc. (Grusak et al., 1999, Vreugdenhil et al., 2004). A powerful technique to study such polygenic traits is quantitative trait loci (QTL). It is based on genetic differences that are present between accessions of one species and is generally performed on genetic segregating populations like $F_2$s or recombinant inbred line (RIL) populations. A QTL is a chromosomal region associated with specific phenotype and identified using flanking molecular markers without the prior knowledge of the genes involved. QTL for the accumulation of minerals (Fe, Zn etc) and the anti-nutrient, phytate have been identified in some plants. The QTLs involved in grain Fe concentration (Gregorio et al., 2000) and Zn tolerance (Wissuwa et al., 2006) in rice; for Fe and Zn in bean (Beebe et al., 2000; Cichy et al., 2005) and for several cation mineral contents in Arabidopsis (Vreugdenhil et al., 2005) have been identified. In addition, five genomic regions affecting the quantity of phytate and inorganic phosphorus in seeds and leaves of Arabidopsis were identified (Bentsink et al., 2003). Since the same segregating population was used for QTL analysis of phytate content and mineral accumulation, loci affecting Fe and Zn accumulation were largely different from those controlling the phytate content of seeds. Thus, it should be possible to breed for low phytate genotypes, without decreasing the levels of Fe and Zn (Ghandilyan et al., 2006). Following the identification of QTL, candidate genes or loci can be resolved through fine mapping and map-based cloning, and this information could be used for gene-based selection or marker-assisted breeding strategies. An advantage of this strategy is that knowledge of the genes and/or chromosome loci controlling mineral content of the shoot or seed in one species could be used in a different target species by exploiting gene homology and/or genome collinearity (Ghandilyan et al., 2006; Broadley et al., 2007).

### 5.2.4. Genetic modification to improve plant mineral content

Transgenic methods for enhancement of mineral content rely on improving mobilization from the soil, uptake from the rhizosphere, translocation to the shoot and accumulation of mineral elements in edible tissues in bioavailable form. Furthermore, because minerals must be transported to various organs, such transgenic strategies should be used which are directed towards enhancing the supply processes, rather than manipulating the receiving organ itself. These strategies have mainly focused on the two most common mineral deficiencies, Fe and Zn. Three different methods have been used to increase these minerals by genetic engineering.

#### 5.2.4.1. Introduction of metal-binding protein genes

Ferritin, a conjugated protein with molecular weight about 540 kDa assembled from 24 subunits into a spherical cell (Laulhere et al., 1988), can store 4500 iron atoms in its central cavity (Theil, 1987, 2004; Harrison and Arosio, 1996). Ferritin with about 23% of its weight as iron, and high stability and survivability in vitro digestion conditions, is thus a major iron storage protein in both plants and animals (Theil, 1987, 2004). Ferritin plays two main roles in living cells, both through modulation of iron content (Theil, 1990). One is to provide iron in bioavailable form (Murray-Kolb et al., 2003). The other is to prevent damage from free-radicals produced by iron/dioxygen interactions. Several workers have investigated the possibility to increase iron in plants by ferritin gene transformation strategies. Transgenic tobacco plants with high iron content have been developed by transferring soybean ferritin cDNA under the control of a CaMV35S promoter (Goto et al., 1998). Goto et al. (1998) reported a positive correlation between ferritin levels and iron accumulation in ferritin-transformed tobacco, raising the possibility that higher iron concentrations in planta could be achieved by increasing the expression of ferritin. In rice and wheat, constitutive expression of soybean ferritin gene by corn ubiquitin promoter increased iron in vegetative organs but not in seeds (Drakakaki et al., 2000). Similarly, transgenic lettuce constitutively expressing ferritin grew larger and faster than the wild type control (Goto et al., 2000). Further Goto and coworkers reported endosperm specific expression of soybean ferritin in rice seeds increased the iron content by three-fold over non-transformed seeds (Goto et al., 1999). The expression of Phaseolus vulgaris ferritin gene under the control of glutelin promoter in Japonica rice endosperm resulted in a more than two-fold increase in Fe content (Lucca et al., 2001). Vasconcelos et al. (2003) introduced soybean ferritin gene under endosperm-specific glutelin promoter in Indica rice and observed increases in both iron and zinc in seeds even after commercial milling (polishing). Recently, Qu et al. (2005) compared the ferritin expression levels and Fe accumulation levels in seeds by generating two kinds of ferritin hyper-expressing lines (double ferritin (DF) - by introducing soybean ferritin

Elemental biofortification of crop plants

gene under the control of strong promoters, rice seed storage glutelin and rice seed storage globulin genes promoters, and one ferritin (OF) - by introducing soybean ferritin gene under the control of glutelin promoter alone). In these lines, ferritin expression (in rice endosperm) and the maximum iron concentrations in seeds were higher by 13-fold and 30%, respectively than previously reported (Goto et al., 1999). The iron concentration in the OF and DF lines was about three-fold higher than in non-transformed control seeds. Qu and coworkers observed that the mean Fe concentration in leaves of high ferritin expressing lines decreased to less than half of the non-transformed plants, and the plants showed chlorosis after flowering, even on an iron-rich medium. The authors (Qu et al., 2005) concluded that accumulation of Fe in seeds of hyper-expression ferritin rice did not depend on the expression level of exogenous ferritin but may also be limited by Fe uptake and transport. It is evident that over-expression of ferritin in rice might double or triple iron intake of populations in areas where rice is a staple food. In transgenic maize, co-expression of soybean ferritin and *Aspergillus* phytase increased the levels of bioavailable iron in endosperm (Drakakaki et al., 2005).

Lactoferrin is another iron storage protein which is highly resistant to proteolytic enzymes and gastric pH. Human lactoferrin binds to specific receptors on the surface of the brush border membrane of the intestinal cell (Lönnerdal and Iyer, 1995) with concomitant iron uptake by the cell (Suzuki et al., 2001). Recently lactoferrin was expressed in potatoes (Chong and Langridge, 2000) but with the objective to study its antimicrobial properties. Nandi et al. (2002) expressed recombinant human lactoferritine (rHLF) in rice endosperm and observed not only 5g rHLF per kg dehusked rice grains, but simultaneously increased iron content about two-fold.

5.2.4.2. Over-expression of native genes for mineral binding proteins

Plant mineral element binding proteins express either at low levels or in a plant tissue that is normally not eaten. The example of the first category is ferritin which is present in the soybean, French bean, rice, but the level of expression is too low to contribute substantially to the iron content of the plants. The conventional breeding (selection and breeding for high-ferritin varieties) or over-expressing the native ferritin gene should substantially increase the ferritin content. Whether the increased ferritin expression is correlated to a concomitant increase in the iron content is likely to be dependent on the plant's capacity to take up iron and transport it to the endosperm (Lönnerdal, 2003).

Another mineral element-binding protein is leghemoglobin which is present in legume root nodules that are not eaten. This protein accumulates iron in more bioavailable form i.e. heme. It remains to be explored whether the expression of leghemoglogin can be directed to the seeds (beans) or whether the gene can be inserted as a storage protein in other plants such as cereals. However, insertion of soybean leghemoglobin-encoding genes into potato tubers resulted in reduced growth and decreased tuber production when compared with wild type plants (Chaparro-Giraldo et al., 2000).

### 5.2.4.3. Reduced anti-nutrients or increased promoters for increasing bioavailability of minerals

Plant food (especially seeds and grains) contain various anti-nutrients (phytate and polyphenols) or promoters (cysteine, β-carotene, ascorbic acid etc.) that can reduce or increase the absorption of non-heme Fe, Zn and other nutrients in humans. The levels of these antinutrients or promoters are influenced by both genetic and environmental factors. Recently genetic engineering approaches have made it possible to either reduce or eliminate antinutrient or increase the levels of promoter substances significantly in staple plant foods (Frossard *et al.*, 2000; Welch, 2001). However, a few studies have suggested that antinutrients are potential antioxidants and protect plants from pests and pathogens (see Brinch-Pedersen *et al.*, 2002), reducing or eliminating these compounds need careful evaluation.

(i) Decreased anti-nutrients that inhibit mineral absorption

**Introduction of phytase gene:** Reduction of phytate content in grain by over-expression of phytase, a phytate degrading enzyme in edible portions can increase iron and zinc absorption (Holm *et. al.*, 2002; Smolin and Grosvenor, 2003; Chiera, 2004). Two transgenic strategies have been adopted to reduce phytate concentrations in food and feed. The first strategy knocks out enzymes in the *myo*-inositol hexa*kis*phosphate (InsP6) biosynthetic pathway. The knock out of *myo*-inositol-1-phosphate synthase (*GmMIPS1*) gene by RNAi in transgenic soybean has reduced phytate (InsP6) content but also inhibited seed development (Nunes *et al.*, 2006). Similar knock out of *myo*-inositol-3-phosphate synthase gene (*RINO1*) by antisense technology in transgenic rice seed has increased available inorganic phosphates which is accompanied by a molar-equivalent decrease in phytic acid, without a reduction in total phosphorus levels (Kuwano *et al.*, 2006). The second strategy is to over-express phytase, phytate-degrading enzyme, in edible portions. Phytase has been commercially produced based on the filamentous fungus, *Aspergillus niger* (Brinch-Pedersen *et al.*, 2002). In humans, when *A. niger* phytase was added to high phytate bread roll prior to consumption, ion absorption increased from 14.3 to 26.1%, suggesting that effective and complete phytate degradation occurred in the stomach (Sandberg *et al.*, 1996; Hurrell *et al.*, 2003). A large number of studies have shown that *A. niger* phytase can be synthesized effectively in transgenic plants such as canola, tobacco and soybean (Brinch-Pedersen *et al.*, 2000). However, *A. niger* phytase is inactivated at temperatures higher than 60°C. Therefore, this phytase is not useful in transgenic cereals for human consumption, because cooking procedures inactivate the enzyme. An alternative has been to introduce thermo-tolerant phytase under a seed-specific promoter to obtain transgenic seeds with heat-stable phytase. Phytase from *Aspergillus fumigatus* has a strong capacity for refolding into an active conformation after thermal denaturation. However, in transgenic rice expressing *A. fumigatus* phytase in endosperm, only slight activity was recovered after cooking (Lucca *et al.*, 2002). For unknown reasons, the *in planta* synthesized phytase is not thermo-tolerant when expressed in fungi.

Elemental biofortification of crop plants

Eventually, the different glycosylation pattern in plants may affect the heat stability (Lucca et al., 2006). The other extremely thermophilic microorganisms may aid in finding such phytases that are not denaturated under high temperatures, will provide higher heat stability *in planta* and will allow phytate degradation upon cereal digestion. However, the potential allergenicity of such species-foreign proteins must be investigated in detail.

On the other hand, the antinutrients such as phytate and polyphenol are potential antioxidants and may play an important and benefial role in human diets by acting as anti-carcinogens or by decreasing the risk of heart diseases or diabetes. Therefore, reducing antinutrient content to increase iron bioavailability remains controversial.

(ii) Increased synthesis of enhancers of mineral absorption

**Amino-acids:** Enhancing diets in cysteine and cysteine-rich peptides has been shown to enhance bioavailability of non-heme Fe and histidine is shown to facilitate Zn absorption (Welch and Graham, 2004). Other organic acids such as fumarate, citrate and succinate can also enhance mineral element absorption.

Low-molecular weight cysteine-rich peptides, metallothionein (MT) are metal binding proteins that are present in animals, fungi, cyanobacteria and plants. Overexpression of a cysteine-rich MT gene in rice increased the cysteine content of soluble seed protein about seven-fold in endosperm (Lucca et al., 2001). However, it has not yet been demonstrated whether cysteine-containing peptides formed during the digestion of the MT enhance iron absorption in man in a similar way as free cysteine, glutathione or the cysteine-containing peptides of meat (Lucca et al., 2006). However, rice seeds are consumed after boiling at 100°C which may oxidize the sulphydryl (SH) groups of cysteine to disulphide (SS) groups and therefore, may hamper the positive effect of cysteine-containing peptides on iron absorption. A recent study has shown that cooking doesn't impair non-heme iron absorption from a phytate-rich meal (Baech et al., 2003).

**Vitamins:** β-carotene (pro-vitamin A) has been shown to enhance the bioavailability of non-heme iron in staple food diets fed to humans (Garcia-Casal et al., 1998) although the mechanism underlying this effect is unknown. β-carotene synthesis in the endosperm of Japonica (Ye et al., 2000) and Indica (Datta et al., 2003) rice (Golden rice) has been demonstrated by introducing three genes, phytoene synthase (from daffodil), phytoene desaturase (from *Erwinia uredovara*) and lycopene β-cyclase (from daffodil) under endosperm-specific promoters. Recently, new golden rice lines have been developed that contain extremely high amounts of β-carotene compared with the original Japonica golden rice variety (Paine et al., 2005). Similarly, ascorbic acid content in Arabidopsis has been increased by increased expression of gene GalUR (D-galacturonic acid reductase) (Agius et al., 2003) or L-Gul (Jain and Nessler, 2000) or L-galactose guanyl transferase in tobacco (Laing et al., 2007). Whether the golden rice and other plants that have high levels of β-carotene or ascorbic acid enhance iron absorption is not yet known. Therefore, combining high iron traits with plants rich in β-carotene or in ascorbic acid could lead to a highly effective, cheap and simple contribution to the relief of major health problems.

5.2.4.4. Modifying uptake and transport pathways of micronutrients

An alternative approach to improve micronutrient content is by genetic engineering. Only small parts of trace minerals present in soil are available for accumulation in the storage organs. In recent years, various genes have been identified which codes for membrane transport proteins of different mineral nutrients. These protein products include membrane transporters, as well as proteins that facilitate the availability of minerals at root-soil interface. In the first category, data are now available for genes encoding various divalent metal transporters for example $Fe^{+2}$, $Zn^{+2}$, $Mn^{+2}$, $Cu^{+2}$, $Ni^{+2}$ and others (Fox and Guerinot, 1998; Eide, et al., 1996; Grotz et al., 1998, see other chapters). Iron uptake and tolerance of Fe-deficient soils in non graminaceous plants can be by over-expressing genes encoding either Fe (III) chelate reductase (FRO2, Samuelsen et al., 1998; Connolly et al., 2003; Vasconcelos et al., 2006) or high affinity Fe-transporting IRT1 protein (belongs to ZIP family) or any of its orthologs (Connolly et al., 2002). In graminaceous plants, iron acquisition is characterized by the synthesis and secretion of iron-chelating phytosiderophore (PS, from the mugineic acid family (MAs) synthesized from L-methionine via nicotianamine) and by a specific uptake system (YS1) for iron (III)-phytosiderophore complexes (PS-FeIII). Nicotianamine synthase (NAS) and nicotianamine amino transferase (NAAT) are the critical enzymes in the biosynthesis of MAs. To improve Fe uptake in graminaceous plants, the production of PSs and the expression of the PS-FeIII transporting YSL proteins is required. Transgenic rice expressing the barley *naatA* and *naatB* genes encoding enzymes nicotianamine aminotransferase involved in the biosynthesis of PSs have shown tolerance to low Fe availability, when grown in calcareous soil. The transformants secreted more amounts of PSs than non-transformants. However, information on mineral content in the seeds was not reported (Takahashi et al., 2001). Over-expression of nicotianamine synthase, *Hv*NAS1 from barley has a positive effect on root to shoot transport of Fe and Zn and doubled the Fe and Zn concentrations in young leaves of tobacco (Takahashi, 2003). Similarly, over-expression of a nicotianamine synthase (*AtNAS1*) also resulted in a marginal increase in shoot Fe (1.5-fold), and Zn (1.2-fold) in transgenic tobacco (Douchkov et al., 2005).

The over-expression of Arabidopsis zinc transporter (*AtZIP1*) in transgenic barley increased seed Fe and Zn contents, although seeds from transformed lines were significantly smaller than seeds from non-transformed plants (Ramesh et al., 2004). Once in the root, both minerals will be transported to the xylem for further transport towards the above ground parts. The heavy metal ATPase (HMA) genes (*At HMA3*, *AtHMA4*) which express in vascular tissue appear to perform this task at least for Zn (Hussain et al., 2004). Over-expression of *AtHMA4* increased the Zn content in the Arabidopsis shoot by two-fold (Verret et al., 2004). However, Zn may end up in the shoot apoplast rather than inside the cells. Although this may be beneficial to the plant to avoid high intracellular Zn concentration, it may be

## Elemental biofortification of crop plants

disadvantageous when trying to load the shoot with Zn for further loading into grain by phloem transport. Moreover, AtHMA4 can transport Cd as well (Mills *et al.*, 2005), also the Cd content is increased, but only when available in substantial amounts. HMAs have not been shown to transport Fe. Therefore, over-expression of YSL transporters, of which same family members transport Fe-NA chelates (Schaaf *et al.*, 2004) seems to be more appropriate.

Transplastomic tobacco plants generated by co-transformation with two vectors, one containing the 16SrDNA marker gene and the other a polyhistidine-tagged large subunit of rubisco (*rbcL* gene) accumulated higher Zn amounts than wild-type when grown on Zn-enriched media. The higher Zn increase observed exceeded the estimated chelating ability of the polyhistidine sequence, indicating a perturbation in intracellular Zn homeostasis (Rumeau *et al.*, 2004).

Crops with enhanced Se accumulation can be developed by genetic engineering. Selenium and sulphur are nutrients with very similar chemical properties. Their uptake and assimilation occurs through common pathways from selenate and sulfate, respectively, by activated ATP sulfurylase. Se accumulation in plant shoots by over-expression of enzymes of rate-limiting steps such as ATP sulfurylase (APS, a key enzyme of sulphur/selenium assimilation involved in the reduction of selenate/sulfate to selenite/sulfite)(Van Huysen *et al.*, 2004), and Se accumulation and volatilization by CGS (cystathionine-γ-synthase, mediates the conversion of Se-cysteine to Se cystathionine, the first enzyme in the Se volatilization as dimethlselenide) (Van Huysen *et al.*, 2004) or SMT (selenocysteine methytransferase methylates selenocysteine to non-protein aminoacid methylselenocysteine which is subsequently converted into dimethyl diselenide that is released harmlessly into the atmosphere) (LeDuc *et al.*, 2004) or selenocysteine lyase (Banuelos *et al.*, 2007) or both APS and SMT (LeDuc *et al.*, 2006) has been recognized as a low cost, environmentally friendly approach to remove Se from contaminated sites. The plant species used for phytoremediation of Se-laden soils on harvest can be blended with animal forage and fed to animals in Se-deficient areas. The sulphate transporters and the enzymes of the S-assimilations pathway in relation to Se accumulation can be further exploited.

## 6. CONCLUSIONS AND FUTURE PROSPECTS

Mineral malnutrition (e.g. iron, zinc and selenium deficiencies) is damaging the health of three billion people (half of the world's population) especially in developing countries. Dietary diversification is the ultimate solution but this is not immediately practical. The agronomic practices, diet supplements and fortification of food have played an important role in tackling the malnutrition problem to a great extent. Unfortunately, these types of programmes do not reach many of those afflicted (especially resource-poor rural families in developing nations) and in many nations they have not proven to be sustainable. Increasing the minerals in edible parts of the major crop plants (biofortification), either through conventional breeding or genetic manipulation, has shown promising results. Although plant breeding has a long gestation period it might provide a more sustainable and cost effective solution in the long run, delivering minerals to the entire population. Availability of natural genetic variations and the knowledge of genetics of these variations in mineral concentrations, are essential to select and breed crop plants for increased concentrations in edible parts without affecting yield and quality. Conventional breeding is now complemented by molecular markers (QTL mapping) and marker assisted selection (MAS) to speed up the breeding efficiency. Identification of accurate genetic markers and the presence or absence of specific alleles of genes known to play a role in mineral uptake and translocation to the grain will help in identification of new genes affecting mineral contents. In order for genetic modification to be efficient and effective to improve the amounts and bioavailability of mineral in the edible part of the crop, a better understanding of molecular biology and regulation of mineral uptake, transportation and redistribution and the biosynthetic pathways of anti-nutrients and promoters is essential. Our knowledge in the area of loading of minerals into phloem for their inward delivery to developing seeds is lacking. Functional genomics approaches will help to identify more or rather all genes involved in mineral transport and biosynthesis of anti-nutrients and promoters but also the cellular and tissue localization of the proteins they encode and their interaction. The transgenic approach in which expression of several genes for enhanced uptake and transport through multiple membrane and tissue systems, storage, and bioavailability will help. The efforts to improve plant mineral status will continue to receive more attention not only for the nutritional value of our food supply but also for the health and reproductive output of our agronomic crops.

## 7. LITERATURE CITED

Administrative Committee on Coordination, Subcommittee on Nutrition, and International Food Policy Research Institute (2000). *Fourth Report on the World Nutrition Situation*, United Nations, Geneva.

Agius, F., Gonzalez-Lamothe, R., Caballero, J.L., Munoz-Blanco, J., Botella, M.A. and Valpuesta, V. (2003) Engineering increased vitamin C levels in plants by overexpression of a D-galacturonic acid reductase. *Nat. Biotechnol.*, **21**: 177-181.

Arthur, J.R. (2003). Selenium supplementation: does soil supplementation help and why? *Proc. Nutr. Soc.*, **62**: 393-397.

Baech, S.B., Hansen, M., Bukhave, K., Jensen, M., Sorensen, S.S., Purslow, P.P., Skibsted, L.H. and Sandstrom, B. (2003) Increasing the cooking temperature of meat does not affect non-haem iron absorption from a phytate-rich meal in women. *J. Nutr.*, **133**: 94-97.

Banuelos, G., LeDuc, D.L., Pilon-Smits, E.A.H. and Terry, N. (2007) Transgenic Indian mustard overexpressing selenocysteine lyase or selenocysteine methyltransferase exhibit enhanced potential for selenium phytoremediation under field conditions. *Environ. Sci. Technol.*, **41**: 599-605.

Banziger, M. and Long, J. (2000) The potential for increasing the iron and zinc density of maize through plant-breeding. *Food Nutr. Bull.*, **21**: 397-400.

Beebe, S., Gonzalez, A. and Rengifo, J. (2000) Research on trace minerals in the common bean. *Food Nutr. Bull.*, **21**: 387-391.

Elemental biofortification of crop plants

Bentsink, L., Yuan, K., Koornneef, M. and Vreugdenhil, D. (2003) The genetics of phytate and phosphate accumulation in seeds and leaves of *Arabidopsis thaliana*, using natural variation. *Theor. Appl. Genet.,* **106**: 1234-1243.

Boccio, J.R. and Iyengar, V. (2003) Iron deficiency-causes, consequences and strategies to overcome the nutritional problem. *Biol. Trace. Elem. Res.,* **94**: 1-31.

Bouis, H.E. (2002) Plant breeding: a new tool for fighting micronutrient malnutrition. *J. Nutr.* **132**: 491s-494s.

Bouis, H.E. (2003) Micronutrient fortification through plant breeding: Can it improve nutrition in man at low cost? *Proc Nutr. Soc.,* **62**: 403-411.

Bouis, H.E., Chassy, B.M. and Ochanda, J.O. (2003). Genetically modified food crops and their contribution to human nutrition and food quality. *Trends Food Sci. Technol.,* **14**: 191-209.

Branca, F. and Ferrari, M. (2002) Impact of micronutrient deficiencies on growth: the stunting syndrome. *Annu. Nutr. Metabolism,* **46**: 8-17.

Brinch-Pedersen, H., Sorensen, L.D. and Holm, P.B. (2002) Engineering crop plants: getting a handle on phosphate. *Trends Plant Sci.,* **7**: 118-125.

Brinch-Pedersen, H., Olesen, A., Rasmussen, S.K. and Holm, P.B. (2000). Generation of transgenic wheat (*Triticum aestivum* L) for constitutive accumulation of an *Aspergillus* phytase. *Mol. Breed.,* **6**: 195-206.

Broadley, M.R., White, P.J., Bryson, R.J., Meacham, M.C., Bowen, H.C., Johnson, S.E. et al. (2006) Biofortification of UK food crops with selenium. *Proc. Nutr. Soc.,* **65**: 169-181.

Broadley, M.R., White, P.J., Zelko, I. and Lux, A. (2007) Zinc in plants. *New Phytol.,* **173**: 677-702.

Caballero, B. (2002). Global patterns of child health: the role of nutrition. *Annu. Nutr. and Metabolism,* **46**: 3-7.

Cakmak, I. (2004) *Triticum dicoccoides*: an important genetic resource for increasing zinc and iron concentrations in modern cultivated wheat. *Soil Sci. Plant Nutr.,* **50**: 1047-1054.

Chaparro-Giraldo, A., Barata, R.M., Chabregas, S.M., Azevedo, R.A. and Silva-Filho, M.C. (2000) Soybean leghemoglobin targeted to potato chloroplasts influences growth and development of transgenic plants. *Plant Cell Rep.,* **19**: 961-965.

Chen, L., Yang, F.M., Xu, J., Hu, Y., Hu, Q.H., Zhang, Y.L. and Pan, G.X. (2002) Determination of selenium concentration of rice in China and effect of fertilization of selenite and selenate on selenium content of rice. *J. Agric Food Chem.,* **50**: 5128-5130.

Cheng, Z.J. and Hardy, R.W. (2003). Effect of extrusion processing of feed ingredients on apparent digestibility coefficient of nutrients for rainbow trout (*Oncorhynchus mykiss*). *Acquaculture Nutr.,* **9**: 77-83.

Chhunjea, P., Dhaliwal, H.S., Bains, N.S. and Singh, K. (2006) *Aegilops kotschyi* and *Aegilops tauschii* as sources for higher levels of grain iron and zinc. *Plant Breeding,* **125**: 529-531.

Chiera, J.M. (2004). Ectopic expression of a soybean phytase in developing seeds of *Glycine max* to improve phosphorus availability. *Plant Mol. Biol.,* **56**: 895-904.

Chong, D.K. and Langridge, W.H. (2000) Expression of full-length bioactive antimicrobial human lactoferrin in potato plants. *Transgenic Res.,* **9**: 71-78.

Cichy, K.A., Forster, S., Grafton, K.F. and Hosfield. G.L. (2005) Inheritance of seed zinc accumulation in navy bean. *Crop Sci,* **45**: 864-870.

Combs, G.F. (2001). Selenium in global food systems. *Br. J. Nutr.,* **85**: 517-547.

Connolly, E.L., Fett, J.P. and Guerinat, M.L. (2002) Expression of the IRT1 metal transporter is controlled by metals at the levels of transcript and protein accumulation. *Plant Cell*, **14**: 1347-1357.

Connolly, E., Campbell, N., Grotz, N., Prichard, C. and Guerinot, M. (2003) Overexpression of the FRO2 ferric chelate reductase confers tolerance to growth on low iron and uncovers post-transcriptional control. *Plant Physiol.*, **130**: 1102-1110.

Dai, J.-L., Zhu, Y.-G., Zhang, M. and Huang, Y.-Z. (2004) Selecting iodine-enriched vegetables and residual effects of iodate application to soil. *Biol. Trace Element Res.*, **101**: 265-276.

Darnton-Hill, I. and Nalubola, R. (2002) Fortification strategies to meet micronutrients needs: successes and failure. *Proc. Nutr. Soc.*, **61**: 231-241.

Datta, K., Baisakh, N., Oliva, N., Torrizo, L., Abrigo, E., Tan, J., Rai, M., Rehana, S., Al-Balili, S., Beyer, P. *et al.* (2003) Bioengineering 'golden' indica rice cultivars with β-carotene metabolism in the endosperm with hygromycin and mannose selection system. *Plant Biotechnol J.*, **1**: 81-90.

Davidson, L., Galan, P., Cherouvrier, F., Kastenmayer, P., Juillerat, M.A., Hercberg, S. and Hurrell, R.F. (1997) Bioavailability in infants of iron from infant cereals: effect of dephytinization. *Am. J.Clin. Nutr.*, **65**: 916-920.

Drakakaki, G., Christou, P. and Stoger, E. (2000) Constitutive expression of soybean ferritin cDNA in transgenic wheat and rice results in increased iron levels in vegetative tissues but not in seeds. *Transgenic Res.*, **9**: 445-452.

Drakakaki G., Marcel S., Glahn R.P., Lund E.K., Pariagh S., Fischer R., Christou P. and Stoger E. (2005) Endosperm-specific co-expression of recombinant soybean ferritin and *Aspergillus* phytase in maize results in significant increases in the levels of bioavailable iron. *Plant Mol. Biol.*, **59**: 869-880.

Douchkov, D., Gryczka, C., Stephan, U.W., Hell, R. and Baumlein, H. (2005) Ectopic expression of nicotinamine synthase genes results in improved iron accumulation and increased nickel tolerance in transgenic tobacco. *Plant Cell Environ.*, **28**: 365-374.

Eichholzer, M. (2003) Micronutrient deficiencies in Switzerland: causes and consequences. *J. Food Eng.*, **56**: 171-179.

Eide, D., Broderius, M., Fett, J. and Guerinot, M.L. (1996). A novel iron-regulated metal transporter from plants identified by functional expression in yeast. *Proc. Natl. Acad. Sci.*, **93**: 5624-5628.

Fairweather-Tait, S. and Hurrell, R.F. (1996). Bioavailability of minerals and trace elements. *Nutrition Research Reviews*, **9**: 295-324.

FAO (2003) The Scourage of 'Hidden Hunger', Global Dimensions of Micronutrient Deficiencies. *Food Nutrition and Agriculture*, FAO, 32.

FAO/WHO (2003). Preliminary report on recommended nutrient intakes. Joint FAO/WHO Expert Consultation on Human Vitamin and Mineral Requirements, FAO, Bangkok, Thailand. Food and Agricultural Organization of the United Nations Rome, Italy and World Health Organization, Geneva, Switzerland.

Farnham, M.W., Hale, A.J., Grusak, M.A. and Finley, J.W. (2007) Genotypic and environmental effects on selenium concentration of broccoli heads grown without supplemental selenium fertilizer. *Plant Breed.*, **126**: 195-200.

Fox, T.C. and Guerinot, M.L. (1998). Molecular biology of cation transport in plants. *Ann. Rev. Plant Physiol.*, **49**: 669-696.

Frossard, E., Bucher, M., Machler, F., Mozafar, A. and Hurrell, R. (2000). Potential for increasing the content and bioavailability of Fe, Zn and Ca in plants for human nutrition. *J. Sci. Food and Agriculture*, **80**: 861-879.

Elemental biofortification of crop plants

Garcia-Casal, M.N., Layrisse, M., Salano, L., Baron, M.A., Arguello, F., Llovera, D, Ramirez, J. Leets, I. and Tropper, E. (1998) Vitamin A and beta-carotene can improve non heme iron absorption from rice, wheat and corn by humans. *J. Nutr.*, **128**: 646-650.

Ghandilyan, A., Vreugdenhil, D. and Aarts, M.G.M. (2006) Progress in the genetic understanding of plant iron and zinc nutrition. *Physiol. Plant.*, **126**: 407-417.

Gibson, C., Lyons, G., Choi, B.-J., Park, S.-G. and Stewart, D. (2007) Selenium enriched barley. Biofortification to improve human health.

Gibson, R.S. (2004). Strategies for preventing micronutrient deficiencies in developing countries. *Asia Pac. J. Clin. Nutr.*, **13**: Suppl. 523.

Golden, M.H.N. (1991). The nature of nutritional deficiency in relation to growth failure and poverty. *Acta Paediatrica Scandinavica*, **374**: 95-110.

Goos, R.J., Johnson, B., Jackson, G. and Hargrove, G. (2004) Greenhouse evaluation of controlled-release iron fertilizers for soybean. *J. Plant Nutr.*, **27**: 43-55.

Goto, F., Yoshihara, T. and Saiki, H. (1998) Iron accumulation in tobacco plants expressing soybean ferritin gene. *Transgenic Res.*, **7**: 173-180.

Goto, F., Yoshihara, T., Shigemoto, N., Toki, S. and Takaiwa, F. (1999) Iron fortification of rice seed by the soybean ferritin gene. *Nat. Biotechnol.*, **17**: 282-286

Goto, F., Yoshihara, T. and Saiki, H. (2000). Iron accumulation and enhanced growth in transgenic lettuce plants expressing the iron-binding protein ferritin. *Theor. Appl. Genet.*, **100**: 658-664.

Graham, R.D., Welch, R.M. and Bouis H.E. (2001) Addressing micronutrient malnutrition through enhancing the nutritional quality of staple foods: principles, perspective and knowledge gaps. *Adv. Agron.*, **70**: 77-142.

Gregorio, G.B., Senadhira, D., Htut, T. and Graham, R.D. (2000). Breeding for trace mineral density in rice. *Food Nutr. Bull.*, **21**: 382-386.

Grotz, N., Fox, T., Connolly, E., Park, W., Guerinot, M.L. and Eide, D. (1998). Identification of a family of zinc transporter genes from *Arabidopsis* that respond to zinc deficiency. *Proc. Natl. Acad. Sci.*, USA, **95**: 7220-7224.

Grusak, M.A. and Cakmak, I. (2005). Methods to improve the crop delivery of minerals of humans and livestock. In: *Plant Nutritional Genomics* (Eds. Broadley, M.R. and White, P.J.), Blackwell Publ., Oxford, pp 265-286.

Grusak, M.A., Della Penna, D. and Welch, R.M. (1999). Physiologic processes affecting the content and distribution of phytonutrients in plants. *Nutr. Rev.*, **57**: S27-S33.

Gupta, U.C. and Gupta, S.C. (2000) Selenium in soils and crops, its deficiencies in livestock and humans: implication for management. *Comm. Soil Sci. Plant Analysis*, **31**: 1791-1807

Harrison, P. and Arosio, P. (1996) The ferritin: molecular properties, iron storage function and cellular regulation. *Biochim. Biophys. Acta*, **1275**: 161-203.

He, X. and Nara, K. (2007) Element biofortification: Can mycorrhiza potentially offer a more effective and sustainable pathway to curb human malnutrition? *Trends Plant Sci.*, **12**: 331-333.

Holm, P.B., Kristiansen, K.N. and Pedersen, H.B. (2002). Transgenic approaches in commonly consumed cereals to improve iron and zinc content and bioavailability. *J. Nutr.*, **132**: 514-516.

Hotz, C. and Brown, K.H. (2004) Guest editors. Assessment of the risk of zinc deficiency in populations and options for control. IZiNCG Technical Documents 1, *Food and Nutrition Bulletin* 25. Supplement 2.

Hunt, J.M. (2002). Reversing productivity losses from iron deficiency: The economic case. *J. Nutr.*, **132**: 794-801.

Hurrell, R.F., Juillerat, M.A., Reddy, M.B., Lynch, S.R., Dassenko, S.A. and Cook, J.D. (1992). Soy protein, phytate and iron absorption in humans. *Am. J. Clin. Nutr.*, **56**: 573-578.
Hurrell, R.F, Reddy, M.B., Juillerat, M.A. and Cook, J.D. (2003) Degradation of phytic acid in cereal porridges improves iron absorption in humans. *Am J. Clin. Nutr.*, **77**: 1213-1219.
Hussain, D., Haydon, M.J., Wang, Y., Wong, E., Sherson, S.M., Young J., Camakris, J., Harper, J.F. and Cobbett, C.S. (2004) P-type ATP are heavy metal transporters with roles in essential zinc homeostasis in Arabidapsis. *Plant Cell*, **16**: 1327-1339.
Jain, A.K. and Nessler, C.L. (2000) Metabolic engineering of an alternative pathway for ascorbic acid biosynthesis in plants. *Mol. Breed.*, **6**: 73-78.
Kennedy, G. (2003). The scourage of "hidden hunger" global dimensions of micronutrient deficiencies. *Food Nutr. Agric.*, **32**: 8-18.
Kuwano, M., Ohyama, A., Tanaka, Y., Mimura, T., Takaiwa, F. and Yoshida, K.T. (2006) Molecular breeding for transgenic rice with low-phytic acid-phenotype through manipulation of myo-inositol- 3-phosphate synthase gene. *Mol. Breed.*, **18**: 263-272.
Laing, W.A., Wright, M.A., Cooney, J. and Bulley, S.M. (2007) The missing step of the L-galactose pathway of ascorbate biosynthesis in plants, an L-galactose guanyltransferase, increases leaf ascorbate content. *Proc. Natl. Acad. Sci. USA*, **104**: 9534-9539.
Laulhere, J.P., Lescure, A.M. and Briat, J.F. (1988) Purification and characterization of ferritins from maize, pea and soybean seeds. *J. Biol. Chem.*, **263**: 10289-10294.
LeDuc, D.L., Tarun, A.S., Montes-Bayon, M., Meija, J., Malit, M.F., Wu, C.P., AbdelSamie, M., Chiang, C.Y., Tagmount, A, desouza, M., Neuhierl, B., Bock, A, Caruso, J. and Terry, N. (2004) Overexpression of selenocysteine methyltranferase in Arbidopsis and Indian mustard increases selenium tolerance and accumulation. *Plant Physiol.*, **135**: 377-383.
LeDuc, D.L., AbdelSamie, M., Montes-Bayon, M., Wu, C.P., Reisinger, S.J. and Terry, N. (2006) Overexpressing both ATP sulfurylase and selenocysteine methyltranferase enhances selenium phytoremediation traits in Indian mustard. *Environ. Pollution*, **144**: 70-76.
Loewus, F.A. (2002) Biosynthesis of phytate in food grains. In: *Food Phytates* (Eds. Reddy A.R. and Sathe S.K.) CRC Press, Boca Raton, Florida, pp 53-62.
Lönnerdal, B. (2003) Genetically modified plants for improved trace element nutrition. *J. Nutr.*, **133**: 1490-1493.
Lönnerdal, B. and Iyer, S. (1995) Lactoferrin: molecular structure and biological function. *Annu Rev. Nutr.*, **15**: 93-110.
Lott, J.N.A. *et al.* (2000) Phytic acid and phosphorus in crop seeds and fruits: a global estimate. *Seeds Sci. Res.*, **10**: 11-33.
Lucca, P., Hurrell, R. and Potrykus, I. (2001) Genetic engineering approaches to improve the bioavailability and the level of iron in rice grains. *Theor Appl. Genet.*, **102**: 392-397.
Lucca, P., Hurrell, R.F. and Potrykus, I. (2002). Fighting iron deficiency anemia with iron-rich rice. *J. Am. Coll. Nutr.*, **21**: 184-190.
Lucca, P., Poletti, S. and Sautter, C. (2006) Genetic engineering approaches to enrich rice with iron and vitamin A. *Physiol. Plant.*, **126**: 291-303.
Lyons, G. and Bouis, H.E. (2003). High-selenium wheat: biofortification for better health. *Nutr. Res. Rev.*, **16**: 45-60.
Lyons, G., Stangoulis, J.C.R. and Graham, R.D. (2003) Nutriprevention of disease with high-selenium wheat. *J. Austr. College Nutr. and Environ. Med.*, **22**: 3-9.

Lyons, G.H., Stangoulis, J.C.R. and Graham, R.D. (2004) Exploiting micronutrient interaction to optimize biofortification programs: the case for inclusion of selenium and iodine in the *Harvest Plus* program. *Nutr. Rev.*, **62**: 247-252.
Lyons, G.H., Judson, G.J., Ortiz-Monasterio, I., Genc, Y., Stangoulis, J.C. and Graham, R.D. (2005) Selenium in Australia: selenium status and biofortification of wheat for better health. *J. Trace Elem. Med. Biol.*, **19**:75-82.
Marero, G.F., Trowbridge, F.L., Yip, R., Sullivan, K.M. and West, C.E. (1991) The antinutritional factors in weaning foods prepared from germinated legumes and cereals. *Lebensmittelwissenchaft Technol.*, **24**: 177-181.
Mason, J.B. and Garcia, M. (1993). Micronutrient deficiency – the global situation. *SCN News*, **9**: 11-16.
Maziya-Dixon, B., Kling, J.G., Menkir, A. and Dixon, A. (2000) Genetic variation in total carotene, iron and zinc contents of maize and cassava genotypes. *Food Nutr. Bull.*, **21**: 419-422.
McGuire, J. (1993). Addressing micronutrient malnutrition. *SCN News*, **9**: 1-10.
Mendoza, C. (2002). Effect of genetically modified low phytic acid plants on mineral absorption. *Int. J. Food Sci. Technol.*, **37**: 759-767.
Mendoza, C., Vitiri, F., Lonnerdal, B., Young, K.A. Raboy, V. and Brown, K.H. (1998) Effect of genetically modified, low-phytic acid maize on absorption of iron from tortillas. *Am. J. Clim. Nutr.*, **68**: 1123-1127.
Meng, F., Wei, Y., and Yang X. (2005) Iron Content and bioavailability in rice. *J. Trace Elem. Med. Biol.*, **18**: 333-338.
Mills, R.F., Francini, A., Ferrlira da Rocha, P.S., Baccarini, P.J., Aylett, M., Krijger, G.C. and Williams, L.E. (2005) The plant $P_{1B-type}$ ATPase At HMAu transports Zn and Cd and plays a role in detoxification of transition metals supplied at elevated levels. *FEBS Lett.*, **579**: 783-791.
Misra, B.K., Sharma, R.K. and Nagarajan, S. (2004) Plant breeding: A component of public health strategy. *Curr. Sci.*, **86**: 1210-1215.
Mora, J.O. (2003). Proposed vitamin A fortification levels. *J. Nutr.*, **133**: 2990S-2993S.
Murray-Kolb, L.E., Welch, R., Theil, E.C. and Beard, J.L. (2003) Women with low iron stores absorb iron from soybeans. *Am J Clin Nutr.*, **77**: 180-184.
Nandi, S., Suzuki, Y.A., Huang, J., Yalda, D., Pham, P., Wu, L., Bartley, G., Huang, N. and Loennerdal, B. (2002) Expression of human lactoferrin in transgenic rice grains for the application in infant formula. *Plant Sci.*, **163**: 713-722.
Nestel, P., Bouis, H.E., Meenakshi, J.V. and Pfeiffer, W. (2006) Biofortification of staple food crops. *J. Nutr.*, **136**: 1064-1067.
Nube, M. and Voortman, R.L. (2006) Simultaneously addressing micronutrient deficiencies in soil, crops, aminals and human nutrition: opportunities for higher yields and better health. *Staff Working Paper, Centre for World Food Studies*, Vrije Universiteit, Amsterdam, The Netherlands.
Nunes, A.C.S., Vianna, G.R., Cuneo, F. *et al.* (2006) RNAi-mediated silencing of the *myo*-inositol-1-phosphate synthase gene (*GmMIPS1*) in transgenic soybean inhibited seed development and reduced phytate content. *Planta* (online)
Ozturk, L., Yazic, M.A., Yucel, C., Torun, A., Cekic, C., Bagci, A., Ozkan, H., Braun, H-J., Sayers, Z. and Cakmak, I. (2006) Concentration and localization of zinc during seed development and germination in wheat. *Physiol. Plant.*, **128**: 144-152.
Paine, J.A., Shipton, C.A., Chaggar, S., Howells, R.M., Kennedy, M.J., Vernon, G., Wright, S.Y., Hinchliffe, E., Adams, J.L., Silverstone, A.L. and Drake, R. (2005) Improving the nutritional value of Golden Rice through increased pro-vitamin A content. *Nat. Biotechnol.*, **23**: 482-487.

Pathak, P. and Kapil, U. (2004) Role of trace elements, zinc, copper and magnesium during pregnancy and its outcome. *Indian J. Pediatrics*, **71**: 1003-1005.

Poirier, K.A. (1994) Summary of the derivation of the reference dose for selenium. In: *Risk Assessment of Essential Elements* (Eds. Mertz, W., Abernathy, C.O. and Olin, S.S.), International Life Sciences Institute Press, Washington DC, pp 157-166.

Poletti, S., Gruissem W. and Sautter, C. (2004) The nutritional fortification of cereals. *Curr. Opin. Biotechnol.*, **15**: 162-165.

Poletti, S. and Sautter, C. (2005) Biofortification of the crops with micronutrients using plant breeding and transgenic strategies. *Minerva Biotech.*, **17**: 1-11.

Prinsen Geerlings, P.D., Brabin, B.J. and Omari, A.A. (2003) Food prepared in iron cooking pots as an intervention for reducing iron deficiency anaemia in developing countries: a systematic review. *J. Hum. Nutr. Dietet.*, **16**: 275-281.

Qu, L.Q., Yoshihara, T., Ooyama, A., Goto, F. and Takaiwa, F. (2005) Iron accumulation does not parallel the high expression level of ferritin in transgenic rice seeds. *Planta*. **222**: 225-233.

Raboy, V. (1996). Cereal low phytic mutants: a "global" approach to improving mineral nutritional quality. *Micronutrients and Agriculture*, **2**: 15-16.

Raboy, V., Young, K.A., Darsh, J.A. and Cook, A. (2001) Genetics and breeding of seed phosphorus and phytic acid *J. Plant Physial.*, **158**: 489-497.

Ramakrishnan, U., Manjrekar, R., Rivera, J., Gonzales-Cossio, T. and Martorell, R. (1999). Micronutrient and pregnancy outcome: a review of the literature. *Nutrition Research*, **19**: 103-159.

Ramesh, S.A., Choimes, S. and Schachtman, D.P. (2004). Overexpression of an Arabidopsis zinc transporter in *Hordeum vulgare* increases short term zinc uptake after zinc deprivation and seed zinc content. *Plant Mol. Biol.*, **54**: 373-385.

Reddy, N.S. and Bhatt, G. (2001) Contents of minerals in green leafy vegetables cultivated in soils fortified with different chemical fertilizers. *Plant Foods for Human Nutrition*, **56**: 1-6.

Rengel, Z. and Graham, R.D. (1995) Importance of seed Zn content for wheat growth on Zn-deficient soil. I. Vegetative growth. *Plant Soil*, **173**: 259-266.

Rengel, Z., Batten G.D. and Crowley D.E. (1999) Agronomic approaches for improving the micronutrients density in edible portions of field crops. *Field Crops Res.*, **60**: 27-40.

Rumeau, D., Becuwe-Linka, N., Beyly, A., Carrier, P., Cuine, S., Genty, B., Medgyesy, P., Horvath, E. and Peltier, G. (2004) Increased zinc content in transplastomic tobacco plants expressing a polyhistidine-tagged Rubisco large subunit. *Plant Biotech. J.*, **2**: 389-395.

Sadana, U.S. and Nayyar, U.K. (2000) Amelioration of iron deficiency in rice and transformation of soil iron in coarse textured soils of Punjab, India. *J. Plant Nutr.*, **23**: 2061-2069.

Samuelsen, A.I. *et al.* (1998) Expression of the yeast FRE genes in transgenic tobacco. *Plant Physiol.*, **118**: 51-58.

Sandberg, A.S., Hulthen, L.R. and Turk, M. (1996) Dietary *Aspergillus niger* phytase increases iron absorption in humans. *J. Nutr.*, **126**: 476-480.

Schaaf, G., Ludewig, U., Erenoglu, B.E., Mori, S., Kitahara, T. and von Wiren, N. (2004) ZmYS1 functions as a proton-coupled symporter for phytosiderophore and nicotianamine-chelated metals. *J. Biol. Chem.*, **279**: 9091-9096.

Schneeman, B.O. (2001). Linking agricultural production and human nutrition. *J. Sci. Food and Agriculture*, **81**: 3-9.

Elemental biofortification of crop plants

Schulte E.E. (2004) Soil and applied iron, *University of Wisconsin-Extension, A3554*, Cooperative Extension Publication.
Shrestha, A.K., Arcot, J. and Paterson, J.L. (2003). Edible coating materials - their properties and use in the fortification of rice with folic acid. *Food Res. Int.*, **36**: 921-928.
Sichert-Hellert, W., Kersting, M., Alexy, U. and Manz, F. (2000). Ten year trends in vitamin and mineral uptake from fortified food in German children and adolescents. *Eur. J. Clin. Nutr.*, **54**: 81-91.
Smith, S.E. and Read, D.J. (1997) *Mycorrhizal Symbiosis* (2$^{nd}$ edn.) Academic Press, New York.
Smolin, L.A. and Grosrenor, M.B. (2003). *Nutrition: Science and Applications*, John Wiley and Sons, New York.
Smrkolj, P., Germ, M., Kreft, I. and Stibilj, V. (2006) Respiratory potential and Se compounds in pea (*Pisum sativum* L.) plants grown from Se-enriched seeds. *J. Exp. Bot,* **57**: 3595-3600.
Suzuki, Y. A., Skin K. and Lonnerdal, B. (2001) Molecular cloning and functional expression of a human fetal small intestinal lactoferrin receptor. *Biochemistry*, **40**: 15771-15779.
Swain, J.H., Newman, S.M. and Hunt, J.R. (2003). Bioavailability of elemental iron powders to rats is less than bakery-grade ferrous sulphate and predicted by iron solubility and particle surface area. *J. Nutr.,* **133**: 3546S-3552S.
Takahashi, M., Nakanishi, H., Kawasaki, S., Nishizawa, N.K. and Mori, S. (2001) Enhanced tolerance of rice to low iron availability in alkaline soils using barley nicotinamine aminotransferase genes. *Nat. Biotechnol.,* **19**: 466-469.
Takahashi, M. (2003) Overcoming Fe deficiency by transgenic approach in rice. *Plant Cell Tiss. Org. Cult.,* **72**: 211-220.
Takatsuji H. (1988) Zinc-finger transcription factors in plants. *Cell Mol. Life Sci.,* **54**: 582-596.
Theil, E.C. (1987). Ferritin: Structure, gene regulation and cellular function in animals, plants and microorganisms. *Annu. Rev. Biochem.*, **56**: 289-315.
Theil, E.C. (1990). Regulation of ferritin and transferrin receptor mRNAs. *J. Biol. Chem.*, **265**: 4771-4774.
Theil, E.C. (2004) Iron, ferritin, and nutrition. *Annu. Rev. Nutr.*, **24**: 327-343.
Timmer, C.P. (2003) Biotechnology and food systems in developing countries. *J. Nutr.* **133**: 3319-3322.
Tucker, G. (2003) Nutritional enhancement of plants. *Curr. Opin. Biotech.*, **14**: 221-225.
Van Huysen, T., Terry, N. and Pilon-Smits, E.A.H. (2004) Exploring the selenium phytoremediation potential of transgenic Indian mustard overexpressing ATP sulfurylase or cystathionine-γ-synthase. *Int. J. Phytoremediation*, **6**: 1-8.
Vasconcelos, M., Datta, K., Oliva, N., Khalekuzzaman, M., Torrijo, L., Krishan, S., Oliveira, M., Goto, F. and Datta, S.K. (2003) Enhanced iron and zinc gene. *Plant Sci.,* **164**: 371-378.
Vasconcelos, M., Eckert, H., Arahana, V., Graef, G., Grusak, M.A. and Clemente (2006) Molecular and phenotypic characterization of transgenic soybean expressing the Arabidopsis ferric chelate reductase gene, *FRO2. Planta*, **224**: 1116-1128.
Velu, G., Rai, K.N., Muralidharan, V., Kulkarni, V.N., Longvah, T. and Raveendran, T.S. (2007) Prospects of breeding biofortified pearl millet with high grain iron and zinc content. *Plant Breeding,* **126**: 182-185.

Verret, F., Gravat, A., Auroy, P., Leonhardt, N., David, P., Nussaume, L., Vavasseur, A. and Richaud, P. (2004). Over-expression of AtHAMA4 enhances root to shoot translocaion of zinc and cadmium and plant metal tolerance. *FEBS Lett.,* **576**: 306-312.

Vreugdenhil, D., Aarts, M.G.M., Koornneef, M., Nelissen, H. and Ernst, W.H.O. (2004) Natural variation and QTL analysis for cationic mineral content in seeds of Arabidopsis thaliana. *Plant Cell and Environ.,* **27**: 828-839.

Vreugdenhil, D., Aarts, M.G.M., Koornneef, M., Broadley, M.R. and White, P.J. (2005) Exploring natural genetic variation to improve plant nutrient content. In: *Plant Nutritional Genomic* (Eds. Broadly, M.R. and White, P.J.), Blackwell Publ., Oxford, pp 201-219.

Underwood, B.A. (2003). Scientific research: essential but is it enough to combat world food insecurities? *J. Nutrition.,* **133**: 1434S-1437S.

United Nations Administrative Committee on Coordination/ Sub Committee on Nutrition (ACC/SCN) (2005). Fourth report on the world nutrition situation. United Nations, Geneva.

Wegmuller, R., Zimmermann, M.B. and Hurrell, R.F. (2003). Dual fortification of salt with iodine and encapsulated iron compounds: stability and acceptability testing in Morocco and Cote d'Ivoire. *J. Food. Sci.,* **68**: 2129-2135.

Welch, R.M. (1986). Effects of nutrient deficiencies on seed production and quality. *Advances in Plant Nutrition,* **2**: 205-247.

Welch, R.M. (1995). Micronutrient nutrition of plants. *Critical Reviews of Plant Science,* **14**: 49-82.

Welch, R.M. (2001). Micronutrients, agriculture and nutrition; linkages for improved health and well being. In: *Perspectives on the Micronutrient Nutrition of Crops* (Eds. Singh, K., Mori, S. and Welch, R.M.) Scientific Publishers, Jodhpur, India, pp 247-289.

Welch, R.M., Combs Jr., G.F. and Duxbury, J.M. (1997). Towards a 'Greener' revolution. *Issues in Science and Technology,* **14**: 50-58.

Welch, R.M., Graham, R.D. and Bouis, H.E. (2001). Addressing micronutrient malnutrition through enhancing the nutritional quality of staple foods: principles, perspectives and knowledge gaps. *Adv. Agron.,* **70**: 77-142.

Welch, R.M. and Graham, R.D. (2002) Breeding crops for enhanced micronutrient content. *Plant Soil,* **245**: 205-214.

Welch, R.M. and Graham, R.D. (2004) Breeding for micronutrients in staple food crops from a human nutrition prospective. *J. Exp. Bot.,* **55**: 353-364.

Welch, R.M., House, W.A., Ortiz-Monasterio, I. and Cheng, Z. (2005) Potential for improving bioavailable zinc in wheat grain (*Triticum* species) through plant breeding. *J. Agric. Food Chem.,* **53**: 2176-2180.

White, P.J. and Broadley, M.R. (2004) Interactions between selenium and sulphur nutrition in *Arabidopsis thaliana. J. Exp. Bot.,* **55**: 1927-1937.

White, P.J. and Broadley, M.R. (2005) Biofortifying crops with essential mineral elements. *Trends Plant Sci.,* **10**: 586-593.

WHO (1999). *Malnutrition worldwide.* Geneva, Switzerland: World Health Organization. http://www.who.int/nut/malnutrition-worldwide.htm.1-13.

WHO (2002). *The world health report 2002. Reducing risks, promoting healthy life.* World Health Organization, Geneva, pp 1-168.

WHO (2004) Iodine status worldwide, *WHO global-database on iodine deficiency* (Eds. de Benoist, B., Andersson, M., Egli, I., Takkouche, B. and Allen, H.). World Health Organization, Geneva.

WHO (2006) Preventing and controlling micronutrient deficiences in populations affected by an emergency, WHO, WFP, Unicef, WHO internet, April, 2006.

Wissuwa, M., Ismail, A.M. and Yanagihara, S. (2006) Effect of zinc deficiency on rice growth and genetic factors contributing to tolerance. *Plant Physiol.*, **142**: 731-741.

World Bank (1994). Enriching lives: overcoming vitamin and mineral malnutrition in developing countries (*Development in Practice Series*). The World Bank, Washington, DC.

Yang, X.E., Chen, W.R. and Feng, Y. (2007) Improving human micronutrient nutrition though biofortification in the soil-plant system: China as a case study. *Environ. Geochem. Health,* **29**: 413-428.

Ye, X., Al Babili, S., Kaloeti, A., Zhang, J., Lucca, P., Beyer, P. and Potrykus, I. (2000). Engineering the pro-vitamin A (β-carotene) biosynthetic pathway into (carotenoid-free) rice endosperm. *Science,* **287**: 303-305.

Zhu, C., Naqvi, S., Gomez-Galera, S., Pelacho, A.M., Capell, T. and Christou, P. (2007) Transgenic strategies for the nutritional enhancement of plants. *Trends Plant Sci.,* **17**: 548-555.

Zimmermann, M.B., Hurrell, R.F. (2002). Improving iron, zinc and vitamin. A nutrition through plant biotechnology. *Curr. Opin. Biotechnol.,* **13**: 142-145.

# SUBJECT INDEX

## ~ A ~

ABA-signaling, 140
ABC (ATP-binding cassette), 214, 226, 283, 291
(ABC) transporter, 294
ABC transporter family, 175
Abscisic acid, 58
ACAs, 68
*Actinida arguta*, 250
*Aegilops tauschii*, 354
Agronomical aspects of potassium, 3
AKT1, 33
AKT2, 13
$Al^{3+}$ tolerance, 141
Alkaloids, 283
Allelopathic substances, 284, 285
*Allium*, 105
*Alonsoa meridionalis*, 242, 245
Amino acid - polyamine - choline (APC), 276
Amino acid permeases (AAPs), 272
Amino acid transport, 268
Amino acid transporter family, 269
Amino acid transporters, 267, 273
Ammonia assimilation, 83
Ammonification, 85
Ammonium, 83
Ammonium compartmentation, 93
Ammonium efflux, 96
Ammonium influx, 95
Ammonium transport, 93, 96
Ammonium uptake, 93, 95
Annexins, 65
Anthocyanins, 290
Anticancer, 285
Antinutrients, 354
Antisense, 273
Antisense strategy, 358
Anti-sigma factor anatgonists, 111, 121
APases, 138
*Apium graveolens*, 242
Apoplast, 111
Apoplastic, 289
Apoplastic invertase, 241
Apple, 249
Aquaporin, 209
Aquaporins, 307
Arabidopsis, 32, 56, 58, 59, 61, 63, 64, 66, 67, 68, 83, 89, 90, 92, 93, 96, 107, 109, 110, 112, 113, 114, 115, 116, 117, 119, 120, 121, 122, 133, 239, 241, 243, 245, 246, 247, 250, 255, 269, 301
Arabidopsis genome, 61, 91, 106, 230
Arabidopsis mutants, 153
*Arabidopsis fumigatus*, 358

*Arabidopsis thaliana*, 89, 90, 93, 111, 121, 179, 214, 242, 355

Aromatic and neutral amino acid transporter, 270

Arsenic transporters, 213, 230

Ascorbic acid, 359

*Aspergillus halleri*, 182, 214, 216

*Aspergillus nidulans*, 90

*Aspergillus niger*, 358

AtFOR2, 153

AtFRD3, 157

AtIRT, 153, 154

ATM, 214, 227

AtNRAMPs, 155

ATP-sulfurylase, 117, 361

AtYSL, 156

Auxin importer, 269

~ B ~

*Bacillus subtilis*, 174

Barley, 57, 93, 161, 253, 255, 289, 301

β-carotene, 359

Berberine, 285

*Beta vulgaris*, 4, 290

bHLH, 158, 187

Bioavailability of iron, 150

Biofortification, 345, 352

Biomolecules, 105

*Botrytis cinerea*, 292

*Bradyrhizobium japonicum*, 290

*Brassica juncea*, 231

*Brassica nigra*, 138

*Brassica oleracea*, 110, 113, 116, 117

BY-Cells, 140

bZIP, 140

~ C ~

$Ca^{2+}/H^+$ exchanger, 56

$Ca^{2+}$-ATPase, 53

$Ca^{2+}$-exchanger, 70

$Ca^{2+}$-movement, 54

$Ca^{2+}$-permeable channels, 53, 58

$Ca^{2+}$-regulated proteins, 52

$Ca^{2+}$-release, 64

Calciome, 55

Calcium, 51, 52

Calcium binding proteins, 65

Calcium channels, 55, 57, 60, 64, 65

Calcium flux, 51

Calcium pump, 56, 65, 66, 67, 70

Calcium reporter system, 57

Calcium reporters, 63

Calcium sensors, 55

Calcium signaling, 51, 55, 64

Calcium signals, 57

Calcium spike, 52, 58

Calcium transport, 51, 55, 72

Calcium transporters, 51, 54, 56, 65, 75

Calmodulin, 52, 66

Calmodulin domain, 61

Carbon fixation, 255

Carbon partitioning, 255

Carotenoids, 288

Catharanthus, 304

*Catharanthus roseus*, 139

Cation Diffusion Facilitator (CDF), 214, 222

Cation exchanger, 70

Cationic amino acid transport (CAT), 269

CAX transporters, 72, 181, 185

CAXs, 72

CCC family, 31

CDF transporters, 182, 185

CDK, 137

cDNA Library, 250

Cellular calcium levels, 57

Cellular compartments, 54

Cellular influx, 105

*Celtis*, 250

Channels, 316

Chenopodium, 248

Chickpea, 117

*Chlamydomonas*, 91

*Chlamydomonas reinhardtii*, 110, 111, 229

Chlorella, 247

Chloronerva mutant, 158

Chloroplast, 243

Chloroplast envelope, 252

Chloroplast genome, 112

Chlorsulfuron, 291

Chromoplast, 243

Cinnamic acid, 291

*Citrus*, 250

*Citrus sinensis*, 242

*Coptis japonica*, 285

*Crambe abyssinica*, 231

$Cu^{2+}$ transporter, 215, 229

Cucumber, 209

Cyanidioschyzon merolae, 229

Cyclic nucleotide-gated (CNG) channels, 25

*Cylindrotheca fusiformis*, 209

Cysteine, 105, 106, 117, 119, 359

Cytochromes, 149

Cytokinin derived signals, 123

~ D ~

*Daucus carota*, 242

Degenerated leaflet (dgl) mutant, 158

*Dioscorea* species, 354

Downstream assimilatory pathway, 120

*Drosophila melonogaster*, 58

Dual affinity transporters, 91

~ E ~

*E. coli*, 16, 118, 174, 252

Endocytosis, 253
Environmental factors, 115
Environmental stress, 54
*Erwinia uredovara*, 359
EST, 283
*Eucalyptus*, 250

~ F ~

Faba bean, 208
Fe starvation, 151
$Fe^{3+}$ reduction, 152
$Fe^{3+}$-chelate reductase, 151
Fe-homeostasis, 159
Ferritin, 158, 356, 357
Ferroprotein, 157
Fertilizer N, 84
Fe-uptake mutants, 217
Flavonoids, 290
Floral tissues, 246
Ford fortification, 345, 351
Functional redundancy, 67
Fungal transporters, 91
Fungicide, 293

~ G ~

Genetic engineering, 123
Genetically modified plants, 74
GFP, 277
GFP-fluorescence, 255

*Gigaspora margarita*, 139
*Glomus versiforme*, 139
GLRs, 63
Glucose transporter, 239, 251, 255
Glucosides, 289
Glutamate binding domain, 63
Glutamate dehydrogenase (GDH), 275
Glutamate receptors, 27, 63
Glutamate synthase (GOGAT), 275
Glutamine synthetase (GS), 275
Glutathione, 105, 106, 117, 119, 289
Glutathione content, 112
Glutathione pcol, 118
Glycine betain, 275
*Glycine max*, 163
Golden rice, 359
*Gossia bidwillii*, 191
GTP-binding proteins, 326
Guard cells, 52, 69, 246
GUS, 277

~ H ~

$H^+$-antiport, 287
$H^+$-ATPase, 17, 151, 302, 311
$H^+$-PPase, 302, 311
$H^+$-symporter, 291
HAK/KT family, 22, 28
*Hansenula polymorpha*, 91
Harvestplus initiative, 354

Heavy metal transporters, 360

Heavy metal uptake transporters, 213

Hemiterpene, 284

Herbivore, 293

Heterologous expression, 56, 271

Hexose transporter-gene, 248

Hexose transporters, 239, 244, 247

Hidden hunger, 347

High affinity ammonium transporters, 83, 93, 95

High affinity nitrate transporters, 83, 85, 87, 88, 89, 90

High affinity Pi transport, 132

High affinity sucrose transporters, 242, 244

High affinity sulfate transporters, 109, 110, 114, 115, 120

High capacity transporters, 245

HKT family, 22, 28

HKT transporters, 28

HKT1, 13, 32, 57

HMA (heavy metal ATPase), 225

*Hordeum vulgare*, 242

Human nutrition, 54, 74, 75

~ I ~

ICP-AES, 194

ICP-MS, 194

*In planta*, 52, 56, 60, 61, 69, 72, 74, 121, 195

Indole alkaloid, 288

Inducible transporters, 118

Intracellular $SO_4^-$ movement, 106

Intracellular transport of amino acids, 277

Ion channels, 9, 18, 301

Ion homeostasis, 302

Iron, 149

Iron chelation, 149

Iron deficiency, 149, 349

Iron fortification, 150

Iron reduction, 149

Iron regulated transporter protein, 214

Iron transport, 149, 153, 162

Iron transport protein, 163

Iron uptake, 150

Iron-regulated protein, 157

IRT, 360

IRTI, 178

~ K ~

$K^+/H^+$ antiporter, 22, 30

$K^+/H^+$ symporter, 19

KAT1, 20

KEA family, 32

KT transporters, 28

KUP, 22, 28

~ L ~

*Lactobacillus plantarum*, 175

Lactoferrin, 357

Late embryogenesis abundant family, 163

LATs, 269

LCT1, 31

Leaching, 84

*Leishmania major*, 230

Lignin, 284

Lipoxygenase, 149

*Lithospermum erythrorhizon*, 292

Lithosphere, 84

Long distance transport of amino acids, 272

*Lotus japonicus*, 290

Low affinity ammonium transporter, 11, 83, 95

Low affinity nitrate transporter, 83, 87, 88, 89, 90

Low affinity Pi transporter, 133

Low affinity sucrose transporter, 244, 245

Low affinity transport, 105, 109, 114, 115

Lupinus, 4

*Lupinus albus*, 138

*Lycopersicon esculentum*, 242

Lysine histidin transporter, 274

~ M ~

Macronutrients, 105

*Magnaportha grisea*, 292

Maize, 93, 162, 209, 246, 247, 290

Mammalian cell lines, 15

Mammalian cells, 56

Manganese accumulation, 173, 177, 190

Manganese deficiency, 173

Manganese toxicity, 173, 174

Manganese transport, 173, 174, 176, 187, 192

Mannitol, 250

Mannitol transport activity, 250

*Marchantia polymorpha*, 112

Marker-assisted selection, 355

Mass spectrometer, 306

MATE, 214, 226, 227

*Medicago*, 248

*Medicago truncatula*, 180, 215

Membrane potential, 112

Membrane proteins, 95, 305

*Mesembryanthemus crystallinum*, 250, 254

Metabolorics, 255

Metal efflux transporters, 213, 222

Metal hyperaccumulators, 216

Metal tolerance protein, 222

Metal transporters, 174

Metal-binding proteins, 356

Metallothionein, 359

Microelectrodes, 4, 10

Micronutrients, 105

Micronutrients accumulation, 350

Micronutrients bioavailability, 350

Micronutrients malnutrition, 348

Micronutrients transport, 360

Micronutrients uptake, 360

Mineral fertilization, 347

Mineral malnutrition, 345

$Mn^{2+}/H^+$ antiport, 177

$Mn^{2+}$-ATPase, 179

Mn detoxification, 174, 177

Mn efflux, 178

Mn homeostasis, 175, 190

Mn hyperaccumulating species, 174, 191

Mn mutants, 193

Mn sequestration, 181

Mn starvation, 176

Mn tolerance, 182

Mn uptake, 177

Monosaccharide sensing proteins, 239

Morphine, 285

MRP, 214

Mugineic acid, 160

Multidrug transporters, 226

Multidrug and Toxin Efflux transporter (MATE), 157

Multidrug resistance (MDR), 286

Multiple mutants, 255

Multiple signaling pathway, 62

Mutant *brz*, 158, 193

Mycorrhiza, 248, 347

*Mycosphaerella graminicola*, 292

myo-inositol transporter, 239, 240, 250

~ N ~

$Na^+/K^+$ antiporter, 70

$Na^+/K^+$ symporter, 52

Natural resistance-associated macrophage protein, 155

Natural variation in minerals, 354

*Neurospora crassa*, 135

Nicotiana, 284

Nicotianamine, 157, 220

*Nicotiana plumbaginifolia*, 88, 91, 288

*Nicotiana tabacum*, 118, 242

Nicotianamine aminotransferase, 160, 161

Nicotianamine synthase, 160, 161

Nicotine, 293

Nitrate, 83

Nitrate assimilation, 83, 84

Nitrate efflux, 92

Nitrate sensor, 92

Nitrate storage, 92

Nitrate transporter, 88, 90

Nitrate uptake, 85, 87

Nitrification, 85

Nitrogen use efficiency, 83, 84, 87, 91, 92, 95, 96

Nitrosolubus, 85

Nitrosomonas, 85

Nitrospora, 85

NMR, 4

Non-selective cation channels, 25

NRAMP, 214, 217
Nramp transporters, 180, 184
Nucleotide gated channel, 61
Nutrient uptake, 83
Nutritional genomics, 74

~ O ~

Oat, 161
Onion epidermis, 255
Organelle phosphate transporters, 133
*Oryza sativa*, 108, 215, 217, 242
Osmoprotectants, 275

~ P ~

*Papaver somniferum*, 285
Parsimony, 282
Patch-clamp technique, 11, 12
Pathogen interactions, 62
Pathogen resistance, 63
PDR, 227
Pea, 273
Peptide antibodies, 251
Petunia, 248
Pharmacognosy, 284
Phase-modulation fluorometry, 8
Phenolic compounds, 283, 289-291
Phi proteins, 140
Phloem loading, 244, 245, 250
Phloem parenchyma, 116

*pho* mutants, 136
Phosphate acquisition, 140
Phosphate fertilizer, 131
Phosphate homeostasis, 137
Phosphate starvation, 137
Phosphate transporters, 131, 134
Phosphate uptake, 132
Phosphate-starvation, 131
Phosphorus, 131
Photo-oxidative damage, 3
Pht family, 133
Phylogenetic analysis, 108
Phylogenetic tree, 241, 242
Physiological aspects of potassium, 3
Phytase, 358
Phytic acid, 355
Phyto alexin, 293
Phytochelatins, 105, 106
*Phytolacca acinosa*, 191
Phytoremediation, 214
Phytosiderophores, 158, 220
Phytosterols, 292
PIB-type ATPase, 224
*Pisum sativum*, 214
Plant breeding, 347
Plant transport, 105
*Plantago major*, 242
Plasma membrane, 93, 105, 106, 243, 252, 239, 243, 244, 246, 251

Plasma membrane influx, 107

Plasma membrane phosphate transports, 133

Plasma membrane proteins, 57, 59

Plasma membrane vesicle, 244, 246

Plasmodesmata, 113

Pleiotropic drug resistance (PDR) Phenylpropanoid, 289

Pollen, 247, 248

Pollen tube, 247

Polyol transport, 249

Polyol transporter genes, 253

Polyol transporter proteins, 253

Polyol transporters, 239, 240, 241, 253, 255

Polyols, 239

Poppy, 287

Post-transcriptional control, 96

Post-transcriptional regulation, 96, 138

Potash, 2

Potassium, 1

Potassium compartmentation, 4, 8

Potassium content of soil, 2

Potassium deprivation, 31, 33

Potassium homeostasis, 5

Potassium transport, 1, 7, 17, 20

Potassium transport mechanism, 6

Potassium transporting proteins, 8

Potassium uptake, 1, 7, 17, 18

Potato, 135

PPi-ase, 17

Primary metabolite, 283

Primary transport, 283

Programmed cell death, 248

Proline, 275

Promoters, 354

Protein kinase, 52

Protein phosphatase inhibition, 246

Proteome, 301

Proteomics, 255

Proton gradients, 51, 55

Proton pumps, 311

Protoplast, 304

ProTs, 275

PsFOR1, 153

*Pteris vitatta*, 230

P-type ATPase, 183, 185

P-type ATPase, 214

~ Q ~

QTL, 355

Quinones, 283

~ R ~

Radiotracer technique, 7

Recombinant inbred lines, 355

Recycling, 20

Remobilization, 20, 83, 91, 92, 95, 96, 115

Reserve N, 84

Resveratrol, 293
Rhizobium, 290
Rhizosphere, 92, 95, 150
Rhizosphere acidification, 151
*Rhus javanica*, 290
Rice, 56, 60, 61, 83, 91, 161, 207
Rice genome, 221
*Ricinus communis*, 242, 245
RNAi, 209, 358
Rosaceae, 249
RT-PCR, 286
Rye, 93

~ S ~

*Saccharomyces cerevisiae*, 13, 52, 55, 69, 174, 214, 217
Saccharum hybrid, 242
S-adenosyl methionine, 160
*Salmonella enterica*, 174
Salt stress, 275
*Schwanniomyces occidentalis*, 16
SDS-PAGE, 305
Secondary metabolite, 283
Secondary multidrug transporters, 228
Selenium deficiency, 349
Shaker-type channels, 14, 21
Si accumulator, 206
Si excluder, 206
Signal molecule, 92

Signal transduction, 52, 55, 119
Silicon, 205
Silicon transport, 205, 209
Silicon transporter, 205
Silicon uptake, 205, 208
Sink tissue, 248
Site-directed mutagenesis, 154
SMR, 214
Snapdragon, 289
Sodium, 4
Soil ecosystem, 84
*Solanum tuberosum*, 242
Sorbitol transporter, 249
Sorbitol transporter gene, 249
Sorghum, 4
Soybean, 208, 248
*Spinacea oleracea*, 242, 354
Spinach, 73, 244
*Spirodela polyrrhiza*, 288
*Staphylococcus aureus*, 175
Stomatal closure, 60
Storage proteins, 302
Strawberry, 208
Stress recognition, 62
*Stylosanthes hamata*, 107, 182, 223
Sucrose binding protein, 240
Sucrose transport activity, 244, 246
Sucrose transporter, 239, 240, 241, 244, 245

Sucrose transporter activity, 253
Sucrose transporter homolog, 254
Sucrose uptake, 253
Sugar beet, 254
Sugar sensors, 244
Sugar storage, 240
Sugar transporter, 239, 240, 243, 254
Sugar transporter protein, 253
Sugarcane, 208, 245, 252
Sulfate distribution, 110
Sulfate permease, 110
Sulfate starvation, 114, 118
Sulfate transport, 114, 115, 117, 123
Sulfate transport activity, 107
Sulfate transporter, 105, 106, 107, 108, 110, 112, 118, 121, 122
Sulfate transporter isoforms, 107
Sulfate uptake, 105, 107, 109, 115, 117, 122, 123
Sulfur availability, 115
Sulfur nutrition, 118
Sulfur responsive element, 120, 123
Sunflower, 209
Switch grass, 231
*Syncchocystis* sp., 175
Systemic signaling, 162

~ T ~

Taproot, 252
T-DNA insertion mutants, 194

Terpenoid, 284, 288
*Teucrium japonicum*, 216, 219
*Thalictrum minus*, 285
*Thlaspi caerulescens*, 214
*Thlaspi goesingense*, 182, 223
*Thlaspi japonicum*, 180
Tobacco, 118, 209, 245, 252
Tomato, 83, 95, 114, 135, 156, 208, 245, 246
Tonoplast, 301
Tonoplast membrane, 93, 112
Transcription factor, 123, 137
Transcription regulator, 158
Transcriptional control, 96
Transcriptional regulation, 96
Transcriptomic, 123
Transcriptomics, 255
Transient expression, 112
Transition metals, 214
Translocation of Mn, 188
Transmembrane domain, 307, 326
Transmembrane heterodimer, 110
Transmembrane sulfate movement, 111
Transmembrane transporter, 107
Transplastomic plants, 361
Transport of secondary metabolites, 292
Transport proteins, 83, 88, 89, 91, 92, 93

Transporter genes, 83, 88, 89, 91, 92, 93, 137

Transporter proteins, 96

*Triticum aestivum*, 242, 354

*Triticum boeoticum*, 354

*Triticum dicoccoides*, 354

*Triticum durum*, 354

Two pore channels, 14, 24

~ V ~

Vacular sulfate, 113

Vacuolar $Ca^{2+}/H^+$ antiporter, 177

Vacuolar channels, 312

Vacuolar efflux, 105

Vacuolar $H^+$-ATPase, 252

Vacuolar membrane proteins, 243, 253, 254

Vacuolar membrane, 60, 72, 239, 252, 253, 255, 303

Vacuolar mutants, 302

Vacuolar sugar, 253

Vacuolar sugar transporters, 239, 255

Vacuolar transporter, 279, 301

Vacuole, 64, 70, 92, 106, 112, 113, 252, 301, 303

Vascular tissue, 113

Vasculature tissue, 69

*Verticillium dahliae*, 116, 122

*Vicia faba*, 242, 273

Vitamins, 105

*Vitis vinifera*, 242

Volatilization, 84

Voltage-clamp technique, 11

~ W ~

Wax, 291

Wheat, 208, 245

~ X ~

Xenobiotics, 289

*Xenopus*, 62, 269, 286

X-ray microanalysis, 4

Xylem parenchyma, 113, 114

Xylem vessels, 113

~ Y ~

Yeast, 52, 54, 55, 56, 58, 62, 67, 70, 91, 107, 240, 249, 255

Yeast complementation, 136

Yellow-stripe 1 like (YSL), 214

Yield, 255

YSL family, 220

YSL/OPT transporters, 186

~ Z ~

*Zea mays*, 242

Zinc deficiency, 349

Zinc transporter, 223

ZIP (Zinc regulated transporter), 154, 214, 360

ZIP transporters, 179, 184